990

Soil Erosion
in the Tropics

Soil Erosion in the Tropics

Principles and Management

Rattan Lal
Department of Agronomy
The Ohio State University

McGraw-Hill, Inc.
New York St. Louis San Francisco Auckland Bogotá
Caracas Hamburg Lisbon London Madrid
Mexico Milan Montreal New Delhi Paris
San Juan São Paulo Singapore
Sydney Tokyo Toronto

To soil conservationists who are committed to using, improving, and restoring the most basic of all resources—the soil

Library of Congress Cataloging-in-Publication Data

Lal, Rattan
    Soil erosion in the tropics: principles and management / Rattan Lal.
       p.   cm.
    Includes bibliographical references.
    ISBN 0-07-036087-1
    1. Soil erosion—Tropics. 2. Soil conservation—Tropics.
I. Title.
S625.T76R37 1990
631.4'5'0913—dc20   90-30142

Copyright © 1990 by McGraw-Hill, Inc. All rights reserved. Printed in the United States of America. Except as permitted under the United States Copyright Act of 1976, no part of this publication may be reproduced or distributed in any form or by any means, or stored in a data base or retrieval system, without the prior written permission of the publisher.

1 2 3 4 5 6 7 8 9 0   DOC/DOC   9 6 5 4 3 2 1 0

ISBN 0-07-036087-1

The sponsoring editor for this book was Jennifer Mitchell, the editing supervisor was Stephen M. Smith, the designer was Naomi Auerbach, and the production supervisor was Suzanne W. Babeuf. It was set in Century Schoolbook by Carol Woolverton.

Printed and bound by R. R. Donnelley & Sons Company.

> Information contained in this work has been obtained by McGraw-Hill, Inc. from sources believed to be reliable. However, neither McGraw-Hill nor its authors guarantees the accuracy or completeness of any information published herein and neither McGraw-Hill nor its authors shall be responsible for any errors, omissions or damages arising out of use of this information. This work is published with the understanding that McGraw-Hill and its authors are supplying information but are not attempting to render engineering or other professional services. If such services are required, the assistance of an appropriate professional should be sought.

For more information about other McGraw-Hill materials, call 1-800-2-MCGRAW in the United States. In other countries, call your nearest McGraw-Hill office.

Contents

Preface   ix

Part 1   Introduction

Chapter 1. Soil Erosion and Sedimentation   3

  Definitions   3
  Agents, Factors, and Causes of Soil Erosion   5
  Types of Erosion   6
  Soil Erosion: Natural vs. Accelerated Process   7
  Soil Erosion and Degradation   9
  Soil Erosion and Civilizations in the Tropics   10
  Soil Erosion in Modern Times   12
  Soil Erosion in the Tropics vs. Temperate Regions   15
  Soil Erosion and Food Production in the Tropics   19
  References   19

Part 2   Basic Processes

Chapter 2. Physical Processes and Climatic Erosivity   23

  Work Involved in Soil Erosion   23
  Common Algebraic Functions Used in Explaining Erosion Processes   25
  Factors Affecting Soil Erosion   26
  Climate   27
  Wind Velocity   52
  Conclusions   55
  References   56

Chapter 3. Soil Properties and Erodibility   60

  Introduction   60
  Mechanical Properties   61
  Strength Properties   65

Hydrologic Properties	67
Rheologic Properties	73
Chemical and Mineralogical Properties	75
Soil Profile Characteristics	81
Dynamic Nature of Soil Properties	82
Conclusions	91
References	92

Chapter 4. Slope and Landforms — 103

Types of Landforms in Relation to Erosion	104
Predominant Types of Landforms in the Tropics	107
Slope and Soil Erosion	111
Slope Modification	126
References	126

Chapter 5. Splash Erosion — 130

Basic Concepts	130
Soil Detachment Mechanism	132
Factors Affecting Soil Splash	132
Factors Affecting Interrill Erosion	148
Conclusions	149
References	149

Chapter 6. Erosion Processes over a Watershed — 154

Types of Flow in Watersheds and Their Properties	155
Sheet or Interrill Erosion	157
Rill Erosion	167
Gully Erosion	171
Pipe or Tunnel Erosion	172
Mass Movements	175
Stream Bank Erosion	177
Coastal Erosion	177
References	177

Part 3 Measurement and Prediction

Chapter 7. Erosion Measurement and Evaluation — 183

Introduction	183
Scales for Measuring Soil Erosion	183
Evaluation of Erosion at Macroscale	185
Evaluation of Erosion at Mesoscale	195
Measurement and Evaluation at Microscale	199
The Scale Problem	213

Laboratory Measurements	216
Conclusions	216
References	218

Chapter 8. Predicting the Erosion Potential — 225

Interrill and Rill Erosion	226
Universal Soil Loss Equation	226
Modifications in USLE	233
Applications of USLE in the Tropics	241
Other Predictive Models	263
Erosion Prediction for Gully and Channel Erosion	267
Physical or Conceptual Models	268
References	272

Part 4 Erosion Control

Chapter 9. Controlling Upland Erosion — 283

Reducing Raindrop Impact	283
Increasing Shearing Resistance of Soil	288
Decreasing the Shear Strength of Overland Flow through Slope Management	302
Conclusions	306
References	306

Chapter 10. Land Use — 309

Introduction	309
Land Use and Soil Erosion	310
Erosion Control in Different Land Use Systems	315
Soil Capability Classification and Erosion Hazard	316
Conclusions	317
References	317

Chapter 11. Crop Management — 320

Introduction	320
Canopy Cover and Soil Erosion	322
Land Use	325
Crop Husbandry	326
Conclusions	347
References	348

Chapter 12. Conservation Tillage — 352

Introduction	352
Conservation Tillage: Basic Concepts and Definitions	353
Erosion Control with Conservation Tillage	360

viii   Contents

Conservation Tillage for Erosion Control on Problem Soils	404
Conservation Tillage for Controlling Wind Erosion	414
Conclusions	415
References	416

Chapter 13. Trees and Soil Erosion — 423

Soil Erosion under Forest	424
Conversion of Natural Forest to Planned Forest	426
Erosion Control in Tree Crops and Planned Forests	431
Afforestation for Erosion Control	449
Agroforestry as an Antierosive Technique	457
Effects of Alley Cropping on Crop Yield	474
Conclusions	477
References	479

Part 5   Special Topics

Chapter 14. Erosion Hazard on Steeplands — 489

Introduction	489
Factors Affecting Soil Erosion	492
Predominant Soil Erosion Processes	496
Magnitude and Extent of the Problem	498
Steep Agricultural Land Technologies	499
Research Needs in Soil Erosion and Its Control	519
Conclusions	527
References	527

Chapter 15. Wind Erosion and Its Control — 531

Introduction	531
Mechanisms of Wind Erosion	532
Particle Movement	535
Forces Opposing Soil Uplift and Entrainment	537
Factors Affecting Wind Erosion	538
Causes of Wind Erosion	539
Wind Erosion Prediction	539
Wind Erosion Hazard	544
Measuring Wind Erosion	547
Control Measures	549
References	559

Author Index   565
Subject Index   575

Preface

Books on soil erosion are typically written by civil engineers, hydrologists, geomorphologists, geographers, or geologists. These books are authoritative in the discipline of their authors, but soil erosion and, particularly, its control cover a broad range of subjects all at once—hydrology, climatology, geology, geomorphology, biology, economics, soil mechanics, soil science, forestry, and agronomy. Few, if any, books encompass so many fields of study. Researchers in soil science and agronomy especially have great difficulty finding reference books that address the broad scope of soil erosion in simple, easy-to-understand language. In addition, most of the vast literature on soil erosion and its ecological and economic consequences describes conditions only in the United States and other temperate-region climates. Therefore, in this book I present the issue of soil erosion from a properly wide frame of reference and emphasize that which occurs in the tropics.

Regardless of geographical or climatological conditions, the physical processes and the principles underlying erosion-control measures do not vary. What differs from region to region is the package of cultural practices required to combat erosion. Conservation-effective cultural practices depend on both biophysical and socioeconomic factors and the latter are dominant in choosing appropriate techniques.

This book addresses the problem of soil erosion, its causes and consequences, and methods to control it. It is a state-of-the-art compendium of the basic principles of soil erosion and technological developments to prevent and control it. Important processes are explained with appropriate examples from different ecological regions. The terminology and definitions used are those adopted by the Soil Conservation Society of America.

The 15 chapters in this book are divided into five parts. Part 1 introduces soil erosion and sedimentation. Part 2 discusses the factors affecting soil erosion and the processes they drive. The five chapters in this part address the subjects of climate, soil, landforms and hydrology, and soil splash or interrill erosion, and the erosion processes over a watershed. Part 3 comprises two chapters and deals with the methods of erosion measurement and prediction. Part 4 discusses the principles and practices of erosion prevention and control in five chapters on controlling upland erosion, land use and farming systems, crop management, conservation tillage, and the role of trees in erosion control. Part 5 addresses special topics and consists of two chapters that deal with the management of steeplands and the problem of wind erosion and its control.

In this book, properties of climate, soil, and landforms relevant to erosion are discussed before predominant processes involved in upland erosion are described. Chapters dealing with the principles and techniques of erosion control naturally follow those describing the cause-effect relations among factors and processes influenced by the principles and techniques.

This book is an up-to-date collation of the subject matter needed for a basic understanding of soil erosion on agricultural lands and its control. The objective has been to review soil erosion problems, processes, and control measures in the global context. This book emphasizes that biological erosion-control measures are more important than those involving engineering techniques. At least two reasons support a bias in favor of biological measures. One, the vegetative cover close to the soil surface effectively decreases soil erosion in many ways. Two, engineering techniques have not been particularly successful, especially in Africa or tropical South America. (Many books provide authoritative and comprehensive coverage of engineering practices for erosion control, so this book deals with them only in passing.)

This book also emphasizes upland erosion by water. This is not to say that wind erosion, coastal and stream-bank erosion, and gully formation and erosion by mass movement are not important. Many books are available by geomorphologists and engineers that treat these subjects at length, so they are given relatively less attention here.

I have drawn heavily on materials and data from many sources, and have taken the liberty of citing data from many colleagues and friends around the world. In this brief preface, it is impossible to list all those who generously helped and contributed in one way or another toward completing this book. The sources of data are acknowledged where the data are cited.

A few colleagues, however, deserve special mention. Several chapters of this book were typed initially by Mrs. Agnes Nwamadi of IITA, Ibadan, Nigeria. Corrections for a number of pages were typed by Mrs. D. Ebenharch. My special thanks are due to Karla Gutheil and Shirley Hall for their help and support in the preparation of this book. Help received from Roshan M. Bajracharya, Brad Ray, William D. Schuster, and M. Luvette Thomas is also gratefully acknowledged.

Rattan Lal

Part 1

Introduction

Chapter

1

Soil Erosion and Sedimentation

Definitions

Soil, the most basic of all resources, is nonrenewable. Once lost, it is difficult to replace within the foreseeable future. New soil formation, development of a biologically productive and economically fertile soil from parent rock, is a slow process measured only on a geological time scale. It takes hundreds to thousands of years to develop the equivalent of, say, a 5-cm layer of fertile soil. In contrast, soil erosion can be drastic and rapid. The equivalent of 1 cm or more of topsoil may be lost in a single rainstorm or windstorm. Literally speaking, the soil formed over hundreds to thousands of years can be blown or washed away in a single climatic event.

Soil erosion

The word *erosion* is derived from the Latin word *erosio,* meaning "to gnaw away." In general terms, soil erosion implies the physical removal of topsoil by various agents, including falling raindrops, water flowing over and through the soil profile, wind velocity, and gravitational pull. *Erosion* is defined as "the wearing away of the land surface by running water, wind, ice or other geological agents, including such processes as gravitational creep" (SCSA, 1982). The process of wearing away by water involves the removal of soluble and insoluble materials. Physical erosion involves the detachment and transport of insoluble soil particles, e.g., sand, silt, clay, and organic matter. The transport may be lateral on the soil surface or vertical within the soil profile through voids, cracks, and crevices. The removal of soluble material as dissolved substances is called *chemical erosion*. Chemical erosion may also be caused by surface runoff or by subsurface flow, where the water moves from one layer to another within the soil profile. Erosion by wind involves processes similar to those by water except that the causative agent in sediment detachment and transport is the wind. The processes of erosion and soil loss have adverse effects on agriculture because they deplete the soil's productive potential and diminish the resource base.

Sedimentation

The soil mass removed from one place is often deposited at another location when the energy of the erosion-causing agent is diminished or too dissipated to transport soil particles. The term *sediment* refers to solid material that is detached from the soil mass by erosion agents and transported from its original place by suspension in water or air or by gravity.

The term *soil erosion* therefore is distinct from soil loss and sediment yield (Wischmeier, 1976; Mitchell and Bubenzer, 1980). *Soil erosion* refers to the gross amount of soil dislodged by raindrops, overland flow, wind, ice, or gravity. *Soil loss* is the net amount of soil moved off a particular field or area, the difference between soil dislodged and sedimentation. *Sediment yield,* in comparison, is soil loss delivered to a specific point under consideration. A field's sediment yield is the sum of soil losses from slope segments minus deposition. The deposition may occur in depressions, at the toes of slopes, along field boundaries, and in terrace channels.

The combined terms *erosion* and *sedimentation by water* embody the process of detachment, transportation, and deposition of sediment by the erosive and transport agents including raindrop impact and runoff over the soil surface (ASCE, 1975). *Detachment* is the dislodging of soil particles from the soil mass by erosive agents. *Entrainment* is the transportation of sediments from their original location.

The term *sedimentation* needs additional clarification. In simple terms, sedimentation implies the process or action of depositing sediments. Sediments dislodged from one location may be deposited at another site and may eventually reach the ocean following repeated cycles of redetachment and reentrainment in rills, channels, streams, river valleys, flood plains, and deltas. The process begins with sediment detachment from uplands and ends with an eventual transport to the ocean. It may take from hundreds or even thousands to millions of years to be completed.

Like erosion, sedimentation has serious environmental and economic implications. Sedimentation decreases the capacity of reservoirs and chokes irrigation canals and tributaries. Sediments are a major source of pollution and eutrophication (the aging of lakes caused by water enrichment). Some researchers, especially engineers, consider sedimentation to be the major process of which erosion is an initial step. Fleming (1981) adopts a broader approach by stating that "the sediment problem may be defined as the detrimental depletion by erosion and transport of soil resources from land surfaces and subsequent accretion by deposition in reservoirs and coastal areas."

Fleming's definition of sedimentation is, in a way, a systems approach whereby the issue of erosion-sedimentation is evaluated in its totality. From this systems approach, the sediment processes involve the denudation chain of erosion—transport, deposition, and coastal circulation. *Erosion* also can be defined in a systems approach in which sedimentation is but an important aspect of the whole interrelated series of processes.

Denudation

The term *denudation* is used by geomorphologists to express erosion in terms of the sediment yield per unit area of watershed. The rate of transport of suspended solids and dissolved solutes from a watershed outlet, expressed per unit area, in geomorphologic terms is referred to as the *denudation rate*. The denudation rate is usually expressed as the volume of material removed per unit area within a given time (cubic meters per square kilometer per year) or depth of material removed per unit of time (millimeters per 1000 years).

Watershed

Watershed refers to a delineated area with a well-defined topographic boundary and a water outlet. Within its topographic boundaries, or water divide, a watershed contains a complex of soils, landforms, land uses, and vegetation. Watershed is a hydrologic unit in which all hydrologic processes are interrelated. The terms *watershed, catchment,* and *basin* are used interchangeably.

Erosion effects

Agriculturally speaking, both erosion and sedimentation affect soil productivity through their respective on-site and off-site effects. Erosion reduces on-site soil productivity by decreasing the rooting depth and depleting nutrient and water reserves. Sedimentation lessens productivity through off-site effects such as decreasing the capacity of water reservoirs and silting of irrigation canals.

Agents, Factors, and Causes of Soil Erosion

Soil erosion is the result of perturbations in the land-vegetation-climate equilibrium. Perturbations may be natural or caused by humans. The magnitude of the effects of perturbations on erosion is influenced by different phenomena. The phenomena or forces are conveniently grouped into agents, factors, and causes of erosion. Agents of erosion are the carriers or the transport system in soil movement. Factors of erosion are those natural or artificial parameters that determine the magnitude of perturbation, e.g., climate, topography, soil, vegetation, and management. Erosion may not occur even when the agents and factors are present. It is the causes of erosion that enhance the effects of agents and factors of erosion and accelerate the various processes involved. The causes of erosion primarily include human activities, e.g., farming practices such as seedbed preparation, deforestation, and cropping systems.

Climatic erosivity

Erosivity refers to the climate's aggressivity. Climatic factors that affect erosivity are precipitation, wind velocity, water balance, mean annual and seasonal temperatures, etc. Temperature and water balance affect the rate at which rainfall is accepted with the onset of rains. Runoff and rainfall are important components of climatic erosivity. The aggressivity of wind-driven rain differs from that of rain in windless conditions.

Soil erodibility

The susceptibility of soil to erosion is called its erodibility. Erodibility is an inherent property of the soil and is influenced by soil characteristics, including texture, structure, permeability, organic matter content, clay minerals, and contents of iron and aluminum oxides. Some climatic factors also influence soil erodibility, e.g., air and soil temperatures and water balance. Soils subjected to extremes of temperature and moisture regimes differ in degree of susceptibility to erosion from those experiencing moderate fluctuations in hydrothermal regimes.

Landforms

Topography affects soil erosion through degree and length of slope, slope shape, and slope aspect. Topographic factors may also influence climatic erosivity by altering the impact of rainfall or wind velocity and runoff and by influencing the microclimate.

Humans

Among the causes of erosion, humans play a major role in soil erosion through their use and abuse of natural resources. Important human activities in relation to erosion are deforestation, grazing, arable land abuse, faulty farming systems, and cropping intensity.

Types of Erosion

Different types of soil erosion can be classified on the basis of major erosion agents. Fluids or gravity is the principal agent of erosion. Wind, rainfall, and running water are the principal agents of soil erosion on arable land in the tropics. Ice as an agent of erosion is relevant in northern latitudes and in highlands. Glacial processes causing soil erosion are discussed by Embleton (1979) and Derbyshire et al. (1981) and are not dealt with in this book. Gravity, as an agent of erosion, is relevant to steeplands. Severe erosion on steepland is observed in South and Central America, parts of south and southeast Asia, and the highlands of tropical Africa.

Different types of erosion on the basis of major agents involved are shown in Fig. 1.1. Water erosion is classified into splash, sheet, rill, and gully erosion on the basis of the principal processes involved. Splash or interrill erosion is caused by raindrop impact (Plate 1.1). Sheet erosion is the removal of a thin, relatively uniform layer of soil particles. Rill erosion is erosion in small channels only a few millimeters wide and deep (Plate 1.2). Rills are transformed to gullies when they cannot be obliterated by normal tillage. Pipe or tunnel erosion is caused by rapid subsurface flow of water. Stream channel erosion and coastal erosion are caused, respectively, by stream flow and ocean waves. Soil movement en masse is caused by gravity.

Soil Erosion: Natural vs. Accelerated Process

Geologic erosion is a natural and inevitable process, and it does not always adversely affect soil or environments. This slow, constructive process is defined as "the normal or natural erosion caused by geological processes acting over long geologic periods and resulting in the wearing away of mountains, the building up of flood plains, coastal plains, etc." (SCSA, 1982). In fact, some of the most fertile valleys of the world have resulted from this slow and natural geologic erosion, e.g., the Nile Valley in Egypt, Indus Valley in Pakistan, Indo-Gangetic alluvium in northern India, Cauca Valley in Colombia. These fertile valleys were the cradles of some of the ancient civilizations. The fertile soils along the major rivers and their deltas were developed by geologic erosion occurring for millions of years over the watershed feeding the major river systems.

Soil erosion, however, becomes serious when the process is accelerated by human activity and its rate exceeds the threshold value equivalent to the counterbalancing and compensatory rate of new soil formation. The threshold value of erosion—the rate at which it starts depleting soil productivity and causing soil degradation—differs for soils developed on different parent materials and in different climatic regions. It is the accelerated erosion that has severe adverse effects on soils and environments. Accelerated erosion is "much more rapid than normal, or geologic erosion, primarily as a result of man's activities

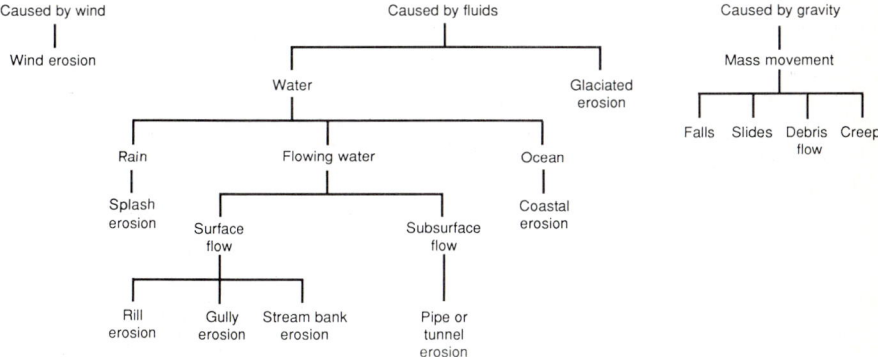

Figure 1.1 Types of erosion.

Plate 1.1 Soil splash or splash erosion is caused by the raindrop impact. (Photo courtesy SWCS, Ankeny, Iowa.)

or, in some cases, of other animals or natural catastrophes that expose bare surfaces, for example, fire" (SCSA, 1982). It is the accelerated soil erosion that menaces, and is the major cause of, soil degradation. Accelerated erosion is caused by perturbations introduced in the soil-vegetation-climate equilibrium mostly by human intervention.

In addition to different rates of denudations, the time scale for two forms of erosion also differs. At geologic time scale, geologic erosion attains a steady state with the rate of land formation. At human time scale, however, accelerated erosion creates an imbalance between the rates of land formation and denudation. Perturbations caused by humans, however, may also alter the processes at the geologic time scale (Fleming, 1981). Henceforth, unless otherwise specified, soil erosion in this book refers to accelerated soil erosion.

Soil Erosion and Degradation

It is important to distinguish among three interrelated but distinctly different phenomena: soil erosion, soil depletion, and soil degradation. *Soil erosion* lessens soil productivity through physical loss of topsoil, reduction in rooting depth, removal of plant nutrients, and loss of water. Soil erosion is a quick process. In contrast, *soil depletion* means loss or decline of soil fertility due to crop removal or removal of nutrients by eluviation from water passing through the soil profile. The soil depletion process is less drastic and can be easily remedied through cultural practices and by adding appropriate soil amendments. *Soil degradation,* however, is an all-encompassing broad term. It implies decline in soil quality through deterioration of the physical, chemical, and biological properties of the soil. Accelerated soil erosion is one of the processes that leads to

Plate 1.2 Rills are small channels only a few millimeters wide and deep. (Photo courtesy SWCS, Ankeny, Iowa.)

soil degradation. Soil degradation may be caused by accelerated erosion, depletion through intensive land use, deterioration in soil structure, changes in soil pH, leaching, salt accumulation, buildup of toxic elements such as aluminum or manganese to toxic levels, or excessive inundation leading to reduced soil conditions and poor aeration.

Still another distinction is that soil erosion can be expressed quantitatively by the depth or weight of soil removed per unit area and unit time. With the present state of knowledge, however, the term *soil degradation* is still ambiguous, vague, and qualitative. The critical limits in relation to crop production for those soil properties that decline with progressive soil degradation are not known. The limits apparently differ for different soils, crops, and climatic conditions. Different soils and crops vary significantly in their critical limits of organic matter content, water and nutrient status, porosity, and compaction beyond which crop growth declines. The time required for a soil to attain these limits also varies for different processes that are responsible for soil degradation. Critical limits for some factors may be attained more quickly by accelerated erosion than by fertility deletion through intensive land use or soil compaction and structural alterations from vehicular traffic. This distinction is often difficult to make when different processes interact with one another and accentuate the problem, as often happens under the natural environment. Human-caused perturbations are the most important factor that accelerates soil erosion and accentuates its interaction with other processes which lead to severe soil degradation.

Soil Erosion and Civilizations in the Tropics

Soil erosion is not just a problem of modern times, although in the past 50 years there has been more awareness of its consequence, more understanding of the processes involved, and more knowledge of its cause-effect relationships than ever before. Soil erosion began with the dawn of agriculture, when people began using the land for settled and intensive agriculture. In fact, soil erosion has been a quiet crisis and has plagued the land since people began practicing agriculture by removing the protective vegetation cover and growing food crops on disturbed soil surface. Some types of soil erosion are not easily observed and may occur unnoticed for a long time. That is often the case with surface wash that can occur unnoticed. By the time it has been noticed, it is often too late to do anything worthwhile to remedy it.

According to some estimates, soil erosion and other degradative processes have destroyed, over the millennia, as much arable land as is now cultivated (Kovda, 1977). It is argued that as many as 2 million ha of arable land is lost annually to severe soil erosion and erosion-induced soil degradation. If these estimates, hard to validate as they are, are anywhere near correct, the consequences are alarming and we are facing a truly enormous challenge.

Soil erosion is considered to have caused some of the once-thriving ancient civilizations to vanish (Olson, 1981). In the Middle East, the cradle of civilization, deforestation of cedar forests from countries surrounding the Mediterranean caused severe corrosion, toppled the Phoenicians, and destroyed granaries of the Roman Empire (Eckholm, 1976). The Phoenicians cultivated steeplands by developing stone terraces; however, severe erosion set in where terraces were not constructed or were poorly maintained. Deforestation and overgrazing of the steep slopes were major causes of severe erosion. Agriculture that was thriving about 10,000 B.C. in Mesopotamia, present-day Iraq, has been converted to desert and shifting sand dunes by deforestation and erosion of the surrounding hills and watersheds that nurtured its ancient irrigation system (Lowdermilk, 1953). The ancient kingdoms of what are now Syria and Lebanon also dwindled because of excessive erosion following deforestation and exposure of soil on steeplands (Beasley, 1972). Negev desert has been inhabited since 10,000 B.C. Its loess plains were fertile and productive, but once again its soils, having succumbed to intensive land use, are now an unproductive desert. Archaeologic evidence indicates that Sardis in western Turkey (now ruins lying amid barren lands) was the capital of the ancient kingdom of Lydia. Recent soil surveys of the regions surrounding the ruined cities indicated that the productivity of once-fertile entisols and inceptisols was depleted by erosion, landslides, inundation, and sediment deposition (Olson, 1981). While uplands suffered erosion because of excessive grazing, the alluvial valleys were destroyed by inundation and sediment deposition (Olson, 1981). The very productive lands of northwestern Africa (Tunisia, Algeria) also were ruined by erosion and desert encroachment (Lowdermilk, 1953).

Another ancient civilization in south Asia suffered a similar fate. Intensive agriculture began in the Indus Valley soon after that in Mesopotamia. The extinction of the Harappan-Kalibangan civilization that flourished in northwest India in the pre-Aryan era is also attributed to erosion, erosion-related soil degradation, and desert encroachment. The Harappan culture flourished about 2000 B.C. along the fertile valleys of Ghaggar, Saraswati, Drishadavati, and old Yamuna in the region of what is present-day Haryana and Rajsthan in India and Sind province in Pakistan. Among other factors, severe wind erosion brought that civilization to ruins (Singh, 1982). The story of Golconda Fort, near Hyderabad, in south central India is written in denuded hills and bare rocks. The prosperity of this region came to an abrupt end in 1650 when the Mughal King Aurangzeb's army seized the fort for almost 40 years. The invading army, along with its horses and elephants, denuded the surrounding hills. Then the high-intensity monsoon rains eventually overpowered the landscape by severely eroding the soil of the steeplands. The result is the ghostly landscape surrounding Hyderabad, characterized by exposed bedrock devoid of soil or vegetation (Plate 1.3).

Soils of South and Central America supported thriving civilizations long before the European settlers discovered the so-called new world. Incas developed elaborate systems of soil management for intensive use of steeplands in the Peruvian Andes. Incas conserved soil and water by constructing stone-walled bench terraces such as those at Machu Picchu, Peru (Plate 1.4). It is estimated that in Peru about 1 million ha of land was terraced, of which about one-third is still in cultivation today (Denevan, 1985). Williams (1987) estimated that these terraces are retained by some 3 billion m^2 or more of rock wall requiring at least 4 million person-years of labor for retaining wall construction alone. This was an heroic endeavor that may have been completed over several generations of continuous work. The thin topsoil was rapidly washed away, however, once maintenance of the terrace system was neglected. Vast areas of barren, un-

Plate 1.3 Barren hills and ghostly rockscape around Hyderabad, central India, are the result of deforestation and overgrazing since the seventeenth century.

productive, and abandoned hillslopes now visible throughout the Andes are testimonials to the effects of misusing land (Plate 1.5).

The collapse of a 1700-year-old Mayan civilization in Guatemala around 900 A.D. is also attributed to accelerated soil erosion. Olson (1981) reported that mollisols developed on limestone bedrock were easily eroded when the forest was cleared. As the population increased, soil depletion set in and the Mayan culture paid the ultimate price (Hardin, 1968).

Soil Erosion in Modern Times

In spite of these concrete examples of the disastrous consequences of land misuse and severe soil erosion, modern humans have not learned their lessons. The magnitude of global soil loss from arable lands around the world now is greater than ever before in human history. The effects of erosion, although temporarily masked by modern technologies, are evident in the high cost of producing and maintaining such civil structures as waterways, reservoirs, and harbors. Soil erosion is severe in all regions, temperate and tropical, wherever the land is used beyond its capability by crop and soil management systems that are ecologically incompatible.

Soil erosion results from human greed regardless of the technological developments. For example, severe erosion occurs in intensively used lands in the midwestern United States, the grain basket of the world. The European settlers adopted farming methods perfected in western Europe, a mild-climate region. Growing open-row crops (corn, cotton, and tobacco) on undulating and steep terrains caused havoc in regions characterized by high-intensity rains. It is estimated that during the last 200 years of exploitative agriculture about one-third of the U.S. cropland has been lost to erosion. About 1 to 2 billion tonnes* of sed-

---

*A tonne is a metric ton, equal to 1000 kilograms. It is abbreviated t.

Soil Erosion and Sedimentation 13

Plate 1.4 The famous stone-walled bench terraces at Machu Picchu, Peru. (Courtesy of Dr. C. Felipe-Morales.)

Plate 1.5 Severe soil erosion of once-terraced hill slopes in the Andes.

iment are washed into U.S. streams every year. Soil erosion is also a serious menace to U.S.S.R. croplands. Dregne (1982) estimated that about 27 percent of arable land in the U.S.S.R. is severely or very severely eroded. Zaslavskiy (1977) reported that erodible lands occupy 11 million km^2 in the European U.S.S.R. The mountainous regions of central Asia are also very severely affected.

Soil erosion is equally, if not more, severe in the tropics and subtropics. Soil erosion is extremely severe in many of the major river basins in the tropics and subtropics. Sediment loads in excess of 14,000 t/(km^2·yr) are not uncommon (Table 1.1).

Soil erosion and the resultant degradation are extremely severe in south Asia. The Himalayan-Tibetan mountain ecosystem is one of the most severely eroded. In Nepal, it is estimated that 63 percent of the Shivalik zone, 86 percent of the middle mountain zone, 48 percent of the transition zone, and 22 percent of the high Himalayas have been reduced to poor and fair watershed conditions (Dent, 1984). In China as much as 46 million ha of the loess plateau that drains into the Yellow River is subject to severe erosion. Soil erosion is equally severe in "red soil" subtropical regions of southern China (Plate 1.6).

Africa is no exception. The Sahel suffers from severe wind erosion during the dry season and accelerated gully erosion during the much-awaited rains (Plate 1.7). The Food and Agriculture Organization (FAO, 1979) reported that in Africa north of the equator 11.6 percent of the total land area is affected by water erosion. High erosion rates are especially prevalent in the coastal regions of northwest Africa. Soil erosion is equally serious in eastern Africa and is particularly menacing in the Ethiopian highlands. The Ethiopian highlands are believed to lose over 1 billion t/yr of topsoil (Brown, 1981). Gully erosion is catastrophic in some parts of southeastern Nigeria (Plate 1.8). Soil erosion is also severe in southern Africa whenever large-scale farming is practiced without appropriate conservation measures.

In the humid tropics, new land is rapidly being brought under the plow by the so-called modern, mechanized system. The results of such techniques, regrettably, are not so modern—the disastrous erosion that has plagued people since they began crop husbandry. The cause of high rates of contemporary erosion in Haiti, Dominican Republic, Ethiopia, Himalayas, or anywhere is the misuse or excessive use of land.

Soil Erosion in the Tropics vs. Temperate Regions

In the context of this book, *tropic* is defined as those regions lying within 23° north and south of the equator. Subtropical regions extend between 23 and 40° north and south of the equator, and temperate regions lie beyond latitude 40°. Tropical regions are further classified into humid, subhumid, semiarid, and arid on the basis of the number of wet months exceeding 10, 5 to 10, 3 to 5, and

TABLE 1.1 Sediment Yields from Some Tropical and Subtropical Catchments

Country	River	Sediment yield, t/(km^2 · yr)	Reference
China	Dali	16,300–25,600	Mou and Meng (1981)
Java, Indonesia	Cilutung	12,000	Hardjowitjitro (1981)
Kenya	Perkerra	19,520	Dunne (1974)
New Guinea	Ause	11,126	Pickup et al. (1981)
Taiwan	—	31,700	Li (1976)

Plate 1.6 Soil erosion and sedimentation in the valley of Yellow River, China.

less than 3, respectively. The climax vegetation of these regions is tropical rain forest, semideciduous rain forest, savanna, and scrub grasses (Lal, 1987). Predominant soils at the humid and subhumid tropic are highly leached oxisols, ultisols, and alfisols and less leached inceptisols. Soils of the semiarid and arid regions are not leached. Most predominant soils of these regions are vertisols, alfisols, inceptisols, entisols, and aridisols (Lal, 1987). A very relevant question often asked is whether soil erosion is more serious in the tropics than in the temperate regions. It is a difficult question to answer. Moreover, the answer is confounded by the fact that severity of erosion and its consequences depend on both the absolute quantity of soil eroded and the depth and quality of soil remaining. The rate of erosion or quantity of soil eroded per unit time and per unit area may not be necessarily greater in the tropics than in temperate regions. However, the consequences of erosion are often drastic in tropical regions. The drastic erosion-caused productivity decline in soils of the tropics is due partly to harsh climate and partly to low-fertility and poor-quality subsoil.

It is because of the low productivity of the exposed subsoil that erosion is considered more severe in the tropics than in the temperate-zone soils. Moreover, severe soil erosion in the tropics is observed wherever the land is misused and shortcuts are taken in land development and subsequent management. Consider these examples:

1. Accelerated erosion has been severe in only those large-scale mechanized schemes in Africa and elsewhere in the tropics that were hastily designed and implemented with utter disregard to soils and climate, without proper planning, and without a careful appraisal of biophysical and socioeconomic factors.

2. Severe erosion and erosion-related degradation observed in Central and South America and in the Caribbeans are attributed to the "hamburgerization" of tropic rain forest into pastures regardless of slope steepness, soil suitability, and appropriate stocking rates.

3. Erosion is indeed serious in Parana and other regions of central Brazil where deep oxisols developed on long, undulating terrains have fallen victim to the thriving international soybean trade.

4. Erosion is observed to be serious on structurally unstable and predominantly low-activity alfisols in west Africa, where the agriculture was utterly neglected during the oil boom era of the mid-1970s and where, regardless of the land's suitability, efforts are being made to produce food by so-called modern techniques.

5. Severe erosion observed on vertisols in central India, Australia, and in eastern Africa is attributed to the lack of appropriate technology to overcome the problem of trafficability on wet, clayey soils to provide the much needed protective ground cover during the monsoons.

Plate 1.7 Severe gully erosion in the Sahel, near Niamey, Niger.

Plate 1.8 Gully erosion in eastern Nigeria.

6. Gully erosion is also severe and even disastrous on soils developed on coastal sediments in southeastern Nigeria. Here the social factors have played havoc with the natural environment. The planning and implementation of some development projects leave much to be desired. Those concerned with construction of new roads have often neglected to provide appropriate runoff outlets on highways and civil structures (Plate 1.8). The mining companies have often abandoned the mines without restoring the land. A majority of the rural population in tropical Africa often walk a mile or two many times a day on steep terrain to fetch water for household use. The sunken footpaths often turn into a disastrous gully virtually overnight. Footpaths turned gullies have devoured houses, buildings, roads, and anything else into the path.

The severity of erosion in the tropics, like anywhere else, can be traced to people's greed, short-sightedness, poor planning, and other avoidable causes and otherwise perfectly manageable factors.

Not all soils of the tropics are shallow, fragile, or structurally unstable. Some soils are as stable and resistant to forces of erosion as anywhere in the world. However, harsh climatic conditions, overpopulation, excessive use of the resources beyond their capability, and the gross imbalance in the soil-food-population equation have made tropical ecosystems extremely vulnerable to soil erosion and erosion-induced soil degradation.

Regardless of the climate and ecological factors, soil erosion can be a severe problem in any region where the land is misused and exploited for quick gain.

Soil Erosion and Food Production in the Tropics

For two decades ending in 1980, food production has grossly lagged population in many tropical countries. In subsaharan Africa, the food staples grew only at 1.7 percent a year, but population increased at 3.2 percent a year (Paulino, 1986). Population growth has also surpassed food production in north Africa and the Middle East. Food production has to be increased substantially to avoid mass starvation in many regions.

Although the perpetual food deficit in Africa cannot be entirely attributed to erosion and erosion-induced soil degradation, there is a disturbing degree of correspondence between the areas affected by severe soil erosion and those prone to gross food deficit. There is some flexibility in bringing new land under production, especially in central Africa, South America, and the outer islands in Indonesia. However, it is hardly justifiable to convert tropical rain forests to new arable lands if we are unable to preserve and enhance the productivity of existing lands. And yet, even slight intensification of agriculture on existing lands can drastically increase the soil erosion risks.

Africa, in particular, faces the greatest challenge of breaking the vicious cycle of erosion-induced soil degradation and the resultant decline in crop productivity. Intensification of agriculture is necessary to increase food production, although this also increases the risk of soil erosion. Soil erosion decreases crop productivity which necessitates even more intensification and bringing even the steep, shallow, and marginal land under cultivation.

References

ASCE. 1975. *Sedimentation Engineering,* ASCE, New York.
Beasley, R. P. 1972. *Erosion and Sediment Pollution Control,* 1st ed., Iowa State University Press, Ames.
Brown, L. R. 1981. Eroding the base of civilization. *J. Soil Water Conserv.* 36:255–260.
Brown, L. R. 1984. The global loss of top soil. *J. Soil Water Conserv.* 39:162–165.
Brunsden, D. 1979. Mass movement. In: *Process in Geomorphology* (C. Embleton and J. Thornes, eds.), pp. 130–186, Edward Arnold Ltd., London.
Butchbaker, A. F. 1969. Hailstorm characteristics in the vicinity of a hail suppression project in southwestern North Dakota during 1966, 1967, and 1968. Agr. Eng. Dept. Rpt., N. Dak. State Univ., Fargo, N.Dak.
Changon, S. A., Jr. 1972. Examples of economic losses from hail in the United States. *J. Appl. Meteor.* 11:1128–1136.
Clark, M. 1979. Marine processes. In: *Process in Geomorphology* (C. Embleton and J. Thornes, eds.), pp. 352–377, Edward Arnold Ltd., London.
Denevan, William. 1985. Terrace abandonment in the Peruvian Andes: Extent, causes and prospects for restoration. *Proc. Consejo Nacional de Ciencia Y Tecnologia,* Seminiario-Taller, Lima, Peru (Resumen).
Dent, F. J. 1984. Land degradation: Present status, training and education needs in Asia and the Pacific. In: *UNEP Investigations on Environmental Education and Training in Asia and the Pacific,* pp. 1–20, FAO Regional Office, Bangkok, Thailand.
Derbyshire, E., Gregory, K. J., and Hails, J. R. 1981. *Geomorphological Processes,* Butterworths, London.
Dregne, H. E. 1982. Historical perspective of accelerated erosion and effect on world civilization. In: *Determinants of Soil Loss Tolerance,* pp. 1–14, ASA Spec. Publ. 45, Madison, Wis.
Dunn, T. 1974. Suspended sediment data for the rivers of Kenya.

Eckholm, E. P. 1976. *Loosing Ground,* Norton, New York.
Embleton, C. 1979. Glacial processes. In: *Process in Geomorphology* (C. Embleton and J. Thornes, eds.), pp. 272–306, Edward Arnold, London.
FAO, 1979. *A Provisional Methodology for Soil Degradation Assessment.* FAO/UNEP/UNESCO, Rome.
Fleming, G. 1981. The sediment problem related to engineering. In: *Proc. Southeast Asian Regional Symposium on Problems of Soil Erosion and Sedimentation* (T. Tingsanchali and H. Eggers, eds.), pp. 3–14, Asian Institute of Technology, Bangkok, Thailand.
Hagen, L. J., and Butchbaker, A. F. 1967. Climatology of hailstorms and evaluation of cloud seeding for hail suppression in southwestern North Dakota. In: *Proc. Fifth Conf. Severe Local Storms,* St. Louis, American Meteorological Society, pp. 336–347.
Hagen, L. J., Lyles, L., and Dickerson, J. D. 1975. Soil detachment from clods by simulated rain and hail. *Trans. ASEA* 18:540–543.
Hardin, G. 1968. The tragedy of the commons. *Science* 162:1243–1248.
Hardjowitjitro, H. 1981. Soil erosion as a result of upland traditional cultivation in Java Island. In: *Proc. Southeast Asian Regional Symposium on Problems of Soil Erosion and Sedimentation* (T. Tingsanchali and H. Eggers, eds.), pp. 173–179, Asian Institute of Technology, Bangkok, Thailand.
Kovda, V. A. 1977. Soil loss: An over-view. *Agro-Ecosystems* 3:205–224.
Lal, R. 1987. *Tropical Ecology and Physical Edephology,* Wiley, New York.
Larson, W. E. 1981. Protecting the soil resource base. *J. Soil Water Conserv.* 36:13–16.
Lee, H. T., 1984. Soil conservation in China's loess plateau. *J. Soil Water Conserv.* 39:306–307.
Li, Y. H. 1976. Denudation of Taiwan Island since the Pliocere Epoch. *Geology* 4:105–107.
Lowdermilk, W. C. 1953. Conquest of the land through seven thousand years. *SCS Agric. Inform. Bull.* 99, Washington.
Mitchell, J. K., and Bubenzer, G. D. 1980. Soil loss estimation. In: *Soil Erosion* (M. J. Kirby and R. P. C. Morgan, eds.), pp. 17–62, Wiley, Chichester, England.
Mou, J., and Meng, Q. 1981. Sediment delivery ratio as used in the computation of watersheds sediment yield. *J. Hydrol.* 20:27–38.
Olson, G. W. 1981. Archaeology: Lessons on future soil use. *J. Soil Water Conserv.* 36(5):261–264.
Paulino, L. A. 1986. Food in the Third World. IFPRI Res. Rpt. 52, Washington.
Pickup, G., Higgins, R. J., and Warner, R. F. 1981. Erosion and sediment yield in Fly River drainage basins. *IAHS Publ.* 132:438–456.
Rapp, A. 1963. The debris slides at Ulvadal, western Norway: An example of catastrophic slope processes in Scandinavia. *Nach. Akad. Wissen. Gottingen, Math-Physik, Klasse* 13:195–210.
SCSA. 1982. *Resource Conservation Glossary,* 3d ed., Soil Conservation Society of America, Ankeny, Iowa, p. 193.
Singh, H. P. 1982. Management of desertic soils. In: *Review of Soil Research in India,* pp. 676–699, ICAR, New Delhi, India.
Thornes, J. 1979. Fluvial processes. In: *Process in Geomorphology* (C. Embleton and J. Thornes, eds.), pp. 213–271, Edward Arnold Ltd., London.
Williams, L. S. 1987. Inca terraces and controlled erosion. *Proc. Conf. of Latin Americanist Geographers,* Merida, Mexico, 6–10 January 1987.
Wischmeier, W. H. 1976. Use and misuse of the universal soil loss equation. *J. Soil Water Conserv.* 31:5–9.
Zaslavskiy, M. N. 1977. Erodible lands in the USSR. *Soviet Soil Sci.* 9:460–464.

Part 2

Basic Processes

Chapter 2

Physical Processes and Climatic Erosivity

Work Involved in Soil Erosion

Soil erosion is physical, fluvial, or geomorphologic "work." Work is involved in detachment, transport, and deposition of soil particles. Performing work involves expending of energy. Energy is supplied by the agents of erosion, e.g., raindrops, water and wind movement, gravity, etc. Depending on the predominant agents of soil erosion, different types of energy are involved in performing the work, e.g., kinetic and potential energy. Chemical energy plays an indirect role, such as in solution weathering. Now we discuss some of the basic physical principles involved in performing erosion work.

Force and work

Agents of erosion are the forces involved in the erosion process. The term *force* denotes an operating agent. It may also be defined as the capacity to persuade. In the context of soil erosion, it is the force of raindrops, wind and water flow, or gravity that causes or persuades soil movement. While causing the soil to move, the agents of erosion perform work. For example, when a force acts on a soil with slope θ and the soil moves through a distance d, the work done in causing soil erosion is given by

$$\text{Work} = F(\cos \theta) d \qquad (2.1)$$

where F is the force, θ is the slope angle, and d is the distance. Both F and d are vectors, but the work performed is a scalar quantity. In SI (Système International) units, the force is measured in newtons; i.e., one newton (1 N) is the force required to give a 1-kg mass of soil an acceleration of 1 m/s².

Potential energy

Potential energy is the energy stored in a system as a result of previous work done on the system, e.g., energy stored by virtue of position or in chemical bonds. The difference in potential energy among two soil particles is given by

$$\Delta \text{ potential energy} = (M_A - M_B)gh \qquad (2.2)$$

where M_A and M_B are the masses of particles A and B, respectively, g is the acceleration due to gravity (9.8 m/s²), and h is the vertical distance between the two particles which is equal to $d \sin \theta$.

The potential energy of soil on steeplands is a major factor of erosion and results from the position of the soil above the earth's surface. Potential energy in landscapes is created by geologic processes. A soil body with high potential energy is in an unstable equilibrium. The body has many forces acting on it. If the shear strength or the resistance force is less than the gravitational pull, the soil will be displaced until it attains a new equilibrium. The SI unit of energy is the *joule* (J), where 1 J is the work done when a force of 1 N moves soil a distance of 1 m. Energy is also measured in calories, where 1 cal is equal to 4.2 J. One calorie is the heat required to raise the temperature of 1 g of water 1°C. The *heat of wetting* soil refers to the energy released when water vapors are absorbed on the soil's surface. The maximum heat of wetting is released when a monomolecular layer of water is formed on the surface of soil particles.

Kinetic energy

Kinetic energy is possessed by a moving body, e.g., falling raindrops, water moving on the soil surface and in the subsurface, flowing water in channels and streams, moving ice as a glacier, blowing wind, etc. The kinetic energy E of a body is

$$E = \tfrac{1}{2}mv^2 \qquad (2.3)$$

where m is the mass of a moving body and v is its velocity in meters per second.

In soil erosion, we are concerned with the terminal velocity of raindrops and the velocity of overland flow. The *terminal velocity* is defined as "the constant equilibrium velocity of a free-falling body." Terminal velocities of bodies falling through the earth's atmosphere are proportional to their weight. For example, the terminal velocity of a raindrop depends on the drop size.

The kinetic energy of flowing water depends on the velocity of water moving on the surface in a stream or a channel. Velocity in a channel is related to discharge and channel characteristics. Manning's equation to calculate velocity in a channel flow is

$$v = \frac{1.009}{n} R^{2/3} S^{1/2} \qquad (2.4)$$

where v is the velocity, S the gradient, R the hydraulic radius, and n the roughness coefficient.

In addition to the velocity of water flow, the shear stress exerted by flowing water depends on the type of flow, i.e., laminar or turbulent. The type of flow depends on the ratio of inertial forces to viscous forces. The inertial forces increase with an increase in velocity. The laminar flow occurs at low velocity so a streak of dye introduced in a laminar flow stays coherent, straight of constant width, and is not distorted. Turbulent flow occurs at high velocity, and a streak of dye is immediately distorted. The ratio of inertial to viscous forces is defined by the Reynolds number R

$$R = \frac{vL}{\eta} \qquad (2.5)$$

where R is dimensionless, v is velocity (m/s), L is a characteristic length or flow depth (m), and η is the kinematic viscosity (m²/s). Turbulent flow has more shear strength to detach and transport particles than laminar flow.

Momentum

The *momentum* P of a body is the product of its mass m and velocity v and is a vector quantity:

$$P = mv \qquad (2.6)$$

Kinetic energy and momentum are related and have the same variables. The law of conservation of momentum has an important application in soil erosion processes. Newton's second law of motion states that the rate of change of momentum of a body is equal to the resultant of all external forces exerted by the body. A similar law was postulated by the French physicist Descartes, who stated that when two bodies collide, the sum of their momentums will not change. In an elastic collision between two bodies, there are no frictional losses, and the resultant momentum of two bodies after collision is exactly equal to that before the collision. This law has an important application in the erosion caused by raindrops colliding with soil particles.

Common Algebraic Functions Used in Explaining Erosion Processes

Five principal mathematical functions [$F(X,Y)$] commonly used to express functional relationships among independent (X) and dependent (Y) variables involved in performing the erosion work:

1. *Linear.* The linear function is the most simple mathematical equation, and it is

$$Y = aX + b \tag{2.7}$$

where a and b are experimentally determined constants.

2. *Logarithmic.* Function $F(X,Y)$ becomes linear when X is plotted on a logarithmic scale and Y on a linear scale. This function is denoted by

$$Y = a \ln X + b \tag{2.8}$$

where ln is the natural logarithm.

3. *Power.* When Y is related to X through a power relation, the function is defined as follows:

$$Y = aX^b \tag{2.9}$$

4. *Exponential.* The exponential function relates Y and X according to

$$Y = ae^{bX} \tag{2.10}$$

where e is the naperian logarithm.

5. *Polynomial.* The polynomial equation is commonly used to describe effect of slope length on erosion and has the form

$$Y = a + bX + cX^2 + \cdots + mX^n \tag{2.11}$$

These functions are widely used to describe functional relations among various erosion processes.

Factors Affecting Soil Erosion

Factors of soil erosion are those ecological parameters that influence the effects of the agent of soil erosion. The most important factors of soil erosion are climate, soil, hydrology, landforms, and humans (Fig. 2.1). People are the most important active factor. Through their activity, people can either accelerate soil erosion by land misuse or curtail it by adopting proper soil and crop manage-

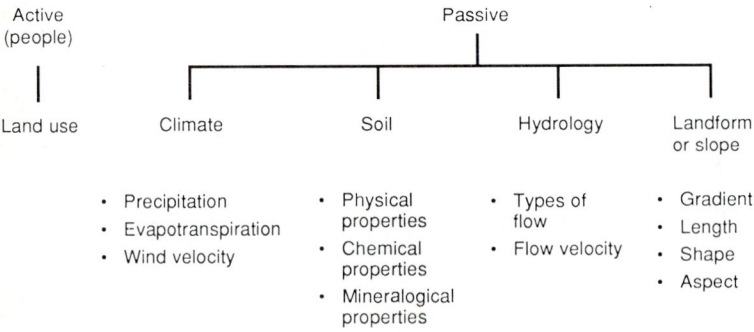

Figure 2.1 Factors of soil erosion.

ment practices. Because the effects of climate, hydrology, landform, and soil properties can be controlled and manipulated through management, these factors may be designated "passive." This chapter explains climate as a factor of soil erosion. Soil properties are discussed in Chap. 3, landforms in Chap. 4, and hydrologic processes over a watershed in Chap. 6.

Climate

In relation to soil erosion, the two most important climatic factors having direct effects on erosion are precipitation and wind velocity. Other climatic factors have an indirect effect on soil erosion, such as water balance, evapotranspiration, temperature, and relative humidity. Indirect factors affect the erosivity of rainfall by altering the soil moisture regime and the proportion of rainfall that may become overland flow.

Precipitation is an all-encompassing term, and it comprises fog, mist, hail, snow, and rain. It is the snow and rain that play major roles in soil erosion. In northern latitudes, snow melt causes high amounts of runoff in spring. Snow melt, however, is not a factor of water erosion in the lowland tropics or subtropics. In the humid tropics, rainfall and associated wind are important climatic factors that affect the potential water erosion of an ecology.

Rainfall

Soil erosion is affected by the character of rainfall, including the amount, distribution, energy load, seasonality, and variability. In general, erosion per unit amount of rain decreases with increasing rainfall. Similarly, the relative potential erosion by wind decreases with increasing rainfall. The amount of rainfall governs the overall water balance and the relative proportion that becomes runoff. In addition to the rainfall amount, intensity and distribution during storm are important characteristics in soil erosion. The distribution of intensity during storm affects erosion by altering the amount and rate of overland flow. The time of the peak intensity period in a rainstorm influences the amount and rate of runoff. Some storms have their highest intensities in the beginning and lowest intensities later. For example, data in Fig. 2.2a show an intensity of 213 mm/h during the first 5 min of the storm. Such storms are called *advanced storms*. They usually generate less runoff because the soil water storage capacity is still unutilized and soil can absorb excessive rain. Other storms begin with medium intensity and reach their peak later (Fig. 2.2b). They are called *delayed storms* and generate considerable runoff. There are also *composite storms* with peak intensities in the beginning and in later stages (Fig. 2.2c). Each intensity distribution pattern presents a different type of soil erosion hazard.

The effects of rainfall characteristics on erosion are not constant but vary with soil type, relief, and predominant vegetation. Over and above the amount of rainfall, it is effective rainfall that influences the amount and rate of surface runoff. In terms of soil erosion, *effective rainfall* refers to the proportion that reaches the soil surface directly, infiltrates the ground, and does not contribute to overland flow. The effectiveness of rain depends on the rainfall amount and

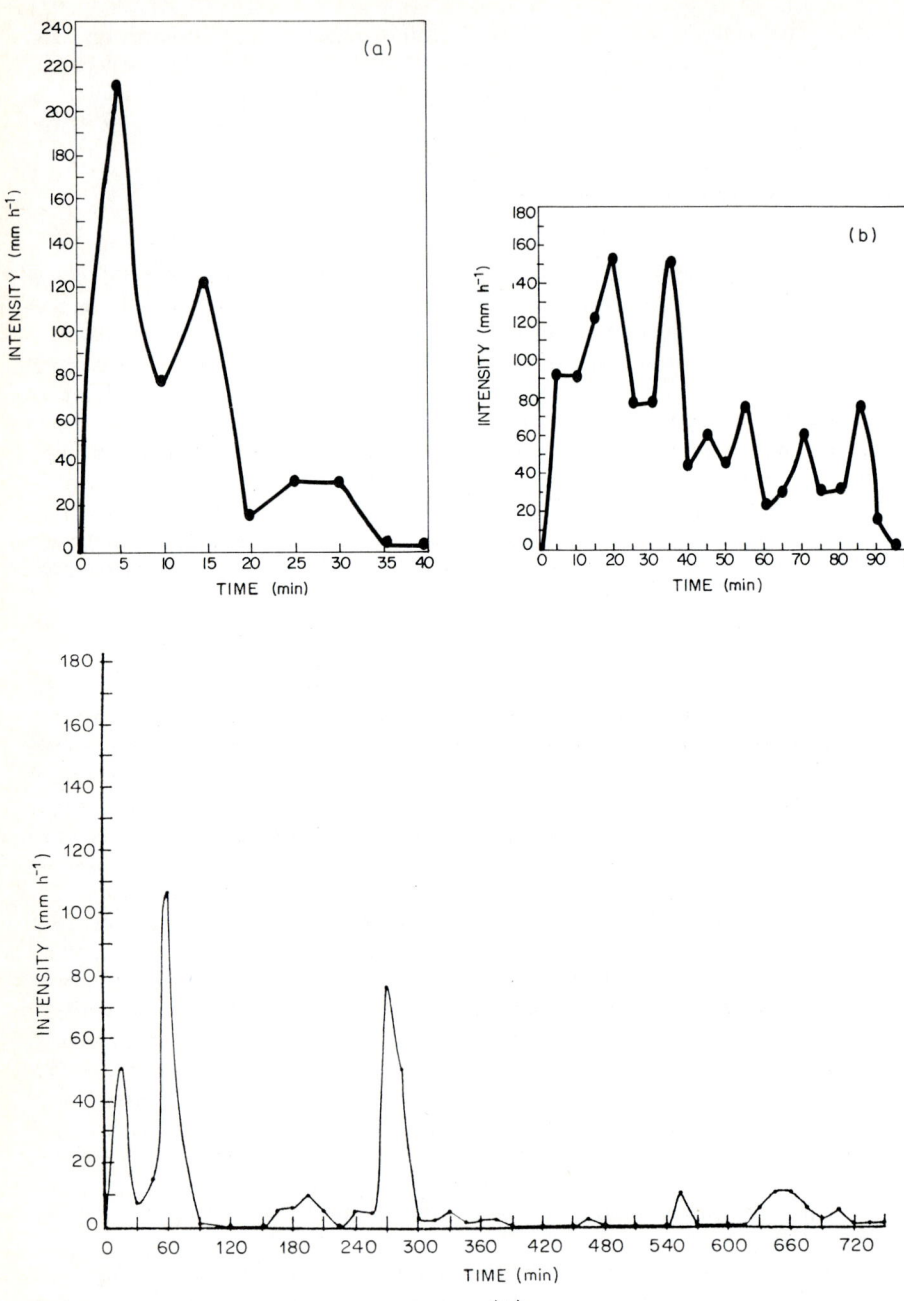

Figure 2.2 (*a*) An advanced rainstorm, (*b*) a delayed rainstorm, (*c*) a composite storm.

Physical Processes and Climatic Erosivity   29

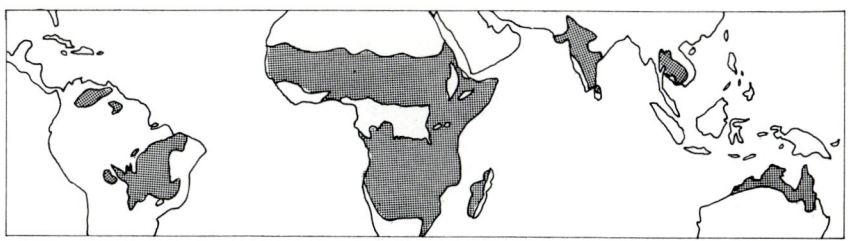

Figure 2.3 Shaded areas show subhumid and semiarid climates with severe problems of soil erosion.

intensity and on other physical factors, i.e., soil, vegetation cover, slope, and evaporation.

The broad category of moisture regimes (i.e., humid, subhumid, semiarid, and arid) depends on relative proportions of precipitation and evapotranspiration. Soil erosion by water is often severe in subhumid and semiarid regions (Fig. 2.3) characterized by a prolonged dry season. After a prolonged dry period, intense downpours fall on soil devoid of any protective vegetation cover and cause severe erosion. Erosion by water becomes severe in humid regions only when the existing vegetation cover is removed (Fig. 2.4). Wind erosion is severe in semiarid and arid climates (Fig. 2.5).

Tropical vs. temperate rains. Characters of tropical rains are influenced by high temperatures and the general nature of atmospheric circulation. In contrast to rains in the middle latitudes, rains in the tropics are not produced by cyclonic circulation and frontal effects, but are mainly caused by convection, which causes air masses to rise from 6000 to 15,000 m where the water vapor condenses into droplets. The droplets serve as nuclei of condensation during their fall and increase in size. Such convectional rains are greatly influenced by relief, e.g., a rapid ascent caused by a mountain range. That is the reason that the effects of the windward and leeward sides on rainfall distribution are drastic.

Rains in the tropics, particularly those caused by thunderstorms, have sharp, high-intensity peaks. Because tropical rains are caused by convection, they are generally accompanied by lightning and thunder, are localized, and are intense. In general, tropical regions receive a high proportion of torrential

Figure 2.4 Shaded areas show humid climates where erosion becomes severe whenever protective vegetation cover is removed.

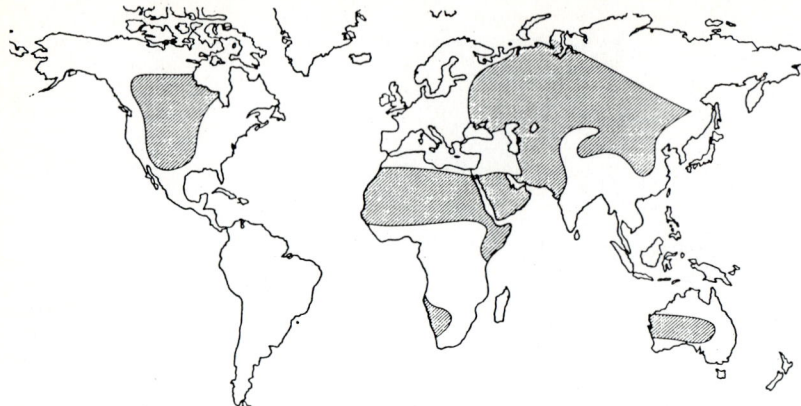

Figure 2.5 Shaded areas show regions particularly susceptible to wind erosion.

rains. These intense rains are partly related to the high temperature. The high temperatures inhibit the formation of raindrops from ice crystals and permit relatively large quantities of water vapor per unit volume of air. The air masses that are cooled during convection release large quantities of water through condensation. The result is intense downpours, high rates of rainfall per unit time, and relatively high drop size.

Rainfall intensity. Both the amount and the rate of rainfall, or its intensity, affect soil erosion. The same amount of rain falling over a short time causes more

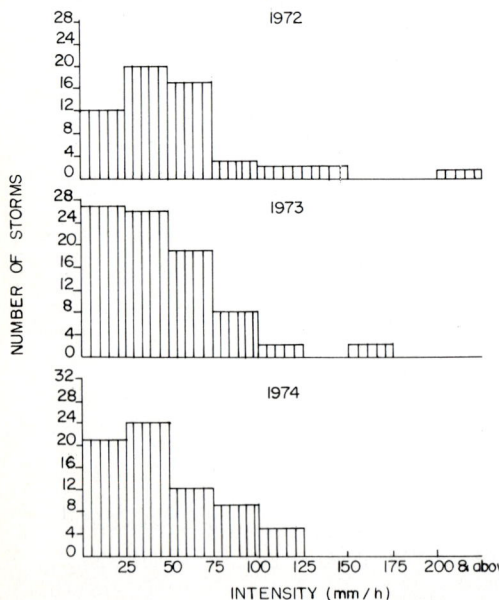

Figure 2.6 Intensity distribution of some tropical rains at Ibadan, Nigeria, from 1972 to 1974.

Physical Processes and Climatic Erosivity

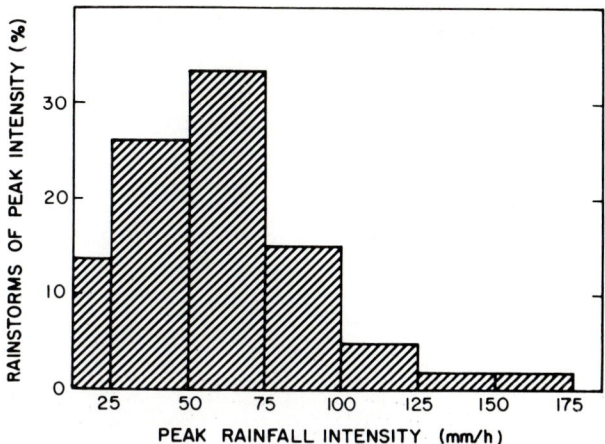

Figure 2.7 Distribution of peak intensities for 60 storms during 1975 in Ibadan.

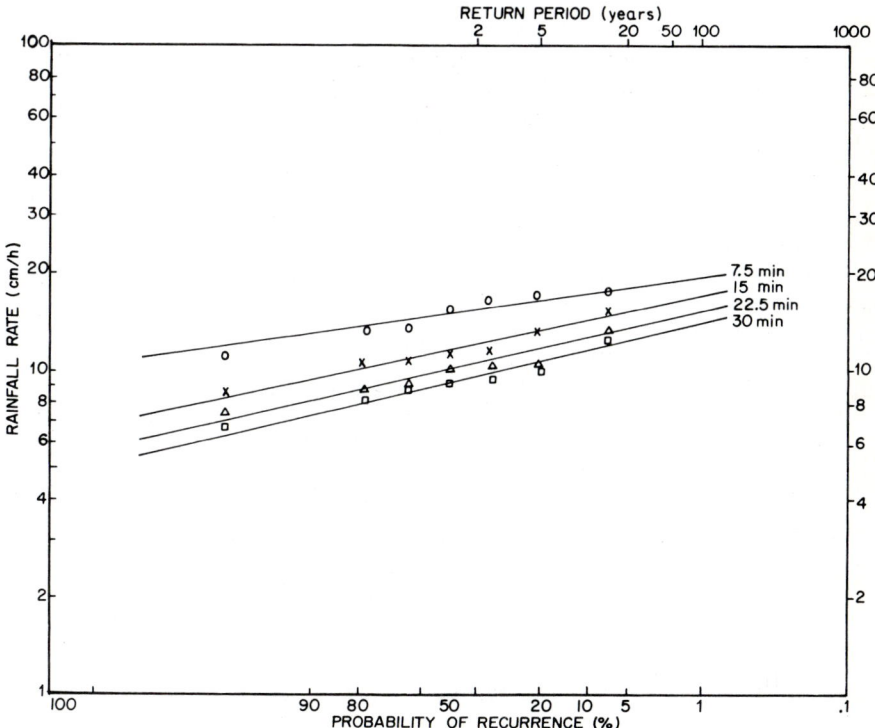

Figure 2.8 Return period of rainstorms received at Ibadan from 1972 to 1977.

Basic Processes

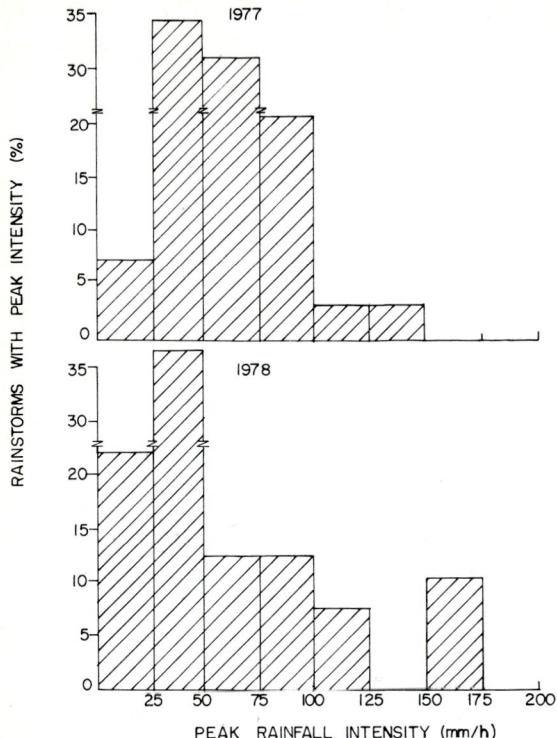

Figure 2.9 Intensity distribution of some tropical rains at Ibadan from 1977 to 1978.

erosion than when it is distributed over a relatively long time and falls as a gentle rain of low intensity.

For the reasons explained above, tropical rains fall at higher intensity than temperate rains. Some tropical rains may attain a short-time (5- to 10-min) in-

TABLE 2.1 Rainfall Intensity Duration Relations for Indore, Central India

Storm duration, min	Storm intensity, mm/h (maxima recorded)			
	1963–1974	1976	1977	1978
5	—	132	125	125
10	—	120	—	81
15	144	88	80	72
30	132	62	68	52
45	94	57	39	44
60	62	46	36	36
75	40	—	—	—

SOURCE: Shaxson et al., 1981.

TABLE 2.2 Maximum Rainfall Intensities in Indonesia

Duration, min	Maximum intensity, mm/h					
	Jakarta	Bogor	Pasuruan	Padeng	Pontianak	Ambon
5	192	192	192	192	228	216
15	140	156	140	172	160	156
30	118	130	128	152	146	96
60	88	110	86	100	80	63

SOURCE: Keersebilik et al., 1972.

tensity of 150 to 200 mm/h (Figs. 2.6 and 2.7). High rainfall intensity is related both to the relatively big drop size and to the number of drops falling per unit area per unit time. Data in Fig. 2.6 show the frequency distribution of rainfall intensity for rainstorms recorded at the International Institute of Tropical Agriculture (IITA), Ibadan, Nigeria, from 1972 to 1974. The 7.5-min maximum intensities of 61 percent of the rainstorms in 1972 and 57 percent of those during 1973 were between 25 and 75 mm/h. The intensity range of 25 to 75 mm/h is commonly observed in temperate regions, e.g., North America and Europe. But intensities exceeding 75 mm/h were recorded for 15 percent of the rainstorms at IITA during 1972 and 10 percent during 1973. Intensities exceeding 100 mm/h were recorded for 7 percent of the storms during 1972 and 3 percent of those during 1973.

Similar records for 1975, 1972–1977, and 1977–1978 in Figs. 2.7 to 2.9 show that 6.4 and 17.1 percent of rainstorms, respectively, had peak intensities exceeding 100 mm/h. In 1978, about 10 percent of the rainstorms had peak intensities exceeding 150 mm/h. The probability of receiving rainstorms at different intensities at Ibadan, based on records for 1972 to 1977, are shown in Fig. 2.8. Records for that short period show that a storm with intensity of 145 mm/h sustained over 7.5 min has a return period of only 2 years. Occurrences of similarly intense rains with sharp peaks were reported for western Nigeria by Wilkinson (1975) and for northern Nigeria by Kowal (1972).

High rainfall intensities are also reported from tropical Asia (Tables 2.1 and

TABLE 2.3 Observed Rainfall Intensities in East Shropshire, England

Date	Time of event	Duration, min	Rainfall, mm	Rainfall intensity, mm/h
9/12/82	1327–1410	43	1.4	1.45
14/3/83	1130–1150	20	0.4	1.2
15/9/83	1415–1420	5	0.4	8.0
10/10/83	1035–1045	10	0.3	1.8
30/1/84	1300–1400	1	1.0	1.0

SOURCE: Fullen, 1985.

2.2). In Taiwan, the highest annual rainfall of 6000 mm was recorded at Ho-Sha-Liao in Keelung area (Lee, 1981). A daily rainfall of 1000 mm, apparently associated with a typhoon, was recorded at Hsin Liao in I-Lan region of Taiwan (Chang and Lee, 1977). Similarly intense rains are reported from other regions of the tropics, e.g., the Caribbeans, southeast Asia, and the Pacific.

Rainfall intensities are generally low in the oceanic temperate-zone climate of Europe. The data in Table 2.3 show extremely low rainfall intensities of 1.0 to 8.0 mm/h computed over 1 to 43 min. Low-energy events are a common feature of the north European climate. Such low-energy erosive events can cause erosion, especially on compacted soils that have low erosivity thresholds (Fullen, 1985). Moreover, high intensities are sometimes observed in the temperate zone, for intensities exceeding 150 mm/h have been observed in Germany. Although such intensive rains may comprise as low as 3 to 10 percent of all rains received, it is the intense storms that cause as much as 90 percent of the total erosion recorded. High rainfall amounts and intensities have also been reported in the United Kingdom (Evans and Nortcliff, 1978; Table 2.4). Rodda (1967) developed a statistical procedure to predict the occurrence of intense rains in the United Kingdom and reported that the relationship between the average annual rainfall and the 1-day maximum is linear. The frequency of intense rainfall in Poland has been reported by Lambour (1967). High-intensity rains are less frequent in temperate Europe than in tropical climates.

Compared with Europe, North America and temperate Australia have more high-intensity rains. Short-time intensities exceeding 100 mm/h are observed in the United States and southeastern Australia. That is one reason why soil erosion is more severe in North America than in western Europe. In Zanesville, Ohio, the 5-min maximum intensity has been reportedly as high as 254 mm/h.

TABLE 2.4 Occurrence of Annual Rainfall Events Exceeding 7.5 mm at Wolterton Park, 1966–1975

	Rainfall events, mm/day		
	7.5–10	>10	>20
1975	7	17	3
1974	8	18	2
1973	6	14	2
1972	7	12	2
1971	11	15	2
1970	7	15	2
1969	13	23	7
1968	12	18	5
1967	9	15	2
1966	16	14	3
Mean	10	16	3

SOURCE: Evans and Nortcliff, 1978.

TABLE 2.5 Rainfall Intensity Measured in Zanesville, Ohio, from 1934 to 1942 and Its Effects on Erosion from a Bare Soil

5-min maximum intensity, mm/h	No. downpours	Average erosion per downpour, t/ha
0–25.4	40	3.75
25.4–50.8	61	5.95
50.8–76.2	40	11.78
76.2–101.6	19	11.44
101.6–127.0	13	34.24
127.0–152.4	4	36.32
152.4–177.8	5	38.72
228.6–254.0	1	47.93

SOURCE: Modified from Fournier, 1967.

The intensity exceeded 100 mm/h for about 23 percent of the downpours recorded over an 8-year period. The data in Table 2.5 show that erosion rates increased 12 to 13 times as the 5-min intensity increased 10 times.

Although the frequency of intense rains is low in temperate regions, the rains of even moderate intensities cause more severe erosion there than similarly moderate-intensity rains in tropical regions. Temperate-zone ecosystems are not adapted to high-intensity rains. That is why the threshold rainfall intensity and amount to cause measurable erosion are higher in the tropics than in temperate regions. For example, Hudson (1976) proposed that rainstorms with an intensity of less than 25.4 mm/h are not erosive. In the United States, however, Wischmeier (1955) considered the threshold to be 12.7 mm/h. In Belgium Bollinne (1985) observed that rainfalls of less than 13.0 mm on bare soil were responsible for as much as 33.5 percent of total runoff. Therefore, in the oceanic temperate climate such as in western Europe, the threshold rainfall of 12.7 mm is considered too high. Instead, Bollinne proposed a threshold of 1 mm.

Drop size distribution. Intense rains are caused by big drops, more drops per unit area, per unit time, or both. The drop size distribution is an important factor affecting rainfall erosivity. Attempts have been made to measure drop size since 1892 when Lowe reported the first recorded measurements (Hudson, 1981). Techniques to measure the size of falling raindrops have been described elsewhere in detail (Hudson, 1981) and are therefore discussed only briefly here.

1. The size of a raindrop is estimated from the size of the splash it makes on some collecting device, e.g., paper, slate. The collecting surface is often sprayed with a suitable dye that changes color upon getting wet and forms a stain. The stain method is one of the earliest used techniques (Wiesner, 1895; Lenard, 1904; Defant, 1905). It has been reviewed by Gillespie (1958) and Hall (1970). The size of the stain is assumed to be proportional to the drop size, and the two are related according to

$$D_r = aS^b_r \qquad (2.12)$$

where D_r is the size of a raindrop, S_r is the size of the stain, and a and b are empirical constants. Advanced techniques involving electronic scanners and image analyzers have been used to measure and count stain size (Attle et al., 1980). In addition to drop size, however, stain size is affected by many other factors, including the fall velocity (Becker, 1907; Neuberger, 1942), absorbent paper thickness, time after application of drops, and porosity of absorbing surface (Gillespie, 1958). The method, therefore, requires careful calibration.

2. Photographs are taken of the falling raindrops, and their size is measured (Rogers et al., 1967; Mutchler, 1971; Mutchler and Hansen, 1970). A modified version of this technique is used to measure the velocity of falling drops (Mache, 1904; Laws, 1941). A radar system is also used to assess drop size distribution at the cloud height (Atlas and Plank, 1953).

3. Raindrops are captured in an oil, and their size is measured with a micrometer. This technique is called the *immersion method* (Fuchs and Petajanoff, 1937; May, 1945). A heavy-grade oil is often used so that drops can easily float on the oil. Both small (less than 0.1 mm) and very large (up to 4.75 mm) drops can be measured by the immersion technique (Eigel and Moore, 1983). Different types of immersion liquids can be used for computing falling drops (McCool et al., 1978; Nawaby, 1970).

4. The size of drops is assessed from the size of pellets made. The flour pellet method is the technique first used by Bentley (1904). The drop weight is related to the weight of the flour pellet formed. The drop size is then related to the weight of the drop according to

$$D_r = 3\sqrt{(6/\pi)W} \qquad (2.13)$$

where D_r is raindrop diameter in millimeters and W is the average drop weight in milligrams. The classical work of Laws and Parsons (1943) was based on the flour pellet technique. It has been described by Hudson (1964) and Carter et al. (1974). The method involves calibrating plain flour by dropping waterdrops of known mass into trays containing about a 25-mm-thick layer of uncompacted flour. The air-dried flour pellets formed are dried at 110°C before sieving. The calibration curve involves a plot of pellet size vs. the mass ratio, i.e., the mass of the waterdrop divided by the mass of the pellet. Aina et al. (1977) used this method to measure the drop size distribution of some rains recorded at Ibadan, Nigeria.

5. The size of the drop is computed from the impact it made. Neal and Baver (1937) measured drop impact by using a torsion-balance technique. Drop size can also be measured from a drop's impact by using the acoustic method (Kinnell, 1972). The design of the distrometer, an impact-measuring device that uses a pressure transducer, is based on drop impact (Joss and Waldvogel, 1967; Schleusener, 1967). Advantages and limitations of this technique have been discussed by Kinnell (1972, 1976).

The drop size of individual rainstorms varies considerably. The upper limit of drops in a natural rainstorm, however, is about 6 mm (Blanchard, 1950).

Drops exceeding 6 mm are unstable and break into smaller sizes. A rainstorm has a wide range of drop sizes, and it is often difficult to characterize the drop size of a rain event. A commonly used criterion to express the drop size distribution of a rain event is to compute its *median volume drop diameter*, or D_{50}. Here D_{50} refers to the drop diameter at which 50 percent of the volume of the rain falls in the shape of drops with a smaller diameter and 50 percent as bigger drops. And D_{50} can be computed on the basis of volume of rainfall or number of drops. The D_{50} computed on the basis of number of drops is usually less than that computed on the volume basis. Although it is a cumbersome and laborious technique, characterizing a rainstorm by expressing its complete range of drop

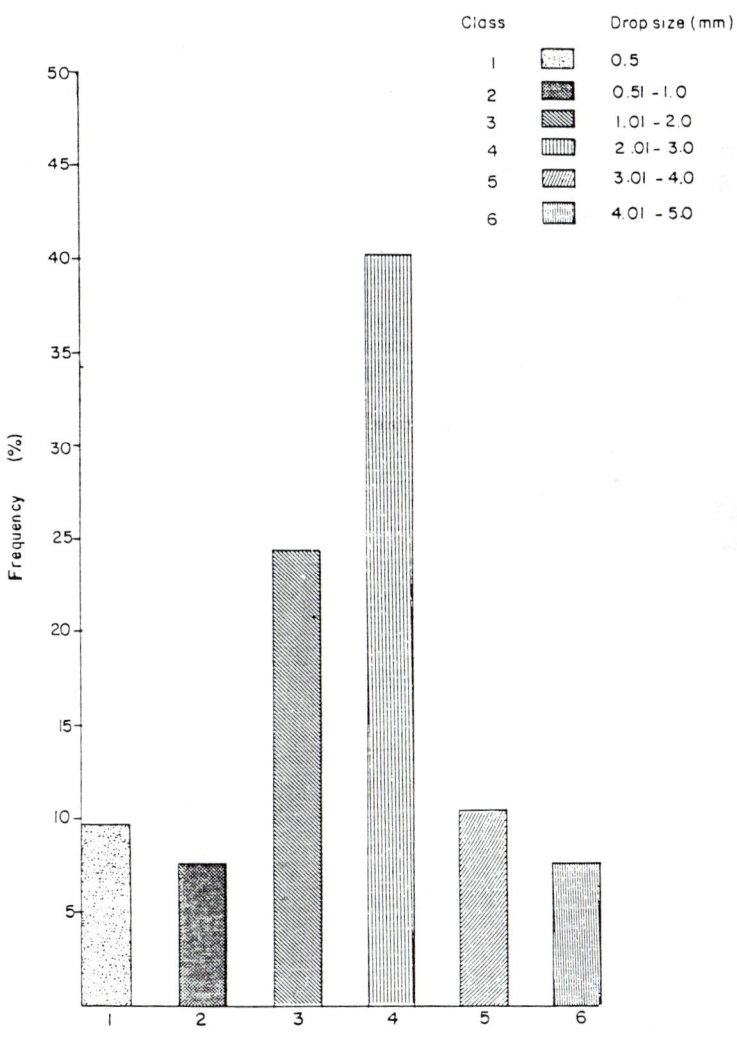

Figure 2.10 Drop size distribution of rains at Ibadan, Nigeria (Lal, 1987).

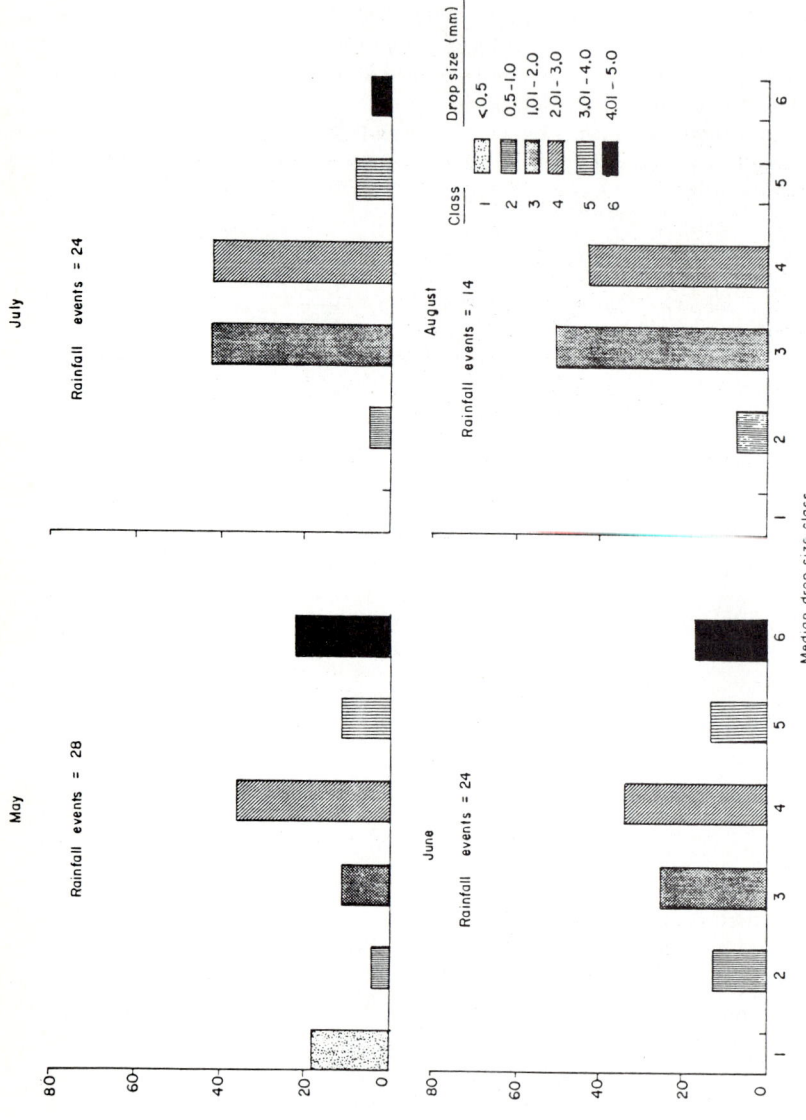

Figure 2.11 Frequency analysis of drop size distribution of rains at Ibadan for different months in the major rainy season (Lal, 1987).

size distribution gives a more complete description of rainfall for a given intensity than does a median drop size (Carter et al., 1974).

On the basis of limited data available from tropical regions, it seems that rains in the tropics are characterized by relatively bigger drops than those in temperate regions. Intense tropical rains of short duration have a relatively high proportion of big drops. Lal (1987) observed that the D_{50} of rainstorms measured at Ibadan ranged from 2 to 4 mm. A comparison of drop size distribution of rain recorded at the IITA, Ibadan, Nigeria (Figs. 2.10 and 2.11), with that at Baton Rouge and Holly Springs in south central United States (Tables 2.6 and 2.7) indicated that rains recorded at Ibadan possess a relatively larger proportion of big drop sizes than those in the United States.

The median drop size is affected by many factors, such as type of rain, amount of rain, duration of rain, rainfall intensity, and wind velocity accompanying the rain. Kowal and Kassam (1976, 1978) measured the drop size distribution of selected rainstorms at Samaru, northern Nigeria. Their data (Fig. 2.12) show tremendous variation in drop size distribution among different types of rains. In general, the larger the rainfall amount per event and the larger the intensity ranges, the bigger the median drop size. Similar observations were made by Baruah (1973) and Carter et al. (1974). The data in Table 2.8, also from Kowal and Kassam (1978), show that for an 18-min, 20-mm rainfall the drop size ranged from 2.34- to 4.86-mm diameter. The predominant diameter was 2.34 mm, while the mean for the rainstorm was 3.42 mm.

High rainfall intensity may result from high median drop size or from more drops per unit area per unit time. For example, D_{50} is related to rainfall intensity (Laws and Parsons, 1943). Median drop size generally increases with increases in rain intensity up to a certain limit according to

$$D_{50} = 2.23 I^{0.182} \tag{2.14}$$

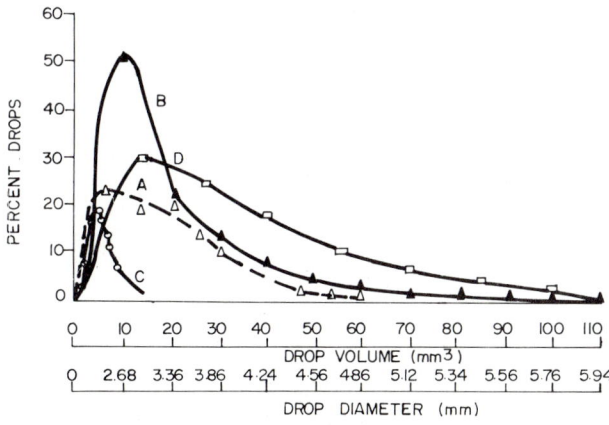

Figure 2.12 Drop size distribution of rains measured at Samaru, Nigeria (Kowal and Kassam, 1978).

TABLE 2.6 Raindrop Size Distribution Typical of the Baton Rouge and Holly Spring Areas in the South-Central United States

Drop diameter, mm	Rainfall intensity, in/h										
	0 to 0.05	0.06 to 0.50	0.51 to 1.00	1.01 to 2.00	2.01 to 3.00	3.01 to 4.00	4.01 to 5.00	5.01 to 6.00	6.01 to 7.00	7.01 to 8.00	8.01 to 9.00	9.01 to 10.0	10.01 and above
	Volume, % of total												
0–0.25	6.7	0.7	0.5	0.5	0.5	0.5	0.6	0.5	0.7	1.5	0.4	0.5	0.4
0.26–0.50	17.3	2.0	1.4	1.3	1.4	1.4	1.9	1.4	2.2	3.2	1.5	1.4	1.9
0.51–0.75	25.2	5.2	2.5	2.1	2.2	2.2	2.8	2.6	3.6	5.0	3.3	2.4	2.3
0.76–1.00	17.5	10.7	3.9	2.9	2.6	2.9	3.1	3.3	3.7	5.8	4.1	2.5	2.6
1.01–1.25	13.1	13.3	5.8	3.9	3.3	3.5	3.6	4.5	4.6	7.2	4.7	2.7	2.8
1.26–1.50	6.0	13.5	8.9	5.4	4.3	4.5	5.0	6.0	5.3	7.7	5.3	3.1	3.3
1.51–1.75	6.2	14.1	9.8	6.1	4.7	4.8	5.2	6.1	5.1	7.8	5.3	3.2	3.3
1.76–2.00	3.2	10.4	9.9	6.5	5.0	5.4	5.9	7.5	5.7	8.6	6.6	3.8	3.8
2.01–2.25	2.6	7.7	10.9	8.0	6.5	6.8	8.2	9.6	7.8	9.7	8.2	4.8	4.8
2.26–2.50	1.2	6.0	9.7	8.0	6.6	7.0	8.2	8.5	8.7	9.5	10.0	5.5	4.9
2.51–2.75	0.6	4.6	8.3	8.4	7.4	7.6	8.8	7.9	8.0	9.2	13.5	6.8	5.7
2.76–3.00	0.4	3.6	6.7	8.0	7.8	8.1	8.7	6.7	8.3	6.9	6.2	7.7	5.8
3.01–3.25	0	2.3	5.4	7.2	7.6	8.2	8.7	7.1	7.9	5.5	6.2	9.4	5.4
3.26–3.50	0	1.7	4.3	6.6	6.9	7.4	7.1	6.5	7.7	4.3	6.1	7.8	6.2
3.51–3.75	0	1.2	2.9	5.5	6.2	6.8	5.6	5.2	4.8	4.1	6.0	7.2	7.8
3.76–4.00	0	1.0	2.8	5.4	6.6	7.3	6.5	4.1	5.9	4.0	12.6	7.7	7.6
4.01–4.25	0	0.7	2.0	5.2	6.0	5.9	5.9	4.3	3.7	0	0	10.4	7.6
4.26–4.50	0	0.7	1.4	3.9	7.0	6.9	3.7	6.0	6.3	0	0	13.1	17.6
4.51–4.75	0	0.2	0.8	2.5	5.1	1.7	0.5	2.2	0	0	0	0	6.2
4.76–5.00	0	0.3	0.9	0.9	1.0	1.1	0	0	0	0	0	0	0
5.01–5.25	0	0.1	0.3	1.3	0.5	0	0	0	0	0	0	0	0
5.26–5.50	0	0	0.9	0.4	0.1	0	0	0	0	0	0	0	0
5.51–5.75	0	0	0	0	0.7	0	0	0	0	0	0	0	0
Number of samples	26	152	90	92	59	37	16	8	6	3	1	2	4

SOURCE: Carter et al., 1974.

TABLE 2.7 Percentages of Total Volume of Rainfall in Various Drop Size Classes and Median Volume Drop Size as Related to Rainfall Intensity

Drop size* class, mm	Rainfall intensity, in/h							
	0.10		0.50		1.00		2.00	
	Urbana	L&P†	Urbana	L&P†	Urbana	L&P†	Urbana	L&P†
0–0.5	0	2.0	0	0.8	0	0.4	0	0.2
0.5–1.0	10.9	16.0	2.1	5.7	1.1	3.8	1.1	3.0
1.0–1.5	57.7	35.4	24.7	18.6	13.0	12.6	9.0	8.8
1.5–2.0	23.0	26.1	29.0	27.6	17.0	22.7	13.1	16.3
2.0–2.5	6.3	12.1	20.3	20.9	19.7	23.3	20.3	21.3
2.5–3.0	1.6	5.2	11.4	13.6	21.8	15.6	20.4	18.7
3.0–3.5	0.5	2.0	7.1	6.8	15.8	10.9	15.8	13.4
3.5–4.0		0.8	2.8	3.0	5.7	5.3	9.6	8.5
4.0–4.5		0.4	1.9	1.8	3.6	2.6	5.8	4.8
4.5–5.0			0.7	0.8	0.8	1.6	0.9	2.4
5.0–5.5				0.4	1.5	0.7	3.2	1.4
5.5–6.0						0.4	0.8	0.8
6.0–6.5						0.1		0.4
Median volume drop size, mm	1.4	1.4	2.0	1.9	2.5	2.2	2.7	2.6

*Data were originally measured in 0.1-mm size intervals but were combined into 0.5-mm size intervals for more direct comparison with the data of Laws and Parsons which were reported in 0.25-mm size intervals.
†Data from Laws and Parsons, *Trans. Am. Geophys. Union*, Part II, pp. 452–459, 1943.
SOURCE: Rogers et al., 1967.

where D_{50} is in millimeters and I is intensity in inches per hour. Best (1950) also observed that D_{50} is related to rainfall intensity through a power function up to a certain maximum intensity of 50 to 75 mm/h according to

$$D_{50} = aI^b \tag{2.15}$$

where I is rainfall intensity and a and b are constants. At high intensity, D_{50} may decrease with increasing intensity. However, there is no single relationship between D_{50} and intensity, and the relationship depends on the type of rain. At Ibadan, Nigeria, Aina et al. (1977) observed that the median raindrop size varied considerably among storms of different intensities (Fig. 2.13). Drop sizes ranged from 1.9 to 4.5 mm with a predominant size of 3 mm. The proportion of larger drops generally increased with an increase in rainfall intensity (r=0.78) and the amount of rainfall per storm (Table 2.9). Some researchers observed that the relationship between drop size and intensity is best expressed

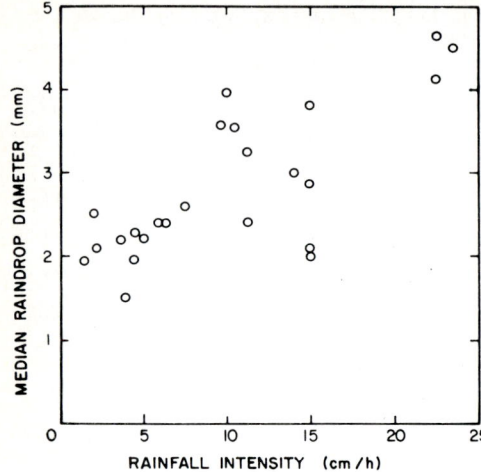

Figure 2.13 Relationship between rainfall intensity and drop size distribution for 12 erosive storms at Ibadan in 1975 (Aina et al., 1977).

by a polynomial rather than a power equation. For southern and central United States, Carter et al. (1974) reported the following polynomial equation:

$$D_{50} = A + BI + CI^2 + DI^3 + \cdots \quad (2.16)$$

Here D_{50} is the medium drop diameter in millimeters; the numerical values of constants A, B, C, and D are 1.63, 1.33, 0.33, and 0.002, respectively; and I is the rainfall intensity in inches per hour. In general, many researchers have ob-

TABLE 2.8 Number and Size of Drops and Kinetic Energy Load Distribution of a 20-mm, 18-min Rainstorm at Samaru, Nigeria, 21 August 1973

Drop size class	Drops per cm^2	Drop no., %	Volume, mm^3/cm^2	Average drop volume, mm^3	Drop diameter, mm	Kinetic energy, ergs/cm^2	Kinetic energy, %
1	21.81	22.7	144.88	6.64	2.34	34,672	5.0
2	17.84	18.5	236.78	13.28	2.94	70,585	10.2
3	19.59	20.3	390.23	19.92	3.36	131,185	18.9
4	14.68	15.2	389.10	26.56	3.70	141,846	20.4
5	9.92	10.3	327.68	33.20	4.00	111,809	16.1
6	9.43	9.8	374.10	39.84	4.24	146,119	21.1
7	2.00	2.2	88.64	46.41	4.46	36,583	5.3
8	0.80	0.8	46.75	53.13	4.66	17,290	2.5
9	0.16	0.2	9.06	59.75	4.86	3,897	0.6
Total	96.31	100.0	2,007.74			693,986	100.0
Average						20.85	3.42

SOURCE: Kowal and Kassam, 1976.

TABLE 2.9 Correlation Coefficients *r* among Median Raindrop Size D_{50}, Rainfall Intensity, and Wind Velocity

Regression	r	r^2
D_{50} (mm) on intensity (cm/h)	0.78	0.56
D_{50} on wind (km/h)	0.40	0.16
Wind on intensity	0.44	0.19

SOURCE: Aina et al., 1977.

served that D_{50} increases with increasing rainfall intensity, then levels off, and falls off slightly with a further increase in intensity (Fig. 2.14).

The effect of wind velocity on drop size varies and depends on other factors. If the general drop size is small, gentle wind may increase drop size through its coalescing effect. The wind effect may be just the opposite if the drop size is relatively large. In Nigeria, Aina et al. (1977) reported little coalescing effect on the median drop size. Effects of wind on rainfall erosivity are discussed in Chap. 6.

In addition to wind velocity, air temperature affects the drop size distribution. Zanchi and Torri (1980) measured raindrop distributions of several rains in central Italy by the flour pellet technique (Table 2.10). They used a power model similar to that of Laws and Parsons (1943) and developed regression equations relating drop diameter corresponding to different volumes of rain intensity. The exponent to I of the equations in Table 2.10 shows significant effects of ambient temperature. Without considering ambient temperature, the coefficient and exponent of the equation relating D_{50} to I were 1.10 and 0.263, respectively. Zanchi and Torri reported that with temperature as a variable, the values were 0.499 and 0.225, respectively (Table 2.10). The importance of tem-

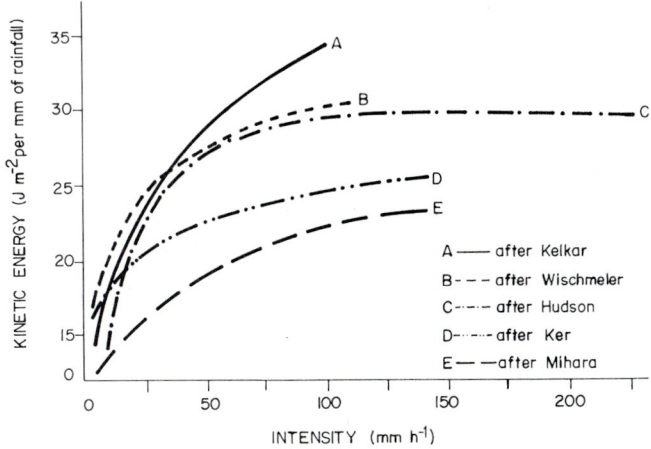

Figure 2.14 General relationships observed between rainfall intensity and drop size distribution for different geographic regions. (Adapted from Stocking and Elwell, 1976; Embleton and Thornes, 1979.)

TABLE 2.10 Equations Relating Intensity and Temperature to Drop Diameters Corresponding to Different Percentages of Cumulative Volume

Equations*	Coefficient of determination†
$D_{12.5} = 0.565 I^{0.133} T^{0.069}$	0.627
$D_{25.0} = 0.586 I^{0.229} T^{0.119}$	0.925
$D_{37.5} = 0.485 I^{0.229} T^{0.245}$	0.963
$D_{50.0} = 0.499 I^{0.225} T^{0.292}$	0.981
$D_{62.5} = 0.666 I^{0.234} T^{0.230}$	0.987
$D_{75.0} = 1.000 I^{0.237} T^{0.135}$	0.989
$D_{87.5} = 3.104 I^{0.267} T^{-0.199}$	0.986
Intensity range: 1.0 I 140.0 mm/h	
Temperature range: 10 T 28°C	

*D = mm; I = mm/h; T = °C.
†All significant, $P \leq 0.01$.
SOURCE: Zanchi and Torri, 1980.

perature in relation to drop diameter distribution is confirmed by the high coefficient of correlation. The correlation coefficient relating D_{50} to I increases from 0.877 when temperature is not considered to 0.981 when temperature is used as a covariable.

The second factor affecting rainfall intensity is drop number. Drop size remaining the same, the higher the intensity, the higher the drop number per unit time. Park et al. (1983) observed that the number of drops is approximately proportional to the square root of rainfall intensity:

$$N_d = 154 I^{0.5} \tag{2.17}$$

where N_d is the total number of drops [number/(m²/s)] and I is rainfall intensity in millimeters per hour.

Terminal velocity. The terminal velocity of fall also depends on drop size. Under natural conditions, the terminal velocity increases with an increase in drop size. Most drops attain terminal velocity in about a 10-m fall under gravity. Laws (1941), Laws and Parsons (1943), and Gunn and Kinzer (1949) established empirical relationships between drop diameter and terminal velocity (Fig. 2.15, Table 2.11). More recent measurements of fall velocity have been reported by Wang and Pruppacher (1977) and Roels (1981). Wang (1972) and Park et al. (1983) computed terminal velocity by using

$$V_T = \frac{(\rho_w - \rho_a) g d}{1.8} \tag{2.18}$$

where ρ_w and ρ_a are water and air density, g is gravitational acceleration, and d is drop diameter.

It has been shown experimentally (Park et al., 1982) that soil detachment

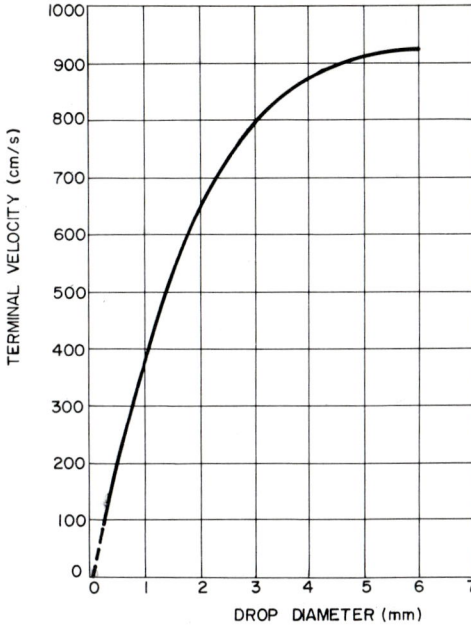

Figure 2.15 Empirical relationship between drop size distribution and terminal fall velocity (Gunn and Kinzer, 1949; Troeh et al., 1980).

and splash increase exponentially with an increase in impact velocity. Impact velocity is, however, greatly influenced by the wind factor. Soil splash is also related to D_{50}. In general, splash erosion increases exponentially with an increase in D_{50}. Data from Ibadan, Nigeria, indicate an increase in sand splash with increasing D_{50}. This subject is discussed at length in Chap. 6.

Kinetic energy. The kinetic energy of a rainfall is equal to the product of the mass of rain falling per unit time and the square of its terminal velocity. To be

TABLE 2.11 Relationship between Drop Size and Terminal Velocity

Drop size, mm	Terminal velocity, m/s	Fall to reach 95% of terminal velocity, m
0.25	1.0	—
0.50	2.0	—
1.00	4.0	2.2
2.00	6.5	5.0
3.00	8.1	7.2
4.00	8.8	7.8
5.00	9.1	7.6
6.00	9.3	7.2

SOURCE: Wischmeier and Smith, 1958.

precise, the kinetic energy E should be computed as a three-way product of each drop size class, the square of its corresponding terminal velocity, the number of drops in each class, multiplied by 0.5 [modified Eq. (2.3)]:

$$E = \sum_{i=1}^{n} \tfrac{1}{2} m_i v_i^2 \qquad (2.19)$$

By more precisely defining and narrowing the size ranges, the energy computation can be made more nearly accurate. For the convenience of computations, however, energy is computed on the basis of median drop size and its corresponding terminal velocity. The latter procedure may lead to considerable error.

Big drop size makes the energy load of tropical rainstorms high. The seasonal distribution of kinetic energy of rains at Samaru, in northern Nigeria, is shown in Figs. 2.16 and 2.17. In this semiarid region, the annual kinetic energy load of 1090-mm annual rainfall was estimated to be 36,000 J/m² (Kowal and Kassam, 1978), which is twice as much as that reported by Elwell and Stocking (1973) for a subtropical region in southern Africa. Lal (1987) reported the seasonal distribution of the energy load for subhumid regions of southwestern Nigeria (Figs. 2.18 and 2.19). The peak energy load occurs in June and September, the most erosive months in the year.

Because kinetic energy cannot be routinely measured, many empirical relationships have been established relating kinetic energy and rainfall intensity. Kinnell (1981) described two separate forms of kinetic energy in relation to rainfall intensity: (1) the expenditure rate of the rainfall kinetic energy E_{RR}, which has units of energy per unit area per unit time, and (2) the amount of rainfall kinetic energy expended per unit quantity of rain E_{RA}, which has units of energy per unit area per unit depth. Thus E_{RR} and E_{RA} are related as follows:

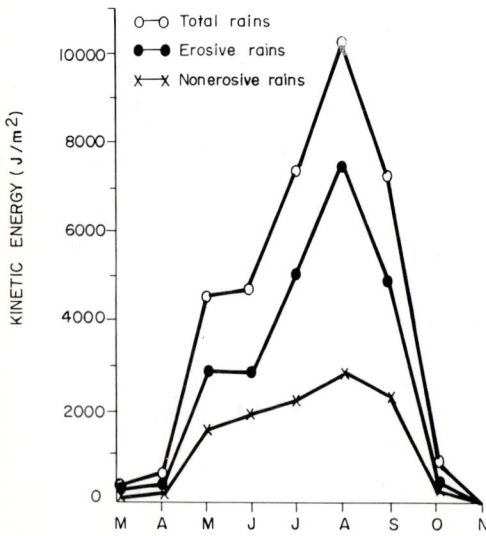

Figure 2.16 Seasonal distribution of rainfall energy at Samaru, northern Nigeria (Kowal and Kassam, 1978).

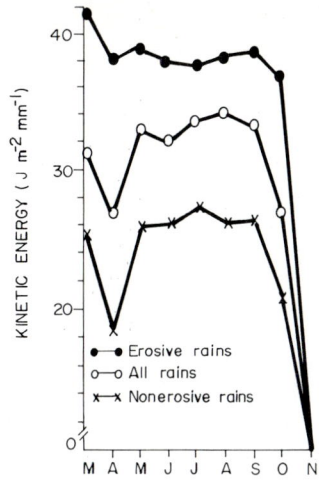

Figure 2.17 Seasonal distribution of rainfall energy at Samaru, northern Nigeria (Kowal and Kassam, 1978).

$$E_{RA} = CE_{RR}I^{-1} \tag{2.20}$$

Here I is the rainfall intensity (depth/time) and C is an empirical constant. Different forms of equations are commonly employed to relate E_{RA} to rainfall intensity:

$$E_{RA} = a + b \log I \tag{2.21}$$

$$E_{RA} = C(b - aI^{-1}) \tag{2.22}$$

$$R_{RA} = bI - a \tag{2.23}$$

Equation (2.17) is used most commonly. Wischmeier and Smith (1958) proposed the following empirical equation to compute kinetic energy from rainfall intensity:

$$E = 210.3 + 89 \log I \tag{2.24}$$

Here E is in tonnes per hectare per centimeter of rain and I is rainfall intensity in centimeters per hour. This equation is valid for I values of up to 76 mm/h. Park et al. (1983) reported an exponential relationship between kinetic energy and rainfall intensity:

$$E = 21{,}107 C_{te} I^{1.16} \tag{2.25}$$

Here E is in joules per unit area and time (ha · h) and C_{te} is the temperature correction factor for the energy relationship. Mutchler and Murphree (1985) proposed some experimentally derived modifications of the equation relating kinetic energy to rainfall intensity:

$$E = 0.273 + 0.217 e^{-0.048I} - 0.413 e^{-0.072I} \tag{2.26}$$

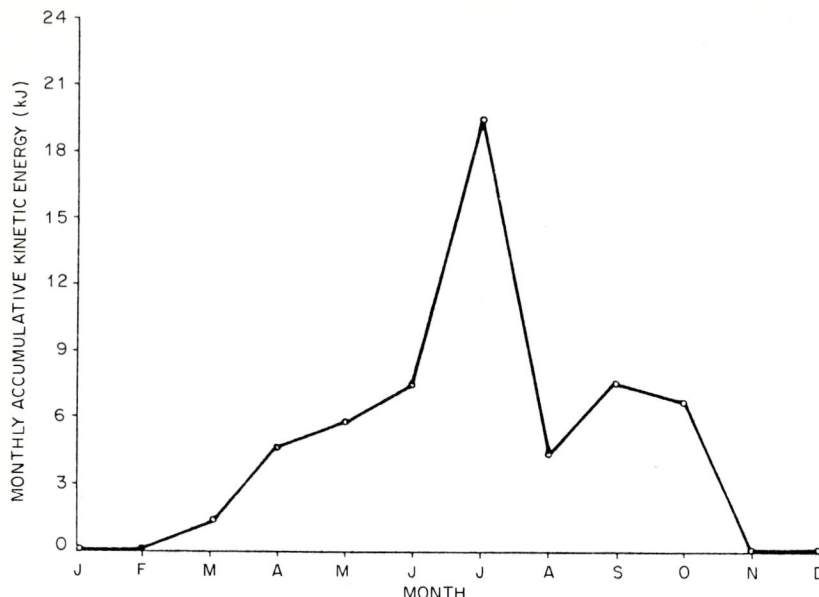

Figure 2.18 Seasonal distribution of rainfall energy for the subhumid tropics at Ibadan, Nigeria (Lal, 1987).

Here E is the kinetic energy per millimeter depth of rain for the corresponding intensity and I is measured in millimeters per hour for short time increments. This equation is valid for intensities exceeding 100 mm/h. On the basis of measurements of drop size distribution in central Italy, Zanchi and Torri (1980) developed regression equations relating rain intensity to kinetic energy with and without considering ambient temperature:

$$E = \begin{cases} 9.81 + 11.25 \log I & r = 0.906 \\ 1.86 + 11.03 \log I + 0.07 \log T & r = 0.999 \end{cases} \quad (2.27)$$

Including temperature improved the coefficient of determination. These equations developed for central Italy differ significantly from Eq. (2.24), developed for the United States.

Some specific relationships relating rainfall intensity and kinetic energy developed for tropical regions are discussed below. Kinnell (1973) related kinetic energy to rainfall intensity as follows:

$$E\,[\text{ergs}/(\text{cm}^2 \cdot \text{s})] = 8.37I - 45.9 \quad (2.28)$$

Here I is the intensity, which ranges from 0 to 300 mm/h. In northern Nigeria, Kowal and Kassam (1976) related kinetic energy to rainfall amount per storm E (ergs/cm^2):

$$E = (41.4R_a - 120.0)10^3 \quad r = 0.99 \quad (2.29)$$

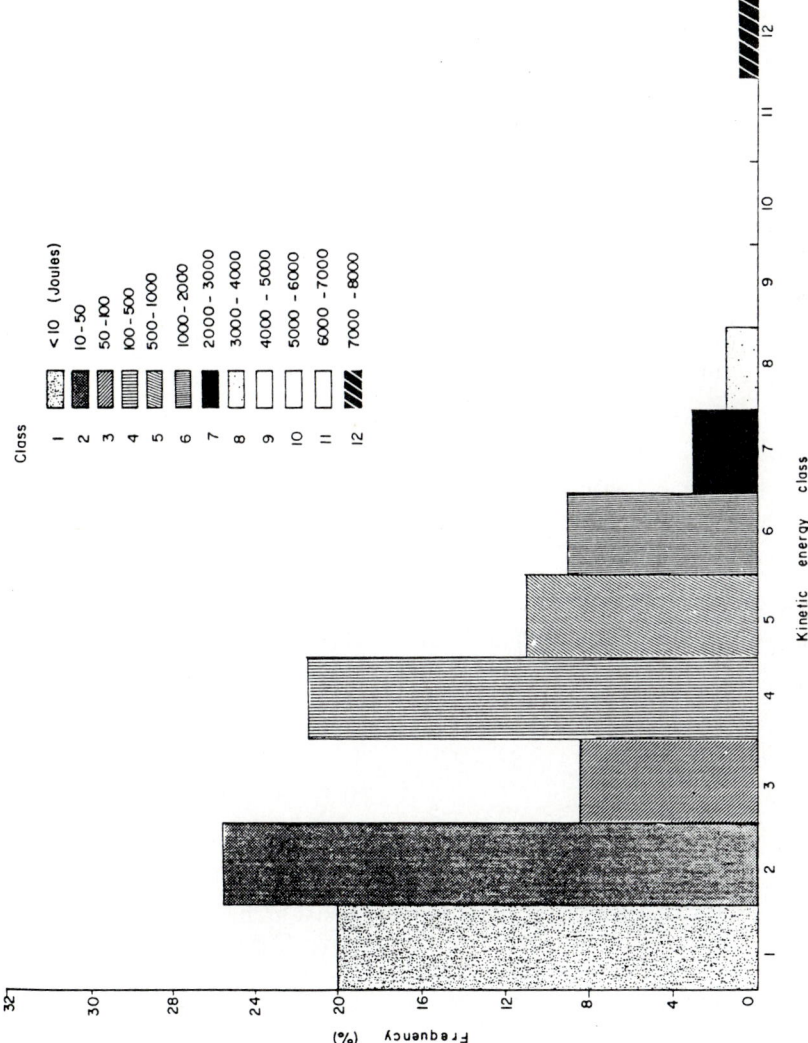

Figure 2.19 Seasonal distribution of rainfall energy for the subhumid tropics at Ibadan, Nigeria (Lal, 1987).

Here R_a is the amount of rainfall per storm in millimeters. In southern Africa, Elwell (1979) also reported a linear relationship between kinetic energy and amount of rainfall

$$E = 18.846 R_a \tag{2.30}$$

where E is in joules per square meter and R_a is the rainfall amount in millimeters. In southwestern Nigeria, Lal (1982) proposed these empirical relationships between kinetic energy and rainfall parameters

$$E = \begin{cases} 24.50 R_a + 27.6 & r = 0.81^{**} \\ 18.18 I_{30} + 18.2 & r = 0.81^{*} \end{cases} \tag{2.31}$$

where E is in joules per square meter, R_a is amount of rainfall in millimeters, I_{30} is the 30-min maximum intensity, r is the correlation coefficient, * indicates significance at the 5% level of probability, and ** indicates significance at the 1% level of probability. Kinnell (1981) reported that the relationship between the kinetic energy per unit quantity of rain E_{RA} and rainfall intensity I for Zimbabwe and Miami is best expressed by equations of the type

$$E_{RA} = Z(1 - Pe^{-hI}) \tag{2.32}$$

where e is the base of natural logarithm and Z, P, and h are empirical constants. The numerical values of these constants vary among different ecological regions.

Momentum. Momentum of rainfall, another characteristic related to soil erosion, is the product of mass and terminal velocity. Like kinetic energy, momentum is computed as the product of median drop size and its corresponding terminal velocity. More accurate computation of the momentum, however, is obtained as a product of the drop size for each class with its corresponding velocity [modified Eq. (2.6)]:

$$P = \sum_{i=1}^{n} m_i v_i \tag{2.33}$$

As with the kinetic energy of rains, few data exist from direct measurements of momentum. It is a capital-intensive undertaking. Therefore, many empirical relationships have been established relating momentum to easily measured rainfall parameters, e.g., amount and intensity. Williams (1969) proposed the following empirical relation for conditions in tropical Australia:

$$\text{Momentum [dynes}/(\text{cm}^2 \cdot \text{h})] = 0.7118 \log I - 1.461 \tag{2.34}$$

Here I is rainfall intensity in millimeters per hour. For Australia, Kinnell (1973) reported a linear relationship between intensity and momentum:

$$\text{Momentum [dynes}/(\text{cm}^2 \cdot \text{s})] = 0.0213 I - 0.62 \tag{2.35}$$

Under tropical conditions at Ibadan, Nigeria, Lal (1981) reported linear relationships of momentum with rainfall intensity and amount:

$$\text{Momentum } [J/(m^2 \cdot s)] = \begin{cases} 6.67 R_a + 9.32 & r = 0.81^{**} \\ 4.79 I_{30} + 8.74 & r = 0.75^{**} \end{cases} \quad (2.36)$$

Here R_a is the rainfall amount in millimeters, I_{30} is the 30-min maximum intensity, and ** indicates significance at the 1% level of probability. On the basis of data for the United States, Park et al. (1985) related momentum to rainfall intensity as

$$\text{Momentum } [kg \cdot m/(s \cdot ha \cdot h)] = 64{,}230 C_{tm} I^{1.07} \quad (2.37)$$

where I is rainfall intensity in millimeters per hour and C_{tm} is the temperature correction factor (dimensionless) for the momentum relationship. The C_{tm} accounts for variations in terminal velocities of different raindrop sizes with the ambient air and raindrop temperatures.

Wind-driven rain. Erosivity of wind-driven rain can differ drastically from that of rain falling under windless conditions. The magnitude of the difference, however, depends on wind velocity, rainfall intensity, and slope characteristics. Disrud et al. (1969) studied the effects of wind on size and shape of raindrops. If waterdrops fall in wind long enough, their shape is altered to oblate spheroid with a flattened area between the bottom and upwind side. The vertical drag velocity decreases with increasing wind velocity. That may decrease the kinetic energy and momentum of wind-driven rain. The effects of wind on rain are complex and confounded by many variables. Wind may increase kinetic energy and momentum by increasing drop size and/or by increasing its impact velocity. Lyles (1977) reported that a wind of 35 km/h at an elevation of 6.1 m could increase the kinetic energy load 3.2 times over that in still air. Wind also alters the angle of raindrop impact, and by adding a horizontal component to drop velocity wind increases its detaching capacity.

Wind has long been recognized as an important factor in erosion caused by wind-driven rain. Wind-driven rains are common even in regions of low winds, e.g., equatorial climates. The data in Table 2.12 show the occurrence of peak wind velocity accompanying rainstorms at Ibadan, Nigeria. Of rainstorms during 1975 50 percent were accompanied by winds exceeding 30 km/h and 12 percent by winds exceeding 40 km/h.

Peak wind velocity does not always coincide with peak rainfall intensity, because of a frequent time lag between the two (Fig. 2.20). Time lags may vary among different types of rains, e.g., advanced or delayed. Highly erosive rains are generally those in which peak intensity and peak wind velocity coincide.

Hails. Hails can be an important factor affecting climatic erosivity in relation to water erosion, especially in the northern latitudes and in tropical highlands. Because of their bigger size and high terminal velocity, hails can cause substantial damage to soil structure and to standing vegetation. Hagen et al. (1975) reported that a considerable area of the United States, extending from northern Texas to western North Dakota, is subjected to hailstorms, especially when

TABLE 2.12 Wind Speed (2 m Aboveground) for Rainstorms Recorded at Ibadan, Nigeria

Maximum wind speed, km/h	Rainstorms, %		
	1975	1976	1977
0–10	0	0	0
10.1–20	25	2	0
20.1–30	27	15	70
30.1–40	38	75	21
40.1–50	9	6	6
50.1–60	3	2	3
Total storms recorded	96	47	34

SOURCE: Lal et al., 1980.

soils are unprotected and devoid of vegetation cover. Hails on bare soil greatly increase the clod breakdown. Most hailstones are less than 0.5 cm in diameter, although some may be as large as 2 cm (Hagen and Butchbaker, 1967; Changnon, 1971). In the United States average hailfall energy ranges from 0.06 to 2.0 cm · N/cm² (Hagen and Butchbaker, 1967). Hailstorms are also reported to affect climatic erosivity in Ethiopian highlands (Hurni, 1981).

Wind Velocity

The kinetic energy supplied by blowing wind causes wind erosion. Moving wind causes a fluid drag τ on the soil surface. This drag (or shear stress, or force per

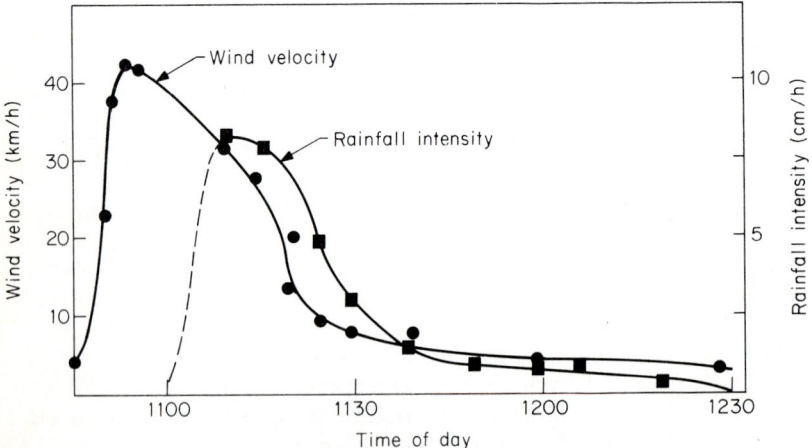

Figure 2.20 Relation between peak rainfall intensity and peak wind velocity at Ibadan, Nigeria (Aina et al., 1978).

Physical Processes and Climatic Erosivity 53

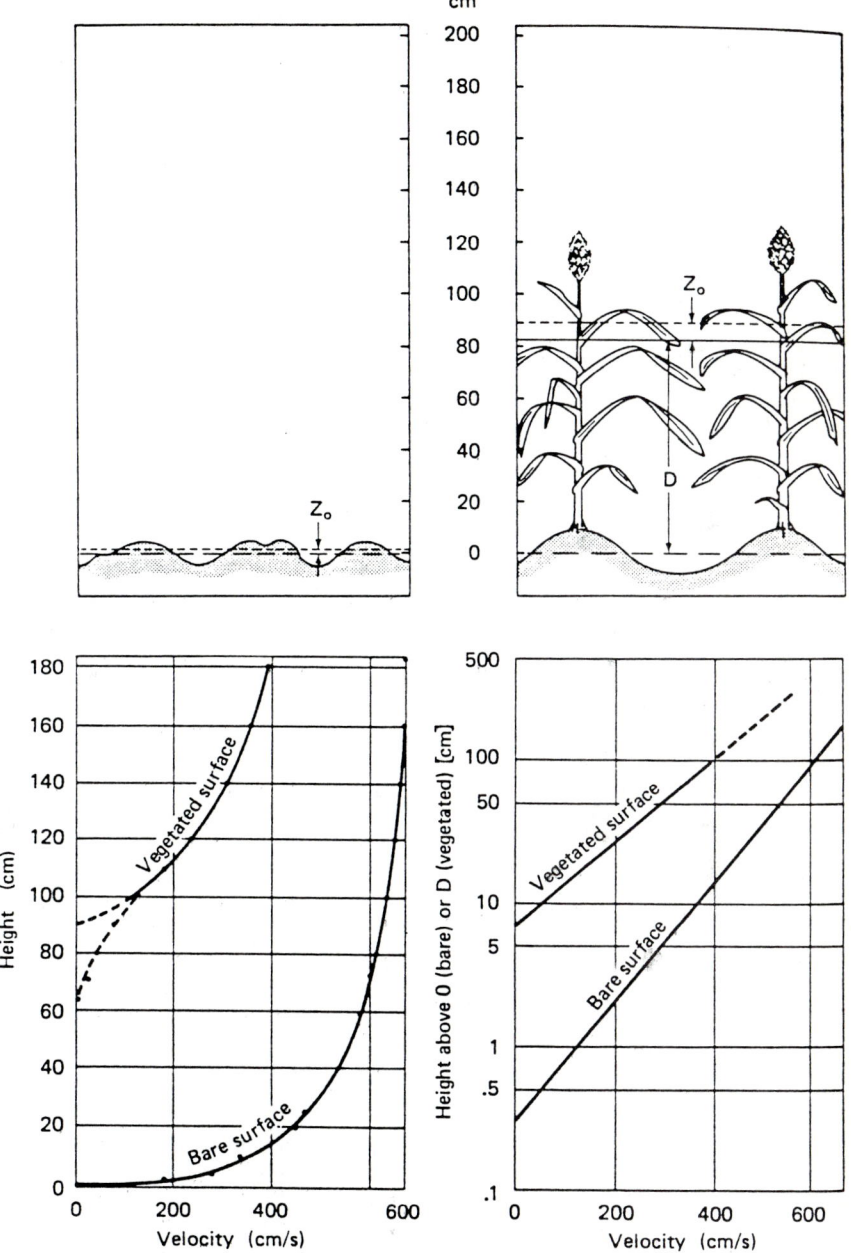

Figure 2.21 Wind profile near the ground, with and without vegetation cover. (Adapted from Troeh et al., 1980.)

unit area) is proportional to the square of the wind velocity V close to the ground surface:

$$\tau = KV^2 \qquad (2.38)$$

Here K is the coefficient that depends on the height at which wind speed is measured and on the roughness of the surface. The wind velocity profile differs for different ground surfaces, such as with and without cover (Fig. 2.21).

Wind velocity has a specific profile. Wind is nearly at rest in contact with the earth's surface because of drag or the frictional forces between the air and soil surface. The average forward velocity of wind near the ground increases with the logarithm of the height above the surface. From velocity records obtained at one height, wind velocity profiles for different heights can be computed via

$$\frac{V_1}{V_2} = \frac{\log Z_1 - \log Z_0}{\log Z_2 - \log Z_0} \qquad (2.39)$$

where V is velocity, Z is height above the surface, subscripts 1 and 2 refer to corresponding velocity and height at a given point, and Z_0 is roughness height.

Zero velocity is somewhere above the average roughness elements of the surface. The height where the wind velocity is zero is called the *roughness height*. The taller the roughness element of the ground, the higher zero velocity is found. Chepil and Woodruff (1963) observed wind velocity profiles for different roughness elements. The data show that wind velocity changes with height above the ground surface. The change in wind velocity with height is called the *velocity gradient*. A mean wind velocity profile equation over a stable surface is

$$V_Z = 5.75 V_* \log \frac{Z_0}{K} \qquad (2.40)$$

where V_Z is the wind speed at height Z, V_* is the friction velocity, and K is the height above the mean aerodynamic surface Z_0, where the wind velocity is zero. Estimated zero velocity is at height $Z_0 + K$, where Z_0 is the aerodynamic surface and K is the height above Z_0 when the velocity is zero. The distance between Z_0 and the average level of the ground surface is the vertical displacement of the velocity gradient by vegetation or other roughness elements of the surface. This displacement is referred to as *zero displacement height D_h* or roughness height (Chepil and Woodruff, 1963). Over small crops and smooth surfaces, K is practically zero. The data show differences in values of K for different ground covers. For impervious stable surface (such as ground), the wind velocity at Z_0 is zero. However, for porous surfaces, such as vegetation cover, the wind velocity is not zero. Rather than having "free flow," flow through the vegetation cover is restricted. It is difficult to define wind velocity for the restricted-flow zone. The equation of wind velocity for the free-flow zone is

$$V_* = \frac{V_Z}{5.75} \log \frac{Z}{K} \qquad (2.41)$$

The friction velocity V_* is the index of the rate of increase of velocity with the logarithm of height. The stronger the wind, the greater the drop velocity. Mean wind drag per unit horizontal area of the ground surface $\bar{\tau}$ is related to drag velocity over vegetation-covered surfaces by

$$\bar{\tau} = a\rho V_*^2 \qquad (2.42)$$

where ρ is the density of air (about 0.0012 g/cm^3 and a is the drag coefficient that varies with the type of cover. If V_* is expressed in centimeters per second, then $\bar{\tau}$ is in dynes per square centimeter. For wind speeds exceeding those required to barely move the soil, the rate of soil movement is proportional to the friction velocity cubed. Wind speeds of less than 5.4 m/s (less than 12 mi/h) at 0.3-m height are considered nonerosive. The capacity of wind to cause soil movement is proportional to friction velocity cubed times the duration of the wind.

The magnitude of the wind erosion force vector r_j is obtained by summing, for all speed groups with wind speed exceeding 5.4 m/s, the product of mean wind speed cubed and a duration factor of specified direction (Skidmore and Woodruff, 1968):

$$r_j = \sum_{i=1}^{n} (\bar{V}_i)^3 f_i \qquad (2.43)$$

Here $(\bar{V}_i)^3$ is the mean wind speed within the ith speed group cubed, and f_i is the duration factor expressed as the percentage of total observations in the jth direction within the ith speed group. The subscript j's indicate direction and take values from 0 to 15, inclusive, representing the 16 principal compass directions. The sum of the magnitude of wind erosion force vectors for all directions gives the total magnitude of wind erosion forces

$$F_T = \sum_{j=0}^{15} \sum_{i=1}^{n} (\bar{V}_{ij})^3 f_{ij} \qquad (2.44)$$

The value of vector F_T indicates the relative capacity of wind to cause soil blowing at a particular location.

Conclusions

Climatic erosivity, or the ability of the climate to cause erosion, is an integrated effect of all climatic variables that contribute to the driving force that causes erosion—kinetic energy. In the context of water erosion, the primary source of kinetic energy is falling raindrops. Raindrop energy is affected by the size, fall velocity, and wind velocity in the case of wind-driven rain. The energy of falling raindrops is directly related to rainfall intensity or amount of rainfall. However, no single formula can explain the relation between drop size and intensity for different geographic regions. Tropical rains are often more intense than temperate rains. However, high-intensity rains also occur in northern latitudes and may cause more severe erosion there than similarly intense rains would in the tropical climates. Another important but less studied source of climatic erosivity is hailstorms. They can be significant in northern latitudes and in tropical highlands.

The major source of energy needed to perform the work involved in wind is the wind velocity. The wind velocity profile is greatly altered by soil surface conditions and especially vegetation cover. Forces generated by wind are greatly altered by soil surface.

References

Aina, P. O., Lal, R., and Taylor, G. S. 1977. Soil and crop management in relation to soil erosion in the rainforest of western Nigeria. In: *Soil Erosion: Prediction and Control* (G. R. Foster, ed.), pp. 75–82, Soil Conservation Society of America, Ankeny, Iowa.

Atlas, O., and Plank, V. G. 1953. Measuring rain with radar. *J. Meteor.* 10:291–295.

Attle, J. R., Oney, D., and Swenson, R. A. 1980. *Applications of Image Analysis,* American Laboratory.

Barnett, A. P., and Rogers, J. S. 1966. Soil physical properties related to runoff and erosion for artificial rainfall. *Trans. ASAE* 9:123–125.

Baruah, P. C. 1973. An investigation of drop size distribution of rainfall in Thailand. Master's thesis 528, Asian Institute of Technology, Bangkok, Thailand.

Becker, A. 1907. Zur Messung der Tropfengroessen bei Regenfaellen nach der Absorptionmethode. *Met. Zs.* 24:247–261.

Bentley, W. A. 1904. Studies of raindrops and raindrop phenomena. *Monthly Weath. Rev.* 32:450–456.

Best, A. C. 1950. The size distribution of raindrops. *Q. J. Roy. Meteorol. Soc.* 76:16.

Blanchard, D. C. 1950. Behavior of water drops at terminal velocity. *Trans. Am. Geophys. Union* 31:836–842.

Bollinne, A. 1985. Adjusting the universal soil loss equation for use in western Europe. In: *Soil Erosion and Conservation* (S. A. el-Swaify, W. C. Moldenhauer, and A. Lo, eds.), pp. 206–213, Soil Conservation Society of America, Ankeny, Iowa.

Bruce-Okine, E., and Lal, R. 1975. Soil erodibility as determined by raindrop technique. *Soil Sci.* 119:149–157.

Bryan, R. B. 1968. The development, use and efficiency of indices of soil erodibility. *Geoderma* 2:5–25.

Carter, C. E., Greer, J. D., Brand, H., and Floyd, J. 1974. Raindrop characteristics in south central United States. *Trans. ASAE* 17:1033–1037.

Chang, T. P., and Lee, S. W. 1977. Watershed management work in Taiwan. Joint Commission on Rural Construction, Taipei, Taiwan.

Changnon, S. A., Jr. 1971. Note on hailstone size distributions. *J. Appl. Meteorol.* 10:168–170.

Chepil, W. S. 1953. Field structure of cultivated soils with special reference to erodibility by wind. *Soil Sci. Soc. Am. Proc.* 18:13–16.

Chepil, W. S., and Woodruff, N. P. 1963. The physics of wind erosion and its control. *Adv. Agron.* 15:211–302.

Defant, A. 1905. Gesetzmaessigkeiten in der Verteilung der verschiedener Tropbengroessen der Regenfaellen. *Sitz Ber. Math. Naturwiss. Klasse Adad. Wiss.* 5:585–646.

De Leenheer, L., and De Boodt, M. 1959. Determination of aggregate stability by change in mean weight diameter. *Meded. Landbouwhogesch Gent.* 24:290–300.

De Vleeschauwer, D., Lal, R., and De Boodt, M. 1978. Comparison of detachability indices in relation to soil erodibility for some important Nigerian soils. *Pedologie* 28:5–20.

Disrud, L. A., Lyles, L., and Skidmore, E. L. 1969. How wind affects the size and shape of raindrops. *Agric. Eng.* 50:617.

Eigel, J. D., and Moore, I. D. 1983. A simplified technique for measuring raindrop size and distribution. *Trans. ASAE* 26:1070–1084.

Elwell, H. A. 1979. Destructive potential of Zimbabwe (Rhodesia) rainfall. *Zimbabwe Agric. J.* 76:227–232.

Elwell, H. A., and Stocking, M. A. 1973. Rainfall parameters for soil loss estimation in a subtropical climate. *J. Agric. Eng. Res.* 18:169–177.

Embleton, C., and Thornes, J. 1979. *Process in Geomorphology.* Edward Arnold, London.

Evans, R., and Nortcliff, S. 1978. Soil erosion in north Norfolk. *J. Agric. Sci. Camb.* 90:185–192.

Falayi, O., and Lal, R. 1979. Effects of aggregate size and mulching on erodibility, crusting and crop emergence. In: *Soil Physical Properties and Crop Production in the Tropics* (R. Lal and D. J. Greenland, eds.), pp. 87–93, Wiley, Chichester, England.

Fournier, F. 1967. Research on soil erosion and soil conservation in Africa. *African Soils* 12:53–96.

Fuchs, N., and Petajanoff, I. 1937. Microscopic examination of fog, cloud and rain-droplets. *Nature* 139:111–112.

Fullen, M. A. 1985. Compaction, hydrological processes and soil erosion on loamysands in east Shropshire, England. *Soil Tillage Res.* 6:17–30.

Gillespie, T. 1958. The spreading of low vapor pressure liquids in paper. *J. Colloid Sci.* 13:32–50.

Gunn, R., and Kinzer, G. D. 1949. The terminal velocity of fall waterdrops in stagnant air. *J. Meteorol.* 6:243–248.

Hagen, L. J., and Butchbaker, A. F. 1967. Climatology of hailstorms and evaluation of cloud seeding for hail suppression in southwestern North Dakota. *Proc. Fifth Conf. Severe Local Storms,* St. Louis American Meteorological Society, Boston, pp. 336–347.

Hagen, L. J., Lyles, L., and Dickerson, J. D. 1975. Soil detachment from clods by simulated rain and hail. *Trans. ASAE* 18:540–543.

Hall, M. J. 1970. Use of the stain method in determining the drop-size distributions of coarse liquid sprays. *Trans. ASAE* 30:33–37, 41.

Henin, S., Monnier, G., and Combeau, A. 1958. Methode pour l'etude de la stabilité structurale de sols. *Ann. Agron.* 1:71–90.

Hoogmoed, W. B., and Stroosnijder, L. 1984. Crust formation on sandy soils in the Sahel. I. Rainfall and infiltration. *Soil Tillage Res.* 4:5–23.

Hudson, N. W. 1964. The flour pellet method for measuring the size of raindrops. *Res. Bull.* 4, Dept. of Conservation and Extension, Salisbury, Rhodesia.

Hudson, N. W. 1976. *Soil Erosion,* B. T. Bastford, London.

Hudson, N. 1981. Instrumentation for Studies of the Erosive Power of Rainfall, pp. 383–390, IAHS Bull. 133, IAHS, Washington.

Hurni, H. 1981. A monograph for the design of labour intensive soil conservation measures in rainfed cultivations. In: *Soil Conservation: Problems and Prospects* (R. P. C. Morgan, ed.), pp. 185–210, Wiley, Chichester, England.

Joss, V. J., and Waldvogel, A. 1967. Ein spectrograph fur niederschlagstropher mit automatischer answertung. *Pure and Appl. Geophys.* 68:240–246.

Keersebilik, N., Muljadi, D., and Sukmena, S. 1972. The possible application of a rainfall simulator method for wet tropical conditions. *Proc. Second ASEAN Soil Conf.,* vol. 2, pp. 32–52, Soil Research Institute, Bogor, Indonesia.

Kinnell, P. I. A. 1972. The acoustic measurement of water drop impacts. *J. Appl. Meteorol.* 11:691–694.

Kinnell, P. I. A. 1973. The problem of assessing the erosive power of rainfall from meteorological observations. *Soil Sci. S. A.* 37:617–621.

Kinnell, P. I. A. 1976. Some observations on the Joss-Walovogel rainfall distrometer. *J. Appl. Meteorol.* 15:499–502.

Kinnell, P. I. A. 1981. Rainfall intensity-kinetic energy relationships for soil loss prediction. *Soil Sci. Soc. Am. Proc.* 45:153–155.

Kowal, J. 1972. Effect of an exceptional storm on soil conservation at Samaru, Nigeria. *Samaru Res. Bull.* 141, pp. 163–172, Inst. Agric. Res., Samaru, Nigeria.

Kowal, J. M., and Kassam, A. H. 1976. Energy and instruments intensity of rainstorms at Samary, northern Nigeria. *Trop. Agric.* 53:185–198.

Kowal, J. M., and Kassam, A. H. 1978. *Agricultural Ecology of Savanna: A Study of West Africa,* Clarendon Press, Oxford, England.

Lal, R. 1981. Analysis of different processes governing soil erosion by water in the tropics. *IAHS Publ.* 133:351–364.

Lal, R. 1982. Temperature profile of soil during infiltration. *Niger. J. Soil Sci.* 2:87–100.

Lal, R. 1987. Soil degradation in relation to climate. Paper delivered at International Symposium on Climate and Food Security, 6–9 February 1987, New Delhi, India.

Lal, R. (ed.) 1988. *Soil Erosion Research Methodology.* Soil Conservation Society of America, Ankeny, Iowa.

Lal, R., Lawson, T. L., and Anastase, A. H. 1980. Erosivity of tropical rains. In: *Assessment of Erosion* (M. de Boodt and D. Gabriels, eds.), pp. 143–153, Wiley, Chichester, England.

Lambour, J. 1967. Frequency of intense rainfall for the territory of Poland. *J. Hydrol.* 5:158–162.

Laws, J. O. 1941. Measurement of fall velocity of water drops and raindrops. *Trans. Am. Geophys. Union* 22:709–721.

Laws, J. O., and Parsons, D. A. 1943. The relationship of raindrop size to intensity. *Trans. Am. Geophys. Union* 24:452–460.

Lee, S. W. 1981. Landslides in Taiwan. In: *Problems of Soil Erosion and Sedimentation* (T. Tingsanchali and H. Eggers, eds.), pp. 195–206, AIT, Bangkok, Thailand.

Lenard, P. 1904. Rain, trans. by R. H. Scott, 1905. *Q. J. Roy. Meteorol. Soc.* 31:62–73.

Luk, S. H. 1979. Effect of soil properties on erosion by wash and splash. *Earth Surface Processes* 4:241–255.

Lyles, L. 1977. Wind erosion: Processes and effects on soil productivity. *Trans. ASAE* 20:880–884.

Lyles, L., Disrud, L. A., and Woodruff, N. P. 1969. Effects of soil physical properties, rainfall characteristics, and wind velocity on clod disintegration by simulated rainfall. *Soil Sci. Soc. Am. Proc.* 33:302–306.

Mache, H. 1904. Ueber die Geschuwindigkeit im Grosse der Regentropfer. *Met. Zs.* 39:278.

May, K. R. 1945. The cassade impactor: An instrument for sampling coarse aerosol. *J. Sci. Instr.* 22:187–195.

McCool, D. K., Robinette, M. J., King, J. T., Molanu, M., and Young, J. L. 1978. Raindrop characteristics in the Pacific Northwest. *Trans. Am. Geophys. Union* 59 (abstract).

Moldenhauer, W. C. 1970. Influence of rainfall energy on soil loss and infiltration rates. II. Effect of clod size distribution. *Soil Sci. Soc. Am. Proc.* 34:673–677.

Moldenhauer, W. C., and Kemper, W. D. 1969. Interdependence of water drop energy and clod size on infiltration and clod stability. *Soil Sci. Soc. Am. Proc.* 33:297–301.

Moldenhauer, W. C., and Koswara, J. 1968. Effect of initial clod size on characteristics of splash and wash erosion. *Soil Sci. Soc. Am. Proc.* 32:875–879.

Mutchler, C. K. 1971. Splash droplet production by water drop impact. *Water Resource Res.* 7:1024–1030.

Mutchler, C. K., and Hansen, L. M. 1970. Splash of a waterdrop at terminal velocity. *Trans. Am. Geophys. Union* 31:836–842.

Mutchler, C. K., and Murphree, C. E., Jr. 1985. Experimentally derived modification of the USLE. In: *Soil Erosion and Conservation* (S. A. El-Swaify, W. C. Moldenhauer, and A. Lo, eds.), pp. 523–527, Soil Conservation Society of America, Ankeny, Iowa.

Nawaby, A. S. 1970. A method of direct measurement of spray droplets in an oil bath. *J. Agric. Eng. Res.* 15:182–184.

Neal, J. H., and Baver, L. D. 1937. Measuring the impact of raindrops. *J. Am. Soc. Agron.* 29:708–709.

Neuberger, H. 1942. Notes on measurement of raindrop sizes. *Bull. Am. Meteorol. Soc.* 23:274–276.

Olson, T. C., and Wischmeier, W. H. 1963. Soil erodibility evaluations for soils on the runoff and erosion stations. *Soil Sci. Soc. Am. Proc.* 27:590–592.

Park, S. W., Mitchell, J. K., and Bubenzer, G. D. 1982. Splash erosion modeling: Physical analysis. *Trans. ASAE* 25:357–361.

Park, S. W., Mitchell, J. K., and Bubenzer, G. D. 1983. Rainfall characteristics and their relation to splash erosion. *Trans. ASAE* 26:795–804.

Rodda, J. C. 1967. A country-wide study of intense rainfall for the United Kingdom. *J. Hydrol.* 5:58–69.

Roels, J. M. 1981. Cited by J. M. Eigel and I. D. Moore (1983), p. 1079.

Rogers, J. S., Johnson, L. C., Jones, D. M. A., and Jones, B. A., Jr. 1967. Sources of error in calculating the kinetic energy of rainfall. *J. Soil Water Conserv.* 22:140–143.

Schleusener, P. E. 1967. Drop size distribution and energy of falling raindrops from a medium pressure irrigation sprinkler. Ph.D. thesis, Michigan State University.

Shaxson, T. F., Brabben, T. E., and Stevenson, E. C. 1981. Rainfall-runoff measurements on basalt soils in Central India. In: *Problems of Soil Erosion and Sedimentation* (T. Tingsanchali and H. Eggers, eds.), pp. 76–84, AIT, Bangkok, Thailand.

Singer, M. J., Blackard, J., Gillogley, E., and Kandiah, A. 1978. Engineering and pedological properties of soils as they affect soil erodibility. California Water Resources Center, Davis, contribution 166.

Skidmore, E. L., and Woodruff, N. P. 1968. Wind erosion forces in the United States and their use in predicting soil loss. USDA, ARS Agric. Handbook 346.

Stocking, M. A., and Elwell, H. A. 1976. Rainfall erosivity over Rhodesia. *Trans. Inst. Br. Geogr.* 1:231–245.

Troeh, F. R., Hobbs, J. A., and Donahue, R. L. 1980. *Soil and Water Conservation for Productivity and Environmental Protection.* Prentice-Hall, Englewood Cliffs, N.J.

Vaneslande, A., Lal, R., and Gabriels, D. 1986. The erodibility of some Nigerian soils: A comparison of rainfall simulator results with estimates obtained from the Wischmeier nomogram. *Hydrol. Process* 1:255–265.

Wang, J. Y. 1972. Methods in agrometeorology, pp. 233–285, *Agricultural Meteorology* (Proceedings of the WMO Seminar), Word Meteorological Org., Geneva.

Wang, P. K., and Pruppacher, H. R. 1977. Acceleration to terminal velocity of cloud and raindrops. *J. Appl. Met.* 16:275–280.

Wiesner, J. 1895. Beitrage zur Kenntniss, der tropi Regens. *K. Akad. Will. Math.— Naturw. Klasse, Sitz Ber.* 104:1397–1434.

Wilkinson, G. E. 1975. Rainfall characteristics and soil erosion in the forest area of western Nigeria. *Exp. Agric.* 11:247–255.

Williams, M. A. J. 1969. Prediction of rainsplash erosion in the seasonally wet tropics. *Nature* 222:763–764.

Wischmeier, W. H. 1955. Punched cards record runoff and soil-loss data. *Agric. Eng.* 36:664–666.

Wischmeier, W. H., and Smith, D. D. 1958. Rainfall energy and its relationship to soil loss. *Trans. Am. Geophys. Union* 39:285–291.

Young, R. A. 1984. A method of measuring aggregate stability under waterdrop impact. *Trans. ASAE* 27:1351–1354.

Young, R. A., and Onstad, C. A. 1982. The effects of soil characteristics on erosion and nutrient loss. *IAHS Publ.* 137:105–113.

Zanchi, C., and Torri, D. 1980. Evaluation of rainfall energy in central Italy. In: *Soil Erosion Assessment* (M. De Boodt and D. Gabriels, eds.), pp. 133–142, Wiley, Chichester, England.

Chapter 3

Soil Properties and Erodibility

Introduction

Soil erosion is a function of two opposing forces, i.e., the driving force of the erosion agent and the resisting force of the soil. Different soils respond differently to the identical kinetic energy of raindrops or the shear stress exerted by moving fluid. Their responses depend on their mechanical makeup and chemical composition. Because of differences in their inherent properties, soils exhibit different degrees of susceptibility (soil erodibility) to the forces generated by erosion agents. All other factors remaining the same, differences in erosion up to 30-fold have been observed due to differences in soil properties (Olson and Wischmeier, 1963). Susceptibility of a soil to erosion is influenced by its physical and hydrologic, chemical and mineralogic, and biological and biochemical properties as well as its soil profile characteristics. Important soil physical properties that affect the resistance of a soil to erosion include texture, structure, water retention and transmission properties, and unconfined compressive and shear strength. The importance of these properties in relation to soil erosion has been reviewed by Bryan (1968, 1976), among others.

Susceptibility to erosion is an integrated response of soil's inherent properties, properties of the eroding fluid and their interaction with climate. There are no simple and measurable soil parameters that can represent the integrated response of the complex variable—soil erodibility. Some variables may indicate different responses on a one-to-one basis than when they are considered as covariables with other properties. The following discussion, therefore, offers a mere guideline to this complex, little understood soil property.

Mechanical Properties

Soil mechanical and physical properties are determined by forces between the particles that compose the soils and the interaction of the particles with liquid and gaseous phases. Soil physical properties in relation to erosion are those determined especially by the constitution of the clay fraction and the interaction between clay and the physicochemical forces generated by erosion.

Texture and particle size distribution

Soil texture implies the visual appearance and feel of a soil. Particle size distribution refers to the diameter of soil particles as determined by laboratory analysis. In relation to soil erosion, the particle size distribution should be characterized according to the system of the International Society of Soil Science: e.g., gravels (greater than 2 mm), coarse sand (2 to 0.2 mm), fine sand (0.2 to 0.02 mm), silt (0.02 to 0.002 mm), and clay fraction (less than 0.002 mm).

Particle size distribution is important in sediment detachment and entrainment. Texture also determines the ease with which a soil can be dispersed. Soils containing low amounts of clay are easily dispersed. The size of soil particles also determines the threshold force required for detachment and entrainment. The bigger the particle, the more force is needed for transport. The size of the particle, primary or secondary, most easily eroded is about 0.1 mm or the equivalent. Soil texture influences soil erosion because coarse particles require a higher fluid drag (wind or water) than small particles. In general, clay and silt-sized particles adhere to form large, heavy aggregates. In some tropical soils, however, the silt-sized fraction is relatively low (Lal, 1979, 1987).

Because of the importance of texture, susceptibility of soil to water erosion has been related to texture-based indices for many soils from different geographic regions around the world. Many indices have been proposed relating texture to soil's susceptibility to erosion, including the dispersion ratio (Middleton, 1930), erosion ratio (Middleton, 1930), and clay ratio (Bouyoucos, 1935).*

In the region of Tziwu-Ling Kansu, China, Tyan and Hwang (1964) observed that intensive cultivation increased the dispersion ratio and erosion ratio as well as soil erosion observed under fluid conditions. The size of the soil particle also determines the threshold force required for detachment and entrainment. The bigger the size, the more force is needed for field conditions. In Taiwan, Wang (1979) observed that the dispersion ratio was a useful criterion to assess soil resistance to erosion. In India, the clay ratio, dispersion ratio, and erosion ratio are related to the extent of erosion measured in field (Chibber et al., 1961;

---

*Dispersion ratio $= \dfrac{\text{silt + clay in undispersed sample}}{\text{silt + clay in dispersed sample}} \times 100$

Erosion ratio $= \dfrac{\text{dispersion ratio}}{\text{colloid content/moisture equivalent}}$

Clay ratio or mechanical ratio $= \dfrac{\text{sand}}{\text{silt + clay}}$

Sharma and Biswas, 1972; Bhola and Jarayam, 1978; Sahi et al., 1977; Jha and Rathore, 1981; Bhatia and Vardani, 1982). From studies of physical properties of five soil types in Utter Pradesh, India, Gupta and Narain (1971) reported that the sum of fine silt and clay can be used as a good diagnostic technique for soil conservation planning and management. In soils of eastern Nepal, Chakrabarti (1969) reported that susceptibility to erosion is significantly related to the dispersion ratio and the clay ratio. In Malaysia, Olofin (1980) observed that a characteristic significantly related to soil erosion was sand/(silt + clay) ratio. In Puerto Rico, Lugo-Lopez (1969) reported that the erosion ratio was an important index of a soil's susceptibility to erosion. In the United States, Sherard et al. (1976) developed a laboratory test to evaluate the amount of dispersive clay in a soil on the basis of the turbidity of water passing through a hole in a soil sample.

Structure

Aggregation. Soil structure and its strength are also important properties that determine the resistance of a soil to dispersion and detachment. In simple terms, soil structure refers to the geomechanical arrangement of soil particles. It is the arrangement of soil particles into easily recognizable geometric shapes that influences the response behavior of the soil to external constraints, e.g., raindrop impact or shearing force of moving water or blowing wind. On the basis of visual observations under field conditions, soil structure is defined according to the packing arrangement of particles to form aggregates. Soil structure may be granular, spherical, platy, prismoidal, rhombohedral, massive, or single-grained. The response to water differs for different types of structure. Soil aggregates are formed by the association of clay particles into domains, domains and silt particles into microaggregates, and microaggregates and sand particles into aggregates. The size of domains is about 5 µm, microaggregates range from 5 to 1000 µm, and aggregates range from 1000 to 5000 µm, or 1 to 5 mm (Greenland, 1971, 1977). A hypothetical model of soil aggregate formation proposed by Williams et al. (1967) is shown in Fig. 3.1.

In the context of soil erosion, however, soil structure should involve the following characteristics: binding of soil particles and resistance to dispersion by water, percentage of water-stable aggregates and mean-weight aggregate diameter, ease of rainfall acceptance and ability to transmit water through the profile, relative proportion of macropores, and pore stability and continuity. In connection with the stable structure, the amount of binding material in the soil is important. Binding material consists of organic matter, the clay and the sesquioxides. Bennett (1926) considered the ratio of silica to sesquioxides [$SiO_2/(Fe_2O_3 + Al_2O_3)$)] an important property in relation to erosion. For some soils in India, susceptibility to erosion is related to the SiO_2/R_2O_3 ratio (Mehta et al., 1963; Haridasan and Chibber, 1971; and Laskar and Govindarajan, 1980).

Erosion by water and wind is also related to the stability of aggregates against abrasive effects of running water or blowing wind. There are many indices of structural stability, e.g., aggregate size distribution and aggregate stability (Peele, 1936; Yoder, 1936; Gerdel, 1937; Peele et al., 1938; Van Bavel, 1950; De Leenheer and De Boodt, 1959; Luk, 1979). In hilly lands in Taiwan, Wang and Lin (1965) observed that the least erodible soils had high aggregate

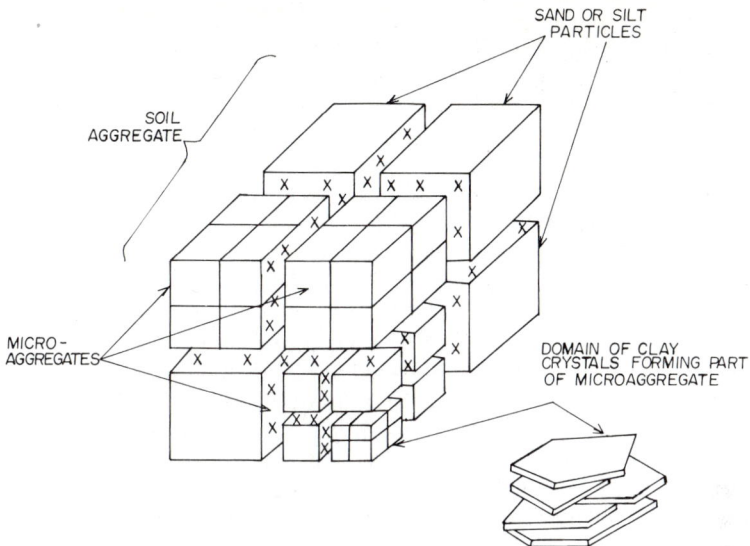

Figure 3.1 A hypothetical model of aggregate formation. X = organic molecules. (Williams et al., 1967.)

stability and that highly erodible soils had high dispersion ratios. For Hungarian chernozems, Major (1973) observed close correlation between water-stable aggregates and susceptibility to erosion. For loess soils in Hungary, Klimes-Szmik (1979) improved soil resistance to erosion by increasing aggregation by applying sokrakol-sodium salt of hydrolyzed polyacrylic acid. In Bulgaria, Krusteva (1977) established a relationship between the erodibility of soil and water stability of its macro- and microstructured aggregates. In general, an inverse correlation exists between the percentage of water-stable aggregates with soil splash (Woodburn and Kozachyn, 1956; Adams et al., 1958; Conaway and Strickling, 1962; Bryan, 1968; Mazurak and Mosher, 1968; Young and Onstad, 1982). Bryan (1976) emphasized the importance of aggregation as an index of soil resistance to erosion. However, Sahi et al. (1976) observed that the degree of aggregation alone was not a sufficient index of soil's ability to resist erosion.

Aggregate size. The size of water-stable aggregates also has a bearing on erosion. The larger the aggregates, the more they resist erosion. Highly structured soils resistant to erosion are those with a high percentage of 0.25- to 5-mm aggregates. Using simulated rainfall on lateritic soils in Uganda, Rose (1961) observed that the size of aggregates markedly affected structural breakdown. In Hawaii, Yamamoto and Anderson (1973) observed that the percentage of water-stable aggregates from 0.25 to 0.5 mm was an important index of soil resistance to erosion. In India, Bhatia and Sarmah (1976) stated that high percentage of water-stable aggregates and a high mean weight diameter are the characteristics of a soil resistant to erosion for alluvial and sedentary soils of Bihar, India.

The ability of a soil to resist erosion is related to both the percentage of aggregation and the distribution of stable aggregates. For soils in Minnesota,

Young and Onstad (1982) observed that the degree of surface soil aggregation and the stability of aggregates to water are both important to resist erosive forces of water. Using simulated rainfall, Alberts et al. (1983) observed that the mean diameter of aggregates eroded was 34 to 44 µm and that the mean diameter of eroded sediments increased with increasing clay content of the soil.

As explained in Chap. 5, the primary force detaching soil particles comes from the impact of raindrops. A useful technique to measure soil resistance to the erosive power of raindrop impact is the waterdrop technique (McCalla, 1944; Smith and Cernuda, 1951; Pereira, 1956; Nearing and Bradford, 1985). Bruce-Okine and Lal (1975) observed that soils that resist erosion require more kinetic energy to disrupt aggregates than soils susceptible to erosion do (Fig. 3.2). Bergsma (1980) used the waterdrop technique to evaluate the erodibility of soils developed on different parent materials. Similarly Al-Durrah and Bradford (1982) used the waterdrop technique to evaluate the susceptibility of nine soils to erosion. Soils resistant to erosion required more kinetic energy to be dislodged than those susceptible to erosion. Aggregate stability as measured by the wet-sieving technique (Yoder, 1936; Russell, 1949; Van Bavel, 1950) reflects soil resistance to the abrasive and slaking forces exerted on soil by flowing water.

Another technique to measure the stability of soil structure is that of De Leenheer and De Boodt (1959). It is a combination of the wet-sieving and waterdrop techniques. Air-dry aggregates are wetted with waterdrops falling from 0.5 m and then are equilibrated over a $1N\ H_2SO_4$ solution in a vacuum desiccator for 24 h. The preequilibrated aggregates then are subjected to wet sieving for 5 min. The *instability index* is defined as the difference (in millimeters) in the mean weight diameter between waterdrop and wet-sieving analyses. Yet another method to determine the structural stability of a soil is that of Henin et al. (1958). With this method, the aggregates are sieved in water, ethanol, and benzene. The stability index H is defined as

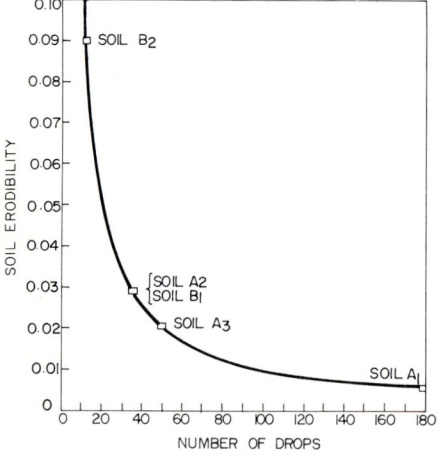

Figure 3.2 Soil erodibility as determined by the raindrop technique. Soil B_2 is the most erodible and A_1 the least. (Bruce-Okine and Lal, 1975.)

$$H = \frac{(A+L)_{\max}}{\text{WS} + \text{ES} + \text{BS}} - 0.9\text{SG} \qquad (3.1)$$

where $(A + L)_{\max}$ is the maximum fraction smaller than 20 μm in filtrate after sieving; WS, ES, and BS refer to percentages of stable aggregates exceeding 200 μm by sieving in water, ethanol, and benzene, respectively; and SG is coarse-sand percentage. Gabriels et al. (1975) and Pauwels et al. (1976), reporting for soils from Ghent, Belgium, stated that Henin's method provided a better index of soil structural stability than either aggregate percentage or change in mean weight diameter.

The applicability of different indices has been evaluated for different soils. Young (1984) observed significant correlation between soil splash and aggregate stability determined by the De Leenheer-De Boodt method. Susceptibility to erosion of some soils in the African tropics is related to the instability index of De Leenheer and De Boodt (1959) and to Henin et al.'s (1958) structural stability index (De Vleeschauwer et al., 1978). Considering the relative merits of all indices, De Vleeschauwer et al. (1978) observed that the choice of a suitable structural index in relation to erosion by water is narrowed to either the De Leenheer-De Boodt index or the Henin et al. index. The index by De Leenheer and De Boodt is applicable for soils from a broad geographic category. Vanelslande et al. (1985, 1987) also concluded that the De Leenheer-De Boodt instability index is valid to evaluate Nigerian soil susceptibility to erosion.

Clod size and soil tilth. Soil tilth, as denoted by surface roughness and cloddiness, is another important factor related to soil erodibility for both wind and water erosion (Lyles and Woodruff, 1961; Lyles, 1975). A rough, cloddy surface has high surface detention capacity and is less vulnerable to wind erosion than a smooth surface (Chepil, 1953). Both clod size and clod density are important soil physical properties that influence resistance to breakdown by rainfall (Moldenhauer and Koswara, 1968; Moldenhauer and Kemper, 1969; Moldenhauer, 1970; Falayi and Lal, 1979). Falayi and Lal (1979) observed that runoff and soil erosion were the least in soil with clod sizes of 10 to 50 mm. A soil with fine tilth is most susceptible to both wind and water erosion. Using simulated rainfall, Sood and Chaudhary (1980) observed that a sandy loam soil's susceptibility to erosion in Himachal Pradesh, India, was related to the initial clod size.

Crusting. Formation of crust and surface seal is a major factor responsible for high runoff rates on soils of the tropics. Structurally unstable soils are readily slaked and form a semipermeable or slowly permeable surface crust (Plate 3.1). The crust formation is particularly severe in soils with low organic matter content, e.g., those in arid and semiarid regions. In the west African Sahel, Valentin (1985) reported that crusting and soil strength are the most important factors affecting the water erosion rate in arid climates.

Strength Properties

Soil bulk density, resistance to penetration, and shear strength are important properties in relation to erosion by water and wind. The number of solid parti-

Plate 3.1 Surface seal or crust formed on structurally unstable soils.

cles per unit volume determines the contact points where forces applied will be dissipated. Soil strength parameters are of particular relevance to erosion by flowing water. The volume weight or soil bulk density determines the total porosity and hence the rate at which rainfall is accepted. The higher the volume weight, the lower, in general, the infiltration rate. Compacted soils, characterized with high bulk density beyond a certain threshold, are more prone to erosion by water than uncompacted soils are. The susceptibility of compacted soils to wind erosion, however, differs from that of erosion by water. Soil compaction may be desirable to control wind erosion, particularly if the compacted soil forms clods on tillage (Lyles and Woodruff, 1961). The bulk density of soils, especially those containing a high proportion of skeletal materials, should be assessed in situ (Lal, 1979).

Penetration resistance is influenced by crusting and skeletal material. Similar to bulk density, compacted soils offer high resistance to penetration. When they are determined under field conditions, both the bulk density and resistance to penetration should be assessed in relation to antecedent moisture content.

Soil is a particulate material. Soil failure, in engineering terms, occurs by rolling and slipping of grains, not by simple tension or compression. Consequently, shear stress is important in the ability of fluids (wind or water) to cause erosion. The property of soil that resists shear stress exerted by fluids is its *shear strength*. Soil's shear strength is particularly important in soil detachment, mass movement, gully erosion, and stream bank erosion.

Soil shear strength is defined by the following equation, called *Coulomb's law*,

$$S = C + \sigma_n \tan \Phi \tag{3.2}$$

where S is shear strength, C is soil cohesion, σ_n is normal stress on critical plane, $\tan \Phi$ is coefficient of friction, and Φ is angle of internal friction. Terzaghi (1925) reported the importance of pore water pressure on shear strength, and

Hvorslev (1937) used laboratory data to prove the effect of pore water pressure on shear strength. A modified form of Eq. (3.2), called the *Coulomb-Hvorslev shear-strength equation*, is

$$S = C' + \sigma' \tan \Phi' \tag{3.3}$$

where C' is the effective soil cohesion (or interparticle attraction effect), σ' is the effective normal stress, and Φ' is the effective angle of normal friction. The effective stress is given by

$$\sigma' = \sigma - U \tag{3.4}$$

where σ' is the intergranular or effective stress, σ is the total stress, and U is the pore water pressure.

Soil erosion is related to shear stress and critical shear stress. Singer et al. (1978) observed significant correlations between shear stress and erosion rates. They concluded that critical shear stress may be used to predict soil erodibility. They proposed a rating system for characterizing soil erodibility on the basis of critical shear values:

Critical shear value	Erodibility
0.0–2.0	Very erodible
2.1–3.0	Fairly erodible
3.1–9.0	Moderately erodible
>9.0	Less erodible

In Sri Lanka, Kandiah (1974) observed that critical shear stress, determined by using a rotating cylinder test apparatus, is a reliable index of a soil's susceptibility to erosion. He also reported a linear relationship between erosion rate and shear stress. These properties are also discussed in Chap. 5.

Hydrologic Properties

Soil hydrologic properties refer to water retention and transmission properties.

Soil water retention

Soil moisture characteristics or p^F curves play a significant role in both water and wind erosion. The energy status of soil water or the pore water pressure influences soil's shear strength [Eq. (3.4)]. Soils with a high percentage of bonding material, organic matter, and clay contents have high water retention capacities. Resistance of soil to fluid drag is also influenced by initial or antecedent moisture content. A drier soil is generally more susceptible to wind and water erosion than a wet soil. Soil moisture provides cohesion between particles and influences the soil strength and infiltration rate. Total soil potential—the me-

68 Basic Processes

chanical work per unit mass of water required to transfer the unit mass against all operative forces from a reference state to the state in question—is an important soil property in relation to erosion by wind and water. Effective strength is the external stress minus soil water potential.

In the west African Sahel, Valentin (1985) reported that soil's mechanical resistance to detachment was related to soil moisture content θ according to a power equation $Y = a\theta^{-b}$. Slonecker et al. (1976) observed that pore water pressure affects the ease with which a raindrop can dislodge soil particles.

The antecedent soil moisture content influences susceptibility to erosion by interacting with other properties. For a catchment in the tropical rain forest region of northeastern Australia, Bonell et al. (1981) observed that saturated overland flow depends on the relationship among temporal variations in rainfall intensity, soil water storage capacity of the upper layer, and permeability of the subsoil layer. In arid United States, Lyles et al. (1974), using simulated rainfall, observed that much less soil was detached from field-moist clods than from air-dry clods. The effects of soil moisture on erosion are discussed at length in Chap. 5.

Soil water transmission

Soil water transmission refers to the ease with which water is transferred from the air-soil boundary into the soil and from one layer to another within the soil. A range of soil properties are used to characterize soil water transmission, including the hydraulic conductivity (both saturated and unsaturated) and infiltration rate. The infiltration rate is an important property in relation to soil erosion by water.

Infiltration. *Infiltration* refers to the downward entry of water from the surface into the soil profile. The *infiltration rate* is the maximum rate at which water can enter a soil when water availability at the surface is not limiting (Horton, 1940). The maximum equilibrium infiltration rate is also referred to as the *infiltration capacity* (Richards, 1952). The term *infiltrability* refers to the infiltration flux when water at atmospheric pressure is made freely available at the soil surface (Hillel, 1971). The difference between the rainfall rate and the infiltration rate, termed the *infiltration index,* is a soil characteristic calculated from field records of runoff and rainfall rates. The infiltration velocity is the instantaneous infiltrate rate; it changes with time. The infiltration velocity is generally high in the beginning and decreases with time until it reaches a constant rate, i.e., the infiltration rate or infiltration capacity. This behavior is explained in a generalized infiltration curve in Fig. 3.3. The decrease in infiltration rate with time results from reduction in moisture potential gradient, change in size distribution of transmission pores, and alterations in soil structure.

Under field conditions, there are two common situations. (1) When the rainfall rate is less than the soil's infiltration rate, infiltration occurs under unsaturated conditions and the rate is governed by water availability at the soil surface, i.e., is flux-controlled, as shown by the curve in Fig. 3.4a. (2) When the rainfall rate exceeds the infiltration rate, the surface layer is saturated and

Soil Properties and Erodibility   69

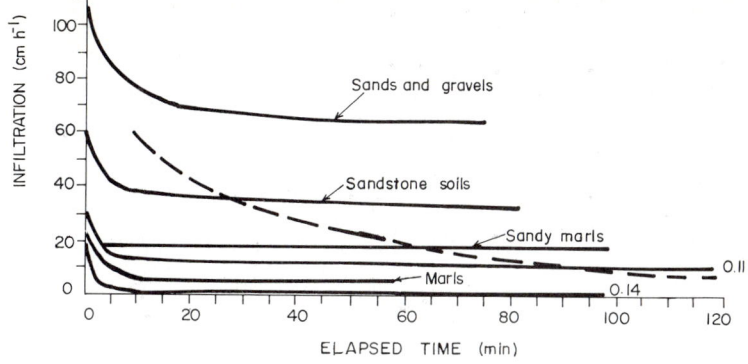

Figure 3.3 The generalized infiltration curve. (Adapted from Thornes, 1976.)

ponding occurs. With surface ponding, the infiltration rate is governed by the soil profile characteristics, i.e., is profile-controlled. The infiltration rate shown by the curve in Fig. 3.4a is under ponded conditions. An example of infiltration rates of some soils in Thailand limited by water availability is shown in Fig. 3.5.

Various authors have described the infiltration process (Childs, 1969; Taylor and Ashcroft, 1972; Swartzendruber and Hillel, 1973; Hillel, 1982; Skaggs and Khaleel, 1982). Green and Ampt (1911) advanced an equation for ponded conditions when water availability is not limiting

$$i = \frac{K_{sat}(H_0 + \rho - H_f)}{\rho} \tag{3.5}$$

where i is the infiltration rate, K_{sat} is the saturated hydraulic conductivity, H_0 is the depth of ponded water, ρ is the vertical depth of saturated zone, and H_f is the soil moisture potential at the wetting front. This original equation has been revised and modified by Swartzendruber and Huberty (1958), Mein and Larson

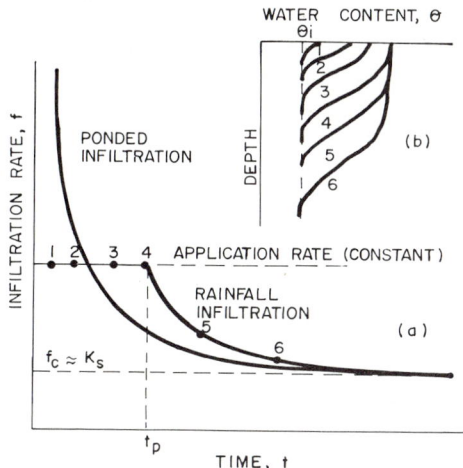

Figure 3.4 The infiltration curve is (a) flux-controlled and (b) profile-controlled. (Adapted from Haan et al., 1982.)

(1973), and Morel-Seytoux and Khanji (1974). The simplified version of the Green and Ampt equation is

$$i = i_c + \frac{b}{I} \qquad (3.6)$$

where I is cumulative infiltration, i is the infiltration rate or volume of water entering a unit soil surface area per unit time, and i_c and b are constants. The constant i_c represents the equilibrium infiltration rate when t and I are both large. Another commonly used empirical equation was advanced by Kostiakov (1932), who described the infiltration rate as an exponential function of time t

$$i = c't^{-\alpha} \qquad (3.7)$$

where c' and α are constants. Horton (1939, 1940) suggested an equation similar to Kostiakov's

$$i = i_c + (i_0 - i_c)e^{-kt} \qquad (3.8)$$

where i_c, i_0, and k are constants. In contrast to Kostiakov's model, however, at time $t = 0$, the infiltration rate is not infinite but equals i_0. Equation (3.8) is

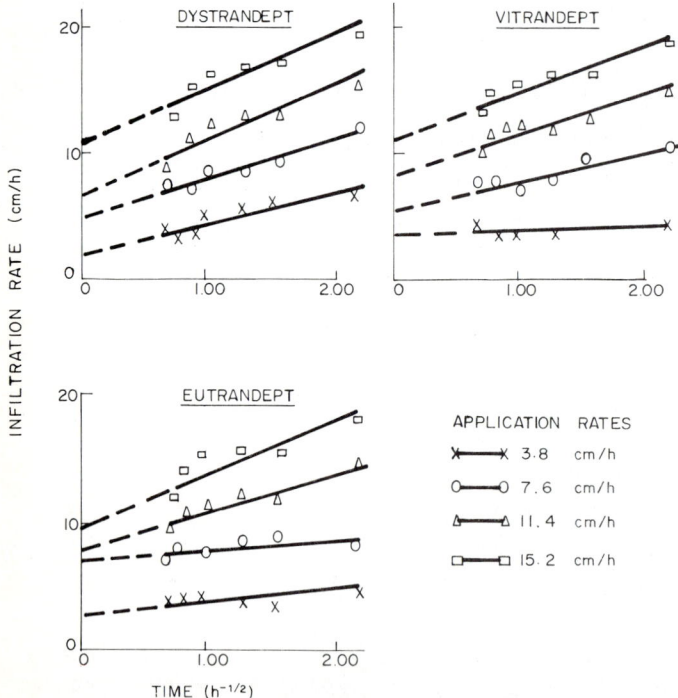

Figure 3.5 Flux-controlled infiltration curves for some soils from the Philippines. (Adapted from Briones, 1982.)

more difficult to use because three constants must be determined experimentally.

The most widely used infiltration model is Philip's (1957)

$$i = \tfrac{1}{2}St^{-1/2} + A \qquad (3.9)$$

where S and A are constants termed the *sorptivity* and *transmissivity*, respectively. Under steady-state conditions, the constant A is approximately equal to the saturated hydraulic conductivity. Equation (3.9) is, however, limited to uniform, homogeneous soils. Holtan (1961) and Holtan and Lopez (1971) described yet another empirical equation based on the storage concept

$$i = (GI)aS^{1.4} + i_c \qquad (3.10)$$

where S is the available storage in the surface layer, GI is the growth index of crops in maturity percentage, a is an index of surface-connected porosity, and i_c is the equilibrium or steady-state infiltration rate. The i_c is estimated from the hydrologic soil group.

Infiltration process. The process of water movement within soil and that related to the advance of the wetting front have been investigated by many. Baver (1938) observed that the wetting front advances under the influence of gravitational and capillary forces. Later Bodman and Colman (1944) postulated that the infiltration process can be defined in five identifiable subzones (Fig. 3.6). (1) The *saturated zone* extends from the surface to a maximum depth of approximately 1.5 cm; (2) the *transition zone* is a region of rapid decrease in soil water content; (3) a *transmission zone* occupies the upper part of the wetted soil and has no moisture gradient but merely conducts water to the wetting zone; (4) the *wetting zone* lies below the transmission zone and has a moisture gradient progressively increasing with depth; and (5) a *wetting front* is rather diffused and irregular and has a high potential gradient. Collis-George and Lal (1971, 1973)

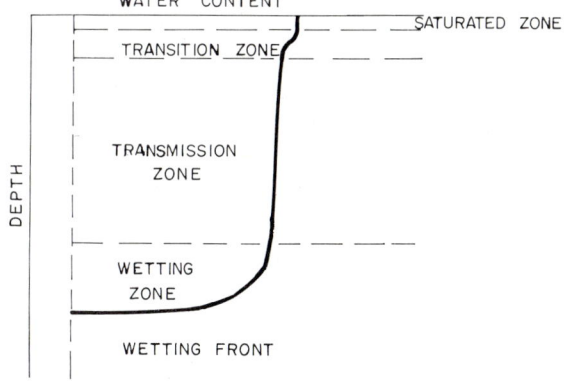

Figure 3.6 Identifiable subzones during the infiltration process as defined by Bodman and Coleman (1944). (Adapted from Haan et al., 1982.)

showed that infiltration is a condensation-evaporation process and that the wetting front can be subdivided into three zones: condensation, evaporation, and liquid wetting front (Fig. 3.7). The condensation-evaporation process controls the rate at which water enters the soil, even if water availability at the surface is not limiting. This implies that infiltration in some structurally unstable soils is profile-controlled even under ponded conditions. It is the profile-controlled infiltration process that initiates and enhances overland flow.

The infiltration rate is best characterized under field conditions (Lal, 1979). Field measurements should be made on large areas, at least 5 m². The most appropriate method would estimate the infiltration rate on a plot or a watershed under natural rainfall. For Central Plain soils in Thailand and a sandy loam soil in the Ord River irrigation area in Australia, respectively, Bridge et al. (1975) and Bridge and Muchow (1982) reported that field infiltration measurements are highly relevant in evaluating water entry into soils. For soils in the Ivory Coast, Moreau (1978) evaluated differences in structural stability of savanna and forest soils by measuring water infiltration rates. Lal (1976) reported for alfisols in Nigeria that soils most susceptible to erosion had low infiltration rates.

Permeability and hydraulic conductivity. The facility of water flow through soils is termed *permeability*. The equilibrium infiltration rate approaches the saturated hydraulic conductivity of the soil. In Darcy's law for saturated flow [Eq. (3.11)], the hydraulic conductivity term K is a measure of soil resistance to water flow:

$$Q = AVt$$

$$V = K\frac{L}{h} = ik$$

$$\therefore K = \frac{QL}{Aht}$$

(3.11)

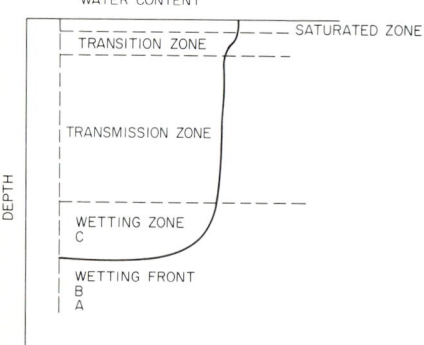

Figure 3.7 Identification of subdivisions of wetting front as observed by Collis-George and Lal (1971, 1973). A = zone of vapor condensation. B = zone of evaporation. C = zone of critical soil moisture potential.

Here Q is the water flux, V is the flow velocity, L/h is the hydraulic gradient, A is the area, t is time, i is the infiltration rate, and K is the hydraulic conductivity. The term *hydraulic conductivity* specifically denotes the porportionality constant in Darcy's law. The term *permeability*, however, is used in general to denote the rate of water movement through soil. According to Terzaghi and Peck (1967), the permeability coefficient K is given by

$$K = \frac{k\gamma_w}{\eta} \qquad (3.12)$$

where γ_w is the density of water, η is the viscosity of water, and k is the property of the medium only.

Permeability of the soil profile is significantly related to properties of different horizons. In a layered profile, permeability is controlled by hydraulic properties of the most restricting layer.

Depending on the permeability rate, soil permeability is conventionally divided into classes as shown in Tables 3.1 and 3.2. Susceptibility of a soil to erosion is related to its permeability. Soils with extremely slow to moderate permeability generate more runoff and are more susceptible to processes governing upland erosion than those with rapid permeability. Use of permeability classes to estimate a soil's susceptibility to erosion is discussed in Chap. 7.

Rheologic Properties

Rheology is the science that describes the behavior of a soil water system in moist to semifluid state. The antecedent soil water content influences soil susceptibility to erosion by affecting cohesion, shear strength, consistency, and plasticity. A simpler agronomic term for those combined properties is *soil tilth*.

Soil consistency—the resistance of soil to deformation or the degree of cohesion and adhesion of the soil mass—has an important effect on the relative importance of processes governing soil erosion. Soil consistency also is highly sig-

TABLE 3.1 Permeability Classes for Saturated Subsoils and Corresponding Range of Hydraulic Conductivity and Intrinsic Permeability

Class	Hydraulic conductivity k		Intrinsic permeability k', 10^{-10} cm^2
	10^{-5} cm/s	in/h	
Very slow	<3	<0.05	<3
Slow	3–15	0.05–0.2	3–15
Moderately slow	15–60	0.2–0.8	15–60
Moderate	60–170	0.8–2.5	60–170
Moderately rapid	170–350	2.5–5.0	170–350
Rapid	350–700	5.0–10.0	350–700
Very rapid	>700	>10.0	>700

SOURCE: O'Neal, 1949.

TABLE 3.2 Permeability Classes for Saturated Subsoils

Class	Hydraulic conductivity k, cm/h	Comments
Extremely slow	<0.0025	So nearly impervious that leaching process is insignificant
Very slow	0.0025–0.025	Poor drainage results in staining; too slow for artificial drainage
Slow	0.025–0.25	Too slow for favorable air-water relations and deep root development
Moderate	0.25–2.5	Adequate permeability
Rapid	2.5–25	Excellent water-holding relations and permeability
Very rapid	>25	Associated with poor water-holding conditions

SOURCE: Smith and Browning, 1946.

nificant in wind erosion. Based on soil water content, Swedish soil scientist Atterberg defined five limits of soil consistency:

1. The *liquid limit* is the water content above which the soil behaves as a viscous liquid and the soil water mixture has no measurable shear strength.
2. The *plastic limit* refers to the water content below which the soil no longer behaves as a plastic material. The plasticity index is the range of moisture content (difference between the liquid limit and the plastic limit) at which the soil behaves as a plastic material.
3. The *shrinkage limit* is the water content below which no more soil volume change occurs with further drying. This property is especially important in water erosion of semiarid and arid region soils.
4. The *sticky limit* refers to the water content at which a soil loses its adhesive property and ceases to stick to other objects. Knowledge of the sticky limit is important in deciding when to plow. The soil's moisture content must be drier than the sticky point so that the soil does not stick to the plow or wheels.
5. The *cohesion limit* is the water content at which the soil particles cease to stick together. Soil dry to the cohesion limit does not form clods on plowing and is therefore highly susceptible to wind erosion.

These five limits apply only to cohesive soils. Coarse-textured upland soils of the humid and subhumid tropics do not exhibit such properties (Lal, 1987). Highly aggregated oxisols and nitosols also behave as cohesionless soils because clay-sized particles are strongly cemented to form stable microaggregates (Ahn, 1979).

As the cohesion limit has effects on wind erosion, the plastic limit is important in tunnel erosion. Initiation of tunnel erosion in semiarid and arid climates is often caused by water flowing rapidly through shrinkage cracks (Stagg, 1978; Burt, 1980; Yair et al., 1980; Anderson and Burt, 1982). Cracking clays with high swell-shrink capacity are more prone to tunnel erosion than soils with low-activity clays. Liquefaction of soil is an important stage before mud flow or viscous flow is initiated (Bryan et al., 1978). Mud flow is attributed to liquefaction of clay-rich material.

Chemical and Mineralogical Properties

Soil's resistance to forces generated by agents of erosion is also affected by its chemical constituents. Most important among these are the amount and nature of colloids, and the composition of the exchangeable cations on the colloid complex. The colloid complex, relevant to soil erodibility, comprises organic content and the clay minerals.

Organic matter

Soil organic matter content significantly affects soil structure and stability. All other factors remaining the same, soils with high organic matter are less susceptible to erosion than those with low organic matter content. Lal (1981) reported that in tropical Africa severely eroded soils contain less organic carbon than less eroded or uneroded soils. De Vleeschauwer et al. (1978) and Vanelslande et al. (1984, 1985) observed that the organic carbon content of some Nigerian soils was significantly negatively related to the soils' susceptibility to erosion.

Although virgin soils under natural vegetation cover have high organic matter content, cultivated soils generally have low organic carbon content. Uncultivated soils of the tropics do not necessarily have lower organic matter content than their counterparts in the temperate zone, but the organic matter content of cultivated soils in the tropics seems to decline more drastically than in the temperate zone. Additionally, organic matter content varies with rainfall amounts. Soils in humid climates have more organic matter than similarly managed soils in semiarid or arid climates.

A wide range of soil physical properties such as structure, pore size distribution, water retention and transmission properties, and aeration are directly or indirectly influenced by soil organic matter. While we consider soil physical properties, it is important to differentiate the effect of humus from that of crop residue or mulch. The latter has pronounced effects on the hydrothermal regime, raindrop impact and splash, crusting, and infiltration. While humus influences these properties by strengthening the bonds that stabilize the structural units, it maintains a favorable balance between retention and transmission pores.

Soils with low organic matter content are more easily compacted than those with high organic matter content under similar climatic conditions. The effects of organic matter content on soil structure differ for soils with different texture and mineralogic compositions. Quirk and Panabokke (1962) reported that, in addition to the amount and kind of organic matter, its distribution within the matrix is an important structure stability factor. For example, it is the concentration of organic matter contents in the microaggregates that makes them resistant to slaking and dispersion (Turchenek and Oades, 1978). Literature reviews (Allison, 1973; Schnitzer and Khan, 1978) have emphasized the important role of organic matter in soil structure, although the information for tropical regions remains sketchy.

Biswas et al. (1964) reported that organic matter increases due to continuous application of farmyard manure increased the percentage of water-stable aggregates and permeability for alluvial soils in northern India. Better structural stability was also observed with an increase in soil organic matter content due to the addition of water hyacinth (Ghani et al., 1967), sugar cane trash (Sandhu and Bhumbla, 1967), and compost (Kandiah, 1976). The favorable influence of increased organic matter on soil permeability, bulk density, and soil erodibility also was reported for some Indian soils (Ghatol and Malewar, 1978). In Puerto Rico, adding organic matter increased the structural stability of some ultisols (Escolar, 1966). But only negligible improvements in structure resulted from adding even large quantities of farmyard manure to sandy soils of Egypt (Abdou and Metwally, 1967) or to clayey vertisols in central India (Venkobarao et al., 1967).

Many researchers have demonstrated significant relationships between organic carbon content and the percentage of water-stable aggregates. Data from Luk (1979) show that the relation between percentage of water-stable aggregates (WSAs) and organic carbon is best described by the nonlinear second-degree polynomial equation

$$WSA = 24.83 + 3.87C - 0.254C^2 \tag{3.13}$$

where C is the percentage of organic carbon content. The type of relationship differs among soils, depending on the nature and amount of clay content. An increase in organic matter content with continuous application of crop residue mulch also increased soil aggregation and stability of an alfisol in southwest Nigeria (Lal et al., 1980). The percentage of water-stable aggregates and the erosion and dispersion ratios were significantly related to the quantity of residue return, which in turn was related to the organic matter content of the soil. The increase in structural stability was associated with an increase in saturated hydraulic conductivity, accumulative infiltration and equilibrium infiltration rates, high soil-water sorptivity and transmissivity, relatively larger percentages of transmission pores, and low soil bulk density. Similar effects were also reported for loess soils of northern Nigeria (Lawes, 1957).

Mechanisms responsible for improving aggregation by soil organic matter are complex (Greenland, 1979). The model by Williams et al. (1967) in Fig. 3.1 shows that forces of organic molecules play an important role in cementing together domains and microaggregates into aggregates. So aggregate stability is closely related to organic matter content. Organic materials act as lining spread over the surfaces of domains and microaggregates. In particular, organic polymers contribute to the stability of aggregates, probably through interdomain or intermicroaggregate bonding (Fig. 3.1). Deshpande et al. (1964) showed a major loss of stability by an ultisol from Katherine, northern Australia, when polysaccharides were chemically removed. The loss of stability was over and above that induced by sorption of anions arising from the treatment. However, stabilization of aggregates by interdomain and microaggregate bonding is almost certainly confined to surface soils containing appreciable organic matter content.

For soils of Martinique, Turenne (1982) emphasized the importance of nitrogen compounds forming humid substances and developing a stable structure. The activity of earthworms, stimulated by fresh, decomposed organic matter, is also important (Lal, 1987). As organic matter undergoes microbial decomposition, microbial slimes and by-products increase the bonding strength among domains and microaggregates. Release of a variety of linear organic polymers, humid substances of low molecular weight, polysaccharides, and polyuronides binds the particles into micro- and macroaggregates (Harris et al., 1966).

Organic matter content may not always increase soil resistance to forces of erosion agents. High contents of organic carbon in some soils may lead to the development of hydrophobic soil characteristics (Krammes and Osborn, 1969; Harris et al., 1966). Meeuwig (1971) observed that water-repellent soils with high organic matter content may be erodible because mutual electrostatic repulsions develop between aggregates.

Clay minerals

Soil structure and its strength are influenced by the amount and nature of clay minerals. Grissinger (1966) noted that soil erodibility varied with clay type and amount. The amount and the nature of clay influence soil structure in at least two ways. First, clay domains and silt-sized particles are cemented together to form microaggregates. Second, the force which holds the domains together is usually electric and depends on the charge properties of the clay minerals. In addition to the organic matter's role in providing bonding between microaggregates, in some soils hydrous oxides create cement between domains and microaggregates. Oriented clay films, along with microbial by-products, add some stability to microaggregates and aggregates as well.

The structural properties of some tropical soils, especially those containing predominantly low-activity clays, require special mention. Low-activity clays (LACs) are those with an effective cation-exchange capacity (ECEC) of 16 meq/100 g clay or less. They are soils in which the clay fraction is composed predominantly of kaolinite and halloysite with hydrous oxides of iron and aluminum. They are mostly alfisols, ultisols, and oxisols. While alfisols and ultisols may have some easily dispersible clay in their soil profiles, the clay in oxisols is highly resistant to dispersion (Table 3.3) because of strong interparticle forces caused by the hydrous oxides of iron and aluminum. Stable aggregation also has been observed in some alfisols and ultisols, especially those belonging to the Pale- and Rhodic great groups such as Rhodic Paleustults formed on basalts in western Nigeria (Moormann, 1981) and in Kenya (Ahn, 1979).

Not all soils in the tropics containing low-activity clays have stable aggregation. Some alfisols and ultisols have less stable structure. They have coarse to medium-textured surface horizons and a sharp transition to clayey B horizons (Plate 3.1). They are structurally unstable, prone to crust (Plate 3.2), and easily

TABLE 3.3 Stable Microaggregation in Oxisols and Nitosols

Soil	Horizon (depth, cm)	Clay content, % of soil			Proportion of clay in stable aggregates
		Before dispersion	After dispersion	Difference	
More stable					
Rhodic Paleudult or nitosol	A 11 (5)	20.5	58.4	37.9	0.65
	A 12 (15)	18.9	51.5	32.6	0.63
Kikuyu friable clay, Kenya	Upper B 2 (30)	4.2	69.8	65.6	0.94
	Lower B 2 (90)	1.8	82.8	81.0	0.98
Alfic Eutrorthox (Akamadon series, Ghana)	A 11 (5)	7.7	45.2	37.5	0.83
	A 12 (13)	7.4	45.2	37.8	0.84
	B 1 (38)	7.2	48.0	40.8	0.85
	B 2.1 (80)	11.3	64.7	53.4	0.83
Less stable					
Plinthic Paleudult (Bekwei series, Ghana)	A 11 (5)	18.6	19.0	0.4	0.02
	A 12 (13)	16.4	20.4	4.0	0.20
	B 21 (30)	19.0	43.2	24.2	0.56
	B 23 (101)	31.9	68.1	36.2	0.53
Plinthic Paleudult (Asuansi series, Ghana)	A 11 (5)	8.8	12.2	3.4	0.28
	A 12 (13)	9.6	12.4	2.8	0.23
	B 1 (28)	10.1	22.2	12.1	0.55
	B 2.3 (82)	21.5	49.7	28.2	0.57

SOURCE: Ahn, 1979.

compacted, and their infiltration rate declines readily with cultivation. In comparison, oxisols and the alfisols and ultisols that have only a gradual increase in clay content down their profiles are relatively stable soils structurally. Oxisols are often more acid than other LAC soils and may have gibbsite and high proportions of oxides and hydrous oxides in their clay fractions (Eswaran and Tavernier, 1980). They also are termed *nitosols*. Together these soils cover a vast part of the subhumid and humid tropics.

The stability of domains and microaggregates is shown by the fact that field texture ascribed to these soils indicates that they are loamy rather than clayey. The stable microaggregates are not easily destroyed even by treating with hydrogen peroxide and prolonged shaking. The particles separate only after the addition of an anionic dispersing agent. That is a reason for their low-erodibility soil compared with those containing 2:1 lattice clays (El-Swaify, 1977).

Soil Properties and Erodibility 79

(a)   (b)

Plate 3.2 Horizonation of tropical soils with a distinct B_{2t} horizon: (a) quartz gravel, (b) hardened plinthite.

Differences in physical behavior among soils with clays composed predominantly of kaolinite or halloysite, oxides, and hydrous oxides of iron and aluminum (latosols) and those with clays consisting predominantly of montmorillonite (margalitic soil) have been explained by Koenigs (1961). He reported that margalitic soils are flocculated more readily by solutions of higher electrolyte concentrations but latosols are dispersed more readily in strong electrolytes. He also observed that a pure kaolinite mineral specimen behaved differently, and so he attributed the anomalous behavior of the latosols to the influence of the oxides and hydrous oxides of iron and aluminum.

It is now known that oxides of iron carry a net positive charge to about pH 7, and oxides of aluminum up to about pH 9. Thus at the pH of many oxisols and ultisols and some alfisols (pH below 6), the oxides of iron and aluminum are strongly positively charged. Therefore, these minerals promote aggregation of clay minerals with a permanent negative charge. This includes many of the kaolin and halloysite particles occurring with the oxides in LACs. Also at a soil pH of 4 to 5 most hydrous oxides of iron and aluminum are close to their zero-point charge, so their dispersive forces are low.

Koenigs (1961) attributed disperson of the latosol clays by strong electrolytes to the specific adsorption of the anion, rendering the oxide surfaces negatively charged and promoting dispersion of the aggregates. This phenomenon has been intensively studied in recent years, and the sorption of phosphate, silicate, and many organic anions is now well understood (Bowden et al., 1980).

The extent to which the oxide and hydrous oxide surfaces are inactivated through anion adsorption probably largely determines the stability of structural elements in field soils. Dispersion and translocation of clay in many alfisols and ultisols are possible because silicate or phosphate or organic anions "deactivate" the clay. A smaller proportion of aggregated clay is normally found in the surface horizons of these soils than in subsoils, which indicates that organic matter is important in deactivation. Deshpande et al. (1968) observed that iron oxides in many red soils influence aggregation or aggregate stability very little. They attributed this to deactivation of oxides by anion sorption.

The more active oxides in nitosols and oxisols apparently are less subject to deactivation, perhaps because the amount and surface extent of the oxides are greater and are associated with a higher silica-removal rate during soil formation (Gallez et al., 1977) and stronger mobilization of aluminum under more acid conditions. Stronger stability certainly appears to be associated with the gibbsite. The portions of aluminum in the lattices of iron oxides and hydrous oxides may also influence their activity.

While organic anions may have a dispersive effect, due to specific sorption by the oxides and hydrous oxides, organic polymers contribute to the stability of aggregates, probably through interdomain or intermicroaggregate bonding. Thus there exists a strong interaction between soil organic matter content and the amount and nature of clay in determining structural stability and susceptibility to erosion. The fine-soil fraction interacts with soil organic matter content to form stable aggregates that resist the dispersive effects of raindrop impact. It is the percentage of stable aggregates larger than 0.25 mm that imparts resistance to erosion. The data in Table 3.4 from Malaysia indicate the importance of interaction between texture and organic carbon content in relation to soil erosion. Serdang series with the maximum measured erosion contains the least organic carbon and is characterized by the fine sandy loam texture. In contrast, Munchang series with the least measured erosion rate contains the highest organic carbon and is characterized by the clay texture. In China DeMing (1986) reported that the percentage of water-stable aggregates was linearly correlated with soil organic content. He also observed that the wash resistivity index, or the soil's resistance to overland flow, increased linearly with an increase in clay content.

TABLE 3.4 Effects of Soil Organic Carbon Content and Texture on Erosion Rates in Malaysia

Soil series	Texture	Organic carbon, %	Aggregates >0.25 mm, %	Soil erosion, t/ha
Munchang	Clay	1.87	83.1	100
Rengam	Sandy clay loam	1.69	59.0	212
Serdang	Fine sandy loam	1.10	55.9	339
Holyrood	Loamy sand	1.35	73.5	252
Sg. Buloh	Loamy coarse sand	2.02	64.6	220

SOURCE: Rubber Research Institute, 1975.

Exchangeable cations

The nature of cations on the exchange complex determines the soil-structure type and strength. Soils containing predominantly bivalent cations (Ca^{2+}, Mg^{2+}) have more stable structure than those containing monovalent cations (Na^+, K^+). Arulanandan et al. (1973) concluded from laboratory experiments using a rotating-cylinder test apparatus that the ionic concentration and composition of pore fluid and eroding fluid, as well as clay type and amount, affect soil particle detachment. Singer et al. (1978) reported that the susceptibility of some soils from California to water erosion increased linearly with an increasing sodium absorption ratio (SAR). In contrast, the vulnerability of soils to erosion decreased exponentially with increasing contents of dithionite iron.

The presence of Na^+ on the exchange complex increases soil dispersibility. In other words, a soil saturated with Na^+ is more easily dispersed than one saturated with Ca^{2+}. Dispersive sodium-rich soils are vulnerable to pipe erosion in semiarid and arid regions (Stocking, 1976; Imeson et al., 1982). Dispersive soils are also susceptible to splash and rill erosion (Wendelaar, 1976). In Gujrat, India, Shah and Patel (1974) observed that the permeability of soils containing high contents of Na^+ was lower than that of those containing calcium. In Sri Lanka, Kandiah (1976) reported an interaction between organic matter content and SAR. With high SAR, organic matter content provides stability and strength to the soil. The effectiveness of organic matter content in forming stable aggregates depends on the flocculating ability of the pore fluid. Kandiah observed that the optimum organic matter content at SAR 2 was about 4 percent.

In New South Wales, Australia, where tunnel erosion is a serious problem, Crouch (1978) reported that the ease of tunnel formation related to the ease with which a soil is dispersed when immersed in water. The main difference between soil from tunneling and nontunneling areas was the dispersion index. The dispersion ratio was related to the exchangeable-sodium percentage.

In soils of the central alluvial tract of Uttar Pradesh, India, the erosion ratio was negatively correlated with the calcium and magnesium saturation percentage ($Ca^{2+} + Mg^+$).

Heat of wetting

Heat of wetting refers to the heat released when water vapors condense on a dry soil surface to form a monomolecular layer. The amount of heat released depends on the amount and nature of clay minerals, the total surface area (internal and external), and the soil moisture potential. Heat released during the wetting process has an important effect on soil structure and ease of particle detachment. This subject is discussed at length in Chap. 5.

Soil Profile Characteristics

Soil profile characteristics influence erosion directly and indirectly. Vegetation growth, an important agronomic factor, affects erosion by providing protective cover on the soil surface and contributing soil organic matter reserves. Better and deeper root distribution in a soil profile favors the structure and lessens erodibility. Over and above the effects on soil fertility, profile characteristics influence both the magnitude and the type of erosion in the following ways:

1. *Profile characteristics influence water flow.* The rate and the type of water flow through the soil profile are influenced by hydrologic properties of different horizons. Abrupt change in hydrologic properties from one horizon to another initiates processes that lead to erosion. For example, sand over clay can cause severe erosion of the top cohesionless sandy material. Initiation of rill, tunnel, and gully erosion is attributed to slowly permeable material underlying a permeable horizon. In general, profiles characterized with tight subsoils are affected more severely with accelerated erosion than those with more friable, permeable subsoil.

2. *Profile characteristics influence vegetative growth.* Soils with shallow topsoil horizon and those close to bedrock are more susceptible to erosion than those with deep A horizon. If subsoil properties are unfavorable to root growth, either from physical impedance or nutritional imbalance, the topsoil is often prone to accelerated erosion. Such soils can support only scanty vegetation that is easily denuded by grazing or other natural factors. Both wind and water erosion become accelerated on denuded surfaces and on soils with low organic matter content and poor structure (Olson, 1949). Low-fertility soils, in both temperate and tropical zones, are more easily eroded and degraded than fertile soils.

Dynamic Nature of Soil Properties

Characterizing soil by its physical and chemical properties provides useful guidelines to its susceptibility to forces generated by agents of erosion. As far as possible, physical and strength parameters of a soil should be measured in situ, under natural conditions as they exist in field situations. While the risk to erosion may be inferred from the soil characteristic, it is important to realize that soil properties are not static. Soil is a dynamic, ever-changing entity $[P = f(t)$, where P is a soil property and t is time]. The rate of change in soil properties depends on the intensity and type of land use and the interaction between management and ecological factors. Susceptibility to erosion may depend on the inherent characteristics, but the inherent characteristics are constantly changing.

Although all soil properties are likely to change, some change faster than others. Also properties of soils in harsh climates (e.g., tropical regions) may change more readily than those in mild climates. The rate of change is also influenced by the antecedent level of the properties considered. Some examples of changes in soil properties that directly bear on soil erosion potential follow.

Mechanical properties

Change in land use and methods of soil and crop management can drastically affect soil texture and structure. Lal (1976) observed drastic changes in textural properties of the surface horizon in only 3 years. The data in Fig. 3.8 show the magnitude of changes among management systems in soil texture during

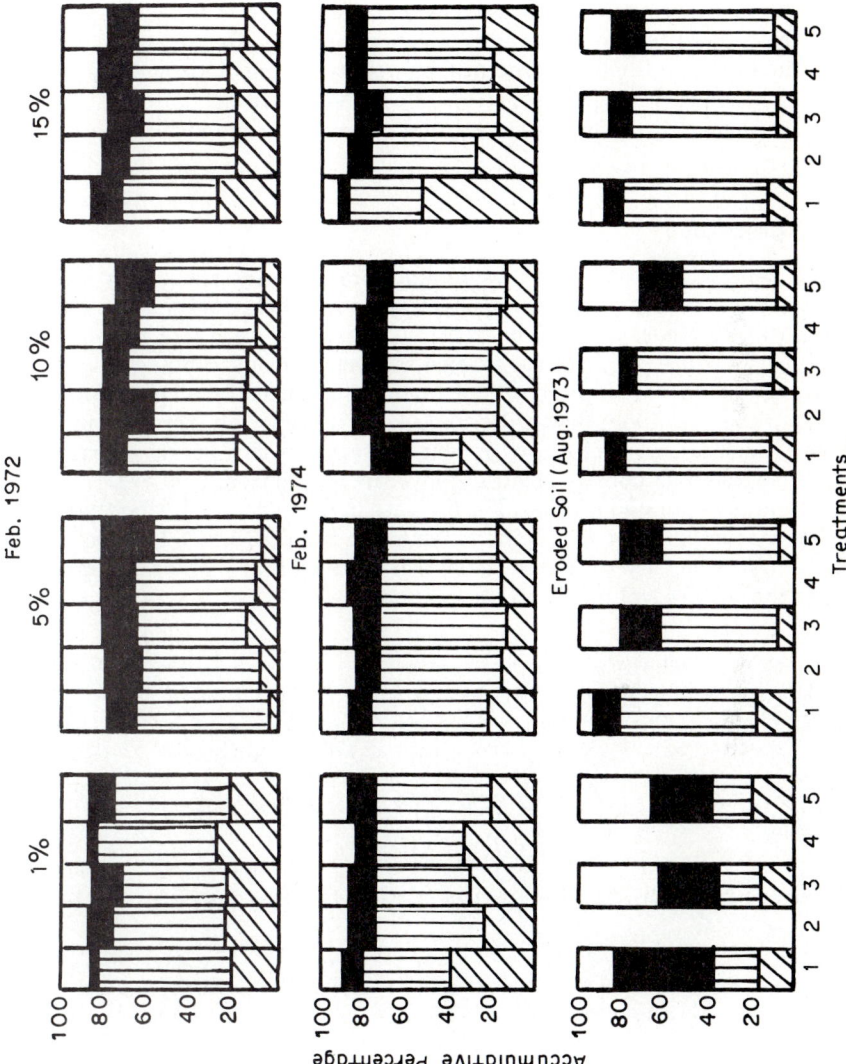

Figure 3.8 Alteration in textural properties of the surface layer due to erosion of an alfisol at Ibadan, Nigeria (Lal, 1976).

84   Basic Processes

3 years. Smallest changes came with mulch and no-till treatments; greatest, with plowing. Bare, fallow plots had a drastic increase in gravel content and a decrease in sand and clay content (Plate 3.3). Increased gravel in the surface horizon can alter a soil's susceptibility to erosion.

Soil structure is more readily altered by management than soil texture. Measurable differences can be observed between before and after plowing. An important factor that influences structural aggregates is the biotic activity of soil fauna, i.e., earthworms, termites, etc. Biotic activity is readily altered by crop residue management, tillage practices, and agricultural chemicals (Plates 3.4 and 3.5).

Strength properties

Soil bulk density, pore size distribution, and total porosity are readily altered by management and by raindrop impact. The kinetic energy of impacting drops can drastically increase soil bulk density and form a surface seal of low porosity. Formation of surface seal involves at least two mechanisms: compaction of surface aggregates and washing in of fine particles produced by dispersion of aggregates (McIntyre, 1958; Bryan, 1973). Luk (1979) reported increases in bulk density of surface layers of soil with poor aggregation. For poorly aggregated prairie soils, the bulk density of surface layer increased from 0.99 to 1.11 g/cm^3 after a 60-min, simulated rainfall. The increase in bulk density during the same period was much greater for alpine forest soils: from 0.85 to 1.15 g/cm^3. The changes in bulk density were reflected in runoff rates that increased appreciably after the surface soil had been compacted.

Plate 3.3 Concentration of gravels in the surface layer of plowed, bare alfisol at Ibadan, Nigeria.

Soil Properties and Erodibility 85

(a)

(b)

Plate 3.4 Casting activity of *Hyperiodrilus africanus* (a) under forest and (b) in burned savanna.

(a)

(b)

Plate 3.5 The activity of termites on soil surface: (*a*) large mound, (*b*) sheeting and feeding galleries.

TABLE 3.5 Effects of Methods of Deforestation on Soil Bulk Density and Penetrometer Resistance of 0- to 5-cm Layer

Treatment	Bulk density, g/cm^3				Penetrometer resistance, kg/cm^2			
	1978	1979	1980	1981	1978	1979	1980	1981
Traditional farming	0.64	1.06	1.07	1.27	0.21	0.96	0.52	1.32
Manual clearing	0.68	1.17	1.17	1.39	0.20	1.40	0.75	1.19
Shear blade clearing	0.70	1.19	1.37	1.38	0.26	1.00	1.84	2.19
Tree pusher/root rake clearing	0.60	1.24	1.32	1.42	0.20	1.30	0.73	1.23

Land cleared in 1979.
SOURCE: Lal, 1984.

Over and above the effect of raindrop impact, vehicular traffic increases soil compaction under field conditions (Raghavan et al., 1977; Soane et al., 1981, 1982; Voorhees et al., 1985), more drastically in tropical than in temperate-zone soils. Mechanized land clearing in the tropics severely compacts soil (Lal, 1984b), and the effects persist for many years (Tables 3.5 and 3.6). As compaction increases, runoff rate and runoff amount also increase. So the increase in compaction is related to the decline in organic matter.

Hydrologic properties

Water retention and transmission properties are also time-dependent. Hydraulic conductivity changes even during one rainstorm, and so does the infiltration rate. The rate of change is faster in soils containing expanding-lattice clays than in those containing predominantly low-activity clays. Using simulated rain, Vanelslande et al. (1987) observed that soil infiltration rates decreased during the test rain (Fig. 3.9). The decrease in infiltration rate during a rainstorm event is partly attributed to increased bulk density, decreased porosity, and formation of surface seal. The infiltration rate decrease was marked by a corresponding increase in runoff.

Under field conditions, the decline in saturated hydraulic conductivity and

TABLE 3.6 Changes in Soil Bulk Density and Gravel Content of 0- to 10-cm Layer after 6 Years of Continuous Corn Cultivation with Mechanized Farm Operations

	Initial data 1975		2 years		4 years		6 years	
	NT	P	NT	P	NT	P	NT	P
Gravel, %	—	—	8.1 ± 10.2	17.8 ± 11.4	7.0 ± 4.0	23.0 ± 3.0	17.0 ± 5.2	22.4 ± 10.1
Bulk density, g/cm^3	1.43 ± 0.12	1.37 ± 0.17	1.42 ± 0.13	1.37 ± 0.30	1.41 ± 0.1	1.51 ± 0.1	1.40 ± 0.13	1.53 ± 0.14

NT = no-tillage; P = plowed.
SOURCE: Lal, 1985.

infiltration rate with time after cultivation started indicates degradation in soil structure. Wang et al. (1985) reported a marked decline in saturated hydraulic conductivity of clay soils in Ottawa, Canada, after 5 years of continuous corn culture. Saturated hydraulic conductivity declined to less than 1 m/s. For tropical alfisols in Nigeria, Lal (1985) found that the infiltration rate declined with mechanized farm operations (Fig. 3.10). Cumulative infiltration 2 h after the test began on no-till and plowed watersheds decreased from 75 to 65 cm in 1976 to 38 and 28 cm in 1978, 28 and 9 cm in 1979, and 12 and 5 cm in 1980, respectively. Such drastic declines in infiltration rate stem from structural collapse and elimination of "transmission pores" caused by vehicular traffic and soil compaction.

The structural collapse indicated by the decline in water transmission properties is also reflected in altered retention pores. For the same watershed that produced the data on infiltration (Fig. 3.10), continuous cultivation also changed soil moisture retention characteristics. Water storage capacity of soil cultivated 6 consecutive years decreased drastically (Table 3.7).

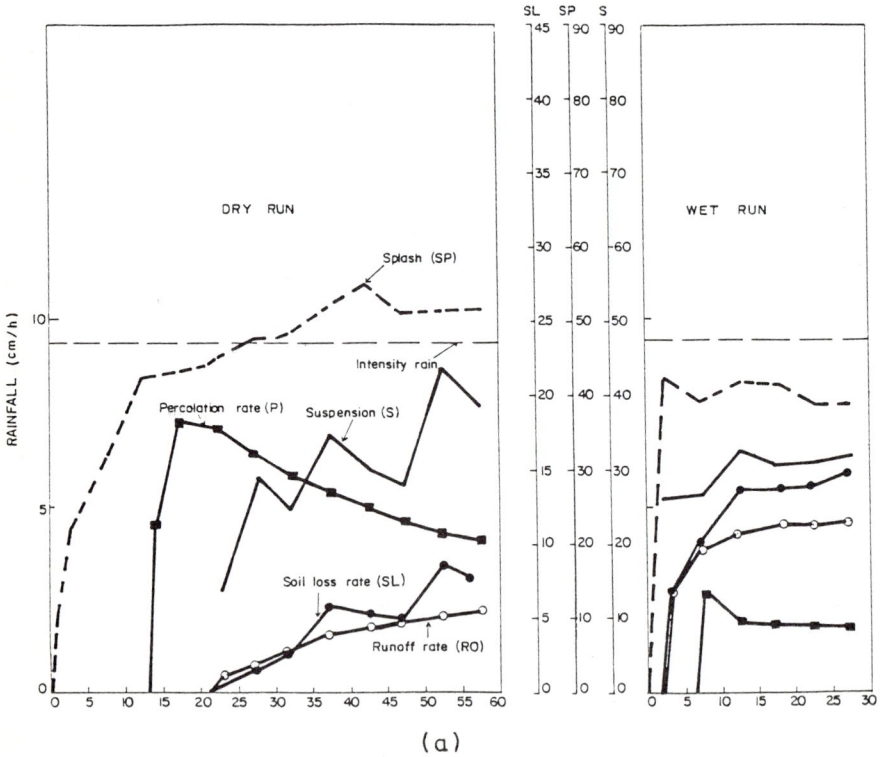

Figure 3.9 Decrease in infiltration rate of some tropical soils subjected to simulated rains: (a) dry run; (b) wet run. Soil loss rate in t/(ha · h); splash in g/5 min; suspension in g/L. (Vanelslande et al., 1987.)

Chemical properties

Soil chemical properties are dynamic and always changing, most notably the organic matter content. With cultivation, soil organic matter declines irrespective of soil or climate. The dynamics of organic carbon in any soil can be treated quantitatively by using a first-order equation (Greenland, 1986)

$$\frac{dc}{dt} = kc + a \qquad (3.14)$$

where c is the mass of carbon per unit area in a fixed mass of soil and a is the addition of carbon per unit time t per unit area.

The organic matter content of cultivated soils in the tropics declines more rapidly than that in soils in temperate regions. So cultivated soils in the tropics generally have less organic matter than comparable soils in northern latitudes (Bates, 1960; Greenland and Kowal, 1960; Jenny and Raychaudhuri, 1960; Klinge, 1962; Bourliere and Hadley, 1970; Bartholomew, 1972; Davidson, 1975; Volkoff, 1977; Gaikwad and Goel, 1977; Kadeba, 1978; Aina, 1979; Smith, 1979). It is important to distinguish among factors that control the rate at which crop residues decompose and those that affect the decomposition of dif-

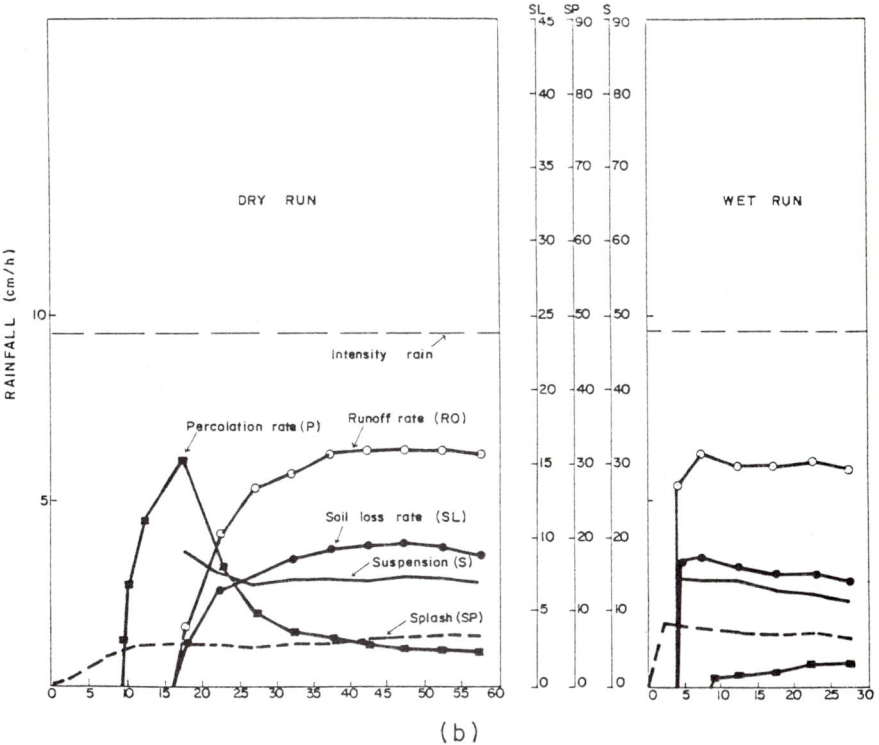

Figure 3.9 (*Continued*)

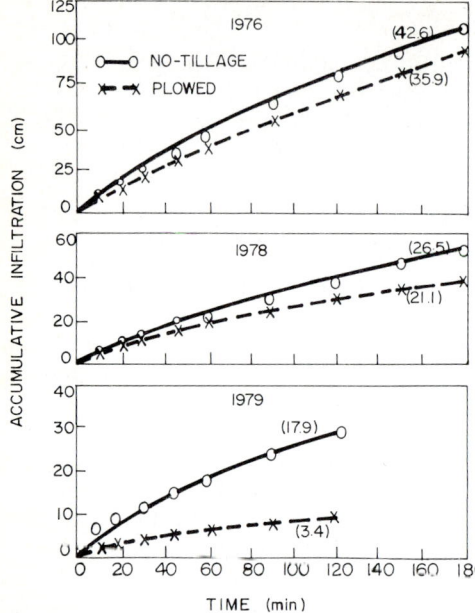

Figure 3.10 Changes in infiltration rate of a tropical soil with length of cultivation period (Lal, 1985).

ferent fractions of soil organic matter. Activity of soil animals, e.g., earthworms and termites, the C:N ratio, and moisture content affect decomposition rates of crop residue (Cook et al., 1979). Several other factors, however, are responsible for rapid decomposition and decline in soil organic matter.

Organic matter content of soils under the bush fallow system is generally higher than under continuous cultivation. Rapid decline in soil organic matter occurs immediately after land clearing. Hoore (1961) and Feller (1979) observed in the Ivory Coast and Senegal, respectively, that clearing and cultivating of alfisols were accompanied by increased release of CO_2, associated with transformation of a fraction of humic acids and humin to fulvic acids and compounds soluble in ethanol and bromoform. Cunningham (1963), working in the forest

TABLE 3.7 Change in Soil Water Storage Capacity (θ, g/g basis) of Plowed Watershed Growing Maize at Ibadan, Nigeria

Metric potential, MPa	Initial data 1975	6 years later
0	39.5 ± 3.5	30.9 ± 7.4
0.003	35.2 ± 2.8	23.1 ± 2.8
0.01	14.7 ± 3.7	13.8 ± 1.8
0.1	8.4 ± 3.3	9.2 ± 1.3
0.3	7.0 ± 3.0	—
1.5	5.4 ± 2.8	—

SOURCE: Lal, 1985.

zone of Ghana, showed that even without cultivation and cropping, deforestation decreases organic carbon from 25 to 57 percent in the top 0- to 5-cm depth and from 17 to 30 percent for the 5- to 15-cm depth in 3 years.

Jones (1973) compared organic matter values of soil samples from virgin woodland or bush and from recently cultivated or fallow fields in the savanna region of Nigeria. He reported mean carbon contents for virgin and cleared lands of 1.03 and 0.58 percent, respectively. Ollagnier et al. (1978) conducted land-clearing experiments on the Ivory Coast and reported that contents of carbon, nitrogen, and humic acids in windrows were higher the first year after felling than in soil of unfelled forest. The high organic matter in windrows was probably from the organic-rich surface layer scraped to the windrows. After 3 years, however, the organic matter content of the top 30 cm of soil was only 60 percent that of soils under forest. Seubert (1975) reported similar adverse effects of mechanized land clearing on organic matter for soils of the Upper Amazon Basin in Peru. These experiments and those conducted in Nigeria at IITA (Lal et al., 1986) indicated that methods of deforestation and land development affect soil organic matter content. Mechanical land clearing with unsuitable attachments and unskilled operators often removes surface soil to windrows and causes rapid losses of soil organic matter content from the remaining cultivated land.

Soil organic matter and mean annual temperature are inversely related (Jenny and Raychaudhuri, 1960). According to Vant Hoff's temperature rule, the rate of decomposition increases 2 to 3 times with every 10°C increase in mean annual temperature. In addition to temperature, annual rainfall and its seasonal distribution affect soil organic matter (Kadeba, 1970; Jones, 1973). In the savanna region of northern Nigeria, Jones (1973) reported a decrease of 0.17 percent in organic matter per 100-mm decrease in mean annual rainfall. Kang et al. (1981a), surveying Nigerian soils, observed large differences in the organic carbon status of soils from various ecological zones: forest (1.3 ± 0.08 percent) > derived savanna (0.98 ± 0.07 percent) > Guinea savanna (0.7 ± 0.06 percent). Jenkinson and Ayanaba (1977) reported the decomposition rate in the humid tropical environment (Ibadan, Nigeria, with mean annual temperature of 26°C) was about 4 times faster than that in the humid temperate region at Rothamsted, United Kingdom, with mean annual temperature of 9°C. This indicates that annual inputs of organic matter in the tropics should be about 4 times greater than those in higher latitudes, to maintain soil organic matter, once steady-state conditions are attained and no other factors are causing depletion. Since the rate of decay depends on organic matter content, the rapid decline the first year after deforestation is markedly reduced in subsequent years (Nye and Greenland, 1964).

Conclusions

Susceptibility of soil to erosion is complex. It is influenced by many soil properties and their interactions with climate and management systems. Processes governing soil erodibility are not well understood, so more research is required to understand the principles influencing it.

Because soil susceptibility to erosion is influenced by always-changing properties, soil erodibility is dynamic. With continuous, intensive cultivation, soil's vulnerability to erosion is likely to increase. The land-use systems recommended should account for possible increases in erosion risks with time.

References

Abdou, F. M., and Metwally, S. Y. 1967. The effect of organic matter, chemical fertilization and rotation soil aggregation. *J. Soil Sci. Union Arab Repub.* 7:51–59.

Ackerman, C. W., and Corinth, R. L. 1962. Empirical equation for reservoir sedimentation. *Int. Assoc. Sci. Hydrol. Publ.* 59.

Adams, J. E., Kirkham, D., and Scholtes, W. H. 1958. Soil erodibility and other physical properties of some Iowa soils. *Iowa St. Coll. J. Sci.* 32:485–540.

Adu, S. V. 1972. Eroded savanna soils of the Navrongo-Bawku area, northern Ghana. *Ghana. J. Agric. Sci.* 5(1):3–12.

Ahmad, N. 1960. Soil erosion by the Indus and its tributaries. *Pak. Geogr. Rev.* 15(2):5–17.

Ahmad, N., and Breckner, E. 1974. Soil erosion on three Tobago soils. *Trop. Agric. (Trin.)* 51:313–324.

Ahn, P. M. 1977. Erosion hazard and farming systems in East Africa. In: *Soil Conservation and Management in the Humid Tropics* (D. J. Greenland and R. Lal, eds.), Wiley, Chichester, England.

Ahn, P. M. 1979. Microaggregation in tropical soils: Its measurements and effects on the maintenance of soil productivity. In: *Soil Physical Properties and Crop Production in the Tropics* (R. Lal and D. J. Greenland, eds.), pp. 75–86, Wiley, Chichester, England.

Aina, P. O. 1979. Soil changes resulting from long-term management practices in western Nigeria. *Soil Sci. Soc. Am. J.* 43:173.

Alberts, E. E., Wendt, R. C., and Piest, R. F. 1983. Physical and chemical properties of eroded soil aggregates. *Trans. ASAE* 26(2):465–471.

Al-Durrah, M. M., and Bradford, J. M. 1982. Parameters for describing soil detachment due to single waterdrop impact. *Soil Sci. Soc. Am. J.* 46(4):836–840.

Allis, J. A. 1962. Comparison of storm runoff volumes from small single-crop watersheds and from a larger mixed-crop watershed. *Agric. Eng.* April.

Allison, F. E. 1973. *Soil Organic Matter and Its Role in Crop Production,* Elsevier, Amsterdam.

Anderson, M. G., and Burt, T. P. 1982. The contribution of throughflow to storm runoff: On evaluation of a chemical mixing model. *Earth Surf.* 7:565–574.

Arulanandan, K., Sargunam, A., Loganathan, P., and Krone, R. B. 1973. Applications of chemical and electrical parameters to prediction of erodibility. In: *Soil Erosion—Cause and Mechanisms, Prevention and Control,* pp. 42–51, Highway Research Board, Spec. Rep. 135, Washington.

Ateshian, K. H. 1976. Comparative costs of erosion and sedimentation control measures. *Proc. Third Fed. Interagency Sediment. Conf.,* Water Resources Council, Denver.

Babu, R., Tejwani, K. G., Agarwal, M. C., and Bhushan, L. S. 1978. Distribution of erosion index and isoerodent map of India. *Indian J. Soil Conserv.* 6(1):1–12.

Barnett, A. P., Carreker, J. R., and Abuna, F. 1972. Soil and nutrient losses in runoff with selected cropping treatment in tropical soils. *Agron. J.* 64:391–395.

Barnett, A. P., and Rogers, J. S. 1966. Soil physical properties related to runoff and erosion from artificial rainfall. *Am. Soc. Agric. Eng. Trans.* 9(1):123–125.

Bartholomew, M. V. 1972. Soil nitrogen supply processes and crop requirement. *Tech. Bull.* 6, International Soil Fertility Evaluation and Improvement Program, Dept. Soil Science, North Carolina State Univ., Raleigh.

Bates, J. A. R. 1960. Studies on a Nigerian forest soil. I. The distribution of organic matter in the profile and in various soil fractions. *J. Soil Sci.* 11:246.

Baver, L. D. 1938. Rainfall characteristics of Missouri in relation to runoff and erosion. *Soil Sci. S.A.* 2:233–536.

Bennett, H. H. 1926. Some comparisons of the properties of humid-tropical and humid-temperate American soils, with special reference to indicated relations between chemical composition and physical properties. *Soil Sci.* 21:349–375.

Bergsma, E. 1980. Method of reconnaissance survey of erosion hazard near Merida, Spain. In: *Assessment of Soil Erosion* (M. De Boodt and D. Gabriels, eds.), pp. 55–66, Wiley, Chichester, England.

Bhatia, K. S., and Sarmah, N. 1976. Studies on the physical properties of some Assan soils in relation to their erodibility. *J. Indian Soc. Soil Sci.* 24(4):369–373.

Bhatia, K. S., and Shanker, H. 1981. Erodibility of the soils of central alluvial tract of Uttar Pradesh. *Indian J. Agric. Sci.* 51(4):244–252.

Bhatia, K. S., and Vardani, B. 1982. Physicochemical and erosional behaviour of red and black soils of Bundlekhaud region of Uttar Pradesh. *J. Indian Soc. Soil Sci.* 30(4):523–527.

Bhola, S. N., and Jarayam, N. S. 1978. Erodibility character of black soils of Bellary. *Mysore J. Agric. Sci.* 12(1):86–90.

Biswas, T. D., Das, B., and Verma, H. K. G. 1964. Effect of organic matter on some physical properties of soil in the permanent manurial experiments. *Proc. Symp. Fertil. Indian Soils 1962, Bull. Natn. Inst. Sci. India* 26:142–147.

Bodman, G. B., and Colman, E. A. 1944. Moisture and energy conditions during downward entry of water into soils. *Soil Sci. S.A.* 8:116–122.

Bonell, M., Gilmour, D. A., and Sinclair, D. F. 1981. Soil hydraulic properties and their effect on surface and subsurface water transfer in a tropical rainforest catchment. *Hydrol. Sci. Bull.* 26(13):1–18.

Bourliere, F., and Hadley, M. 1970. The ecology of tropical savannas. *Ann. Ecol. Systematics* 1:125.

Bouyoucos, G. J. 1935. The clay ratio as a criterion of susceptibility of soils to erosion. *J. Am. Soc. Agron.* 27:738–741.

Bowden, J. W., Nagarajah, S., Barrow, N. J., Posner, A. M., and Quirk, J. P. 1980. Describing the adsorption of phosphate, citrate and selenite on a variable-charge mineral surface. *Aust. J. Soil Res.* 18:49–60.

Brams, E. A. 1971. Continuous cultivation of West African soils: Organic matter diminution and effects of applied lime and phosphorus. *Pl. Soil* 35:401.

Bridge, B. J., Boonyoi, S., and Arromratan, U. 1975. Properties affecting water entry in Central Plain soil, Thailand. *J. Agric. Sci. (Thailand)* 8(4):117–193.

Bridge, B. T., and Muchow, R. C. 1982. Soil water relationships for cunnurra clay and Ord sandy loam in the Ord river irrigation area. *Trop. Agro. Tech. Memo.*, CSRO, Australia, no. 30.

Briones, A. A. 1982. Characteristics and fertilization of andepts in the Philippines. *Trop. Agric. Res. Ser.*, No. 15, pp. 251–264, Department of Soil Science, University of the Philippines, Los Baños.

Bruce-Okine, E., and Lal, R. 1975. Soil erodibility as determined by raindrop technique. *Soil Sci.* 119:149–157.

Bryan, R. B. 1968. Development, use and efficiency of indices of soil erodibility. *Geoderme* 2:5–26.

Bryan, R. B. 1973. Survey crusts formed under simulated rainfall on Canadian soils. *Consiglin Nazionale Pisa Conf.* 2:1–30.

Bryan, R. B. 1976. Considerations on soil erodibility indices and sheetwash. *Catena* 3(1):99–111.

Bryan, R. B. 1977a. Methodology used to determine the maximum potential average annual soil loss due to sheet and rill erosion in Morocco. *FAO Soils Bull.* 34:39–48.

Bryan, R. B. 1977b. Predicting soil losses due to sheet and rill erosion. In: *FAO Conservation Guide*, FAO, Rome, Italy. Vol. 1:121–149.

Bryan, R. B., Yair, A., and Hodges, W. K. 1978. Factors controlling the initiation of runoff and piping in Dinosaur Provincial Park Badlands, Alberta, Canada. *Z. Geomorphol.* 29:151–168.

Bryan, R. B., Yair, A., and Hodges, W. K. 1979. Influence of slope angle on soil entrainment by sheet wash and rainsplash. *Earth Surface Processes* 4(1).

Burt, A. P. 1980. *Rainfall in the Southern Pennines*. Hnudersfield Polytechnic, Dept. of Geography, Occasional Paper 8.

Burt, A. P., and Anderson, M. G. 1980. Soil moisture conditions on an instrumented slope,

March–October 1976. In: *Atlas of Drought in Great Britain* (J. C. Doornkamp and K. J. Gregory, eds.), vol. 44.
Carson, M. A., and Kirkby, M. J. 1972. *Hillslope Form and Process.* Oxford University Press, Cambridge, England.
Chakela, Q. K. 1981. *Soil Erosion and Reservoir Sedimentation in Lesotho.* UNGI Rep. No. 54 Upps.
Chakrabarti, D. C. 1969. Investigation on erodibility and water stable aggregates of certain soils of eastern Nepal. *J. Indian Soc. Soil Sci.* 17:465–470.
Chepil, W. S. 1953. Factors that influence clod structure and erodibility of soil by wind. II. Water stable structure. *Soil Sci.* 76(5):389–399.
Chepil, W. S. 1955a. Factors that influence clod structure and erodibility of soil by wind. IV. Sand, silt and clay. *Soil Sci.* 80(2):155–162.
Chepil, W. S. 1955b. Factors that influence clod structure and erodibility of soil by wind. V. Organic matter at various stages of decomposition. *Soil Sci.* 80(5):413–421.
Chibber, R. K., Ghos, P. C., and Satyaharayana, K. V. S. 1961. Studies on the physical properties of some of the Himadral Pradesh soils formed on different materials in relation to their erodibility. *J. Indian Soc. Soil Sci.* 9:187–192.
Childs, E. C. 1969. *The Physical Basis of Soil Water Phenomena,* Wiley, London.
Collis-George, N., and Lal, R. 1971. Infiltration and structural changes as influenced by initial moisture content. *Aust. J. Soil Res.* 9:107–116.
Collis-George, N., and Lal, R. 1973. The temperature profiles of soil columns during infiltration. *Aust. J. Soil Res.* 11:93–105.
Conaway, A. W., Jr., and Strickling, E. 1962. A comparison of selected methods for expressing soil aggregate stability. *Soil Sci. S.A.* 26:426–430.
Cook, A. G., Critchley, B. R., Critchley, U., Perfect, T. J., Russel-Smith, A., and Yeadou, R. 1979. The effects of soil treatment with DDT on the biology of a cultivated forest soil in the sub-humid tropics. *Pedobiologia* 19:279.
Crouch, R. J. 1978. Variation in the structural stability of soil in a tunnel-erosion area. In: *Modification of Soil Structure* (W. W. Emerson, R. D. Bond, and A. R. Dexter, eds.), Wiley, Chichester, England.
Cunningham, R. K. 1963. The effect of clearing a tropical forest soil. *J. Soil Sci.* 14:334.
Davidson, T. 1975. Soil organic matter fluctuations in some tropical wetland soils in Turrialba. *Turrialba* 25:183.
De Leenheer, L., and De Boodt, M. 1959. Determination of aggregate stability by the change in mean weight diameter. *Int. Symp. Soil Structure. Meded. Landbw.* 24:290–300.
De Ming, Shi. 1986. Soil erosion and its control in red soil regions of China. In: *Proc. Int. Symp. on Red Soils,* Elsevier, Holland, pp. 678–699.
De Ploey, J. 1972. A quantitative comparison between rainfall erosion capacity in a tropical and a middle-latitude region. *Geogr. Pol.* 23.
Deshpande, T. L., Greenland, D. J., and Quirk, J. P. 1964. Role of iron oxides in the bonding of soil particles. *Nature* 201:107–108.
Deshpande, T. L., Greenland, D. J., and Quirk, J. P. 1968. Changes in soil properties associated with the removal of iron and aluminum oxides, *J. Soil Sci.* 19:108–122.
De Vleeschauwer, D., Lal, R., and De Boodt, M. 1978. Comparison of detachability indices in relation to soil erodibility for some important Nigerian soils, *Pedologie* 28(1):5–20.
Douglas, I. 1967. Man, vegetation and the sediment yield of rivers. *Nature.* 215:925–928.
Douglas, I. 1968. Erosion in the Sungei Gombak catchment, Selangor, Malaysia. *J. Trop. Geogr.* 26:1–16.
Douglas, I. 1969. *Sediment Sources and Causes in the Humid Tropics of Northeast Queensland, Australia.* 27–39, Br. Geomorph. Res. Group. Occas. Pap. 5.
Douglas, I. 1973. *Rates of Denudation in Selected Small Catchments in Eastern Australia.* Occ. Pap. in Geogr., 21. 128. University of Hull, Australia.
Dragoun, F. J. 1962. Rainfall energy as related to sediment yield. *J. Geophys. Res.* 64(4):1495–1505.
Dragoun, F. J. 1981. A soil loss estimation technique for southern Africa. In: *Soil Conservation: Problems and Prospects* (R. P. Morgan, ed.), Wiley, Chichester, England.
Ekern, P. S. 1953. Problems of raindrop impact erosion. *Agric. Eng.* 34:23–25.

Ellison, W. D. 1947. Soil erosion studies, parts I to VII. *Agric. Eng.* 28:145, 197, 245, 297, 349, 402, 442.

El-Swaify, S. A. 1977. Susceptibilities of certain tropical soils to erosion by water. In: *Soil Conservation and Management in the Humid Tropics* (D. J. Greenland and R. Lal, eds.), pp. 71–80, Wiley, Chichester, England.

Elwell, H. A., and Stocking, M. A. 1973. Rainfall parameters for soil loss estimation in subtropical climate. *J. Agric. Eng. Res.* 18:169–177.

Emerson, W. W. 1967. A classification of soil aggregates based on their coherence in water. *Aust. J. Soil Res.* 5:47–57.

Emmett, W. W. 1978. Overland flow. In: *Hillslope Hydrology* (M. J. Kirkby, ed.), pp. 145–176, Wiley, Chichester, England.

Epstein, E., Grant, W. J., and Struchtmeyer, R. A. 1966. Effects of stones on runoff, erosion and soil moisture. *Soil Sci. Soc. Am. Proc.* 30(5):638–640.

Escolar, R. P. 1966. Stability of soil aggregates treated with distillery shops or blackstrap molasses. *J. Agric. U.P.R.* 50:174–185.

Eswaran, H., and Tavernier, R. 1980. Classification and genesis of oxisols. In: *Soils with Variable Charge* (B. K. G. Theng, ed.), pp. 427–442, New Zealand Society of Soil Science, Lower Hutt, New Zealand.

Falayi, O., and Lal, R. 1979. Effect of aggregate size and mulching on erodibility, crusting, and crop emergence. In: *Soil Physical Conditions and Crop Production in the Tropics* (R. Lal and D. J. Greenland, eds.), pp. 87–94, Wiley, Chichester, England.

FAO, UNEP, and UNESCO, 1979. *A Provisional Methodology for Soil Degradation Assessment.* FAO, Rome.

Farmham, C. W., Beer, C. E., and Heinemann, H. G. 1966. Evaluation of factors affecting reservoir sediment deposition. *Int. Assoc. Sci. Hydrol. Publ.* 71:747–758.

Farmham, C. W., Foster, G. R., Meyer, L. D., and Onstad, C. A. 1967. La recherche en erosion et conservation des sols sur le continent Africain. *Sols Afr.* 12(1):5–53.

Farmham, C. W., Foster, G. R., Meyer, L. D., and Onstad, C. A. 1977a. An erosion equation derived from basic erosion principles. *Trans. Am. Soc. Agric. Eng.* 20(4):678–682.

Farmham, C. W., Foster, G. R., Meyer, L. D., and Onstad, C. A. 1977b. A runoff erodibility factor and variable slope length exponents for soil loss estimates. *Trans. Am. Soc. Agric. Eng.* 20(4):683–687.

Feller, C. 1979. Evolution des sols de defriche recente dans la region des Terres Neuves (Senegal Oriental). *Cah. ORSTOM Ser. Pedologie* 15(3):291.

Free, G. R. 1960. Erosion characteristic of rainfall. *Agric. Eng.* 41(7):447–449, 455.

Freire, O., and Pessotti, J. E. S. 1974. *Erodibility of Sao Paulo Soils.* Anais da Escola superior de Agricultura "Luiz de queiroz," Departamento de Solos e Geologia, ESALQ, Brazil, 31:333–340.

Gabriels, D., Verdouck, O., de Boodt, M., and Minjamu, W. 1975. The (average) value of the wet-aggregate distribution as an erodibility index for conditioned soils. *Mededelingen van de Faculteit landbouwwetenschappen Rijks universiteit Ghent* 40(314):1351–1357.

Gaikwad, S. T., and Goel, S. K. 1977. Statistical characterization of soil profile trends. *J. Indian Soc. Soil Sci.* 25:298.

Gallez, A., Herbillon, A. J., and Juo, A. S. R. 1977. The indexes of silica saturation and silica reactivity as parameters to evaluate the surface properties of tropical soils. *Soil Sci. Soc. Am. J.* 41:1146–1154.

Gavand, M. 1968. Les sols bien drainés sur materiaux sableux du Niger: essai de systematique regionale. *Cah. ORSTOM, Ser. Pedol.* 6:277.

Gerdel, R. W. 1937. Reciprocal relationships of texture, structure, and erosion on some residual soils. *Soil Sci. Soc. Am. Proc.* 2:537–545.

Ghani, M. O., Hasan, K. A., and Talukder, H. A. 1967. Effect of cowdung and water hyacinth on the aggregation of a red soil of Dacca. *Pakist. J. Soil Sci.* 3(1):26–31.

Ghatol, S. G., and Malewar, G. U. 1978. Vertical distribution of nitrogen and phosphorus in relation to organic carbon content in profiles of mau campus, Parbhani. *Res. Bull. Marathurada Agric. Univ.* 2(3):30–33.

Green, W. H., and Ampt, G. A. 1911. Studies on soil physics. *J. Agr. Sci.* 4(1):1–24.

Greenland, D. J. 1971. Changes in the nitrogen status and physical condition of soils under pasture. *Soils Fertil.* 34:237–251.
Greenland, D. J. 1977. Soil damage by intensive arable cultivation: Temporary or permanent? *Phil. Trans. Roy. Soc. (Lond.)* B281:193–208.
Greenland, D. J. 1979. Structural organization of soils and crop productivity. In: *Soil Physical Properties and Crop Production in the Tropics* (R. Lal and D. J. Greenland, eds.), pp. 47–56, Wiley, Chichester, England.
Greenland, D. J. 1986. Soil organic matter in relation to crop nutrition. *Int. Conf. Management and Fertilization of Upland Soils,* Nanjing, China, 7–11, September 1986.
Greenland, D. J., and Kowal, J. M. L. 1960. Nutrient content of the moist tropical forest of Ghana, *Pl. Soil* 12:154.
Grissinger, E. H. 1966. Resistance of selected clay systems to erosion by water. *Water Resources Res.* 2:131–138.
Gupta, R. N., and Narain, B. 1971. Investigations on some physical properties of alluvial soils of Utter Pradesh related to conservation and management. *J. Indian Soc. Soil Sci.* 19(1):11–22.
Haan, C. T., Johnson, H. P., and Brakensiek, D. L. (eds.). 1982. *Hydrologic Modeling of Small Watersheds.* ASAE, St. Joseph, Mich.
Hadley, R. F., and Lusby, G. C. 1967. Runoff and hillslope erosion from a high intensity thunderstorm. *Water Resources Res.* 3(1):139–143.
Haridasan, M., and Chibber, R. K. 1971. Effect of physical and chemical properties on the erodibility of some soils of the Malwa Plateau. *J. Indian Soc. Soil Sci.* 19(3):293–298.
Harris, R. F., Chesters, G., and Allen, O. N. 1966. Dynamics of soil aggregation. *Adv. Agron.* 18:107–170.
Henin, S., Monnier, G., and Combeau, A. 1958. Methode pour l'étude de la stabilité structurale des sols. *Ann. Agron.* 9:71–90.
Hillel, D. 1971. *Soil and Water: Physical Principles and Processes,* Academic Press, New York.
Hillel, D. (ed.) 1982. *Advances in Irrigation,* vol. 1, Academic Press, New York.
Holtan, H. N. 1961. *A Concept for Infiltration Estimates in Watershed Engineering,* U.S. Dept. Agr., ARS 41–51.
Holtan, H. N., and Lopez, N. C. 1971. USDAHL-70 model of watershed hydrology. *U.S. Dept. Agr. Tech. Bull.* 1435.
Hoore, J. D. 1961. Tropical soils and vegetation. *Proc. Abidjan Symp. 1959,* pp. 49–55.
Horton, R. E. 1939. Analysis of runoff plot experiments with varying infiltration capacity. *Trans. Am. Geophys. Union* 4:693–694.
Horton, R. E. 1940. An approach toward a physical interpretation of infiltration-capacity. *Soil Sci. S.A.* 16:85–88.
Hundson, N. W. 1958. Erosion control research. Progress report on experiment at Henderson Research Station, 1953–56. *Rhod. Agric. J.* 54(4):297–323.
Hvorslev, M. J. 1937. Physical properties of remolded cohesive soils. Ph.D. thesis translated and published (1969) as Translation No. 69-5 by Waterways Experiment Station, Vicksbury, Miss.
Imeson, A. C., Kwaad, F. J. P. M., and Verstrataen, J. M. 1982. The relationship of soil physical and chemical properties to the development of badlands in Morocco. In: *Badland Geomorphology and Piping* (R. B. Bryan and A. Yair, eds.), pp. 47–70, Geobooks, Norwich, Conn.
Ives, N. C. 1951. Soil and water runoff studies in a tropical region. *Turrialba* 1:240–244.
Jenkinson, D. S., and Ayanaba, A. 1977. Decomposition of carbon-14 labeled plant material under tropical conditions. *Soil Sci. Soc. Am. J.* 41:912–915.
Jenkinson, D. S., and Johnston, A. E. 1977. Soil organic matter in the hoosfield continuous barley experiment, pp. 87–101, Report Rothamsted Experimental Station for 1976, Part 2.
Jenny, H., and Raychaudhuri, S. P. 1960. Effect of climate and cultivation on nitrogen and organic matter reserves in Indian soils. ICAR, New Delhi, India.
Jha, M. N., and Rathore, R. K. 1981. Erodibility of soil in shifting cultivation areas of Tripura and Orissa. *Indian Forester* 107(5):310–313.

Jones, M. J. 1973. The organic matter content of the savanna soils of West Africa. *J. Soil Sci.* 24:42.
Joshua, W. D. 1977. Soil erosion power of rainfall in the different climatic zones of Sri Lanka. *Proc. Paris Symp.*, July 1977, IAHS-AISH Publ. 122.
Jungerius, P. D. 1975. The properties of volcanic ash soils in dry parts of the Colombian Andes and their relation to soil erodibility. *Catena* 2(1/2):69–80.
Kadeba, O. 1970. Organic matter and nitrogen status of some soils from the savanna zone. *Proc. Conf. Forestry Assoc. Nigeria,* Ibadan, Nigeria.
Kadeba, O. 1978. Organic matter status of some savanna soils of northern Nigeria. *Soil Sci.* 125:122.
Kandiah, A. 1974. Critical shear stress approach in the evaluation of hydraulic erodibility of cohesive soils. *J. Nat. Agric. Soc. Ceylon* 11–12:18–26.
Kandiah, A. 1976. Influence of organic matter on the erodibility of saturated illitic soil. *Mededelingen van de Faculteit Landbouwwetenschappen, rijksuniversiteit Gent* 41(1):397–406.
Kang, B. T., Okoro, E., Acquaye, D., and Osiname, O. A. 1981a. Sulfur status of some Nigerian soils from the savanna and forest zones. *Soil Sci.* 132:220–227.
Kang, B. T., Wilson, G. F., and Sipkens, L. 1981b. Alley cropping maize (*Zea mays* L.) and leucaena (*Leucaena leucocephala* Lam.) in southern Nigeria. *Pl. Soil* 63:165.
Klimes-Szmik, A. 1979. Erodibility of Hungarian soils developed on loess. *Agrokemia es Talajtan* 28(1–2):3–14.
Kline, H. 1962. Contributions to the knowledge of tropical soils, IV and V. *Z. PflErnahr. Dung.* 97:40–51, 106–118.
Klinge, H. 1962. Beitrage zur Kenntris tropischer Boden V. Uber Gesamtkohlenstaff und Stickstoff in Boden des braziliznischen Amazonasgebietes. *Z. Pflanzn. B.* 97:106–118.
Koenigs, F. F. R. 1961. The mechanical stability of clay soils as influenced by the moisture conditions and some other factors. *Versl. Landb. Onderz. Onderz. Wageningen Nr. 67.7.*
Kostiakov, A. N. 1932. On the dynamics of the coefficient of water-percolation in soils and on the necessity for studying it from a dynamic point of view for purposes of amelioration. *Trans. Sixth Comm. Internat. Soc. Soil Sci.,* Pt. A, pp. 17–21.
Kowal, J. M., and Kassam, A. M. 1977. Energy load and instantaneous intensity of rainstorms at Samaru, northern Nigeria. In: *Soil Conservation and Management in the Humid Tropics* (D. J. Greenland and R. Lal, eds.), pp. 57–70, Wiley, Chichester, England.
Krammes, J. S., and Osborn, J. 1969. Water repellent soils and wetting agents as factors influencing erosion. *Proc. Symp. Water-Repellent Soils,* Univ. California, Riverside, pp. 177–178.
Krusteva, V. S. 1977. Relationship between soil structure and erodibility. *Pochvoznanie i Agroklimiya* 12(4):72–80.
Lal, R. 1976. Soil erosion problems on an alfisol in western Nigeria and their control. *IITA Monogr.* Ibadan, Nigeria.
Lal, R. 1979. Physical characteristics of soils of the tropics: Determination and management. In: *Soil Physical Properties and Crop Production in the Tropics* (R. Lal and D. J. Greenland, eds.), pp. 7–46, Wiley, Chichester, England.
Lal, R. 1981. Soil erosion problems on alfisols in western Nigeria. VI. Effects of erosion on experimental plots. *Geoderme* 15:215–228.
Lal, R. 1984a. Mechanized tillage systems effects on soil erosion from an alfisol in watersheds cropped to maize. *Soil Tillage Res.* 4:349–360.
Lal, R. 1984b. Compaction, erosion and soil mechanical problems on tropical arable lands, pp. 160–164. *Proc. Int. Workshop on Soils,* 12–16 September 1983, Townsville, Qld, Australia.
Lal, R. 1985. Mechanized tillage systems effects on properties of a tropical alfisol in watersheds cropped to maize. *Soil Tillage Res.* 6:149–161.
Lal, R. 1986. Effects of eight tillage treatments on a tropical alfisol. I. Maize growth and yield. *J. Sci. Food & Agric.* 37:1073–1082.
Lal, R. 1987. *Tropical Ecology and Physical Edaphology,* Wiley, Chichester, England, p. 732.

Lal, R., De Vleeschauwer, D., and Malafa Nganje, R. 1980. Changes in properties of a newly cleared tropical alfisol as affected by mulching. *Soil Sci. Soc. Am. J.* 44:827.

Lal, R., and Kang, B. T. 1982. Management of organic matter in soils of the tropics and subtropics. *XIII Congress ISSS,* New Delhi, India, 3:152–178.

Lal, R., Sanchez, P. A., and Cummings, R. W., Jr. 1986. *Land Clearing and Development in the Tropics,* Balkema, Rotterdam.

Laskar, S., and Govindarajan, S. V. 1980. Erodibility of soils of Tripura. *Indian J. Agric. Sci.* 50(2):161–167.

Lawes, D. A. 1957. Preliminary report on soil-water-crop relationships at Samaru, northern Nigeria, 1956. *Emp. Cott. Gr. Rev.* 34:155–160.

Lugo-Lopez, M. A. 1969. Prediction of the erosiveness of Puerto Rican soils on basis of the percentage of silt and clay when aggregated. *J. Agric. Univ. Puerto Rico* 53:187–190.

Luk, S. H. 1979. Effect of soil properties on erosion by wash and splash. *Earth Surface Processes* 4:241–255.

Lyles, L. 1975. Possible effects of wind erosion on soil productivity. *J. Soil Water Cons.* 30:279–283.

Lyles, L., Dickerson, J. D., and Schmeidler, N. F. 1974. Soil detachment from clods by rainfall: Effects of wind, mulch cover and initial soil moisture. *Trans. ASAE* 17:697–700.

Lyles, L., and Woodruff, N. P. 1961. Surface soil cloddiness in relation to soil density at time of tillage. *Soil Sci.* 91:178–182.

Major, I. 1973. The use of aggregation data on the estimation of soil erodibility. *Kiserletugyi kozlemenyek, Novenytermesztes* 6(1–3):19–33.

Mazurak, A. P., and Mosher, P. N. 1968. Detachment of soil particles in simulated rainfall. *Soil Sci. Soc. Am. Proc.* 32:716–719.

McCalla, T. M. 1944. Water drop method of determining stability of soil structure. *Soil Sci.* 58:117–123.

McGuinness, J. L., Harrold, L. L., and Edwards, W. M. 1971. Relation of rainfall energy and stream flow to sediment yield from small and large watersheds. *J. Soil Water Conserv.* 26:233–235.

McIntyre, D. S. 1958. Soil splash and the formation of surface crust by raindrop impact. *Soil Sci.* 85:261–266.

Meeuwig, R. 1971. Soil stability on high-elevation rangeland in the intermountain area. U.S. Forest Service Res. Paper, Int. 94.

Mehta, K. N., Sharma, V. C., and Deo, P. G. 1963. Erodibility investigations of soils of eastern Rajasthan. *J. Indian Soc. Soil Sci.* 11:23–31.

Mein, R. G., and Larson, C. L. 1973. Modeling infiltration during a steady rain. *Water Resources Res.* 9:384–394.

Meyer, L. D., Foster, G. R., and Romkens, M. J. M. 1975. *Source of Soil Eroded by Water from Upland Slopes.* U. S. Dept. Agric., Agric. Res. Serv. (ARS-S-40).

Meyer, L. D., and Momken, E. J. 1965. Mechanics of soil erosion by rainfall and overland flow. *Trans. Am. Soc. Agric. Eng.* 8:572–577.

Meyer, L. D., and Wischmeier, W. H. 1965. Mathematical simulation of the process of soil erosion by water. *Trans. Am. Soc. Agric. Eng.* 12:754–758, 762.

Middelburg, E. A. 1952. De Kalkbehoefte van tropicale roodaarden. *Bergcultures* 21:126–130.

Middleton, P. R. 1930. Properties of soils which influence soil erosion. *U.S.D.A. Tech. Bull.* 178.

Moldenhauer, W. C. 1970. Influence of rainfall energy on soil loss and infiltration rates. II. Effect of clod size distribution. *Soil Sci. Soc. Am. Proc.* 334:673–677.

Moldenhauer, W. C., and Kemper, W. D. 1969. Interdependence of water drop energy and clod size on infiltration and clod stability. *Soil Sci. Soc. Am. Proc.* 34:673–677.

Moldenhauer, W. D., and Koswara, J. 1968. Effect of initial clod size on characteristics of splash and wash erosion. *Soil Sci. Soc. Am. Proc.* 32:875–879.

Moore, T. R. 1979. Rainfall erosivity in East Africa. *Geogr. Ann.* 61A(3–4):147–156.

Moormann, F. R. 1981. Representative toposequences of soils in southern Nigeria and their pedology. In: *Characterization of Soils in Relation to Their Classification and Management for Crop Production in the Tropics* (D. J. Greenland, ed.), pp. 10–29, Oxford University Press, London.

Moreau, R. 1978. Influence of mechanical mellowing and water infiltration on the structural stability of a ferrallitic soil in central Ivory Coast. *Cah. OSTOM, Ser. Predol.* 16:413–424.
Morel-Seytoux, H. J., and Khanji, J. 1974. Derivation of an equation of infiltration. *Water Resources Res.* 10:795–800.
Morgan, M. A. 1969. Overland flow and man. In: *Water, Earth and Man* (R. J. Chorley, ed.), pp. 239–255, Methuen, London.
Morgan, R. P. C. 1973. The influence of scale in climatic geomorphology: A case study of drainage density in west Malaysia. *Geogr. Ann.* 55A(2):107–115.
Morgan, R. P. C. 1980. Field studies of sediment transport by overland flow. *Earth Surface Processes* 5:307–316.
Murray-Rust, H. 1972. Soil erosion and reservoir-sedimentation in a grazing area west of Arusha, northern Tanzania. *Geogr. Ann.* 54A:3–4.
Musgrave, G. W. 1947. The quantitative evaluation of water erosion: A first approximation. *J. Soil Water Conserv.* 2:133–138.
Mutchler, C. K., and Young, R. A. 1975. *Soil Detachment by Raindrops.* U.S. Dept. Agric., Agric. Res. Serv. (ARS-S-40).
Nearing, M. A., and Bradford, J. M. 1985. Single waterdrop splash detachment and mechanical properties of soils. *Soil Sci. Soc. Am. J.* 49:547–552.
Nye, P. H., and Greenland, D. J. 1964. Changes in the soil after clearing tropical forest. *Pl. Soil* 21:101.
Ollagnier, M., Lanzeral, A., Olivin, J., and Ochs, R. 1978. Evolution des sols sans palmeraie après defrichement de la fôret. *Oleagineux* 33(11):537.
Olofin, E. A. 1980. The determination and significance of indices of soil erodibility, Ulu Langat district, Selangor—A case study. *Malaysian J. Trop. Geog.* 2:26–34.
Olson, O. C. 1949. Relations between soil depth and accelerated erosion on the Wasatch Mountains. *Soil Sci.* 67:417–451.
Olson, O. C., and Wischmeier, W. H. 1963. Soil erodibility evaluations for soils on the runoff and erosion stations. *Soil Sci. Soc. Am. Proc.* 27:590–592.
O'Neal, A. M. 1949. Soil characteristics significant in evaluating permeability. *Soil Sci.* 67:403–409.
Onstad, C. A., and Foster, G. R. 1975. Erosion modelling on a watershed. *Trans. Am. Soc. Agric. Eng.* 18(2):288–292.
Pauwels, J. M., Gabriels, D., Eeckhout, G. 1976. Evaluation of different criteria to assess the stability of the soil surface. Dept. of Soil Physics, Faculty of Agricultural Sciences, State University of Ghent, Belgium.
Peele, T. C. 1936. The effect of calcium on the erodibility of soils. *Soil Sci. Soc. Am. Proc.* 1:47–51.
Peele, T. C., Beale, O. W., and Latham, E. E. 1938. The effect of lime and organic matter on the erodibility of Cecil clay. *Soil Sci. Soc. Am. Proc.* 3:289–295.
Pereira, H. C. 1956. A rainfall test for structure of tropical soils. *J. Soil Sci.* 7:68–75.
Philip, J. R. 1957. The theory of infiltration: 4 Sorptivity and algebraic infiltration equation. *Soil Sci.* 84:257–264.
Quantin, P., and Combeau, A. 1962. The relationship between erosion and structural stability of the soil. *C.R. Acad. Sci. Paris* 254:1855–1857.
Quirk, J. P., and Panabokke, C. R. 1962. Pore volume-size distribution and swelling of natural soil aggregates. *J. Soil Sci.* 13:71–81.
Raghavan, G. S. V., McKyes, E., and Beaulieu, B. B. 1977. Prediction of soil clay compaction. *J. Terramech.* 14:31–38.
Ramaiah, R., and Sreenivas, G. N. 1975. Rainfall erosion index of Bangalore agro-climatic area. *Mysore J. Agric. Sci.* 9(3):448–453.
Rapp, A. 1974. Physiographical research and teaching on developing countries, pp. 53–446. *Symp. Appl. Phys. Geogr.*, UNGI Rep. 34.
Rapp, A., Axelsson, V., Berry, L., and Murray-Rust, H. 1972. Soil erosion and sediment transport in the Morogoro River catchment, Tanzania. *Geogr. Ann.* 54(9A, 3–4):125–155.
Rentro, G. W. 1975. *Use of Erosion Equations and Sediment Delivery Ratios from Predicting Sediment Yield.* U.S. Dept. Agric., Agric. Res. Serv. (ARS-S-40).

Rentro, G. W. 1978. Soil erosion in central Europe. *Pedol.* 28:145–160.
Richards, L. A. 1952. Report of the Subcommittee on Permeability and Infiltration, Committee on Terminology, Soil Science Society of America. *Soil Sci. S.A.* 16:85–88.
Rose, C. W. 1960. Soil detachment caused by rainfall. *Soil Sci.* 89(1):28–35.
Rose, C. W. 1961. Rainfall and soil structure. *Soil Sci.* 91:49–54.
Rubber Research Institute. 1975. Annual Report, Kuala Lumpur, Malaysia.
Russell, M. B. 1949. Methods of measuring soil structure and aeration. *Soil Sci.* 68:25–35.
Sahi, B. P., Pandey, R. S., and Singh, S. N. 1976. Studies on water stable aggregates in relation to physical constants and erosion indices of alluvial and sedentary soils of Bihar. *J. Indian Soc. Soil Sci.* 24(2):123–128.
Sahi, B. P., Singh, S. N., Sinha, A. C., and Acharya, B. 1977. Erosion index—A new index of soil erodibility. *J. Indian Soc. Soil Sci.* 25(1):7–10.
Sandhu, B. S., and Bhumbla, D. R. 1967. Effect of addition of different organic materials and gypsum on soil structure. *J. Indian Soc. Soil Sci.* 15:141–147.
Schnitzer, M., and Khan, S. U. (eds.). 1978. *Soil Organic Matter*, Elsevier, Amsterdam.
Seubert, C. A. 1975. Effects on land clearing methods on crop performance and changes in soil properties in an ultisol of the Amazon jungle in Peru. M.Sc. thesis, North Carolina State Univ., Raleigh.
Shah, R. K., and Patel, N. A. 1974. Note on the effect of different salt solutions on permeability of the soil. *Indian J. Agric. Res.* 8(2):117–118.
Sharma, R. R., and Biswas, N. R. D. 1972. Erodibility of hill soils of Sutlej catchment area in Himachal Pradesh. *Indian J. Agric. Sci.* 42(2):161–169.
Sherard, J. L., Dunnigan, L. P., Decker, R. S., and Steele, E. F. 1976. Pinhole tests for identifying dispersive soils. *J. Geotech. Eng. Div. ASCE* 102:69–85.
Singer, M., Ugoliri, F. C., and Zachara, J. 1978. In situ study of podzolization of tephra and bedrock. *Soil Sci. So.* 40:105–111.
Singer, M. J., Jaintzky, P., and Blackard, J. 1982. The influence of exchangeable sodium percentage on soil erodibility. *Soil Sci. Soc. Am. J.* 46(1):117–121.
Singh, G., Dayal, R., and Bhola, S. N. 1967. Soil and water loss studies under different vegetative covers on 0.5 and 1% slope at Kota. *J. Soil Water Conserv. India* 15(3/4):17–23.
Skaggs, R. W., and Khaleel, R. 1982. Infiltration. In: *Hydrological Modeling of Small Watersheds* (C. T. Haan, H. P. Johnson, and D. L. Brakensieck, eds.), pp. 119–166. American Society of Agricultural Engineers, St. Joseph, Mich.
Slonecker, L. L., Olsen, T. C., and Moldenhauer, W. C. 1976. Effect of pore water pressure on sand splash. *Soil Sci. Soc. Am. J.* 40(6):948–951.
Smith, D. D., and Wischmeier, W. H. 1962. Rainfall erosion. *Adv. Agron.* 14:109–148.
Smith, O. L. 1979. An analytical model of the decomposition of soil organic matter. *Soil Biol. Biochem.* 11(6):585.
Smith, R. M. and Browning, D. R. 1946. The influence of evacuation upon the laboratory percolation rates and wetting of undisturbed soil samples. *Soil Sci.* 62:243–253.
Smith, R. M., and Cernuda, C. F. 1951. Some applications of water-drop stability testing to tropical soils of Puerto Rico. *Soil Sci.* 71–337–346.
Soane, B. D., Blackwell, P. S., Dickson, J. W., and Painter, D. J. 1981. Compaction by agricultural vehicles: A review. I. Soil and wheel characteristics. *Soil Tillage Res.* 1:207–238.
Soane, B. D., Cickson, J. W., and Campbell, D. J. 1982. Compaction by agricultural vehicles: A review. III. Incidence and control of compaction in crop production. *Soil Tillage Res.* 2:3–36.
Sood, M. C., and Chaudhary, T. N. 1980. Soil erosion and runoff from a sandy loam soil in relation to initial clod size, tillage-time moisture and residue mulching under simulated rainfall. *J. Indian Soc. Soil Sci.* 28:24–27.
Stagg, M. J. 1978. Rill patterns derived from air photographs of the Grwyne Fechan catchment, Black Mountains. *Cambria* 5:22–36.
Stocking, M. A. 1972. Relief analysis and soil erosion in Rhodesia using multi-variate techniques. *Z. Geomorph, N.F.* 16(4):432–443.
Stocking, M. A. 1976. Tunnel erosion. *Rhod. Agric. J.* 73:35–39.

Stocking, M. A., and Elwell, H. A. 1973. Soil erosion hazard in Rhodesia. *Rhod Agric. J.* 70(4):93–101.
Stromquist, L., and Johansson, D. 1978. Soil erosion and sediment transport in the Mtera Reservoir Region. *Z. Geomorph, N.F.* Suppl. 29:43–51.
Swartzendruber, D., and Hillel, D. 1973. The physics of infiltration. In: *Ecological Studies.* vol. 4, *Physical Aspect of Soil Water and Salts in Ecosystems* (A. Hadas, D. Swartzendruber, P. E. Rijtema, M. Fuchs, and B. Yaron, eds.), pp. 3–15, Chapman and Hall, London.
Swartzenruber, D., and Huberty, M. R. 1958. Use of infiltration equation parameters to evaluate infiltration differences in the field. *Trans. Am. Geophys. Union* 39:84–93.
Taylor, S. A., and Ashcroft, G. L. 1972. *Physical Edaphology: The Physics of Irrigated and Nonirrigated Soils,* W. H. Freeman, San Francisco.
Temple, P. H. 1972. Measurements of runoff and soil erosion at an erosion plot scale with particular reference to Tanzania, *Geogr. Ann.* (9A, 3–4):203–220.
Temple, P. H., and Rapp, A. 1972. Landslides in the mgeta area, western Uluguru mountains, Tanzania. *Geogr. Ann.* 54:157–193.
Temple, P. H., and Sundberg, A. 1972. The Rufiji river, Tanzania, hydrology and sediment transport. *Geogr. Ann.* 54(9A, 3–4):345–368.
Terzaghi, K. 1925. Principles of soil mechanics. II. Compressive strength of clays. *Eng. News-Record* 95(20):799.
Terzaghi, K., and Peck, R. B. 1967. *Soil Mechanics in Engineering Practice,* 2d ed., Wiley, London.
Thornes, J. B. 1976. *Semi-Arid Erosional Systems.* London School of Economics, Department of Geography, Occas. Paper No. 7.
Thornthwaite, C. W., Sharpe, C. F. S., and Dosch, E. F. 1942. Climate and accelerated erosion in the arid and semi-arid Southwest, with special reference to the Polcacca Wash drainage basin, Ariz. *U.S. Dept. Agric. Tech. Bull.* 808.
Turchenek, L. W., and Oades, J. M. 1978. Organo-mineral particles in soils. In: *Modification of Soil Structure* (W. W. Emerson, R. D. Bond, and A. R. Dexter, eds.), pp. 137–144, Wiley, Chichester, England.
Turemie, J. F. 1980. Soil structure stability and the organic system in heavy montmorillonite clays. *Trop. Agric. (Trinidad and Tobago)* 59(2):157–161.
Turenne, J. F. 1982. Soil structure stability and the organic system in heavy montmorillonite clays. *Trop. Agric.* 59:157–161.
Tyan, C. Y., and Hwang, Y. D. 1964. Investigation on physical properties of soil in relation to the index of soil resistance to erosion in the region of Tziwu-ling, Kansu. *Acta. Pedol. Sin.* 12:286–296.
Valentin, C. 1985. Organisations pelliculaires superficielles de quelques sols de region sub desertique (Agadez-Niger) dynamique de formation et consequences sur l'economie en eau. *Etudes et Theses,* ORSTOM, Paris.
Van Bavel, C. H. M. 1950. Mean weight-diameter of soil aggregates as statistical index of aggregation. *Soil Sci. S.A.* 14:20–23.
Vanelslande, A., Lal, R., and Gabriels, D. 1985. Erodibility of some Nigerian soils, pp. 51–56. *Proc. Int. Symp. On Erosion, Debris Flow and Disaster Prevention,* 3–5 September, 1985, Tsukuba, Japan.
Vanelslande, A., Lal, R., and Gabriels, D. 1987. The erodibility of some Nigerian soils: A comparison of rainfall simulator results with estimates obtained from the Wischmeier nomogram. *Hydrol. Processes* 1:255–265.
Vanelslande, A., Rousseau, P., Lal, R., Gabriels, D., and Ghuman, B. S. 1984. Testing the applicability of a soil erodibility nomogram for some tropical soils. *IAHS Publ.* 144:463–473.
Venkobarao, K., Nair, P. K., Rao, S. B. P., and Chattopadhyay, S. 1967. Ineffectiveness of farmyard manure in improving soil aggregation in blacksoil of Bellary. *Ann. Arid Zone* 6:138–145.
Virgo, K., and Munro, R. N. 1978. Soil and erosion features of the central plateau region of Tigrai, Ethiopia. *Geoderma* 20(2):131–158.
Volkoff, B. 1977. La matière organique des sols ferrallitiques du nordeste du Brésil. *Cah. ORSTOM Ser. Pedologie* 15:275.

Voorhees, W. B., Evans, S. D., and Warnes, D. D. 1985. Effect of pre-plant wheel traffic on soil compaction water use, and growth of spring wheat. *Soil Sci. Soc. Am. J.* 49:215–220.

Wang, C., McKeagne, J. A., and SwitzerHowse, K. D. 1985. Saturated hydraulic conductivity as an indicator of structural degradation in clayey soils of Ottawa area, Canada. *Soil Tillage Res.* 5:19–31.

Wang, S. 1979. Erodibility of slopeland soils in Taiwan. *J. Agric. Assoc. China* 105:39–42.

Wang, S. T., and Lin, T. H. 1965. Relative erodibility of soils on hilly lands in Taiwan. I. Properties of the soils affecting soil erosion. *J. Taiwan Agric. Res.* 14(4):33–42.

Wendelaar, F. E. 1976. Field identification of sodic soils. *Rhod. Agric. J.* 73:77–86.

Wilkinson, G. E. 1975. Rainfall characteristics and soil erosion in the rainforest area of western Nigeria. *Exp. Agric.* 11(4):247–255.

Williams, A. R. K., and Morgan, R. P. C. 1976. Geomorphological mapping applied to soil erosion evaluation. *J. Soil Water Conserv.* 31(4):164–168.

Williams, B. G., Greenland, D. J., and Quirk, J. P. 1967. The effect of poly (vinyl alcohol) on the nitrogen surface area and pore structure of soils *Aust. J. Soil Res.* 5:77–83.

Williams, J. R., and Berndt, H. D. 1972. Sediment yield computed with universal equation. *Proc. Am. Soc. Civ. Eng., J. Hydraul. Div.* 98(HY 12):2087–2098.

Williams, J. R., and Berndt, H. D. 1976. Sediment yield prediction based on watershed hydrology. 1976 Winter Meet, Am. Soc. Agric. Eng., Chicago. Pap. No. 76-2535.

Williams, M. A. J. 1969. Prediction of rainsplash erosion in the seasonally wet tropics. *Nature* 222:763–764.

Wischmeier, W. H. 1962. Rainfall erosion potential. *Agric. Eng.* 43:212–215, 225.

Wischmeier, W. H., Johnson, L. B., and Cross, R. V. 1971. A soil erodibility nomograph for farmland and construction sites. *J. Soil Water Conserv.* 26:189–193.

Wischmeier, W. H., and Smith, D. D. 1958. Rainfall energy and its relationship to soil loss. *Trans. Am. Geophys. Un.* 39(2):285–291.

Wischmeier, W. H., and Smith, D. D. 1965. Predicting rainfall erosion losses from cropland east of the Rocky Mountains—Guide for selection of practices for soil and water conservation. U.S. Dept. Agric., *Agric. Res. Serv. Handbook* 282, Washington.

Wischmeier, W. H., Smith, D. D., and Uhland, R. E. 1958. Evaluation of factors in the soil-loss equation. *Agric. Eng.* 39(8):458–462, 474.

Woodburn, R., and Kozachyn, J. 1956. A study of relative erodibility of a group of Mississippi gully soils. *Trans. Am. Geophys. Union* 37:749–753.

Yair, A., Bryan, R. B., Lavee, H., and Adar, E. 1980. Runoff and erosion processes and rates in the Zin Valley badlands, northern Negev, Israel. *Earth Surface Processes* 5:205–255.

Yamamoto, T., and Anderson, H. W. 1973. Splash erosion related to soil erodibility indexes and other forest soil properties in Hawaii. *Water Resources Res.* 9(2):430–432.

Yoder, R. E. 1936. A direct method of aggregate analysis of soils, and a study of the physical nature of erosion losses. *J. Am. Soc. Agron.* 28:337–351.

Young, R. A. 1984. A method of measuring aggregate stability under waterdrop impact. *Trans. ASAE* 27:1351–1354.

Young, R. A., and Onstad, C. A. 1982. The effect of soil characteristics on erosion and nutrient loss. *IAHS Publ.* 137:105–113.

Young, R. A., and Wiersma, J. L. 1973. The role of rainfall impact in soil detachment and transport. *Water Resources Res.* 9(6):1620–1636.

Zingg, A. W. 1940. Degree and length of land slope as it affects soil loss in runoff. *Agric. Eng.* 21:59.

Chapter

4

Slope and Landforms

Landform is a feature of the earth's surface with distinctive form characteristics which can be attributed to the dominance of particular processes or structures in the course of its development and to which the features can be clearly related (Savigear, 1965). *Slope* refers to the gradient of landscape with reference to a baseline and affects the amount and quality of sediments and the type and velocity of water flow. Both landform and slope play an important role in the origin, entrainment, and deposition of entrained sediment. Because of its prominent role in accelerated erosion and land degradation, management of landform and slope has been a preoccupation of land use planners for millennias toward an attempt to control erosion from steeplands (Chap. 1). Slope management is both labor- and capital-intensive, and some grandiose projects on erosion control have failed (Sanders, 1986) because of poor planning, faulty techniques used in slope modification, lack of subsequent maintenance of engineering structures installed, or poor acceptability by farmers of the "improved techniques" of slope management. Considerable frustration and wasteful use of limited resources can be avoided by understanding landforms and their effects on erosion-related processes.

The branch of geology that studies processes leading to the development of landforms and slopes is called *geomorphology*. Slopes and landforms are dynamic and ever-changing entities. However, predominant slopes and landforms at a given time influence important hydrologic processes that affect movement of water over, on, and below the ground surface. Geomorphologic processes responsible for landform development and fluvial processes influenced by predominant slopes and landforms are interdependent. Over and above these nat-

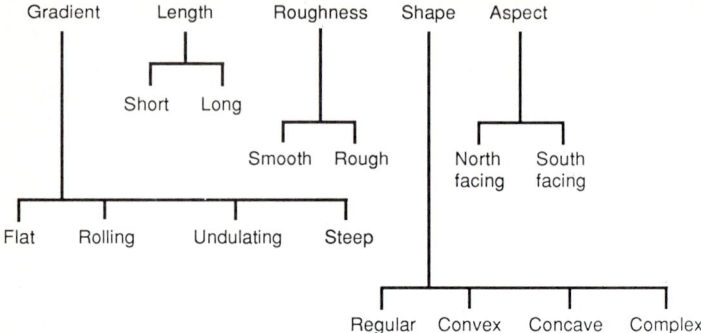

Figure 4.1 Slope characteristics that influence sediment yield from a watershed.

ural processes, human perturbations accelerate some of the processes, complicate the already complex system, and destabilize the landscape until it reaches another state of equilibrium through sediment transport and deposition.

Types of Landforms in Relation to Erosion

Both the source and sinks for sediments and the type of agent involved in sediment entrainment are influenced by slope characteristics. A number of controls influence the rate of fluvial processes on a slope. Slope characteristics that influence sediment transport out of the watershed are shown in Fig. 4.1. Slopes are classified differently on the basis of different criteria. This subject is reviewed in many books on geomorphology (e.g., Carson and Kirkby, 1971; Young, 1972; Gregory and Walling, 1973) and so is described only briefly here.

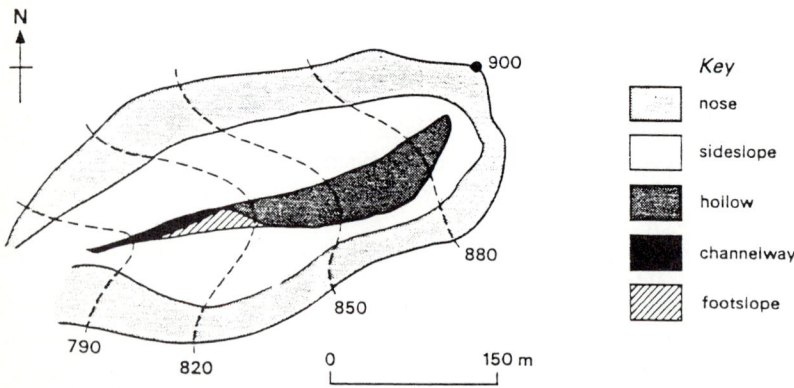

Figure 4.2 Classification of slope areas. (Adapted from Hack and Goodlet, 1960; Gerrard, 1981.)

Hydrologic characteristics

A hillslope landform is characterized on the basis of its slope—angle, length, form, shape, and aspect. The slope angle affects the volume, velocity, and type of water flow as well as the agent of erosion. The slope length also influences flow characteristics and the type of erosion, e.g., interrill, rill, or gully erosion. The sediment originating upslope may be deposited at the footslope depending on the shape. The effect of slope gradient and length is greatly modified by slope shape. Smooth and regular slopes generate more overland flow than rough slopes with large surface detention capacities. The alterations in microclimate by slope aspect influence water balance and therefore the volume of overland flow. Because soil wetness and antecedent moisture content are altered by the slope aspect, soil erodibility is different on north-facing vs. south-facing slopes.

Position in the landscape

The most obvious landform in relation to erosion is the river valley. On the basis of interaction between landform and overland flow, Hack and Goodlet (1960) classified slopes within a watershed as nose, sideslope, hollow, channelway, and footslope (Fig. 4.2). They observed that the overland flow on different slope types is influenced by physical characteristics that distinguish them. For example, runoff on the noseslope is proportional to a function of the radius of curvature of the contours. Runoff is proportional to a linear function of slope length on the sideslopes and to a power function of slope length in the hollows. In the channelway, runoff is presumably proportional to a power function of channel length. In the footslope, runoff is transitional between sideslope and channelway.

Predominant geomorphologic processes in slope evolution

Slopes are also classified on the basis of predominant fluvial processes involved in their evolution, e.g., erosion or deposition, vertical or lateral movement of

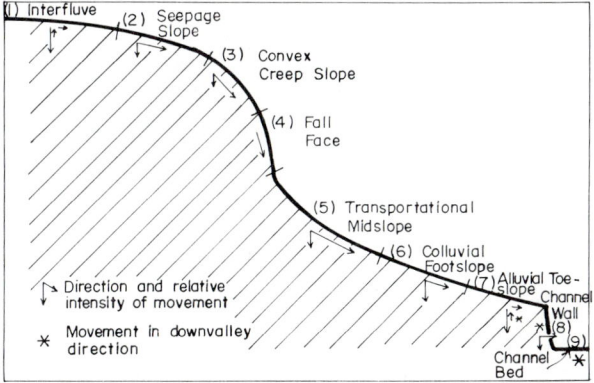

Figure 4.3 A hypothetical nine-unit landscape model. (Adapted from Dalrymple et al., 1968; Derbyshire et al., 1979; Gerrard, 1981.)

water, solid movement with or without water, water movement with or without dissolved and suspended loads, rate of movement, and nature of the sediments. Comprehensive reviews on this subject have been made by Thomas (1974) and Derbyshire et al. (1979), among others. Different control mechanisms govern the predominance of different processes. Dalrymple et al. (1968) proposed a nine-unit land surface model in an attempt to relate processes to the form of slopes (Fig. 4.3). Each slope is related to a specific set of geomorphologic processes responsible for slope evolution. Although specific types of slope would be expected within a broad climatic pattern (e.g., humid tropic, semiarid temperate, humid temperate, semiarid tropics), local variations can occur due to differences in mesoclimate, e.g., those related to aspect.

Channel character and type of water flow

Different types of water flow occur at different locations along the valley axis projected to the divide (Fig. 4.4; Derbyshire et al., 1979). There is a one-to-one

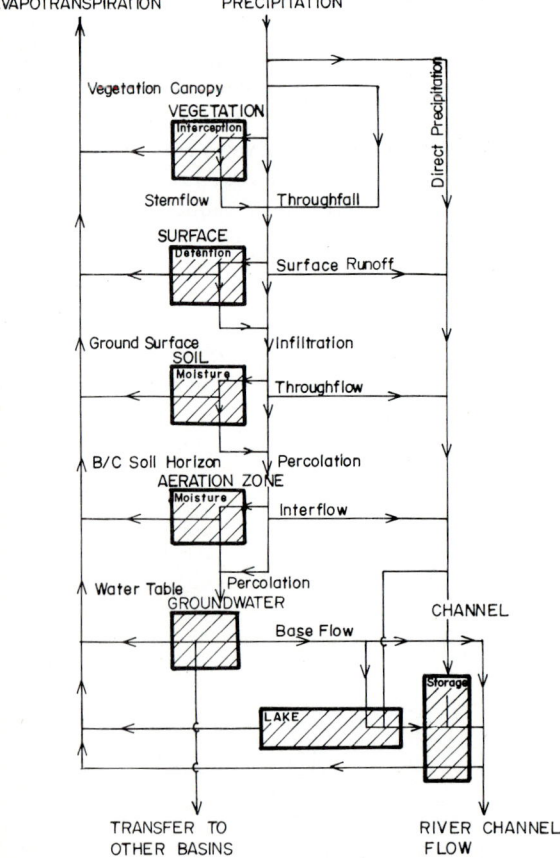

Figure 4.4 Relationship between landscape and predominant flow (Derbyshire et al., 1979).

correspondence between the slope type shown in Fig. 4.3 and the predominant flow associated with it shown in Fig. 4.4. Clearly not all elements may be present in all climates.

Predominant Types of Landforms in the Tropics

Although most landscapes fit into the broad categories described above and are in accord with the present-day climate, some landscapes were developed under the influence of past climate. It is the combination of relief and drainage characteristics that provides specific information on terrain features related to erosion. Drainage density depends on parent material, amount and regolith type, relief, and climate and is greater in humid mountainous regions than in sub-humid or semiarid regions of low relief (Peltier, 1962; Eyles, 1971). The analysis of drainage characteristics, however, is incomplete if the information on drainage density is not accompanied by supporting data on the dimensions of stream channels, vegetation, land use, maximum discharge, and representative hydrographs. Doornkamp and King (1971) analyzed more than 100 third-order watersheds from Uganda. Drainage densities ranged from 0.62 to 6.25 km/km^2. In Nigeria, low drainage densities (0.5 to 1.5 km/km^2) have been reported over basement complex regardless of the climax vegetation, e.g., rain forest, semideciduous forest, or savanna (Wigwe, 1966; Jeje, 1970, 1972; Thorp, 1970). Low drainage densities exist in regions with soils of high infiltration rate.

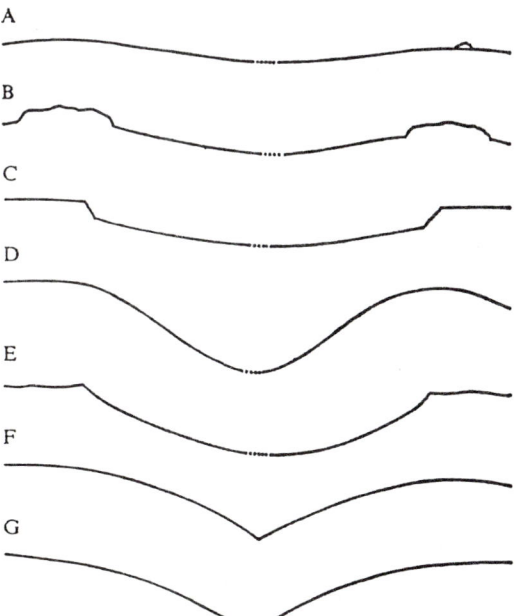

Figure 4.5 Schematic diagrams showing typical valley forms in Tanzania (Louis, 1964).

Slope types and valley forms depend on the rainfall regime and the parent material or lithology (Thomas, 1974). Louis (1964) described different types of valleys in Tanzania (Fig. 4.5). The saucer-shaped valleys (types A, B, and C) predominate in savanna regions with mean annual rainfall ranging between 500 and 1000 mm. These valleys have broad, flat, alluviated floors and gently concave slopes underlaid by deep red soils. The valley sideslopes are generally of less than 3° inclination and have considerable lengths. Groove-shaped valleys are formed in regions of higher rainfall (types D and E). These valleys possess alluviated concave floors confined within steep valley sideslopes. Incised valleys with convex slopes also occur in regions of high rainfall (type F). Some of these valleys may have a flat floor (type F). In the transition zone of Nigeria, Wigwe (1966) also described the occurrence of saucer-shaped valleys. He also reported "bowl-shaped valleys" possessing extensive ramp slopes.

In the Mato Grasso area of Brazil, Young (1970) observed the predominance of convex slopes in the rain forest region. Young proposed five major classes of slope forms (Fig. 4.6). Convex slopes are particularly abundant in the forest-savanna transitionary zone.

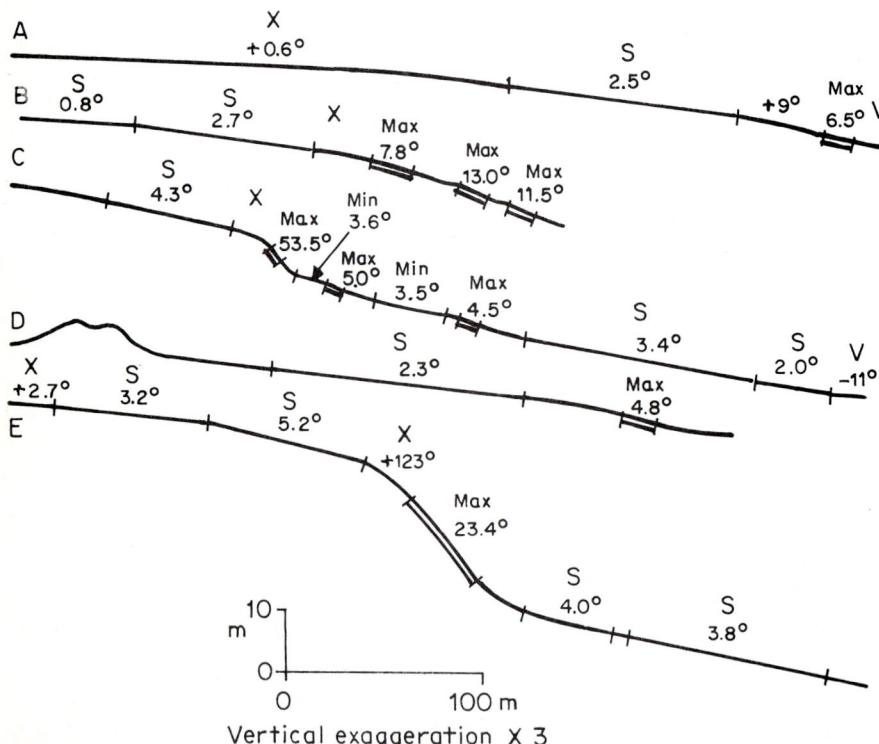

Figure 4.6 Predominant slope forms in the Xarentina-Cachimbo area of Brazil. S = segment, Max = maximum segment, Min = minimum segment, X = convex element, V = concave element. (Young, 1970.)

Plate 4.1 Hardened plinthite in the profile shallow depth.

In seasonably wet or semiarid regions characterized by the presence of hardened plinthite or laterite terrain (Plates 4.1 and 4.2), valleys are characterized by wide hydromorphic floors bordered by steep, spring-sapped slopes. Such valleys have been described in northeastern Brazil by Vann (1963), in Malaysia by Eyles (1971), in central Africa (called Dambos) by Webster (1965), in Malawi by Young (1969a), and in west Africa by Ledger (1969). In the west African Sahel,

Plate 4.2 Hardened plinthite on the surface: large areas in the west African Sahel.

wide saucer-shaped valleys are bordered by laterites. Only the inner 20 to 30 percent of the land surface is cultivable (Figs. 4.7 and 4.8).

Predominant tropical terrain

The slope steepness and slope shape depend on the rainfall regime and parent material. Highly dissected steep slopes are found in the humid regions, and undulating to rolling terrain is found in semiarid and arid regions. On the basis of the study of land relief in semiarid regions of Madras, India, by Budel (1965)

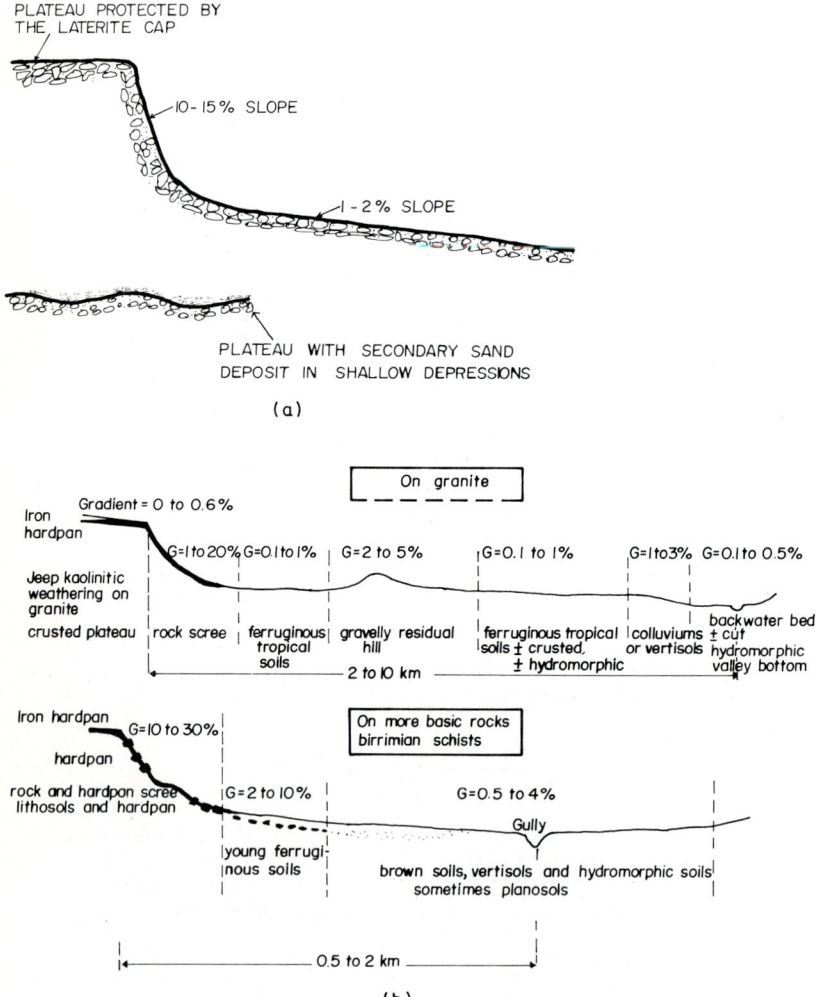

Figure 4.7 (a) Landforms in the Sahel region of Niger. (b) Typical catenas on the Mossi Plateau of Niger. (Roose, 1988.)

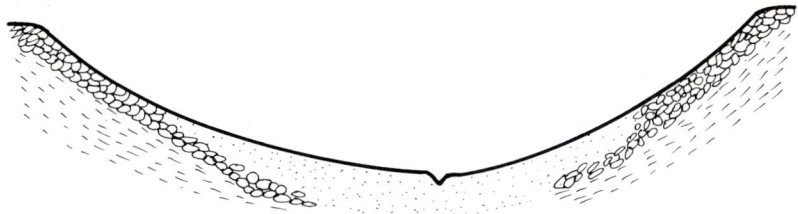

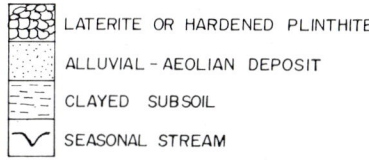

Figure 4.8 Presence of lateritic crust and hardened plinthite in relation to landforms in Niger.

and in Tanzania by Harpum (1963), predominant relief types in the tropics can be classified as follows:

1. *Plain.* The slopes have generally low values of 1 to 2 percent but are surrounded by isolated high inselbergs and bordered by radial pediment slopes of 3.5 to 4.0 percent. This type of relief may occur in coastal regions or at plateau with mean altitude of less than 1000 m.
2. *Mountain.* This relief is characterized by a high frequency of groups of hills or inselbergs leading toward an escarpment dissected by deep, narrow valleys.
3. *Ridge.* This relief appears above the escarpment and consists of narrow, groovelike valleys.

Similar patterns of terrain have been described for Uganda by Doornkamp (1968, 1970), for humid tropical terrain developed over granites by Ruxton and Berry (1961), and for Africa by D'Hoore (1964).

Slope and Soil Erosion

Slope is an important variable that affects erosion processes for all types of soil erosion, e.g., splash, sheet, rill, and gully erosion. Both shearing and transport capacity of flowing water are influenced by slope. Important slope characteristics in relation to erosion are slope steepness, length, and shape.

Slope steepness

Erosion increases with an increase in slope steepness because of the increased downslope component of gravity. Slope steepness has different magnitudes of

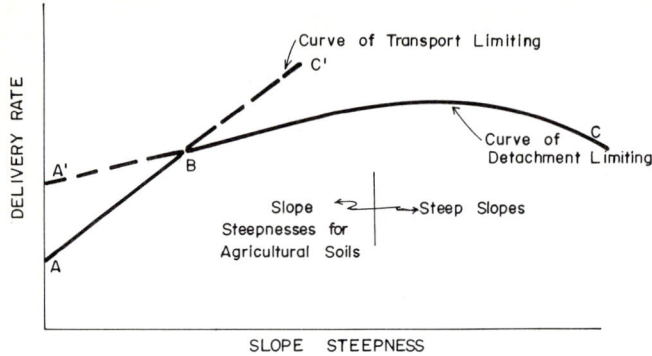

Figure 4.9 Effect of slope steepness on delivery rate of sediments (Haan et al., 1982).

effect on rill and interrill or splash erosion. An increase in slope steepness will increase rill erosion more than interrill erosion. The effect of slope steepness generally levels off at a slope of about 20 percent (Foster and Martin, 1969). That the slope steepness has slight effect on splash is shown by the data of Meyer et al. (1975) (Fig. 4.9). Therefore, the expected increase in erosion with increasing slope steepness is caused by an increase in rill erosion. Field plot experiments conducted at the International Institute of Tropical Agriculture (IITA) have shown that under natural slopes erosion increased with an increase in slope angle (Table 4.1). Similar results have been reported from experiments in Zimbabwe by Hudson and Jackson (1959) (Table 4.2) and in Ivory Coast by Roose (1972).

Many empirical relations have been proposed relating slope steepness to erosion potential. Zingg (1940) proposed a simple equation relating soil loss to slope characteristics. Using simulated rainfall under field conditions, he observed that doubling the degree of slope increased soil loss 2.61 to 2.8 times. His

TABLE 4.1 Effect of Slope Steepness on Erosion from a Plowed Bare Soil Surface of an Alfisol in Western Nigeria

	Soil erosion, t/ha					
	1972		1973		1974	
Slope, %	I	II	I	II	I	II
1	4.0	1.0	7.5	3.7	6.6	2.8
5	32.1	11.1	80.4	75.8	105.5	28.8
10	45.5	13.4	152.9	79.7	115.6	21.3
15	101.0	15.1	155.3	73.9	71.7	23.8
Rainfall, mm	663.2	160.4	526.3	580.6	492.8	248.7

I = first rainy season (March–July), II = second rainy season (August–November).
SOURCE: Lal, 1976.

TABLE 4.2 Soil Erosion (in tons/acre) under Maize on Three Slopes in Zimbabwe

Rain year	Slope, %		
	3.0	4.5	6.5
1953–1954	2.8	2.5	4.0
1954–1955	1.1	0.6	1.7
1955–1956	1.0	1.5	4.2
1956–1957	2.2	4.3	6.7
1957–1958	0.2	0.3	0.7
1958–1959	2.1	4.2	7.0

SOURCE: Hudson and Jackson, 1959.

data showed that the exponential function was a satisfactory empirical model relating slope steepness to erosion potential

$$A = aS^m L^{n-1} \qquad (4.1)$$

where A is the average soil loss per unit area from a land slope of unit width, a is a constant, S is the degree of land slope, L is horizontal length of land slope, and m and n equal 1.49 and 1.6, respectively. Musgrave (1947) proposed a similar equation; however, the values of m and n were 1.35 and 1.37, respectively. In comparison, Van Doren and Bartelli (1956) observed that the average exponent for steepness of slope was 1.5. Kirkby (1969) reported the values of m and n to be 1.35 and 1.72, respectively. Many other researchers have reported the value of the slope exponent, ranging from 0.7 (Neal, 1938), 1.4 in Papua, New Guinea (Humphreys, 1984), and 2.02 in Zimbabwe (Hudson and Jackson, 1959). On the basis of 63 plot year data from Zimbabwe, Hudson and Jackson (1959) observed that the mean value of the exponent was 1.63. They suggested that for all practical purposes the soil loss is satisfactorily estimated by

$$A = KS^{1.63} \qquad (4.2)$$

The exponent values also vary with slope steepness and with management (Roose, 1972; Lal, 1976). In Nigeria, Lal (1976) observed that the power equation was a valid model relating erosion to slope steepness on plowed bare soil only. The power equation did not apply to the mulched or no-till plot. The numerical value of the slope exponent for the Nigerian data ranged from 0.74 to 1.26 (Table 4.3). Furthermore, the correlation coefficient was more significant for high (greater than 25 mm) than low (less than 25 mm) rains.

Smith and Wischmeier (1962) established the polynomial relationship between slope steepness and soil erosion

$$A = 0.43 + 0.30S + 0.43S^2 \qquad (4.3)$$

where A is soil loss in tons per acre and S is the slope in percent. By isolating the effects of single variables Wischmeier and Smith (1978) observed that ero-

TABLE 4.3 Regression Equations Relating Slope Steepness to Erosion from Plowed Bare Soil According to a Power Model

Year	Rainfall	Regression equation	Correlation coefficient
1972	A	$Y = 6.0S^{1.11}$	0.75
1972	B	$Y = 1.6S^{0.82}$	0.30
1973	A	$Y = 9.6S^{1.21}$	0.86
1973	B	$Y = 4.26S^{1.26}$	0.48
1974	A	$Y = 11.8S^{1.13}$	0.81
1974	B	$Y = 11.2S^{0.74}$	0.38

A = rains with amount exceeding 25 mm; B = rains with amount less than 25 mm; S = slope, in percent.
SOURCE: Lal, 1976.

sion rate varied with 1.3 power of the slope angle. Conclusions drawn from field experiments reported above have been validated by the data obtained by Bryan (1979) under more controlled conditions using simulated rainfall for slopes varying from 3° to 30°. His data supported the conclusion of Wischmeier (1959) that when high slope angles are involved, a polynomial function is a better model than a power function.

In contrast to Wischmeier's and Bryan's conclusions, Lal (1976) observed that the polynomial relationship was not as valid for tropical alfisols at Ibadan as the power function was. Lal's data showed that correlation coefficients with polynomials were often low and statistically not significant (Table 4.4). The variability in erosion explained by the polynomial equations was low and ranged from 13 to 44 percent.

Whereas sediment transport is related to slope steepness, the amount of overland flow does not necessarily follow a similar relationship. Under natural field conditions, soil characteristics also change with slope steepness. Soil characteristics that have evolved as a function of slope steepness and that also affect infiltration and runoff include soil texture, clay mineralogy, moisture regime,

TABLE 4.4 Polynomial Regression Equations Relating Slope Steepness to Erosion Potential on Plowed Bare Soil

Year	Rains	Equation	Correlation coefficient R^2
1972	A	$Y = -2.1 + 11.2S$	0.41
	B	$Y = -0.6 + 7.6S$	0.13
1973	A	$Y = -33.5 + 40.2S - 1.65S^2$	0.42
	B	$Y = -37.9 + 49.8S - 2.6S^2$	0.31
1974	A	$Y = -38.8 + 51.97S - 2.57S^2$	0.44
	B	$Y = -33.1 + 49.06S - 2.93S^2$	0.34

A = rainfall exceeding 25 mm; B = rainfall less than 25 mm; S = slope, in percent.
SOURCE: Lal, 1976.

etc. Lal (1976) observed that water runoff from plowed bare soil was not related to slope steepness in the way that soil erosion was (Table 4.5). Whereas soil erosion increased, the water runoff decreased with increasing slope steepness. The water runoff from cropped and mulched plots, however, generally increased with increasing slope steepness (Lal, 1976).

Slope length

Slope length is defined as the distance from the point of overland flow to the point where either the slope gradient decreases enough that deposition begins or the runoff enters a well-defined channel that may be part of a drainage network or a constructed channel (Wischmeier and Smith, 1978). The effect of slope length on erosion potential is not as clearly defined as that of slope steepness. The data base relating slope length to runoff and erosion is also narrow. The slope length has little, if any, effect on the amount and velocity of runoff, but slope length affects both detachment and sediment transport by overland flow. The rill erosion is greater on long slopes than short ones (Foster et al., 1977; Laflen et al., 1978). Although the volume of flow per unit area may be less, the total discharge increases with increasing slope length from the water divide. This implies that the upper part of the slope, closer to the divide, has little, if any, sediment detachment and transport due to overland flow. This was the basis of Horton's (1945) model showing a zone of no erosion near the crest. However, Horton's concept has been questioned because of other fluvial processes that may lead to the development of concentrated flow even on the upper parts of the slope (Yair, 1973).

The effects of slope length on erosion have been described by the linear power of polynomial functions. A few experiments have shown that erosion increases linearly with an increase in slope length (Laflen et al., 1978). The empirical models proposed by Zingg (1940), Musgrave (1947), and Kirkby (1969) indicate, however, that erosion is related to slope length through a power function. Similarly, Horton (1945) observed that erosion varies with 0.6 power of the distance from the slope crest. Van Doren and Bartelli (1956) reported that the exponent n of slope length varied with slope steepness and was 1.42 for 5 percent slope and 1.35 for 9 percent slope.

TABLE 4.5 Effect of Slope Steepness on Water Runoff from Plowed Bare Soil Surface of an Alfisol in Western Nigeria

Slope, %	1972, mm		1973, mm		1974, mm	
	I	II	I	II	I	II
1	225.6	22.7	315.7	191.7	283.2	128.5
5	261.6	39.0	347.3	195.8	345.9	137.1
10	259.1	27.6	311.0	193.1	218.5	84.4
15	214.1	25.5	316.5	185.4	294.3	80.4

I = first rainy season (March–July), II = second rainy season (August–November).
SOURCE: Lal, 1976.

Smith and Wischmeier (1962) proposed a topographic factor LS described by the polynomial equation

$$LS = \frac{L^{0.5}(0.76 + 0.53S + 0.076S^2)}{100} \quad (4.4)$$

where L is slope length in feet and S is slope in percent. Wischmeier (1974) later modified Eq. (4.4) to take into account changes at specific slope gradients and to the unit plot length of 72.6 ft (see Chap. 7):

$$LS = m\,\frac{\lambda}{72.6}\,\frac{430 \sin^2 \theta + 30 \sin \theta + 0.43}{6.56} \quad (4.5)$$

Here λ is the desired slope length, and factor m varies with slope steepness and other irregularities (Foster and Wischmeier, 1974; Wischmeier, 1974). The value of m is 0.5 if the slope is steeper than 4 percent, 0.4 for 4 percent slopes, and 0.3 for slopes of 3 percent or less.

Under field conditions in the tropics Lal (1976, 1983, 1984) observed runoff and soil loss in relation to slope length for a toposequence on tropical alfisols. The data in Table 4.6 indicate an interaction between slope gradient and slope length on runoff. For plots on 1 percent slopes with soil having a tendency to crust, runoff increased with an increase in slope length between 5 and 15 m. A further increase in slope length beyond 15 m resulted in a drastic decrease in water runoff. Taking the average of all slope gradients, however, showed that runoff decreased with an increase in slope length. A regression model of the

TABLE 4.6 Effect of Slope Length and Steepness on Annual Runoff (Measured in mm) from Tropical Alfisol

Slope length, m	Slope, %				Mean
	1	5	10	15	
1977 data					
5	92.3	269.0	310.6	207.0	219.7
10	120.3	198.4	187.8	253.8	190.1
15	228.7	225.1	166.9	85.8	176.6
20	71.6	146.1	243.1	143.5	151.1
Mean	128.2	209.7	227.1	172.5	
1978 data					
5	187.9	578.5	508.0	403.3	419.4
10	245.3	288.8	302.7	265.7	275.6
15	188.2	231.7	189.9	205.9	203.9
20	96.4	165.7	160.3	164.8	146.8
Mean	179.5	316.1	290.2	259.9	

SOURCE: Lal, 1983.

type $W = aL^b$ was fitted to the original data. The regression equation based on combined data for 1977 and 1978 rains was

$$W = 773L^{-0.53} = 0.99 \qquad (4.6)$$

where W is water runoff in millimeters per year and L is slope length in meters. Regression equations of this type were different for different slope gradients (Table 4.7). Polynomial functions were also computed relating runoff to slope length and slope steepness:

$$W = 357.1 + 12.5S - 11.2L - 0.7LS \qquad r = 0.81 \qquad (4.7)$$

$$W_1 = 0.3 + 0.01S - 0.01L - 0.0006LS \qquad r = 0.81 \qquad (4.8)$$

Here W is the mean annual runoff, W_1 is the runoff/rainfall ratio, L is slope length in meters, and S is slope steepness in percent. The correlation coefficients with polynomial equations were less than those with the power equation [Eq. (4.6)]. The data of runoff in relation to slope length from another experiment conducted in 1986 at IITA, Ibadan, are shown in Table 4.8. Once again, the amount of runoff decreased with an increase in slope length. Wischmeier (1966) and Wischmeier and Smith (1978) observed that either slope length has a negligible effect on runoff amount or runoff decreases with an increase in slope length. Wischmeier (1966) reported from his investigations on 21 locations that for 18 sites the total growing season runoff per unit area was greater on short than long slopes. Total dormant season runoff was found to be greater on longer slopes for 11 locations, but it was equal to or greater than runoff on the short slopes for the other 10 sites. Borst and Woodburn (1942) observed no significant effect of slope length on runoff.

The effect of slope length on erosion or sediment entrainment is influenced by runoff amount and its velocity. The available information indicates that, in general, the runoff velocity increases with an increase in slope length. Kramer and Meyer (1969) compared the runoff velocities on slope lengths of 40, 70, and 100 ft and observed that longer slopes had higher runoff velocities and therefore more erosion than shorter slopes. Free and Bay (1969) also observed that although runoff decreased on long compared with short slopes, the reverse was

TABLE 4.7 Regression Equations Relating Mean Annual Runoff, mm, to Slope Length ($y = aL^b$)

Slope, %	1977 plot-year data		1978 plot-year data	
	Coefficient	Exponent	Coefficient	Exponent
1	86.3	0.15	281.3	−0.20
5	441.2	−0.33	2162.1	−0.87
10	466.4	−0.32	1723.0	−0.80
15	454.4	−0.46	1070.0	−0.63

SOURCE: Lal, 1983.

TABLE 4.8 Effect of Slope Length on Water Runoff from an Alfisol on about 10% Slope under Maize-Cowpea Rotation during 1986 at IITA, Ibadan

Slope length, m	Runoff, mm	
	No-till	Plowed
First season: Maize		
60	147	749
50	45	703
40	42	860
30	113	784
20	326	638
10	622	824
Second season: Cowpea		
60	19.1	85.6
50	20.8	48.8
40	15.2	75.7
30	30.8	68.2
20	100.1	68.2
10	156.5	235.6

SOURCE: Unpublished data from Lal.

true for erosion. The magnitude of increase in erosion with an increase in slope length, however, depends on many other factors, including slope gradient and soil and crop management systems.

Lal (1984) monitored erosion on field runoff plots of varying lengths established on different slope gradients. The data in Table 4.9 from plowed bare soil surface show that slope lengths between 5 and 20 m had less effect on soil erosion per unit area than slope steepness. In some cases erosion even decreased with an increase in slope length. There also existed an interaction between slope length and slope gradient with erosion. For steep slopes of 10 and 15 percent, an increase in slope length from 5 to 15 m increased soil erosion per unit area. During 1978, soil erosion on 10 percent slope was in the order of 1, 1.05, and 1.08 for 5-, 10-, and 15-m slope lengths, respectively. Similarly, for 15 percent slopes the soil erosion was in the order of 1, 1.11, 1.51, and 1.60 for 5-, 10-, 15-, and 20-m slope lengths, respectively. For gentle slopes of 1 and 5 percent, however, there was only a slight effect or no effect of slope length on soil erosion per unit area.

Multiple and polynomial regression equations were also developed relating slope length to soil erosion for different slope gradients (Tables 4.10 and 4.11). Polynomial equations of the type shown in Eqs. (4.4) and (4.5) were better for these data than power models. The effects of variable slope length on erosion under field conditions for land of about 10 percent slope under maize-cowpea rotation at Ibadan, Nigeria, are shown in Table 4.10. In plowed treatments, soil

TABLE 4.9 Effect of Slope Steepness and Length on Erosion [Measured in t/(ha · yr)] from an Alfisol in Western Nigeria

Slope, %	Slope length, m				Mean
	5	10	15	20	
1977 data					
1	2.8	1.3	8.3	1.1	3.4
5	92.2	58.0	67.2	21.5	59.7
10	82.1	101.9	134.9	107.0	106.5
15	52.6	72.8	84.6	100.4	77.6
Mean	57.4	58.5	73.8	57.5	
1978 data					
1	4.5	2.8	6.5	2.2	4.0
5	143.4	94.5	117.4	52.0	101.8
10	219.1	229.6	235.8	163.5	212.0
15	190.7	212.4	288.5	306.0	249.5
Mean	139.4	134.8	162.1	130.9	

SOURCE: Lal, 1984.

erosion was generally more from longer than shorter slope lengths. In São Paulo, Brazil, Bertoni and Lombardi (1985) measured runoff and soil erosion from field plots of variable slope length. The slope steepness ranged between 6.5 and 7.5 percent, and the mean annual rainfall of the region is about 1300 mm. Their data show that while the percentage of runoff decreased, soil erosion increased with increasing slope length (Table 4.12).

It is generally accepted that soil erosion per unit area is greater on longer than shorter slope lengths. The exact nature of the mathematical relation, however, depends on soil type, slope steepness, soil and crop management, and rainfall characteristics. For example, in the United States Mutchler and Green (1980) reported that the magnitude of the slope length exponent depends on the

TABLE 4.10 Regression Equations* Relating Soil Erosion with Slope Length for 1978 Data

Slope, %	Soil loss, t/(ha · yr)	Soil loss/rainfall, t/(ha · yr · mm)	Soil loss/runoff, t/(ha · yr · mm)
1	$A = 5.675 L^{-0.148}$	$A = 0.005 L^{-0.149}$	$A = 0.021 L^{0.052}$
5	$A = 305.125 L^{-0.467}$	$A = 0.269 L^{-0.467}$	$A = 0.196 L^{0.257}$
10	$A = 279.996 L^{0.117}$	$A = 0.246 L^{0.117}$	$A = 0.220 L^{0.570}$
15	$A = 96.974 L^{0.385}$	$A = 0.085 L^{0.385}$	$A = 0.076 L^{1.079}$

*$A = aL^b$.
SOURCE: Lal, 1984.

TABLE 4.11 Polynomial Regression Equations Relating Soil Erosion to Slope Length and Slope Steepness for 1978 Data

Dependent variable	Regression equation	Correlation coefficient R
Soil loss, t/(ha · yr)	$A = 44.921 + 10.633S - 2.947L + 0.656LS$	0.93
Soil loss/rainfall, t/(ha · yr · mm)	$A = 0.037 + 0.009S - 0.003L + 0.006LS$	0.93
Soil loss/runoff, t/(ha · yr · mm)	$A = 0.071 - 0.004S - 0.007L + 0.007LS$	0.95

SOURCE: Lal, 1984.

slope gradient. For low slope gradients of 0.5 percent and less, the value of the exponent was about 0.15. The data in Table 4.11 also show that the increase in erosion on longer slopes is greater on plowed soil than on no-till soil.

The effect of the slope gradient on the slope length exponent is related to the amount of deposition on gentle slopes. On long slopes of gentle gradient, there is a possibility of deposition of large particles, resulting in less erosion from long than shorter slope lengths (Long, 1964; Young, 1980). The deposition on footslopes is further encouraged in the case of irregular slopes, e.g., concave slopes. That is why in Trinidad Georges (1977) observed, from an experiment with sugarcane, that longer plots produced significantly less erosion than shorter plot lengths. The effect of slope length on runoff and erosion is, therefore, greatly influenced by slope shape.

Effects of slope length on runoff per unit area and on soil erosion are greatly altered by land use, cropping systems, and tillage methods. Method of seedbed preparation and crop residue mulch affect runoff rate, runoff velocity, and sediment origin and transport. Field experiments conducted in Nigeria showed that slope length had negligible effect on water runoff on plowed plots (Fig. 4.10a). In the no-till plots, however, runoff decreased with an increase in slope length on plowed plots (Fig. 4.10b). In contrast, however, soil erosion decreased with an increase in slope length on no-till plots (Fig. 4.11b). The no-till system of seedbed preparation with crop residue mulch decreases soil splash and reduces sediment transport. These erosion-preventing techniques are not operative in the plow-based system of seedbed preparation.

TABLE 4.12 Effect of Slope Length on Runoff and Soil Erosion in São Paulo, Brazil

Slope length, m	Runoff, % of rainfall	Soil erosion, t/ha
25	13.6	13.9
50	10.7	19.9
100	2.6	32.5

SOURCE: Bertoni and Lombardi, 1985.

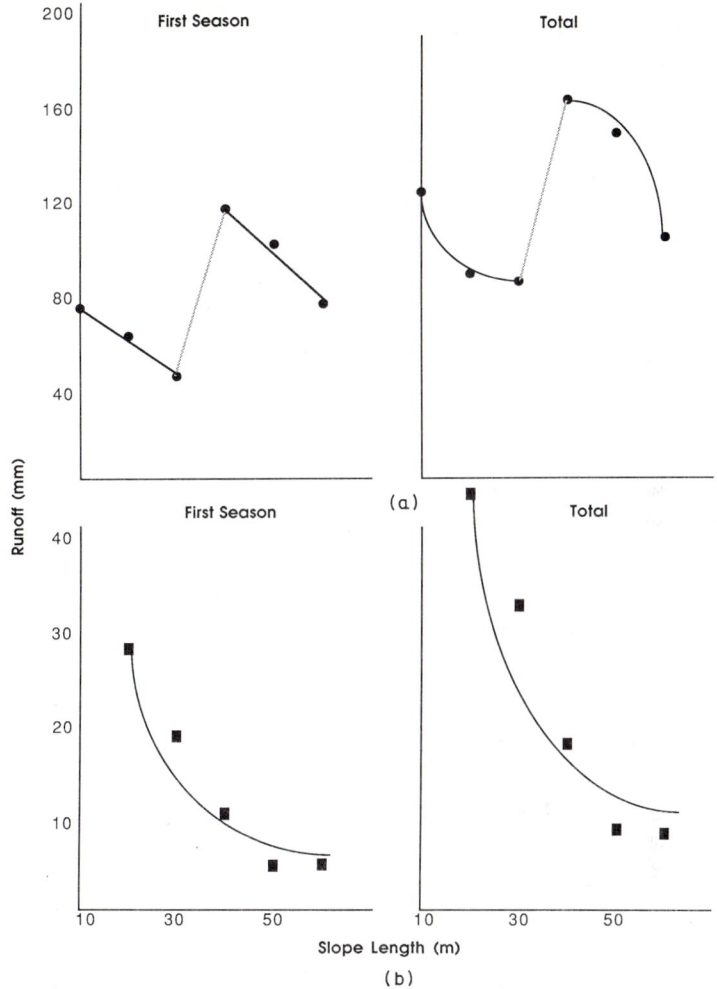

Figure 4.10 Effects of slope length and tillage methods on soil erosion from field plots in Nigeria. (*a*) Plowed; (*b*) no-till. (Unpublished, Lal, 1985.)

Slope shape

The effects of average slope angle and length are drastically altered by the slope of short segments that change the slope shape. Slope shape affects soil erosion by influencing the amount and velocity of overland flow. The slope shape may be uniform, convex, concave, or complex (Fig. 4.12). Whereas convex slopes increase the velocity of overland flow, thereby increasing its detaching and transport capacity, the velocity is decreased on concave slopes that cause deposition. Young and Mutchler (1969a) observed deposition on footslopes in concave slopes and higher erosion rates on convex slopes. Using simulated rainfall, Li

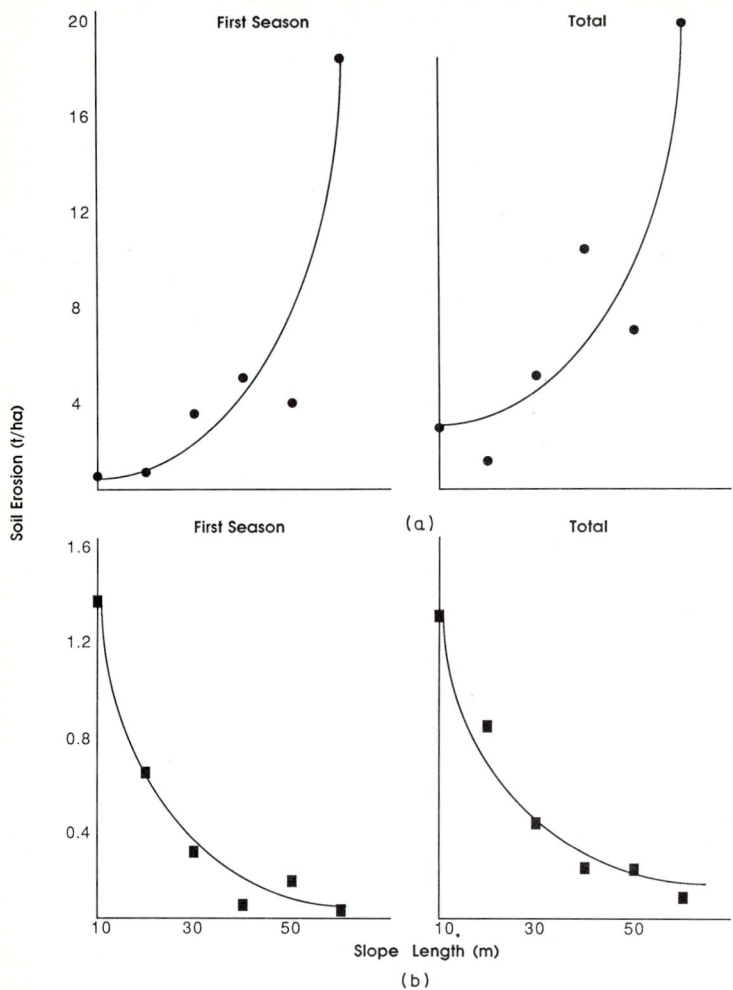

Figure 4.11 Effects of slope length and tillage methods on soil erosion from field plots in Nigeria. (*a*) Plowed; (*b*) no-till. (Unpublished, Lal, 1985.)

et al. (1976) observed that erosion rates were about 5 times greater on convex than on uniform slopes (Fig. 4.13). Similar results were obtained by Kramer and Meyer (1969) (Fig. 4.14).

Young and Mutchler (1969) conducted field experiments on complex topography including concave, uniform, and complex slopes. Although the average slope on all plots was 8.5 percent, the slope on each of the concave and convex plots ranged from 2 to 14 percent. Runoff and soil erosion were evaluated for three surface covers: corn, oats, and cultivated fallow. The data on soil erosion and water runoff are shown in Tables 4.13 and 4.14, respectively. Soil losses from the uniform and convex slopes did not differ significantly. The erosion from concave plots, however, was about 20 t/ha less than from uniform plots.

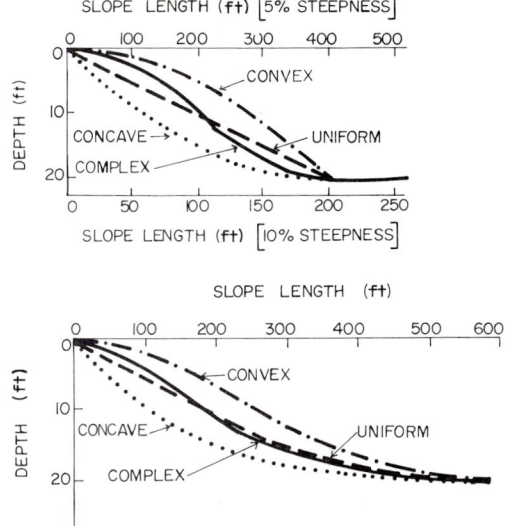

Figure 4.12 Slope shape in relation to depth and length (Meyer and Kramer, 1969).

The flattened slope at the lower end of the concave plots caused considerable deposition. Soil in the convex plots, however, did not erode toward the upper part of the slope. But the runoff loss was greatest on convex slopes (Table. 4.14). Young and Mutchler (1969a) concluded that it is generally the slope of the segment next to the point of measurement that is the control segment and determines the magnitude of runoff and soil losses.

Field measurements of runoff and soil erosion on irregular slopes were reported for a tropical alfisol by Lal (1976). Plots established on natural slopes were plowed, uncropped and maintained weed-free. The data in Table 4.15 show that during 1972, 2.15 times as much soil was lost from the 12.5-m plot of 10 percent regular slope as from the 12.5-m plot of 19.2 percent concave slope. During 1973, 54 percent more soil was lost from the regular slope than from the steeper concave slope. Also during 1972, 83 percent more soil was lost from the 37.5-m plot of 9.3 percent convex slope than from the 37.5-m plot of 13.4 percent complex slope. The complex slope was convex in the upper half and concave in the lower half. Similar data for 1973 indicate that 34 percent more soil was lost from the convex slope than from the complex slope.

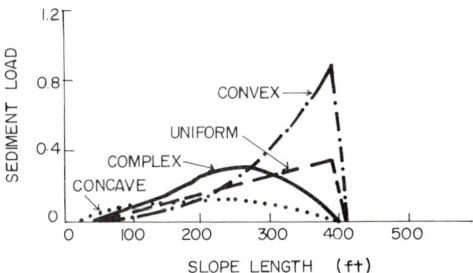

Figure 4.13 Sediment load in relation to slope shape (Meyer and Kramer, 1969).

124 Basic Processes

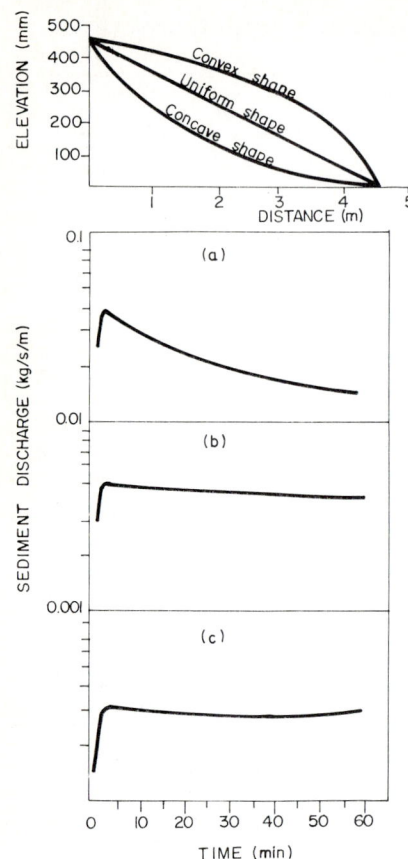

Figure 4.14 Effect of slope shape on sediment yield: (*a*) convex shape, (*b*) uniform shape, (*c*) concave shape. (Adapted from Li et al., 1976; Kirkby and Morgan, 1980.)

The data in Table 4.15 show therefore that the shape of the slope can be more important than the length of the slope. Furthermore, soil erosion is in the order of convex > regular > complex > concave slopes.

Many researchers have developed analytical procedures to evaluate soil erosion from irregular slopes (Meyer and Kramer, 1969). Wischmeier (1974) and Wischmeier and Smith (1978) outlined a procedure for estimating the topo-

TABLE 4.13 Average Soil Losses, tons/acre, from Slope-Shape Erosion Plots

Crop	Slope shape			Average
	Concave	Uniform	Complex	
Corn	6.9	16.8	18.1	13.9
Oats	2.5	6.5	7.2	5.4
Fallow	4.6	15.1	15.7	11.9
Average	4.8	12.8	13.7	10.4

SOURCE: Young and Mutchler, 1969a.

TABLE 4.14 Average Water Losses (in Percent) from Slope-Shape Erosion Plots from Application of 6.35 cm Water in 1 h

	Slope shape			
Crop	Concave	Uniform	Complex	Average
Corn	3.07	3.30	3.63	3.33
Oats	3.20	2.97	3.70	3.30
Fallow	2.67	3.22	3.53	3.14
Average	2.97	3.18	3.63	3.25

SOURCE: Young and Mutchler, 1969a.

graphic factor on irregular slopes. The weighted-average procedure is based on the assumption that the irregular slope can be divided into a small number of equal-length segments in such a manner that the gradient within each segment, for all practical purposes, can be considered uniform.

Regardless of the progress made in understanding the erosion processes in-

TABLE 4.15 Effect of Contour Length on Runoff and Soil Loss from Irregular Slopes of an Alfisol in Nigeria

	12.5 m long		37.5 m long	
Variable	10.0% regular	19.2% concave	9.3% convex	13.4% complex
First season 1972				
Runoff, mm	230.2	214.3	133.6	125.8
Soil erosion, t/ha	21.2	7.9	11.4	5.9
Second season 1972				
Runoff, mm	18.8	21.7	9.1	7.7
Soil erosion, t/ha	2.7	3.2	1.4	1.1
First season 1973				
Runoff, mm	193.1	203.0	88.5	80.5
Soil erosion, t/ha	31.3	22.8	21.5	15.8
Second season 1973				
Runoff, mm	88.1	55.1	25.7	28.0
Soil erosion, t/ha	5.2	0.9	1.5	1.4
First season 1974				
Runoff, mm	302.7	260.4	175.6	157.3
Soil erosion, t/ha	77.3	34.6	114.3	68.6
Second season 1974				
Runoff, mm	162.4	140.7	52.3	52.7
Soil erosion, t/ha	32.3	14.0	40.2	26.8

SOURCE: Lal, 1976.

volved on irregular slopes, erosion on the complex slopes remains difficult to model. The fact remains, however, that the effects of slope length and gradient on erosion are drastically influenced by slope shape.

Slope Modification

Terraced agriculture is by no means a modern invention. Because of an early realization that slope length and gradient influence erosion potential, in the past attempts have been made by many ancient civilizations to cultivate hillslopes by altering slope characteristics through constructing terraces. Terraces are built to alter slope characteristics such that steep slopes otherwise highly susceptible to erosion can be easily managed. Ancient civilizations in Asia and South and Central America developed very elaborate systems of hillslope agriculture. The strength and weakness of terraced agriculture, however, lie in the belief that erosion is mainly influenced by slope gradient and slope length. The terraced system was developed long before we realized the significant role that raindrop impact or soil splash plays in causing soil erosion. Consequently, terraced agriculture is designed to minimize the rill erosion by altering slope characteristics. Terraces have virtually no effect on interrill erosion.

Elaborate reviews have been made of types of terraces and methods of their construction (Hudson, 1981). Terrace systems have been designed on the basis of soil types and slope characteristics, to suit the need for mechanized farming operations. Different terms are used to describe a wide range of terrace systems available. The fact remains, however, that no terrace system is effective unless it is constructed properly and maintained regularly. Regular maintenance is crucial to its success in curtailing soil erosion. Severe gully erosion is caused on hillslopes where the terraces have not been maintained.

References

Bertoni, J., and Lombardi, F. Neto. 1985. *Conservacao do solo*. Livroceres Ltda, Brasil.
Borst, H. L., and Woodburn, R. 1942. The effect of mulching and method of cultivation on runoff and erosion from Muskingam silt loam. *Agric. Eng.* 23:19–22.
Bryan, R. B. 1979. The influence of slope angle on soil entrainment by sheetwash and rainsplash. *Earth Surface Processes* 4:43–58.
Budel, J. 1965. Die relieftypen der Flachenspulzone: Sud-Indiens am Ostabfall Dekans gegen Madras. *Colloq. Geogr.* 8, Bonn.
Butley, B. E. 1967. Soil periodicity in relation to landform development in Southeastern Australia. In: *Landform Studies from Australia and New Guinea*, pp. 231–255 (*Geogrl. Abst.* 68A/41).
Carson, M. A., and Kirkby, M. J. 1971. *Hillslope Form and Process*. Cambridge University Press, London.
Churchward, H. M. 1970. Erosional modification of a laterized landscape over sedimentary rocks: Its effect on soil distribution. *Austr. J. Soil Res.* 8:1–19.
Dalrymple, B., Conacher, A. J., and Blong, R. J. 1968. A hypothetical nine-unit land surface model. *Z. Geomorph.* 12:60–76.
Dau, J. 1965. Soil-catena relationships as evidenced from erosional cycles and changes of base level during the Pleitocere in Israel. *Israel J. Earth Sci.* 13:82–87.
Derbyshire, E., Gregory, K. J., and Hails, J. R. 1979. *Geomorphological Processes*, Butterworths, London.

De Swardt, A. M. J., and Trendall, A. F. 1969. The physiographic development in Uganda. *Overseas Geol. Miner. Resources* 10:241–288 (Geol. Surv. Zambia, Lusaka).

D'Hoore, J. L. 1964. Soil map of Africa, scale 1:5,000,000. Explanatory monograph. C.C.T.A. Publ. 93, Lagos.

Doornkamp, J. C. 1968. The role of inselbergs in the geomorphology of southern Uganda. *Trans. Inst. Br. Geogr.* 44:151–162.

Doornkamp, J. C. 1970. *The Geomorphology of the Mbarara Area.* Geographical Report 1, Dept. of Geography, University of Nottingham, England.

Doornkamp, J. C., and King, C. A. M. 1971. *Numerical Analysis in Geomorphology,* Edward Arnold, London.

Eyles, R. J. 1971. A classification of West Malaysian drainage basins. *Ann. Ass. Am. Geogr.* 61:460–467.

Foster, G. R., Meyer, D. L., and Onstad, C. A. 1977. A runoff erosivity factor and variable slope length exponents for soil loss estimates. *Trans. ASAE* 20:683–687.

Foster, G. R., and Wischmeier, W. H. 1974. Evaluating irregular slopes for soil loss prediction. *Trans. ASAE* 17:305–309.

Foster, R. L., and Martin, G. L. 1969. Effects of unit weight and slope on erosion. *Proc. Am. Soc. Civil Eng.* (Irrigation and Drainage Div.) 95(IR4):551–561.

Free, G. R., and Bay, C. E. 1969. Tillage and slope effects on runoff and erosion. *Trans. ASAE* 12:209–211, 215.

Freeze, R. A. 1972. Role of sub-surface flow in generating sub-surface runoff in upstream areas. *Water Resources Res.* 8:1272–1283.

Gard, L. E., and Van Doren, C. A. 1950. Soil losses as affected by cover, rainfall and slope. *Soil Sci. Soc. Am. Proc.* 14:374–378.

Georges, J. E. W. 1977. Soil erosion in cane fields in hilly lands in Trinidad. *Int. Sugar. J.* 81:147–157.

Gerrard, A. J. 1981. *Soils and Landforms.* George Allen and Unwin, London.

Gregory, K. J., and Walling, D. E. (eds.). 1973. *Drainage Basin: Form and Process,* Edward Arnold, London.

Haan, C. T., Johnson, H. P., and Brakensiek, D. L. (eds.). 1982. *Hydrologic Modeling of Small Watersheds.* ASAE, St. Joseph, Mich.

Hack, J. T., and Goodlet, J. G. 1960. *Geomorphology and Forest Ecology of a Mountain Region in the Central Appalachians.* U.S. Geological Survey, Professional Paper, no. 347.

Harpum, J. R. 1963. Evolution of granite scenery in Tanganyika. *Geol. Surv. Tanganyika Rec.* 10:39–46.

Horton, R. E. 1945. Erosional development of streams and their drainage basins: Hydrophysical approach to quantitative morphology. *Bull. Geol. Soc. Am.* 56:275–370.

Hudson, N. W. 1981. Social, political and economic aspects of soil conservation. In: International Council of Scientific Unions, 1981, pp. 45–54.

Hudson, N. W., and Jackson, D. C. 1959. Results achieved in the measurement of erosion and runoff in southern Rhodesia. *Third Inter-African Soils Conference, Dalaba.* Africa Soil Bureau, vol. 2, pp. 575–584.

Humphreys, G. S. 1984. *The Environment and Soils of Chimbu Province, Papua, New Guinea, with Particular Reference to Soil Erosion.* Department of Primary Industries, Research Bulletin, no. 35, Port Moresbg.

Jamison, V. C., and Peters, D. B. 1967. Slope length of clay pan soil affects runoff. *Water Resources Res.* 3:471–480.

Jeje, L. K. 1970. Some aspects of the geomorphology of southwestern Nigeria. Ph.D. thesis, University of Edinburgh, Scotland.

Jeje, L. K. 1972. Landform development at the boundary of sedimentary and crystalline rocks in south western Nigeria. *J. Trop. Geogr.* 34:25–33.

Kirkby, M. J. 1969. Infiltration, throughflow and overland flow. In: *Water, Earth and Man* (R. J. Chorley, ed.), pp. 215–228, Methuen, London.

Kirkby, M. J., and Morgan, R. P. C. (eds.). 1980. *Soil Erosion.* John Wiley and Sons, Chichester, U.K.

Kramer, L. A., and Meyer, L. D. 1969. Small amount of surface mulch reduces soil erosion and runoff velocity. *Trans. ASAE* 12:638–641, 645.

Laflen, J. M., Baker, J. L., Hartwig, R. O., Buchelle, W. F., and Johnson, H. D. 1978. Soil and water loss from conservation tillage system. *Trans. ASAE* 21:881–886.
Laflen, J. M., and Saveson, I. L. 1970. Surface runoff from graded lands of low slopes. *Trans. ASAE* 13:340–341.
Lal, R. 1976. *Soil Erosion Problems on Alfisols in Western Nigeria and Their Control.* IITA Monograph 1, IITA, Ibadan, Nigeria.
Lal, R. 1983. Effects of slope length on runoff from alfisols in western Nigeria. *Geoderma* 31:185–193.
Lal, R. 1984. Effect of slope length on soil erosion from an alfisol in western Nigeria. *Geoderma* 33:181–189.
Ledger, D. L. 1969. Dry season flow characteristics of west African rivers. In: *Environment and Landuse in Africa* (M. F. Thomas and G. Whittington, eds.), Methuen, London.
Li, R. M., Simons, D. B., and Carder, D. R. 1976. Mathematical modelling of overland flow for soil erosion. In: *National Soil Erosion Conference,* 25–26 May 1976, Purdue University, Lafayette, Ind.
Linden, P. Van Der. 1978. *Serayu Valley Project.* Final report, vol. 3. *Contemporary Soil Erosion in the Sanggreman River Basin Related to the Quaternary Landscape Development. A Pedogeomorphic and Hydro-geomorphological Case Study in Middle-Java, Indonesia.* Netherlands Universities Foundation for International Cooperation (NUFFIC).
Lixandru, G. 1968. Determination of critical slope length in relation to erosion. *Stiinta Sol* 6:12–18.
Long, D. C. 1964. The size and density of aggregates in eroded material. M.Sc. thesis, Iowa State Univ., Ames.
Louis, H. 1964. Uher rampfflachen-und talbuilding in en wechselfenchten tropen besonders nach studien in Tanganyika. *Z. Geomorph.* 8:43–70.
Meyer, L. D., Foster, G. R., and Romkens, J. M. 1975. Relationship of soil loss to slope steepness on short inter-rill areas. In: *Present and Prospective Technology for Predicting Sediment Yield and Sources,* pp. 177–189, USDA, ARS-S-40, Washington.
Meyer, L. D., and Kramer, L. A. 1969. Erosion equations predict land slope development. *Agric. Eng.* 50:522–523.
Musgrave, G. W. 1947. The quantitative evaluation of factors in water erosion, a first approximation. *J. Soil Water Conserv.* 2:133–138.
Mutchler, C. K., and Green, J. D. 1980. Effect of slope length on erosion from low slopes. *Trans. ASAE* 23:866–869, 876.
Neal, J. H. 1938. Effect of degree of slope and rainfall characteristics on runoff and soil erosion. *Agric. Eng.* 19:213–217.
Oyegun, R. O. 1982. Erosion-active surfaces on a pediment slope. *Trop. Agric.* 60(1):53–55.
Peltier, L. C. 1962. Area sampling for terrain analysis. *Prof. Geogr.* 14:24–28.
Petit, M., and Bourgeat, F. 1965. Lavakas in Madagascar: A natural agent in slope development. *Bull. Assoc. Geogr. Fr.,* nos. 332–333:29–33.
Puvaneswaran, P., and Conacher, A. J. 1983. Extrapolation of short-term process data to long-term landform development: A case study from southwestern Australia. *Catena* 10(4):321–337.
Roose, E. J. 1972. Contribution à l'étude de la résistance à l'érosion de quelques sols tropicaux. ORSTOM, Adiopodoume, Abidjan, Nigeria.
Roose, E. J. 1988. Soil and water conservation lessons from steep-slope farming in French-speaking countries of Africa. In: *Conservation Farming on Steeplands* (W. C. Moldenhauer and N. W. Hudson, eds.), Soil and Water Conservation Society, Ankeny, Iowa, pp. 129–139.
Rougerie, G. 1965. Lavakas and slope development in Madagascar. *Bull. Ass. Geogr. Fr.,* nos. 332–333: 15–28(F.) *Bull. Bibliophique Pedol.* ORSTOM 15 (Ped. 65-b.49).
Ruxton, B. P., and Berry, L. 1961. Weathering profiles and geomorphic position on granite in two tropical regions. *Rev. Geomorph. Dyn.* 12:16–31.
Sanders, D. W. 1986. Sloping land: Soil erosion problems and soil conservation requirements. In: *Land Evaluation for Landuse Planning and Conservation in Sloping Areas* (W. Siderius, ed.), pp. 40–50, ILRI, Wageningen, The Netherlands.

Savigear, R. A. G. 1965. A technique of morphological mapping. *Ann. Assoc. Am. Geogr.* 55:513–538.
Smith, D. D., and Wischmeier, W. H. 1962. Rainfall erosion. *Adv. Agron.* 14:109–148.
Speight, J. G. 1968. Parametric description of land form. In: *Land Evaluation* (G. A. Stewart, ed.), pp. 239–250, Macmillan, Australia.
Swan, S. B., St. C. 1972. Land surface evolution and related problems with reference to a humid tropical region, Johor, West Malaysia. *Z. Geomorph.* 16(2):160–181.
Thomas, M. F. 1974. *Tropical Geomorphology: A Study of Landform Development in Warm Climates.* Macmillan, London.
Thorp, M. D. 1970. Land forms. In: *Zaria and Its Regions* (M. J. Mortimore, ed.), pp. 13–32, Dept. of Geography Occasional Paper, ABU, Zaria, Nigeria.
Van Doren, C. A., and Bartelli, L. J. 1956. A method of forecasting soil loss. *Agric. Eng.* 37:335–341.
Vann, J. H. 1963. Developmental processes in laterite terrains in Amampa. *Geogr. Rev.* 53:406–417.
Walker, P. H. 1962. Terrace chronology and soil formation on the south coast of N.S.W. *J. Soil Sci.* 13:178–186.
Webster, R. 1965. A catena of soils on the Northern Rhodesia Plateau. *J. Soil Sci.* 16:31–43.
Wigwe, G. A. 1966. Drainage composition and valley forms in parts of northern and western Nigeria. Ph.D. thesis, University of Ibadan, Nigeria.
Wischmeier, W. H. 1959. A rainfall erosion index for a universal soil-less equation. *Soil Sci. Soc. Am. Proc.* 23:246–249.
Wischmeier, W. H. 1966. Relation of field-plot runoff to management and physical factors. *Soil Sci. Soc. Am. Proc.* 30:272–277.
Wischmeier, W. H. 1974. New developments in estimating water erosion. In: *Proceedings of 29th Annual Meeting of the Soil Conservation Society of America*, Ankeny, Iowa, pp. 133–141.
Wischmeier, W. H. 1975. *Estimating the Soil-Loss Equation's Cover and Management Factor for Undisturbed Areas.* USDA, ARS-S-40, Washington, pp. 118–124.
Wischmeier, W. H., and Smith, D. D. 1978. *Predicting Rainfall Erosion Losses; Guide to Conservation Farming.* USDA Handbook 282, Washington.
Wischmeier, W. H., Smith, D. D., and Uhland, R. E. 1958. Evaluation of factors in the soil loss equation. *Agric. Eng.* 39:458.
Yair, A. 1973. Theoretical considerations on the evolution of convex hillslopes. *Z. Geomorph. N. F. Suppl. Bd.* 18:1–9.
Young, A. 1969a. Natural resource survey in Malawi: Some considerations of the regional method in environmental description. In: *Environment and Landuse in Africa* (M. F. Thomas and G. Whittington, eds.), Methuen, London.
Young, A. 1969b. Present role of land erosion. *Nature* 224:851–852.
Young, A. 1970. Slope forms in Xarantina-Cachimbo Area. *Geogr. J.* 136:383–392.
Young, A. 1972. *Slopes,* Longmans, London.
Young, R. A. 1980. Characteristics of eroded sediments. *Trans. ASAE* 23:1139–1142, 1146.
Young, R. A., and Mutchler, C. K. 1969a. Effect of slope shape on erosion and runoff. *Trans. ASAE* 12:231–233, 239.
Young, R. A., and Mutchler, C. K. 1969b. Soil and water movement in small tillage channels. *Trans. ASAE* 12:543–545.
Zingg, R. W. 1940. Degree and length of land slope as it affects soil loss in runoff. *Agric. Eng.* 21:59–64.

Chapter 5

Splash Erosion

Falling raindrops are the major agent responsible for initiating soil erosion, i.e., causing soil detachment and displacement from its original position. The impact of raindrops on soil erosion is evidenced by pedestals or columns of soil protected by stones or leaf litter while the surrounding soil has been removed (Plate 5.1). It was classical research by Ellison in the mid-1940s that indicated the relative importance of falling raindrops vis-à-vis the overland flow in dislodging soil particles. That realization changed the emphasis on methods of controlling erosion, e.g., preventing and minimizing raindrop impact to curtail the source or origin of sediments rather than trying to minimize the amount and velocity of overland flow.

Basic Concepts

Splash means to strike or dash about a liquid or semiliquid substance and cause it to spatter. (See Plate 1.1.) By corollary, to *splash soil* means to detach and spatter soil particles by impacting raindrops. Soil splash is also called *interrill erosion* (Foster and Meyer, 1972). It is defined as "the spattering of small soil particles caused by the impact of raindrops on wet soils." The loosened and spattered particles may or may not be removed subsequently by surface runoff. Soil splash occurs in two stages: detachment of soil and soil transport. *Detachment* means "the removal of transportable fragments of material from a soil mass by an eroding agent, usually falling raindrops" (SCSA, 1982). The detached particles are transported by two mechanisms: displacement caused by physical impact and particle entrainment caused by overland flow. Depending on their size, cohesive strength, and impact energy, particles are splashed into the air to various heights and deposited at some distance from

Splash Erosion   131

Plate 5.1 Pedestals formed on soils eroded to different depths of topsoil list.

their original location (Plate 5.2). Splashed particles may reach a maximum height of 60 to 70 cm or more and often move horizontally 1.5 m or more on level surfaces (Ellison, 1944). Ellison (1947a, b) observed that splash erosion by raindrop impact may be as much as 225 t/ha by a heavy rain beating on an easily detached bare soil. He showed that the main detaching agent, falling raindrops, influences soil erosion in three ways, by

1. Breaking clods and soil aggregates into individual particles or smaller aggregates
2. Displacing soil particles from their original site
3. Creating turbulence in shallow, overland flow

The three processes constitute erosion by falling raindrops—splash.

 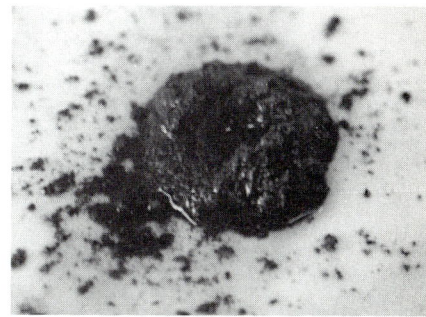

Plate 5.2 Soil splash.

Soil Detachment Mechanism

The mechanism of soil detachment by raindrops is complex and involves changes in energy level of the soil water system. Water from a raindrop acts both as an energy source and as a wetting agent. The complex wetting process occurs in three well-defined stages (Yariv, 1976): (1) dry soil, (2) soil water mixture or fluidized soil, and (3) soil cum overland flow. The energy required for soil detachment differs for various stages of wetness.

Physical analysis of the splash erosion process in the drop solid domain is discussed by Park et al. (1982). The schematics in Fig. 5.1 show different forces acting on a soil particle. The magnitude of particle displacement can be computed from Newton's second law of motion

$$W_d \cos \alpha = W_s V_s \cos \theta + W_d V_R \cos \beta \tag{5.1}$$

where W_s is the soil particle mass, W_d the drop mass, V_R the drop impact velocity, V_s the sediment incipient velocity, θ the slope angle, α the impact angle, and β the repulse angle. The impact angle α is equal to α_1 plus θ, where α_1 is the inclined angle of drop for the vertical axis and is related to the wind direction. Angle β is equal to $\pi/2 - 2\Phi - \alpha$, where Φ equals the impact angle to the tangential line at the point of impact. With the knowledge of variables involved, Eq. (5.1) can be used to compute the splash downslope. Splash can also be computed by using the equation of continuity and the law of conservation of momentum.

Factors Affecting Soil Splash

Soil splash is influenced by many factors, including antecedent soil properties, landform, rainfall characteristics, properties of overland flow, and vegetation cover. Various aspects of these factors are illustrated in Figs. 5.2 and 5.3.

Soil moisture potential and wetting process

The collision between dry soil particles and raindrops is essentially a collision between two elastic bodies. The energy of the raindrop is transmitted to the aggregate or soil clod, but the clod may still retain its shape. The clod progressively gets wet, its soils moisture potential increases, its soil strength de-

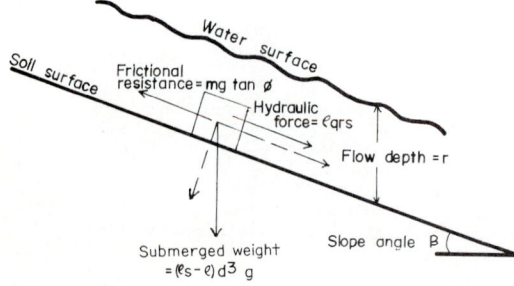

Figure 5.1 Schematics of different forces acting on a submerged soil particle. $s = \tan \beta$. (Park et al., 1982.)

Soil properties (resistance to scour)	Landform	Rainfall characteristics	Overland flow	Vegetation cover
• Soil moisture potential • Particle size distribution • Soil structure	• Slope steepness • Slope shape and aspect • Slope length	• Mass, size, shape, and impact velocity of raindrop • Kinetic energy and momentum • Intensity • Wind velocity	• Depth • Type of flow, for example, laminari's turbulence	• Canopy cover • Foliage distribution

Figure 5.2 Factors affecting soil splash.

creases, and its particles are detached and spattered about by drops later impacting on it.

Quick wetting of dry clods affects their detachability in two ways: through the pressure of air entrapped and by releasing the heat of wetting.

Entrapped air. The entrapped air on quick wetting can virtually explode, breaking the clod and spattering soil particles into the air. Energy in the form of air pressure within the soil contributes substantially to detachment at the soil surface. Badrashi et al. (1981) showed that entrapped air increased soil detachment 21 percent. Quirk and Panabokke (1962) reported significant effects of entrapped air on slaking when a dry soil is rapidly immersed in water.

Heat of wetting. The wetting process also influences soil detachment by altering the energy status of the soil water system. In addition to facilitating a gradual escape of air, slow wetting changes the energy status of the soil water system. The heat of wetting, or the energy released when soil water potential changes from one energy state to another, plays a significant role in detachment of relatively dry soil. The drier the soil and the greater the surface area, the

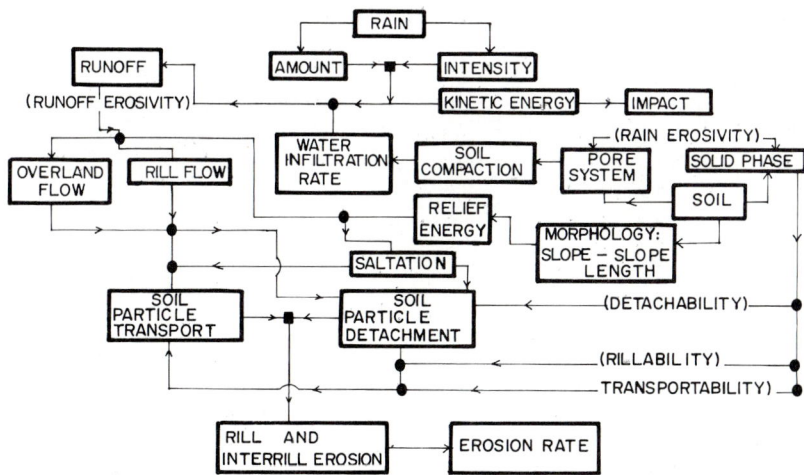

Figure 5.3 Interaction between rainfall, overland flow, and terrain on splash and sediment transport (Chisci, 1981).

more heat of wetting is released, the greater the change in soil strength, and the more soil is detached per unit energy expended by the raindrop impact. In addition to the quantity of heat released, the rate at which it is released and dissipated is important. Fast release causes more soil detachment and splash. An example of heat release during the wetting process is shown for two Australian soils in Table 5.1.

Many researchers have observed that some soils are more easily detached when dry than when moist (Young, 1984). For example, Bruce-Okine and Lal (1975) observed that fewer drops were required to disrupt a dry clod than a wet clod (Table 5.2). Furthermore, the effect of soil moisture potential on detachment depends on the surface area and hence the particle size distribution. In general, clayey soils are more easily detached when at low (more negative) than at high (less negative) moisture potential. The high detachability of clayey soils at low moisture potential is attributed partly to the heat of wetting and partly to entrapped air. The amount of heat of wetting released can be computed from the changes in entropy and enthalpy of the soil water system.

The importance of the heat of wetting on soil detachment is more relevant in tropical soils that contain predominantly high-activity clays, i.e., vertisols, inceptisols, and other clayey soils. They are dry and hot at the end of the long, dry season of 5 to 8 months. Because the soils are devoid of any vegetation cover, high-intensity monsoon rains cause severe soil splash and interrill erosion. With decreased soil water potential, soil detachability by impacting drops may increase up to a certain critical moisture potential. This critical potential differs among soils of different chemical, physical, and mineralogic compositions (de Ploey, 1971; Sloneker et al., 1976).

TABLE 5.1 Heat Balance Assessments of Heat of Wetting ΔH_{exp} Compared with Thermodynamic Values
Values in cal/g

Thermistor depth, cm	Kraznozem		Chernozemic soil		
	60% R.H.	O.d. soil	60% R.H.	O.d. soil, 4.44 cm	O.d. soil, 10 cm
2.5	−2.6	−5.0	−8.0*	−23*	−28*†
10	−2.9	−5.0	−2.4	−4.7	−13.3† ⎫
17.5	−2.7	−4.0	n.d.	n.d.	−5.1† ⎬ ∼−36‡
25	−2.5	−3.8	n.d.	n.d.	n.d. ⎭
40	−2.1	−4.1	n.d.	n.d.	n.d.
Average	−2.6	−4.4			
Theoretical	−3.1	∼−23	−7.4	∼−35	∼−35

n.d. = not determined.
*See text, corrected for expansion of soil.
†For 1, 5, 9, and 16.5 cm, respectively.
‡See text, after correction for radiation losses.
SOURCE: Collis-George and Lal, 1973.

TABLE 5.2 Effects of Soil Moisture Potential on Number of Raindrops Required to Detach an Aggregate

Soil textural class	Horizon	pF	Average no. drops required to detach an aggregate at waterdrop temperature		
			30°C	40°C	50°C
Clay	A_1	4.44	194	173	143
Clay	A_2	4.44	86	65	34
Clay	A_3	4.44	48	37	36
Sandy clay loam	B_1	4.44	23	23	17
Sandy clay loam	B_2	4.44	11	11	9
Clay	A_1	7.0	21	19	11
Clay	A_2	7.0	14	11	11
Clay	A_3	7.0	17	14	11
Sandy clay loam	B_1	7.0	25	17	15
Sandy clay loam	B_2	7.0	19	11	10

SOURCE: Bruce-Okine and Lal, 1975.

Shear strength. Once the monomolecular layer is formed, a further increase in pore water pressure influences detachability by changing the soil strength. Pore water pressure influences soil shear strength, as explained by Terzaghi's effective stress equation

$$\sigma' = \sigma - \psi_m \tag{5.2}$$

where σ' is the effective stress, σ is the externally applied stress, and ψ_m is the pore water pressure. Negative pore water pressure contributes positively to the effective stress. Cruse and Larson (1977) developed an empirical relation between shear strength and detachability:

$$(D \times 10^4)^{1/2} = 6.4337 - 0.098\sigma' + 0.004(\sigma')^2 \quad R^2 = 0.86 \tag{5.3}$$

Al-Durrah and Bradford (1981) observed that splash decreased exponentially with increasing shear strength. These authors observed a linear relationship between soil splash and the ratio of drop kinetic energy to soil shear strength

$$D = 0.36 + \frac{0.007E}{\sigma'} \quad R^2 = 0.97 \tag{5.4}$$

where D is soil splash in milligrams per drop, E is the kinetic energy, and σ is the shear strength.

Soil shear strength, being influenced by soil texture and particle size distribution, differs among soils developed on different parent materials. Working on some Japanese soils, Ezaki (1985) reported measurable differences in splash among soils developed on coarse-grained decomposed granite, fine-grained decomposed granite, and a red soil. Ezaki proposed the following equation relating splash to rainfall intensity and soil properties:

$$\frac{E}{\gamma d_{50}^2} = \frac{Ai^B}{d_{50}} - C \tag{5.5}$$

Here E is the sediment yield per unit width, γ is the soil density, i is the maximum rainfall for a 10-min period, d_{50} is the median grain size, and A, B, and C are constants. The value of exponent B ranged from 1.60 to 2.06. Ability to form crust is another soil factor that governs splash (McIntyre, 1958).

Slope gradient

In general terms, soil detachment is independent of slope. Soil splash is, however, affected by slope steepness and the direction of the rainfall vector in relation to the direction of slope. Soil particles splattered in air may move horizontally as much as 1.5 m from the original location. During the landing under the force of gravity, the amount of material splashed downslope is often more than that splashed upslope (Fig. 5.4). This means that soil can move downslope even if there is no runoff. However, in windless rain falling on a flat soil, the net soil movement from an area should be zero. All factors being the same, net splash from an area may increase with increasing slope steepness.

The amount of soil carried downslope depends on many factors including slope steepness, drop size, wind velocity, and soil surface conditions, e.g., vegetation cover or mulch. These factors and their interaction are responsible for a controversy in the literature regarding the effects of slope steepness on the amount of splash carried downslope.

Some researchers have observed increased splash downslope with increasing slope gradient. For example, Ellison (1944) established that the amount of splash downhill is approximately equal to 50 percent plus the land slope percentage. Foster and Martin (1969) observed that, depending on the soil and its bulk density, splash downslope increased as the slope increased up to 33 percent but decreased on steeper slopes.

The rate of splash is also time-dependent, even during a rain event of constant erosivity. De Ploey and Savat (1968) observed an increase in the proportion of material landing downslope from drop impact with an increase in slope angle. However, they reported a negligible increase between slope angles of 30° and 40°. Bryan (1979) made similar observations and reported that the quan-

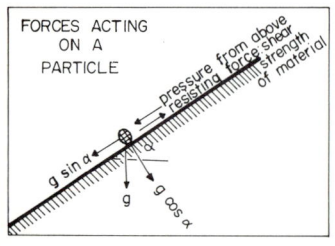

Figure 5.4 Relative partitioning of soil splashed upslope and downslope as a function of slope gradient. (Adapted from Derbyshire et al., 1979.)

tity of splash downslope peaked between 15° and 18° slope angle. He developed a regression equation relating splash downslope with slope angle

$$D = 0.079 + 0.028S - 0.0007S^2 \quad R^2 = 0.63 \quad (5.6)$$

where D is splash downslope and S is slope angle in degrees. Savat (1981) concluded that splash increases as $(\sin \alpha)^{1.9}$, where α is the slope angle. This means that splash downslope increases as $(\sin \alpha)^{0.9}$. However, the value of the exponent varies with changes in soil properties and antecedent moisture content. It has been reported to be 0.69 by Moeyersons and de Ploey (1976) and 0.8 by Mosley (1974).

Many other researchers also have observed either little or no effect of slope gradient on splash downslope. Farmer (1973) observed only a small effect of slope steepness on splash even with slope ranges of 2, 18, and 32 percent. There was approximately 0.1 percent increase in creep per unit increase in percentage of slope. Morgan (1978) observed no relationship between percentage of downslope splash and slope steepness because only a small portion of the rainfall energy contributed to splash erosion. He concluded that splash is so small that its major role is to detach soil particles before overland flow removes them. Quansah (1981) also observed a negligible effect of slope steepness on splash (Table 5.3). According to his data, the kinetic energy of rain accounted for 79 to 80 percent of variations in amounts of soil detached, with the residual effects attributable to many variables, including slope gradient. The exponent relating splash detachment to kinetic energy ranged from 0.84 to 1.35. The corresponding values for slope exponents were 0.13 to 0.27. Lewis (1981), making observations on field runoff plots on alfisols at the International Institute of Tropical Agriculture (IITA) in western Nigeria, concluded that most rain events caused soil particle movement by rain splash. But he found no correlation between the distance that soil moved downslope and slope steepness. The data in Table 5.4 show that both the coefficient and the slope intercept were maximum for middle

TABLE 5.3 Effects of Kinetic Energy E and Slope Steepness S on Splash Detachment D_d and Transport D_t

Soil	Regression equation	R^2		
		E	S	Equation
Standard sand	$D_d = 0.0002E^{1.06}S^{1.0}$	0.84	0.06	0.90
Sand	$D_d = 0.0003E^{0.84}S^{0.13}$	0.72	0.08	0.81
Clay loam	$D_d = 0.003E^{1.16}S^{1.25}$	0.66	0.14	0.79
Clay	$D_d = 0.00001E^{1.35}S^{0.27}$	0.74	0.14	0.88
Standard sand	$D_t = 0.00005E^{0.97}S^{0.75}$	0.84	0.06	0.90
Sand	$D_t = 0.001E^{0.92}S^{0.92}$	0.23	0.61	0.85
Clay loam	$D_t = 0.004E^{0.75}S^{1.15}$	0.20	0.67	0.88
Clay	$D_t = 0.001E^{0.90}S^{1.37}$	0.19	0.62	0.81

D_d = splash detachment (kg/m^2), D_t = net splash transport (kg/m^2); S in percent.
SOURCE: Quansah, 1981.

TABLE 5.4 Movement of Soil Materials with Slope and Erosivity Relations

Slope category	Average r	Movement/soil equation	Maximum r	Movement/soil equation
1%	0.68	$M = 6.1AI + 141.2$	0.70	$M = 10.9AI + 197.6$
5%	0.79	$M = 36.9AI + 599.6$	0.76	$M = 41.5AI + 983.8$
10%	0.72	$M = 31.5AI + 497.9$	0.73	$M = 38.0AI + 751.3$
15%	0.78	$M = 15.9AI + 248.1$	0.75	$M = 19.6AI + 358.3$

All are significant at the 0.05 level using a t test. M = movement in millimeters; AI = the product of the 7½-min maximum rainfall intensity and the total rainfall.
SOURCE: Lewis, 1981.

slopes, that is, 5 percent. The maximum movement of soil particles for any storm was 4005 mm on a 5 percent slope. The corresponding values were 1501 mm on a 15 percent slope, 3045 mm on a 10 percent slope, and 900 mm on a 1 percent slope. These differences in movement on slopes of various steepness also relate to differences in soil texture—in this experiment, an increasing coarseness of soil texture with increasing slope steepness. Fine-textured materials on middle slopes splash long distances, but coarse-textured material only moves short distances (Quantin and Combeau, 1962).

In spite of the controversy, it is apparent that soil splash down steep slopes exceeds that on gentler slopes. The amount of splash downslope, however, is influenced by such other factors as drop size distribution, soil texture, vegetation cover, and soil surface conditions.

Rainfall characteristics

Soil splash is related to rainfall amount and intensity. For a given amount of rainfall, high-intensity rain produces more splash than rain at lower intensity. Many researchers have shown that rainfall intensity is more important than rainfall amount (Nichols and Saxton, 1932; Tamhane et al., 1959; Meyer and Wischmeier, 1969; Foster and Meyer, 1972, 1975; Foster et al., 1977; Meyer, 1981). They developed many empirical regression equations relating soil splash to rainfall intensity. But these equations are site-specific. The effect of intensity I on erosion has often been expressed as a power function of I ($A = aI^b$ with exponent b ranging from 1.5 to 2.2) (Neal, 1938; Ekern, 1954; Ezaki, 1985; Park et al., 1985). In comparison Inoue (1985) observed a logarithmic relation between I and soil splash.

The numerical value of the exponent is also related to soil properties and canopy cover (Meyer, 1981). Meyer (1981) reported that the exponent decreased with an increase in clay content:

$$\text{Exponent } b = 2.1 - (0.01)(\% \text{ clay}) \tag{5.7}$$

On the basis of the data available in the literature, Gilley and Finkner (1985) proposed the rainfall detachment factor for rainfall intensity

$$R_{K(c)t} = (1.299 \times 10^{-5})(I^{1.366}) \qquad (5.8)$$

$$R_{K(c)t} = 4.71 I^{0.36} \quad \text{per mm rainfall} \qquad (5.9)$$

where $R_{K(c)t}$ is the raindrop detachment factor [kg/(m^3 · s^2)] and I is rainfall intensity (mm/h). One reason for site specificity of empirical relations relating soil splash to rainfall intensity is the considerable spatial and temporal variation in rainfall intensity.

In general, tropical rains fall at higher intensity than temperate rains. Some tropical rains can attain a 5- to 10-min intensity of 150 to 200 mm/h. High intensity over short time intervals is known to cause severe erosion in the tropics. In Colombia, Suarez de Castro (1980) observed that both runoff and soil erosion were affected by the maximum amount of rain received in a 5-min interval (Table 5.5). High erosion rates due to short-term but intense rains have been widely reported in the tropics (Hutchinson et al., 1958; Wilkinson, 1975).

High rainfall intensity is related to relatively big drop size and numbers per unit area per unit time. Lal (1981) reported a significant correlation coefficient between sand splash and rainfall intensity and rainfall amount:

$$D_s = 17.6 I_{30} + 1.64 \qquad r = 0.84^{**} \qquad (5.10)$$

$$D_s = 22.7 r_a + 19.73 \qquad r = 0.84^{**} \qquad (5.11)$$

Here D_s is the sand splash in grams per square meter, I_{30} is the 30-min maximum intensity, r_a is the rainfall amount in millimeters, and ** indicates significance at the 1% level of probability. Rainfall intensity and rainfall amount are easily monitored, so they are quite practical parameters in estimating soil splash and erosion by water.

Drop size distribution. Soil splash depends on the kinetic energy of the impacting raindrop, and hence on its size. (See Chap. 2.)

Kinetic energy. The physical characteristics of impacting raindrops affect the quantity and nature of soil material detached. The kinetic energy is influenced

TABLE 5.5 Relation between Rainfall Intensity and Soil Erosion at Chichina, Colombia

Rainfall, mm	Maximum 5-min rain, mm	Runoff, mm	Erosion, t/ha
20.6	7.9	6.8	7.35
21.4	5.0	11.1	1.74
18.0	4.5	7.8	1.06
21.8	2.2	4.5	0.47
20.0	1.9	0.8	0.12
22.0	1.0	n.d.	0.06

Runoff and erosion were measured on a 20 percent slope; n.d. = not determined.
SOURCE: Suarez de Castro, 1980.

by the drop size and terminal velocity. The size of the soil particle displaced depends on the terminal velocity of the impacting drops. There is a threshold impact velocity below which soil particles (W_s) are not displaced by raindrop (W_d) impact. One of the earliest models relating splash to drop size was proposed by Ellison (1947a). He related soil detachment to drop diameter, terminal velocity, and rainfall intensity by

$$A = KV^{4.33}d^{1.07}I^{0.65} \qquad (5.12)$$

where A is the relative amount of soil detached, K is a constant characteristic of soil properties, V is the velocity of raindrops (ft/s), d is the diameter of raindrops (mm), and I is the rainfall intensity (in/h). Bisal (1960) developed similar relationships.

It is widely accepted that the kinetic energy of impacting raindrops is the predominant factor responsible for soil splash (Ekern, 1950; Mihara, 1951; Rose, 1960; Bubenzer and Jones, 1971). Young and Wiersma (1973) reported that decreasing the rainfall impact energy by 89 percent, without decreasing the rainfall intensity, decreased soil losses 90 percent or more, indicating that the impact energy of a rainfall is the major factor that initiates soil detachment. Ghadiri and Payne (1977, 1979) also concluded that the breakdown of crumbs was related to the drop diameter and to the square of its velocity, i.e., the kinetic energy.

Research by these and other workers has shown that splash detachment is related to kinetic energy by

$$\text{Splash detachment} = aE^b \qquad (5.13)$$

where the value of exponent b ranges from 0.8 for sandy soils to 1.8 for clays. Quansah (1981) recorded a value of about 1.0. The kinetic energy may, however, be expressed per unit of drop area or per unit of drop circumference.

Meyer (1965) suggested that the kinetic energy per unit of drop area was a potentially important factor in soil erosion. Al-Durrah and Bradford (1982) identified the kinetic energy per unit of drop circumference as an important factor initiating soil erosion. Gilley and Finkner (1985) observed that rainfall parameters which provided the best statistical fit comprised kinetic energy times the unit of drop circumference

$$D_{K(c)*} = \frac{\sum_{i=1}^{n} a_i d_{i*}^4 V_{i*}^2}{\sum_{i=1}^{n} a_i} \qquad (5.14)$$

where a_i = no. drops of particular diameter d_i
c = raindrop circumference (length)
d_i = drop diameter of particular size class (length)
d_{i*} = relative drop diameter of particular size class, $d_{i*} = d_i/d_0$
d_0 = normalized equivalent drop diameter (length)

V_i = impact velocity of drop with diameter d_i (length/time)
V_0 = normalized drop impact velocity (length/time)
V_{i*} = relative drop impact velocity of drop with diameter d_i, $V_{i*} = V_i/V_0$
$D_{K(c)*}$ = relative raindrop detachment induced by kinetic energy times unit of drop circumference

Some researchers have argued that soil splash is related more to the momentum than to the kinetic energy of rain. (See Chap. 2.)

The importance of momentum over kinetic energy in soil detachment and splash was put forward by Rose (1960), Williams (1969), and Kinnell (1973). They argued that momentum is more logically related to splash because it is a measure of pressure or force per unit area and has the nature of mechanical stress. Park et al. (1980) argued that splash erosion is significantly affected by such rainfall parameters as rainfall momentum and number of drops by

$$A = a(M_0)^b(N_{dt})^c \qquad (5.15)$$

where A is splash erosion (g/cm^2), M_0 is rainfall momentum per unit area [kg · m/(s · cm^2)], N_{dt} is the number of raindrops per unit area (N/cm^2), and a, b, and c are constants.

The controversy regarding the relative importance of kinetic energy vs. momentum seems trivial because both are related to the same variables M and V and neither is entirely independent of the other.

Depth of overland flow

The ability of a raindrop to cause detachment and soil splash differs when overland flow is present or absent (Kinnell, 1973, 1976; Morgan, 1980; Meyer, 1985; Poesen, 1981). Walker et al. (1978) also observed that raindrop impact in runoff flow is important in soil detachment. The impacting raindrops in overland flow increase turbulence, which increases both detachment and spattering of soil particles. Raindrop impact transforms hydraulic patterns in sheet flow, generates local turbulence, and may even retard flow velocity (Yoon and Wenzel, 1971). There are different types of overland flow depending on the velocity, depth, and changes in flow patterns with time (Chow, 1959; Embleton and Thornes, 1979).

An interaction exists between the drop diameter and depth of overland flow. The diameter that causes maximum detachment depends on the depth of overland flow. Palmer (1963) reported that soil splash generally increased with increased depth of overland flow up to a threshold approximately equal to the diameter of the impacting raindrop. Park et al. (1982) used Palmer's (1963) concept and computed splash in reaction to the depth of overland flow. Splash increased slightly with increasing depth of overland flow to a critical water depth approximately equal to one drop diameter; splash then decreased sharply as water depth increased. Mutchler and Larson (1971) observed that splash produced by waterdrop impact varied with waterdrop diameter D and overland flow depth d. Splash increased from 0 for $d/D = 0$ to a maximum of $d/D = 0.14$ and $d/D = 0.20$ for D values of 0.559 and 0.296 cm, respectively. When the water depth was equal to or exceeded 3D, it had relatively less effect on splash. Mutchler and Young (1975) observed maximum detachment when the depth of water film was between one-third and one-fifth of drop diameter.

In addition to soil splash, raindrop impact in the overland flow may increase the transport capacity of the overland flow. Walker et al. (1978) observed that solids discharged in overland flow impacted by raindrops were approximately 5 times those in overland flow for the same water discharge. The size of particle detached is also influenced by the overland flow. The shift of the peak is generally markedly toward larger particle size in the presence of a thin water film rather than without it (Farmer, 1973). Ellison (1944) reported that pebbles as large as 10 mm in diameter were moved by raindrops when the pebbles were partly submerged in overland flow.

Wind-driven rain. Soil detachability by a wind-driven rain differs from that by a windless rain (Lal et al., 1980). The data in Table 5.6 show that up to 73 percent more soil detachment occurred at a wind velocity of 13.4 m/s than with windless rain. By adding a horizontal component, wind alters the vertical component of the vector. A raindrop falling vertically through wind gradually gains horizontal velocity until it reaches a state where its horizontal velocity component is equal to the wind velocity (Umback and Lembke, 1966; Morrison et al., 1985). The raindrop approach angle from vertical is given by

$$\text{Angle} = \arctan \frac{W}{V} \quad \text{rad} \tag{5.16}$$

where W is the wind-induced horizontal component of velocity and V is the vertical component of velocity, assumed to be the terminal fall velocity. Data in Table 5.7 are based on Eq. (5.16) which relates horizontal raindrop velocities and kinetic energies for an angular fall of a 2.2-mm-diameter raindrop. The wind velocity also influences the size and shape of impacting raindrops, as well as their terminal velocity, by altering the resistance of air or by breaking large drops into smaller droplets (Lyles et al., 1969; Disrud et al., 1969). Smith and Wischmeier (1962) observed that the velocity of wind-driven rain can be computed by multiplying the drop's terminal velocity in still air by the secant of the angle between vertical and the direction of fall in the wind. Lyles (1977) reported that the kinetic energy of a 2-mm drop in a 32 km/h wind was 2.75 times

TABLE 5.6 Effect of Wind Velocity on Soil Detachment

Wind velocity, m/s	Soil detachment (arbitrary units) at different intensities, cm/h		
	1.6	2.84	5.61
0	56	93	97
6.7	95	98	100
13.4	97	100	100

SOURCE: Lyles et al., 1969.

TABLE 5.7 Horizontal Raindrop Velocities and Kinetic Energies for Angled Fall of 2.2-mm-Diameter Raindrops

Terminal velocity V, m/s	Horizontal component of velocity W		Angle from vertical,* deg	Energy E, µJ
	m/s	mi/h		
6.9	0	0	0	132
6.9	1.8	4.1	15	142
6.9	3.9	8.9	30	177
6.9	6.9	15.4	45	265
6.9	11.9	26.7	60	531

*Angle from vertical = arctan (W/V) rad.
SOURCE: Morrison et al., 1985.

that of a similar drop falling in still air. The kinetic energy of wind-driven rain can be computed from the following equation, derived by Morrison et al. (1985):

$$E = \tfrac{1}{2}mq^2 \quad \text{J} \tag{5.17}$$

Here m is the mass of the drop, and q is the resultant linear velocity. In terms of terminal velocity and rain angle P_A, the equation to compute the kinetic energy becomes

$$E = \tfrac{1}{2}mV^2[(\tan P_A)^2 + 1.0] \quad \text{J} \tag{5.18}$$

For examples at $A = 45°$, E is double the value for vertical rainfall in still air.

Disrud and Krauss (1971) reported that soil detachment from clods exposed to wind-driven rain was greater than that caused by similar rainfall intensities without wind. Lyles et al. (1974) observed rainfall detached 2.68 times more soil when accompanied by a 25 mi/h wind than by no wind. Many tropical rains are accompanied by heavy winds that affect their erosivity. The data in Figs. 5.5 and 5.6 show the peak wind velocity in relation to the peak rainfall intensity for two rains recorded at Ibadan, Nigeria. The erosivity of wind-driven rain is most affected when there is no phase difference between the peak rainfall intensity and the peak wind velocity.

Vegetation cover

The effect of canopy cover on rainfall erosivity and splash erosion has long been recognized (Haynes, 1940). Vegetation cover may dissipate raindrop impact and protect the soil against splash. Vegetation cover alters the volume, drop size distribution, impact velocity, and kinetic energy of rainfall reaching the ground. A striking example of the protective effect of a cover in dissipating raindrop energy came from an experiment in southern Africa by Hudson and Jackson (1959). They laid out field plots to determine the relative importance of raindrop splash and surface flow by imposing two treatments over clean, cultivated, bare soil: a control with no cover and a wire gauze suspended about 15 cm above

144   Basic Processes

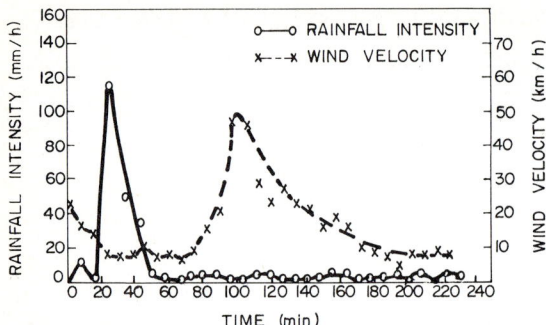

Figure 5.5 Phase difference between the maximum wind velocity and peak rainfall intensity (Lal et al., 1980).

the soil surface to break the force of raindrops but allow the rain to fall through as a fine spray. The results, shown in Table 5.8, indicate that there was severe soil erosion in a region of subtropical convective thunderstorms with high intensities and large drop size where there was no protective cover and that erosion was effectively controlled by a cover. Average soil erosion over 6 years from the bare, unprotected plot was about 123 times that from the plot with a gauze suspended above it. Hudson and Jackson also observed that a third treatment with a dense grass sward (*Digitaria swazilandensis*) also effectively controlled erosion. Their data, and similar experiments conducted by many others, indicate two important principles: (1) Raindrop splash plays a predominant role over and above the cutting and abrasive effect of runoff in causing severe erosion, and (2) any cover is effective in completely controlling soil erosion, even in extremely erosion-prone environments. The results of their experiment would have been quite different if the gauze had been suspended 5 or 10 m, rather than 15 cm, above the soil surface. The effect of vegetation cover on soil splash depends on many factors, and the height of the canopy cover is important.

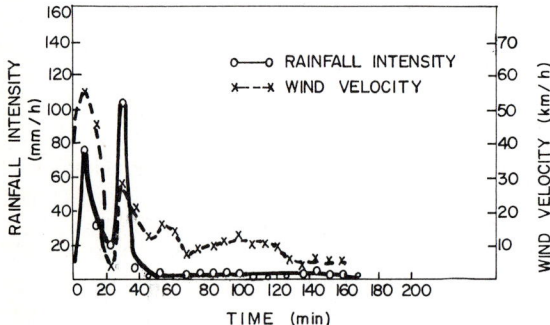

Figure 5.6 The peak rainfall intensity and maximum wind velocity coincide (Lal et al., 1980).

TABLE 5.8 Erosion from Bare and Gauzed Plots for a Soil in Southern Africa

Observation period	Bare plot, t/ha	Gauzed plot, t/ha
1953–1954	156.2	0
1954–1955	573.5	2.3
1955–1956	153.7	5.1
1956–1957	139.7	0.3
1957–1958	56.1	0.3
1958–1959	227.8	2.8
Average	217.8	1.8

SOURCE: Modified from Hudson and Jackson, 1959.

Crop canopy. Effects of canopy cover on soil splash vary among crops, depending on foliage characteristics, canopy height, and ground cover percentage. High rates of soil splash from bare soil and reduced splash by crop cover are reported even from climates of mild erosivity. For example, in Belgium, Bolline (1978) observed splash rates of 9 to 40 kg/($m^2 \cdot$ yr) on bare soil, 2 to 18 kg/($m^2 \cdot$ yr) under sugar beet, and 1 to 4 kg/($m^2 \cdot$ yr) under winter wheat for Brussels Sands series at Hesbaye. The crop cover reduced splash erosion by a factor of 2 to 9. In Poland, splash rates of 2.0 to 8.3 kg/m^2 from a bare soil and 6.2 to 16.2 kg/m^2 from a compacted cart track were reported on a loamy soil over 137 days (Froehlich and Slupik, 1980). In the United Kingdom, Morgan (1982) recorded splash rates of 29.0 to 36.5 kg/($m^2 \cdot$ yr) for bare sandy soil and 18.3 and 22.6 kg/($m^2 \cdot$ yr) for sandy loam soil under cereals. He later (1985a) used simulated rainfall to evaluate canopy cover effects on soil splash. His data indicated that detachment rates under corn generally increased with increasing canopy cover, but generally decreased under soybean (Table 5.9).

In Nigeria, Lal (1983) reported soil splash under a range of tropical crops of various canopy characteristics. Splash rates over 4 months ranged from 60 t/ha under cassava plus sweet potato to 440 t/ha on bare ridged soil (Table 5.10). The effects of different crops contrasted widely. Differences in crops stem from differences in leaf area index, height and density of canopy, foliage characteristics, and ground cover under different crops. The effect of canopy characteristics on throughfall and its erosivity also depend on the position of drop impact on the leaf, the angle of inclination of the leaf, the condition of the leaf surface, and the impact angle between drops and the leak (Kitanosoni, 1972). The canopy throughfall can be direct (drops not intercepted) or indirect (drops intercepted) penetration. Indirect penetration occurs from coalesced drops that fall from leaf surfaces (Schottman, 1978). Quinn and Laflen (1983) reported that corn canopy effectively reduced the rainfall kinetic energy. Lal (1983) computed regression equations relating soil splash under different crops to rainfall characteristics. Splash erosion correlated significantly with leaf area index, rainfall amount, and 30-min maximum intensity (Table 5.11). The kinetic energy of falling raindrops leads to soil compaction and a decrease in its hydraulic conductivity. The compaction effect of the impacting raindrop is, however, modified by the vege-

TABLE 5.9 Mean Soil Particle Detachment Rate

Crop	Canopy	Detachment rate (g/m^2) per 10 min for design intensities (mm/h)			
		40	50	75	100
Corn	0	163	200	909	776
	8	176	204	869	762
	42	253	428	808	812
	62	224	469	936	1097
	88	312	476	1304	1083
Soybean	0	296	534	566	1267
	10	202	586	662	1312
	35	260	418	631	1152
	58	783	436	351	747
	90	552	324	295	254

SOURCE: Modified from Morgan, 1985a.

tation cover. The vegetation close to the ground surface offers more protection than that high above.

Rainfall intensity. The effect of cover on soil splash also depends on the rainfall intensity and canopy height. For high rain intensities, there may be little difference in drop size distribution between the intercepted and unintercepted rain. The energy reduction by canopy is, however, proportional to the rainfall

TABLE 5.10 Effect of Crop Cover and Methods of Seedbed Preparation on Soil and Sand Splash under Different Cropping Systems from June 14 to October 12, 1982

Cropping system	Soil splash		Sand splash	
	Absolute, t/ha	Relative	Absolute, t/ha	Relative
Cassava (ridges)	186.8	42.4	61.3	65.8
Maize	111.0	25.2	53.5	57.5
Yam (mounds)	203.6	46.3	81.2	87.2
Sweet potato	237.2	53.9	86.1	92.5
Cassava + sweet potato (ridges)	60.2	13.7	62.3	66.9
Sweet potato + maize	140.0	31.8	76.5	82.2
Yam + maize (mounds)	175.0	39.8	74.2	79.7
Cassava + maize (ridges)	65.4	14.9	68.0	73.0
Ridges (bare)	440.1	100.0	93.1	100.0

SOURCE: Unpublished data of S. Huke and R. Lal; Lal, 1983.

TABLE 5.11 Regression Equations Relating Soil Splash with Leaf Area Index, Rainfall Amount, and 30-min Maximum Intensity

Cropping system	Regression equation	r
Cassava	$E = 0.56 + 0.70I_{30} + 0.43A - 0.46\text{LAI}$	0.93
Yam	$E = 0.08 + 0.88I_{30} + 0.51A - 0.18\text{LAI}$	0.76
Maize	$E = 0.15 + 0.06I_{30} + 0.53A - 0.05\text{LAI}$	0.77
Sweet potato	$E = 0.91 + 0.30I_{30} + 0.38A - 0.91\text{LAI}$	0.53
Cassava + sweet potato	$E = 0.45 + 0.18I_{30} + 0.08A - 0.28\text{LAI}$	0.63
Maize + sweet potato	$E = 0.31 + 0.03I_{30} + 0.54A - 0.07\text{LAI}$	0.58
Yam + maize	$E = 0.31 + 0.03I_{30} + 0.65A - 0.07\text{LAI}$	0.80
Cassava + maize	$E = 0.02 + 0.11I_{30} + 0.23A - 0.01\text{LAI}$	0.87

E = splash (kg/m^2); I_{30} = maximum 30-min intensity (in/h); LAI = leaf area index; A = rainfall amount per storm (in).
SOURCE: Unpublished data of S. Huke and R. Lal; Lal, 1983.

intercepted (Morgan, 1982). At low intensities, in contrast, the canopy intercepts small drops that coalesce on the leaves and fall to the ground as large drops with much more kinetic energy than the unintercepted rain.

Canopy height. When coalesced drops fall from large trees, they often reach terminal velocity and have high kinetic energy. Chapman (1948) and Schottman (1978) observed that the kinetic energy of throughfall under a pine forest was greater than that in an open field. In Malaysia, Maene and Chong (1979) observed that rain falling under oil palm consisted predominantly of large drops 3.0 to 3.5 mm in diameter. The large drops fall regardless of rainfall intensity and are obviously formed by small drops' coalescing. At low intensities and high canopy, therefore, the rainfall energy reduced by interception is partly offset by the increased energy from larger drops. Relatively large drops under cotton canopy also have been observed by McGregor and Mutchler (1978), although fewer drops reached the soil surface.

Various models have been proposed to account for canopy cover effects on throughfall erosivity and soil splash. Wischmeier (1975) proposed a canopy subfactor as the ratio of rainfall erosivity with a crop canopy to rainfall erosivity without a crop canopy. In southern Africa, Elwell (1981) proposed an exponential decrease in the rate of detachment with an increased percentage of intercepted rainfall energy. The exponential relationship is applicable to all covers in direct contact with the soil surface, e.g., crops residue mulch (Laflen and Colvin, 1981; Hussein and Laflen, 1982), stone cover (Van Asch, 1980), pasture (Lang and McCaffrey, 1984), and ponded water on the soil surface (Foster, 1982). Because of the variable effects of canopy cover on splash, Foster (1982) proposed an equation to allow for changes in the drop size distribution and fall velocity of raindrops temporarily intercepted by a canopy:

$$D_s = Ki^2\left[G_A + (1-G_A)\left(\frac{M_{ca}V_{ca}^2}{mV^2}\right)\left(\frac{i_{ca}}{i}\right)\right] \quad (5.19)$$

Here D_s is the rate of detachment of soil particles from the soil mass on bare ground, G_A is the fraction of ground area that receives direct throughfall, M_{ca} is the mass of drops falling as temporarily intercepted throughfall, V_{ca} is their impact velocity, m and V are the mass and impact velocity of drops in direct throughfall, and i_{ca}/i is the fraction of rainfall at time t falling from the canopy as transformed drops.

Morgan (1985a, b) used simulated rainfall to evaluate the effects on soil splash of canopy covers of soybean and corn. He observed an exponential decline in splash with increasing canopy cover of the type shown in

$$D_s = K_1 E(e^{-a(\text{INCEP})})^b \quad (5.20)$$

where a ranges in value from 0.03 to 0.15 and INCEP is the percentage of rainfall contributing to permanent interception and stem flow (Morgan et al., 1984).

Factors Affecting Interrill Erosion

A comprehensive review of factors affecting rill erosion has been done by Foster (1982). A brief discussion of some factors important to tropical environments follows.

Interrill erosivity

The kinetic energy of rain, as influenced by drop size distribution and intensity, is the main driving force. Foster and Meyer (1975) proposed that, for intensities between 10 and 250 mm/h, the interrill detachment is proportional to $i^{2.14}$, where i is intensity in millimeters per hour. For tropical regions, Hudson (1981) confirmed that sediment detachment is proportional to rainfall intensity squared.

Interrill erodibility

The susceptibility of soils to erosion depends on soil properties, e.g., texture, structure, organic matter content, oxides of iron and aluminum, and predominant clay minerals. Combining the erodibility and erosivity terms, Foster (1982) proposed a basic equation for interrill detachment

$$D_1 = 0.0138 k_q i^2 \quad (5.21)$$

where D_1 is the detachment rate [kg/(m^2 · h)], k_q is the soil erodibility factor for detachment by raindrop impact [kg · h/(N · m^2)], and i is the rainfall intensity (mm/h). The factor K_1 is appropriately evaluated by the raindrop technique (Bruce-Okine and Lal, 1975; Young, 1984).

Slope

The slope of the interrill area affects the velocity of sheet flow to rills and therefore its transport capacity. The slope factor for interrill areas is (Meyer et al., 1975; Lettanzi et al., 1974; Foster, 1982)

$$S_i = 2.96(\sin \theta)^{0.79} + 0.56 \quad (5.22)$$

where S_i is the interrill slope steepness factor and θ is the slope angle. Combining Eqs. (5.21) and (5.22) yields an equation that represents the maximum amount of interrill detachment:

$$D_1 = 0.0138 K_1 i^2 [2.96(\sin \theta)^{0.79} + 0.56] \quad (5.23)$$

Cover and management

The maximum possible detachment as computed by Eq. (5.23) is greatly influenced by the soil and by crop management. The rainfall erosivity factor is greatly modified by crop canopy, mulch cover, and other factors discussed in Chap. 2. Similarly, the canopy factor influences the interrill detachment. Putting these factors into Eq. (5.23) yields the interrill detachment equation:

$$D_1 = 0.038 K_1 i_{\text{eff}}^2 [2.96(\sin \theta)^{0.79} + 0.56] C_i \quad (5.24)$$

The term i_{eff}^2 is the crop canopy effect on rainfall intensity, and C_i is the cover management factor, which affects the interrill flow velocity as well as the rainfall erosivity. Other relevant management practices include tillage systems, mulch rate, cropping systems, and agroforestry. Specific examples of these practices are cited in Chap. 3.

Conclusions

The literature reviewed indicates that cover in the vicinity of soil surface decreases soil splash. Soil splash decreases exponentially with increasing area covered. The effect of canopy cover at various distances above the ground surface on soil splash varies widely. Canopy cover may increase soil splash, decrease it, or have no effect, depending on a range of other interacting factors, e.g., height, foliage characteristics, rainfall intensity, and drop impact on foliage.

References

Al-Durrah, M., and Bradford, J. M. 1981. New methods of study of soil detachment due to water drop impact. *Soil Sci. Soc. Am. J.* 45:949–952.
Al-Durrah, M. M., and Bradford, J. M. 1982. The mechanism of raindrop splash on soil surfaces. *Soil Sci. Soc. Am. J.* 46:1086–1090.
Badrashi, B., Janett, A. R., and Hoover, J. R. 1981. *The Role of Escaped Soil Air on Erosion in Sand,* ASAE Paper 81-2022, St. Joseph, Mo.
Bisal, F. 1960. The effect of raindrop size and impact velocity on sand splash. *Can. J. Soil Sci.* 40:242–245.

Bolline, A. 1978. Study of the importance of splash and wash on cultivated loamy soils of Hesbaye (Belgium). *Earth Surface Processes* 3:71–84.

Bruce-Okine, E., and Lal, R. 1975. Soil erodibility as determined by raindrop technique. *Soil Sci.* 119:149–159.

Bryan, R. B. 1979. The influence of slope angle on soil entrainment by sheetwash and rainsplash. *Earth Surface Processes* 4:43–58.

Bubenzer, G. D., and Jones, A. B., Jr. 1971. Drop size and impact velocity effects on the detachment of soils under simulated rainfall. *Trans. ASAE* 14:625–628.

Chapman, G. 1948. Size of raindrops and their striking forces at the soil surface in a red pine plantation. *Trans. Am. Geophys. Union* 29:664–670.

Chisci, G. 1981. *Upland Erosion: Evaluation and Measurement.* IAHS Publication No. 133, Wallingford, U.K., pp. 331–349.

Chow, V. T. 1959. *Open-Channel Hydraulics.* McGraw-Hill, New York, 680 pp.

Collis-George, N., and Lal, R. 1973. The temperature profiles of soil columns during infiltration. *Aust. J. Soil Res.* 11:93–105.

Cruse, R. M., and Larson, W. E. 1977. Effect of soil shear strength on soil detachment due to raindrop impact. *Soil Sci. Soc. Am. J.* 41:777–781.

de Ploey, J. 1971. Liquefaction and rain wash erosion. *Z. Geomorph.* 15:491–496.

de Ploey, J., and Savat, J. 1968. Contribution à l'étude de l'erosion par le splash. *Z. Geomorph.* 12:174–193.

Derbyshire, E., Gregory, K. J., and Hails, J. R. 1979. *Geomorphological Processes.* Dawson, Westview Press.

Disrud, L. A., and Krauss, R. K. 1971. Examining the process of soil detachment from clods exposed to wind-driven simulated rainfall. *Trans. ASAE* 14:90–92.

Disrud, L. A., Lyles, L., and Skidmore, E. L. 1969. How wind affects the size and shape of raindrops. *Agric. Eng.* 50:617.

Ekern, P. C. 1950. Raindrop impact as the force initiating soil erosion. Ph.D. thesis, University of Wisconsin, Madison.

Ekern, P. C. 1954. Rainfall intensity as a measure of storm erosivity. *Soil Sci. Am. Proc.* 18:212–216.

Ellison, W. D. 1944. Studies of raindrop erosion. *Agric. Eng.* 25:131–136, 181–182.

Ellison, W. D. 1947a. Soil erosion studies. II. Soil detachment hazard by raindrop splash. *Agric. Eng.* 28:197–201.

Ellison, W. D. 1947b. Soil erosion studies. V. Soil transportation in the splash process. *Agric. Eng.* 28:349–351.

Elwell, H. A. 1981. A soil loss estimation technique for southern Africa. In: *Soil Conservation: Problems and Prospects* (R. P. C. Morgan, ed.), pp. 281–292, Wiley, Chichester, England.

Ezaki, T. 1985. Basic research on soil erosion and conservation of slope. *International Symposium on Erosion, Debris Flow and Disaster Prevention,* Tsukuba, Japan. Erosion Control Engineering Society, pp. 45–50.

Farmer, E. E. 1973. Relative detachability of soil particles by simulated rainfall. *Soil Sci. Soc. Am. Proc.* 37:629–633.

Foster, G. R. 1982. Modeling the erosion process. In: *Hydrologic Modeling of Small Watersheds* (C. T. Haan, H. P. Johnson, and D. L. Brakensiek, eds.), ASAE Monograph 5, St. Joseph, Mich., pp. 297–380.

Foster, G. R., Lane, L. J., Nowlin, J. D., Laflen, J. M., and Young, R. A. 1977. An erosion equation derived from basic erosion principles. *Trans. ASAE* 24:1253–1262.

Foster, G. R., and Meyer, L. D. 1972. A closed-form soil erosion equation for upland areas. In: *Sedimentation* (H. W. Shen, ed.), pp. 1–12, 19, Colorado State University, Fort Collins.

Foster, G. R., and Meyer, L. D. 1975. Mathematical simulation of upland erosion by fundamental erosion mechanics. In: *Present and Prospective Technology for Predicting Sediment Yields and Sources,* ARS-S-40, USDA, Washington, pp. 190–207.

Foster, R. L., and Martin, G. L. 1969. Effects of unit weight and slope on erosion. *Proc. Am. Soc. Civil Eng. (Irrigation and Drainage Div.)* 95(IRU):551–561.

Froehlich, W., and Slupik, J. 1980. Importance of splash in erosion process within a small Flysch catchment basin. *Stud. Geomorph. Carpatho-Balcanica* 14:77–112.

Ghadiri, H., and Payne, D. 1977. Raindrop impact stress and the breakdown of soil crumbs. *J. Soil Sci.* 28:247–258.

Ghadiri, H., and Payne, D. 1979. Raindrop impact and soil splash. In: *Soil Physical Properties and Crop Production in the Tropics* (R. Lal and D. J. Greenland, eds.), pp. 95–103, Wiley, Chichester, England.

Gilley, J. E., and Finkner, S. C. 1985. Estimating soil detachment caused by raindrop impact. *Trans. ASAE* 28(1):140–146.

Haynes, J. L. 1940. Ground rainfall under vegetative canopy of crops. *J. Am. Soc. Agron.* 32:176–184.

Hudson, N. W. 1981. *Instrumentation for the Study of the Erosive Power of Rainfall.* IAHS Publication No. 133, Wallingford, U.K., pp. 383–390.

Hudson, N. W., and Jackson, D. C. 1959. Results achieved in the measurement of erosion and runoff in southern Rhodesia. *Proc. Third Inter-African Soils Conference*, Dalaba, Africa, vol. 2, pp. 575–583.

Hussein, M. H., and Laflen, J. M. 1982. Effects of crop canopy and residue on rill and interrill soil erosion. *Trans. ASAE* 25:1310–1315.

Hutchinson, J., Manning, H. L., and Earbrother, H. G. 1958. On the characterization of tropical rainstorms in relation to runoff and percolation. *Q. J. Roy. Met. Soc.* 84:250–258.

Inoue, S. 1985. Mechanism of soil erosion by raindrop impact. *International Symposium on Erosion, Debris Flow and Disaster Prevention*, Tsukuba, Japan. Erosion Control Engineering Society, pp. 35–39.

Kinnell, P. I. A. 1973. The problem of assessing the erosive power of rainfall from meteorological observations. *Soil Sci. Soc. Am. Proc.* 37:617–621.

Kinnell, P. I. A. 1976. Some observations of the Joss-Waldrogel rainfall distometer. *J. Appl. Met.* 15:499–502.

Kitanosono, T. A. 1972. Interception of rainfall at different growth stages by canopy of tobacco plants. *Proc. Crop Sci. Soc. Jpn.* 41:38–43.

Laflen, J. M., and Colvin, T. S. 1981. Effect of crop residue on soil loss from continuous cropping. *Trans. ASAE* 24:695–709.

Lal, R. 1981. Analysis of different processes governing soil erosion by water in the tropics. *IAHS Publ.* 133:351–364.

Lal, R. 1983. Soil erosion in the humid tropics with particular reference to agricultural land development and soil management. *IAHS Publ.* 140:221–239.

Lal, R., Lawson, T. L., and Anastase, A. H. 1980. Erosivity of tropical rains. In: *Assessment of Erosion* (M. De Boodt and D. Gabriels, eds.), pp. 143–151, Wiley, Chichester, England.

Lang, R. D., and McCaffrey, L. A. H. 1984. Ground cover, its effect on soil loss from grazed runoff plots, Gunnedah. *J. Soil Conserv. New South Wales* 40:56–61.

Lettanzi, A. R., Meyer, L. D., and Baumgardner, M. F. 1974. Influence of mulch rate and slope steepness on inter-rill erosion. *Soil Sci. Soc. Am. Proc.* 38:946–950.

Lewis, L. A. 1981. The movement of soil materials during a rainy season in western Nigeria. *Geoderma* 25:13–25.

Lyles, L. 1977. Soil detachment and aggregate disintegration by wind driven rain. In: *Soil Erosion: Prediction and Control*, pp. 152–159, SCSA, Ankeny, Iowa.

Lyles, L., Dickerson, J. D., and Schmeidler, N. F. 1974. Soil detachment from clods by rainfall: Effects of wind, mulch, mulch cover, and initial soil moisture. *Trans. ASAE* 33:302–306.

Lyles, L., Disrud, L. A., and Woodruff, N. P. 1969. Effects of soil physical properties, rainfall characteristics, and wind velocity on clod disintegration by simulated rainfall. *Soil Sci. Soc. Am. Proc.* 33:302–306.

Maene, L. M., and Chong, S. P. 1979. Drop size distribution and erosivity of tropical rainstorms under the oil palm canopy. Lapuran Penyelidikan Jabatan Sains Tanah 1977–78, Universiti Pertanian Malaysia, Serdang, pp. 81–93.

McGregor, K. C., and Mutchler, C. K. 1978. *The Effect of Crop Canopy on Raindrop Size Distribution and Energy.* Annual Report, U.S. Sedimentation Laboratory, Oxford, Miss.

McIntyre, D. S. 1958. Soil splash and the formation of surface crusts by raindrop impact *Soil Sci.* 85:261–266.

Meyer, L. D. 1965. Simulation of rainfall for soil erosion research, *Trans. ASAE* 8:63–65.
Meyer, L. D. 1981. How rain intensity affects interrill erosion. *Trans. ASAE* 24:1472–1475.
Meyer, L. D. 1985. Interrill erosion rates and sediment characteristics. In: *Soil Erosion and Conservation* (S. A. El-Swaify, W. C. Moldenhauer, and A. Lo, eds.), pp. 167–177, Soil Conservation Society of America, Ankeny, Iowa.
Meyer, L. D., Foster, G. R., and Romkens, M. J. M. 1975. Source of soil eroded by water from upland slopes. In: *Present and Prospective Technology for Predicting Sediment Yield and Sources*. ARS 40, USDA, Washington, pp. 177–189.
Meyer, L. D., and Wischmeier, W. H. 1969. Mathematical simulation of the process of soil erosion by water. *Trans. ASAE* 12:754–758, 762.
Mihara, Y. 1951. *Raindrops and Soil Erosion*. National Institute of Agricultural Science, Tokyo, Japan, series A.
Moeyersons, J., and de Ploey, J. 1976. Quantitative data on splash erosion, simulated on unvegetated slopes. *Z. Geomorph. Suppl. Bd.* 25:120–131.
Morgan, R. P. C. 1978. Field studies of rainsplash erosion. *Earth Surface Processes* 3:295–299.
Morgan, R. P. C. 1980. Field studies of rain splash erosion. *Earth Surface Processes* 5:307–316.
Morgan, R. P. C. 1982. Splash detachment under plant covers: Results and implications of a field study. *Trans. ASAE* 25:987–991.
Morgan, R. P. C. 1985a. Effect of corn and soybean canopy on soil detachment by rainfall. *Trans. ASAE* 28:1135–1140.
Morgan, R. P. C. 1985b. Establishment of plant cover parameters for modelling splash detachment. In: *Soil Erosion and Conservation* (S. A. El-Swaify, W. C. Moldenhauer, and A. Lo, eds.), pp. 377–383, Soil Conservation Society of America, Ankeny, Iowa.
Morgan, R. P. C., Morgan, D. D. V., and Finney, H. J. 1984. A predictive model for the assessment of soil erosion risk. *J. Agric. Eng. Res.* 30:243–253.
Morrison, J. E., Jr., Gerik, T. J., and Bartek, L. A. 1985. Standing stubble protection of soil from wind driven raindrop impacts. *Trans. ASAE* 28:484–488.
Mosley, M. P. 1973. Rain splash and the convexity of badland divides. *Z. Geomorph. Suppl. Bd.* 18:10–25.
Mosley, M. P. 1974. Experimental study of rill erosion. *Trans. ASAE* 17:909–913, 916.
Mutchler, C. K., and Larson, C. L. 1971. Splash amounts from waterdrop impact on a smooth surface. *Water Resources Res.* 7:195–200.
Mutchler, C. K., and Young, R. A. 1975. Soil detachment by raindrops. In: *Present and Prospective Technology for Predicting Sediment Yields and Sources*, ARS 40, USDA, Washington, pp. 113–117.
Neal, J. H. 1938. The effect of degree of slope and rainfall characteristics on runoff and soil erosion. *Mo. Agric. Expt. Sta. Res. Bull.* 280.
Nichols, M. L., and Saxton, H. D. 1932. A method of studying soil erosion. *Agric. Eng.* 13:101–103.
Palmer, R. S. 1963. *The Influence of a Thin Water Layer on Water Drop Impact Forces*, pp. 141–148, IAHS, Publication no. 65, Washington.
Park, S. W., Mitchell, J. K., and Bubenzer, G. D. 1980. *An Analysis of Splash Erosion Mechanics*, ASAE Paper 80-2502, St. Joseph, Mo.
Park, S. W., Mitchell, J. K., and Bubenzer, G. D. 1982. Splash erosion modeling: Physical analyses. *Trans. ASAE* 25:357–361.
Poesen, J. 1981. Rainwash experiments on the erodibility of loose sediments. *Earth Surface Processes Landforms* 6:285–307.
Quansah, C. 1981. The effect of soil type, slope, rain intensity and their interactions of splash detachment and transport. *J. Soil Sci.* 32:215–224.
Quantin, P., and Combeau, A. 1962. Erosion et stabilité structurale du sol. *IAHS Bull.* 59:124–130.
Quinn, N. W., and Laflen, J. M. 1983. Characteristics of raindrop throughfall under corn canopy. *Trans. ASAE* 26:1445–1450.
Quirk, J. P., and Panabokke, C. R. 1962. Pore volume-size distribution and swelling of natural soil aggregates. *J. Soil Sci.* 13:71–81.

Rose, C. W. 1960. Soil detachment caused by rainfall. *Soil Sci.* 89:28–35.
Savat, J. 1981. Work done by splash: Laboratory experiments. *Earth Surface Processes Landforms* 6:275–283.
Schottman, W. R. 1978. Estimation of the penetration of high energy raindrops through a plant canopy. Ph.D. thesis, Cornell University, Ithaca, N.Y.
SCSA. 1982. *Resource Conservation Glossary,* Soil Conservation Society of America, Ankeny, Iowa.
Sloneker, L. L., Olson, T. C., and Moldenhauer, W. C. 1976. Effect of pore water pressure on sand splash. *Soil Sci. Soc. Am. J.* 40:948–951.
Smith, D. D., and Wischmeier, W. H. 1962. Rainfall erosion. *Adv. Agron.* 14:109–148.
Suarez de Castro, F. 1980. *Conservacion de Suelos,* Instituto Interamericano de Ciencias Agricolas, San Jose, Costa Rica.
Tamhane, R. V., Biswas, T. D., and Das, B. 1959. Effects of intensity of rainfall on soil loss. *J. Indian Soc. Soil Sci.* 7:231–235.
Umback, C. R., and Lembke, W. D. 1966. Effects of wind on falling water drops. *Trans. ASAE* 9:805–808.
Van Asch, Th. W. J. 1980. Water erosion on slopes and landsliding in Mediterranean landscape. *Utrechtse Geograf. Stud.,* no. 20.
Walker, P. H., Kinnell, P. I. A., and Green, P. 1978. Transport of non-cohesive sandy mixture in rainfall and runoff experiments. *Soil Sci. Soc. Am. J.* 42:793–801.
Wilkinson, G. E. 1975. Rainfall characteristics and soil erosion in the rainforest area of Western Nigeria. *Exp. Agric.* 11:247–255.
Williams, M. A. 1969. Prediction of rainfall splash erosion in the seasonally wet tropics. *Nature* 222:763–765.
Wischmeier, W. H. 1975. Estimating the soil loss equation's cover and management factor for undisturbed areas. In: *Present and Prospective Technology for Predicting Sediment Yields and Sources,* ARS 40, USDA, Washington, pp. 118–124.
Yariv, S. 1976. Comments on the mechanism of soil detachment by rainfall. *Geoderma* 15:393–399.
Yoon, Y. N., and Wenzel, H. G. 1971. Mechanics of sheet flow under simulated rainfall. *J. Hydraul. Div. ASCE HYG* 97:1367–1386.
Young, R. A. 1984. A method of measuring aggregate stability under water drop impact. *Trans. ASAE* 27:1351–1354.
Young, R. A., and Wiersma, J. L. 1973. The role of raindrop impact in soil detachment and transport. *Water Resources Res.* 9:1629–1639.

Chapter 6

Erosion Processes over a Watershed

The concept of a *watershed* as a hydrologic unit, advanced in the early twentieth century by K. G. Gilbert (Derbyshire et al., 1979), has been widely adopted by hydrologists and geomorphologists to study fluvial processes. The three widely and interchangeably used terms to describe a hydrologic unit are *watershed, catchment,* and *drainage basin.* The term *watershed* is used in this book. Some important watershed characteristics that influence hydrologic processes include relief, parent material, soil type, vegetation, and land use. Hydrologically, a watershed may be conceptualized as having overland, channel, and subsurface flows. Overland flow and channel areas are the major components as far as erosion and sedimentation are concerned (Foster, 1982). Pioneering work by Horton and others in the 1940s paved the way for the development of models that helped relate water and output to precipitation inputs and water characteristics.

The quantitative determination of parameters of hydrologic balance over a drainage basin in relation to watershed characteristics is a major step toward understanding processes involved in the origin and transport of sediments. Precipitation received over a watershed is partitioned into components including interception by vegetation, infiltration, surface detention, overland flow, subsurface flow, evapotranspiration, and groundwater recharge. A simplified version of different routes of water flow over a watershed is shown in Fig. 6.1. Infiltration, surface detention, overland flow, and subsurface water flow are important soil erosion components of the hydrologic cycle.

Different types of erosion as caused by specific fluvial processes over a water-

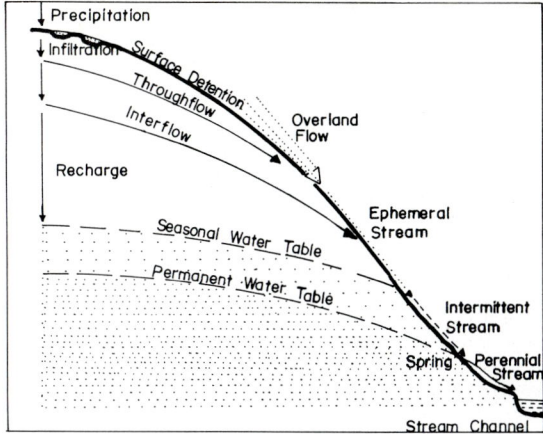

Figure 6.1 Interrelationship between different components of the hydrologic cycle over a watershed (Derbyshire et al., 1979).

shed. For example, any erosion caused by concentrated flow in small channels or rills is called *rill erosion,* and erosion on areas between the rills is *interrill erosion* (Meyer et al., 1975). Both rill erosion and interrill erosion are caused by overland flow and are called *upland erosion.* Upland erosion comprises soil detachment and transport by many interacting erosion processes, i.e., splash, sheet wash, rill erosion, and distinctly channelized fluvial erosion.

Types of Flow in Watersheds and Their Properties

There are a variety of flow types (Table 6.1). Overland flow is an important agent of water erosion that influences detachment, entrainment, and deposition of soil particles. Surface flow is influenced by the rainfall intensity and infiltration rate. Infiltration water refers to water that flows into sinkholes, open cavities, and porous materials and disappears into the ground. In addition to surface and groundwater flow, subsurface flow forms an important component of the hydrologic cycle. Subsurface flow includes pipe flow, an important agent of water erosion. Commonly observed velocities of various types of flow are shown in Table 6.2. The relationship between different components of the hydrologic cycle is shown in Fig. 6.1. While sheet, rill, and gully erosions are caused by different types of surface flow, tunnel erosion is caused by pipe flow—a form of subsurface fluvial.

The different types of flow described in Table 6.2 may be (1) turbulent or laminar, (2) steady or unsteady, and (3) uniform or nonuniform. Turbulent flow is the most relevant to soil erosion. In turbulent flow, the water moves in highly irregular paths, causing an exchange of momentum from one portion of water to another. The turbulence increases shear stresses throughout the fluid. In laminar flow, water moves in smooth paths in laminars or layers, with one layer

TABLE 6.1 Types of Flow in Headwater Areas of Watersheds

Type of flow	Character	Location
Quick Flow		
Overland flow	Surface flow of water because rainfall intensity exceeds infiltration rate. Referred to as *Horton overland flow* or *infiltration overland flow* by some workers.	Semiarid areas where rainfall intensities are high and vegetation cover is sparse. In humid areas may occur adjacent to stream channels or in topographic hollows where water converges.
Saturated overland flow	Surface flow of water that occurs because soil is saturated and infiltration capacity has not been exceeded.	Locations usually close to stream channels or hollows where water table rises rapidly to surface during storm event.
Return Flow		
Throughflow	Movement of water downslope in soil profile usually under unsaturated conditions. Referred to as *unsaturated throughflow* by some workers.	Slopes with well-drained soils often encouraged by discontinuities in soil profile. Lateral flow will occur in soil if the flow meets less resistance than vertical percolation of water.
Saturated throughflow	Lateral flow in soil under saturated conditions.	During storm a saturated wedge will extend upslope in soil profile, and saturated throughflow occurs immediately above.
Translatory flow	Lateral flow in soil occurring by displacing stored water due to addition of "new" water.	Slope of soil of saturated zone.
Interflow	May be used synonymously with throughflow. Some workers describe lateral flow above water table but below soils as interflow which could thus be through unsaturated rock or regolith.	Slopes having permanent water table at depth and any lithological discontinuities may encourage lateral flow of water as interflow.
Saturated interflow	Interflow occurring under saturated conditions.	Affected by extension of saturated wedge beneath surface in upslope direction.
Pipe flow	Flow through subsurface network of interconnected anastomosing pipes or tubes, larger than other soil voids and may be up to 1 m in diameter.	Variety of areas including steep slopes, where erodible layer lies above less permeable layer, or on flood plains marginal to channel banks.
Delayed Flow		
Groundwater flow	Water that has infiltrated into ground, has reached groundwater, and is discharged to surface from spring or seepage at rate determined by hydraulic head.	Areas where groundwater storage is possible due to character of subsurface materials.

SOURCE: Derbyshire et al., 1979.

gliding smoothly over an adjacent layer. In turbulent flow both viscosity and turbulence contribute to shear stress:

$$\tau = (\mu + \eta)\frac{dv}{dx} \tag{6.1}$$

where τ is the shear stress (or force per unit area), μ is the absolute or the dynamic viscosity of the water, η is the roughness coefficient, v is the velocity, and x is the distance.

Steady flow occurs when conditions (velocity, density, pressure, and temperature at any point in water) do not change with time ($\partial v/\partial t = 0$). The flow is unsteady when conditions at any point change with time ($\partial v/\partial t \neq 0$). Uniform flow occurs when the velocity vector at every point is identical (in magnitude and direction) for any given instant ($\partial v/\partial x = 0$). In uniform flow, the velocity vector does not change in any direction throughout the fluid at any instant. In nonuniform flow, the velocity vector varies from place to place at any instant ($\partial v/\partial x \neq 0$).

Sheet or Interrill Erosion

The Soil Conservation Society of America (SCSA) (1982) defines *sheet erosion* as "the removal of a thin fairly uniform layer of soil from the land surface by runoff water." *Sheet erosion* is, however, a misnomer and is attributed to its principal agent, the unchanneled overland flow, which is also called *sheet flow*. But under natural field conditions, there is no such thing. Soil removal by overland flow varies even over short distances. Soil variability and the presence of numerous microdepressions and microhillocks preclude removal of a uniform soil layer even from a seemingly homogeneous field. It is important to realize, however, that some, if not most, soil particles transported by sheet flow were already detached and moved by soil splash caused by the raindrop impact. Although soil detachment by splash precedes transport by sheet flow, both processes may occur simultaneously during a natural rain event. Overland flow represents two components: (1) the difference between the rainfall rate (or intensity) and the infiltration rate and (2) the direct flow or the amount of rain directly con-

TABLE 6.2 Flow Velocity for Different Types of Flow

Type of flow	Velocity of water flow
Overland flow	3 to 15 cm/s on slope of 0.40; less than 0.1 cm/s on low slopes with thick vegetation cover
Vertical percolation	Less than 7.5 cm/day in Whitehall watershed, Georgia
Saturated	20 cm/h; 0.2 to 37.2 cm/h saturated hydraulic conductivity values collected from various field measurements
Throughflow	80 cm/day in B horizon in East Twin Brook catchment, Sommerset; 50 cm/day in B/C horizon
Pipe flow	10 to 20 cm/s in Nant Gerig catchment, central Wales
Stream channel flow	Average 45 cm/s

SOURCE: Derbyshire et al., 1979.

158   Basic Processes

tributing to overland flow. Some rainfall becomes overland flow by falling directly into the sheet flow or channeled flow. The major portion of overland flow during a rain event, however, is the difference between the amount of rain received and infiltration.

Infiltration

The amount, rate, and time of runoff initiation are determined by the infiltration rate. The infiltration rate refers to the rate at which the rainwater is accepted and transmitted through the soil. Theoretical principles governing the infiltration process are described in Chap. 3. The infiltration rate depends on many factors: crusting, heat of wetting, hard setting, and soil compaction.

Crusting.   Surface sealing and crusting are severe in soils that contain predominantly low-activity clays and have low soil organic matter content. The structure of these soils slakes readily on quick wetting. Dispersed soil forms a slowly permeable seal on drying (Plate 6.1). Dispersion is enhanced by raindrop impact on bare soil surface (Thompson and James, 1965). Weakly formed aggregates are easily dispersed. The fine-clay particles move in the interparticle voids and reduce the proportion of transmission or macropores, thereby restricting the infiltration rate. Tropical alfisols are particularly prone to crusting. Vertisols also slake readily, and their infiltration rate is limited by profile characteristics.

The data in Fig. 3.10 show a very rapid decline in infiltration rate of a newly developed alfisol in western Nigeria. The rapid decline in the infiltration rate of plots recently cleared from a forest vegetation stems from crust development and surface seal. Soil compaction sets in with intensive cultivation, especially with motorized farm operations.

Plate 6.1 A slowly permeable seal on the soil surface.

Heat of wetting. The heat of wetting, its magnitude, and the rate of release play important roles in governing infiltration into clayey soils. (See Chap. 3.) The phenomenon is particularly relevant to clayey soils of the tropics and subtropics that are subject to long periods of hot and dry weather before the monsoons. Some heavy-textured vertisols are structurally unstable and slake readily. Like the heat of wetting, slaking is caused by the disruptive effects of air entrapped within the aggregates (Robinson and Page, 1950). Many workers have reported better aggregation and structural stability in soil samples previously moistened (Nijhawan and Olmstead, 1947; Panabokke and Quirk, 1957). Over and above the effect of entrapped air, the heat of wetting is important in regulating the rate at which water is accepted into the soil.

It has been experimentally established that the drier the soil, hence the more negative the soil water potential, the more the surface aggregate disintegrates and the more the infiltration rate is lowered (Collis-George and Lal, 1970, 1971). Collis-George and Lal (1971) reported that the wetter the soil is initially, the higher the infiltration rate and the less slaking and swelling (Figs. 6.2 and 6.3). Lal (1981) made similar observations for soils from west Africa (Figs. 6.4 and 6.5). Measurements of soil temperature before a liquid wetting front into a dry clayey soil indicated a change in temperature of 20 to 40°C (Figs. 6.6 and 6.7). Using the temperature profile of a soil column during the movement of the wetting front, Collis-George and Lal (1973) refined the concept of wetting front as proposed by Bodman and Coleman (1944) and Anderson and Linville (1962). The wetting front zone is now subdivided into two zones: the zone of the liquid wetting front and the zone of vapor condensation or the temperature front. The temperature front was further subdivided into three zones: (a) a zone of vapor condensation and temperature rise, (b) an intermediate zone where temperature falls and heat is lost by conduction to the liquid wetting front, and (c) a zone starting at the liquid wetting front and with temperature decreasing to ambient (Fig. 3.7). With a temperature change of as much as +40°C in dry clayey soil, the total quantity of heat involved is indeed substantial. The quan-

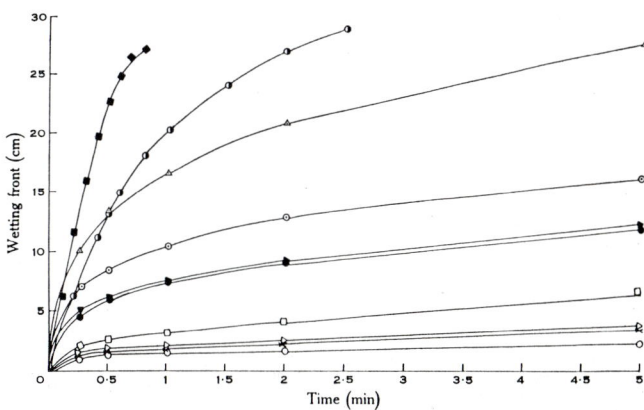

Figure 6.2 Effects of initial soil moisture potential on wetting front movement (Collis-George and Lal, 1971, 1973).

tity of heat evolved Q during the infiltration process can be experimentally determined by using the procedure outlined by Collis-George and Lal (1973):

$$Q = C_\rho \int_{t=\alpha}^{t_{\text{LWF}}} V_t \Delta T_t \, dt \tag{6.2}$$

where t_{LWF} is the time at which the liquid wetting front (LWF) reaches the thermistor, V_t is the infiltration velocity at time t, and C_ρ is the thermal capacity [cal/(°C · cm³)] which takes into account moisture content and bulk density behind the liquid wetting front. The heat-balance assessment of ΔH for two Australian soils is shown in Table 5.1. Soil structure is preserved in those soils where mechanisms exist to dissipate heat energy as quickly as it is produced.

Hard setting and soil compaction. Some soils that set hard when dry are called *hard-setting soils* (Northcote et al., 1975), and this property is related to

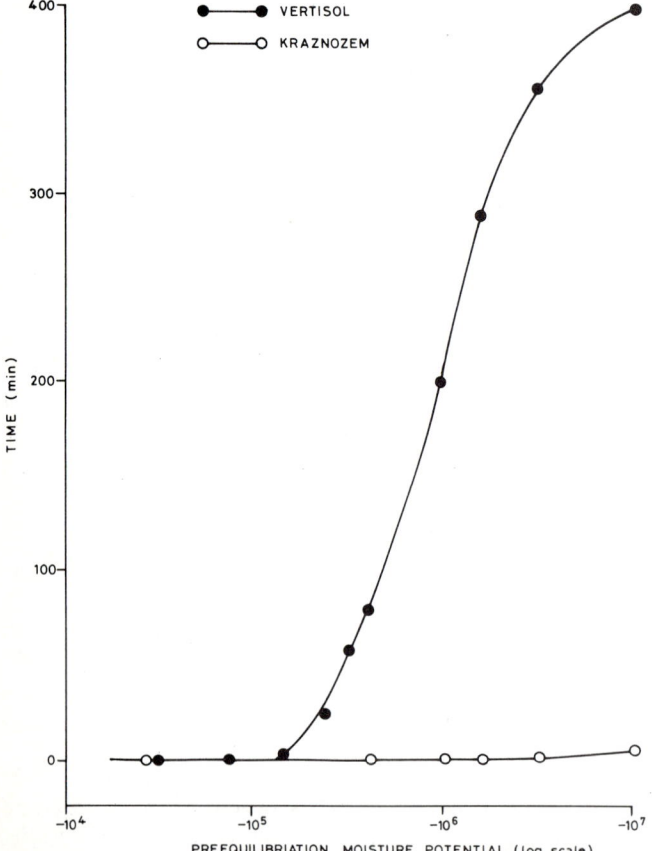

Figure 6.3 Effects of initial soil moisture potential on time required for wetting front to penetrate 25 cm in some Australian soils (Collis-George and Lal, 1971, 1973).

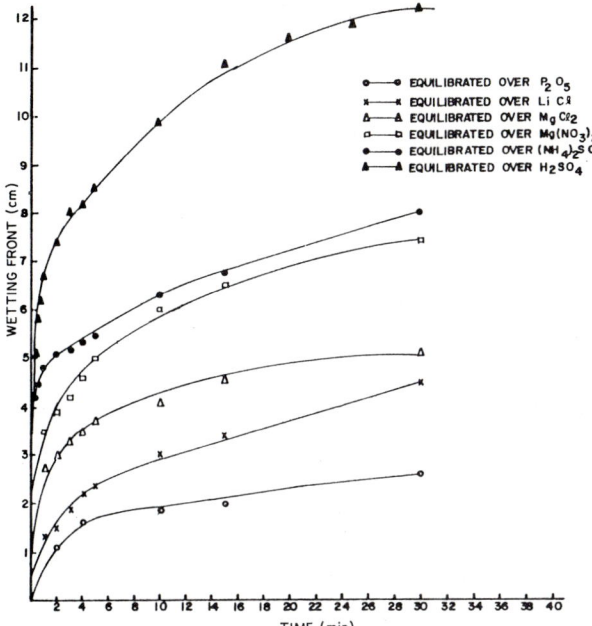

Figure 6.4 Effects of soil moisture potential on wetting front advancement into a clayey soil from Nigeria (Lal, 1981).

ultra desiccation due to excessive drying and high temperatures. Surface seal and crust formation are not necessarily associated with hard setting. Hard-setting soils are widely distributed in tropical Africa (Charreau and Nicou, 1971; Jones and Wild, 1975; Sinclair, 1985; McDonald et al., 1985). These structurally inert soils have a low quantity of low-activity clays. Their structural properties are progressively degraded by the repeated cycle of cultivation followed by drying and hard setting in the prolonged dry seasons. Infiltration rates of these soils progressively decline so that a large proportion of the rainfall received is lost as overland flow.

In addition to surface seal and hard setting, tropical soils also are easily compacted, especially where intensive land use and mechanized farm operations are seen. With continuous cultivation, the infiltration rate of some alfisols may decline by several orders of magnitude in only 3 or 4 years (refer to Fig. 3.10).

Overland flow

Overland flow is water that flows over the land surface en route to a stream channel. Such flow occurs when rainfall exceeds both the infiltration rate and the depression storage capacity of the land surface. Overland flow is the initial phase of surface runoff that eventually becomes a major agent of sediment detachment, entrainment, and deposition. Although overland flow is visualized as a broad sheet flow, it includes many shallow but easily definable channels.

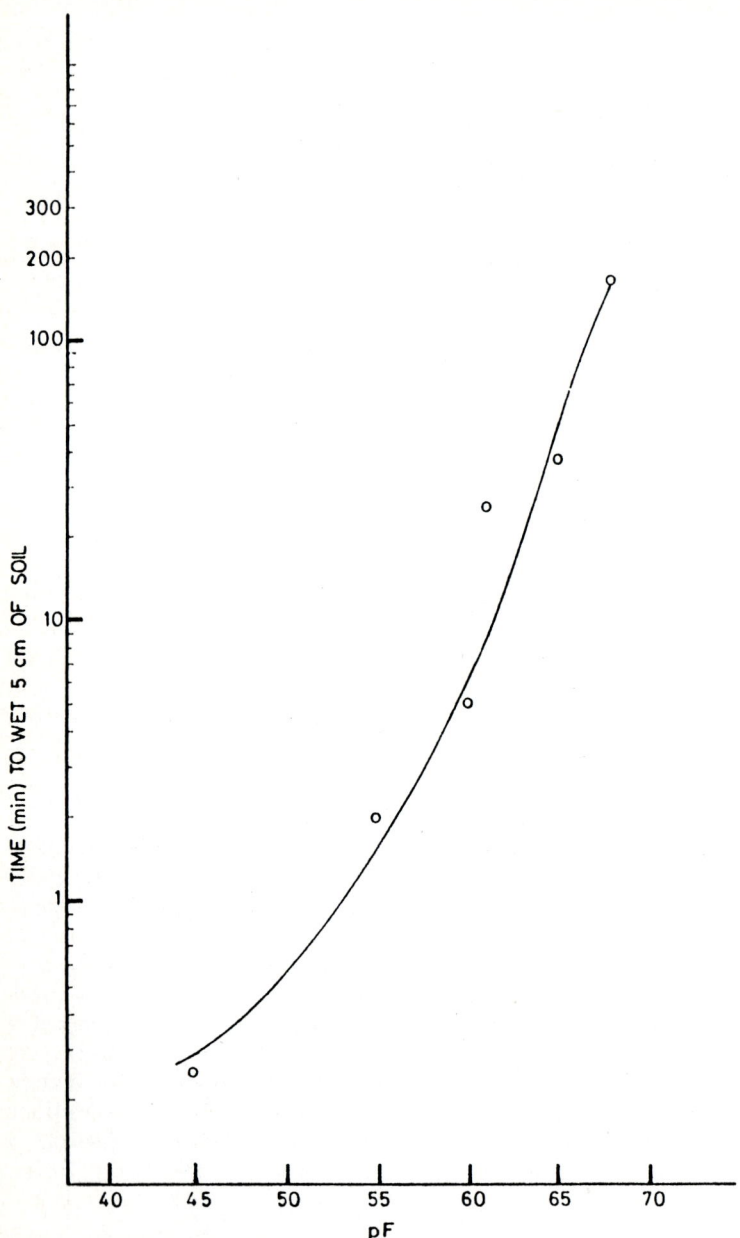

Figure 6.5 Effects of antecedent moisture potential on time needed for wetting front to penetrate 5 cm into a clayey soil from Nigeria (Lal, 1981).

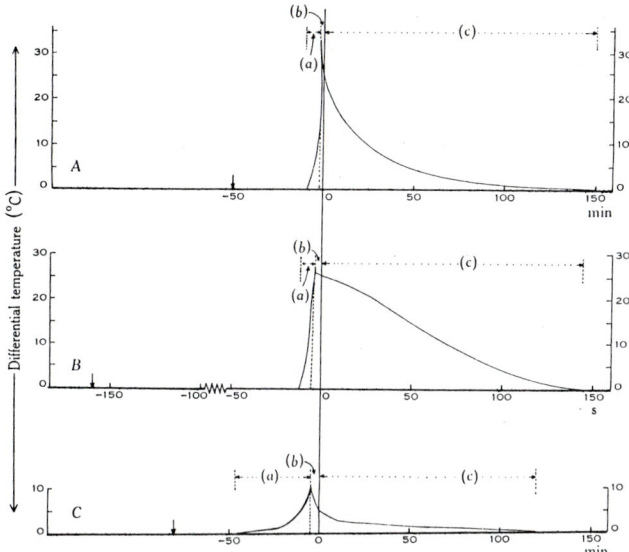

Figure 6.6 Temperature fluctuations at wetting front during infiltration into a vertisol from New South Wales, Australia. A, B, and C are three positions of the thermistor within the column. (a) is a zone of vapor condensation, (b) an intermediate zone, and (c) a zone of the wetting front. (Collis-George and Lal, 1971.)

Origin of overland flow. Excess rainfall over infiltration is first used to fill all the depression storage, which may range from 2.5 cm for smooth-surface clay to 5.0 cm for sandy soils (Chow, 1964). Depression storage may be far greater in soil with stubble and vegetation cover than in bare soil. A detailed description of this process is given by Gerrard (1981). Horton (1945) proposed the concept of the development of overland flow on steeplands. Basically, Horton visualized two main types of water supply to a stream in the drainage basin, i.e., overland flow and groundwater flow. He postulated that at some critical distance downslope from the natural water divide the overland flow accumulates with a shear stress component adequate to detach and transport soil particles. On the basis of Horton's concept of overland flow, Cook (1946) described the sequence of events that leads to the generation of overland flow:

1. A thin water layer forms on the surface, and downslope surface flow is initiated.
2. The flowing water accumulates in surface depressions.
3. When full, these depressions begin to overflow.
4. Overland flow enters microchannels that coalesce to form rivulets which discharge into small gullies. This continues until discharge into a major channel occurs.
5. Along each microchannel, there is lateral inflow from the land.

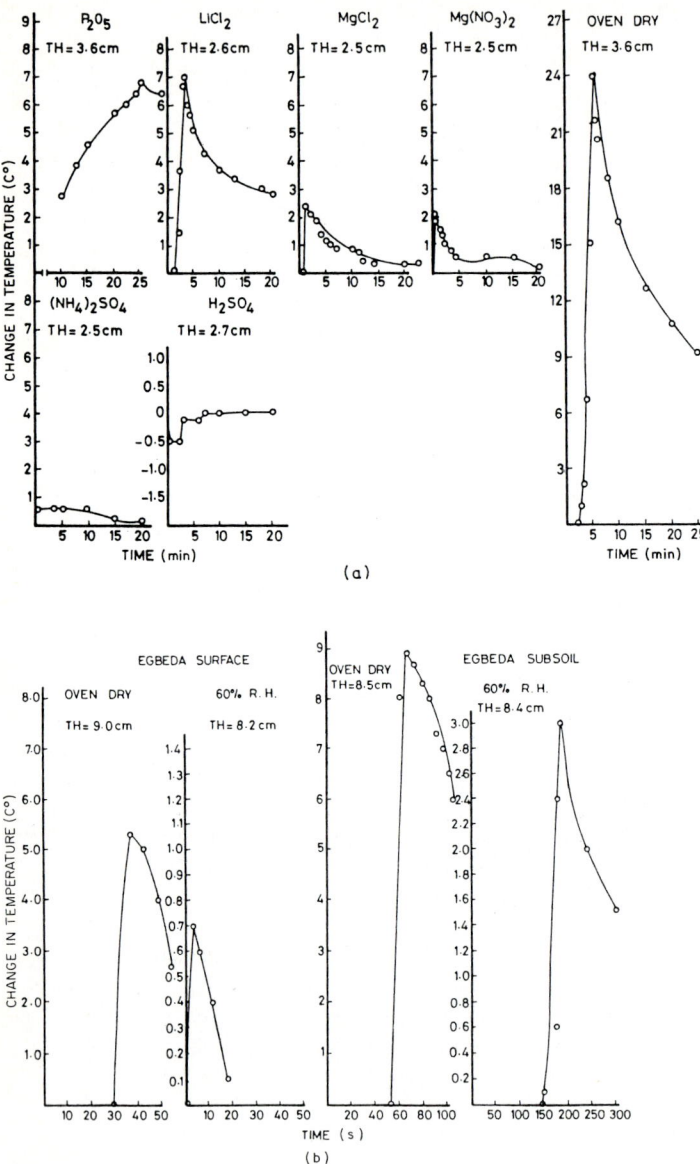

Figure 6.7 Temperature fluctuations at wetting front for some soils in Nigeria: (a) effects of equilibration at relative humidities created by different salt solutions, (b) differences in surface and subsoil of an oxic Paleustalf. TH = thermistor (installed at different depths below the soil surface), R.H. = relative humidity. (Lal, 1981.)

The realization that base flow can occur even when there is no overland flow led to the development of the throughflow model. It is now widely accepted that throughflow is probably the most important mode of water flow on hillsides in many humid regions, especially the humid tropics. A distinction has also been made between throughflow and interflow. *Throughflow* refers to flow in the soil horizons, especially above such relatively impermeable layers as the junction of the A and B horizons. In comparison, *interflow* refers to lateral flow that occurs in the aeration zone above the level of permanent saturation but below the A and B soil horizons. The amount and exact location of subsurface flow depend on site-specific factors, e.g., the presence of impeding layers in the soil profile.

Many geomorphologists now visualize a drainage basin as an extended plane rather than a section. The plane visualization is the basis of the partial-area concept and the variable-source-area model of water flow. The plane concept recognizes regions close to the drainage channel that contribute to water runoff. As the rainy season progresses, the saturated zone expands. The saturated zone that contributes to runoff is dynamic (always changing) in relation to the rainfall regime and antecedent soil moisture status.

Gregory and Walling (1973) explained various components of watershed runoff based on the water balance equation. Their model outlines the contribution of many components; e.g., the quick return flow Q_t refers to the flow in the soil layers, the delayed return flow Q_i is the flow in the aeration zone, and the prolonged return flow Q_b refers to the flow from the saturated zone or the groundwater flow. The magnitude and time when these components occur are influenced by watershed characteristics and antecedent precipitation and soil moisture regime.

Sediments are contributed by various subsystems, i.e., upland or hillslope and channel sides. Overland flow is also called *sheet flow* because water supposedly moves as a thin sheet until it reaches the nearest rill or channel. Under natural field conditions, however, sheet flow rarely, if ever, occurs. For example, Emmett (1978), who conducted field experiments using simulated rain on different slope forms within a watershed, reported that flow rarely occurred as a uniform sheet of water. Regions with crop residue developed a series of puddles—surface water moved downslope in several lateral concentrations of flow. Literally, overland flow ceases to be sheet flow by the time it enters into microchannels and rivulets. The depth of unimpeded prechannel flow rarely exceeds 3 or 4 mm. A major cause of soil detachment and transport in sheet flow is raindrop impact and its interaction with the thin film of overland flow. Processes involved in raindrop impact are discussed in Chap. 2.

Types of overland flow. Shallow flow may be laminar, turbulent, or both. Areas of turbulent flow are often interspersed with areas of laminar flow. Turbulence in the flow creates local forces capable of dislodging soil particles.

Turbulence is caused by falling raindrops and wind-driven rain. The flow may also exhibit rain waves under high-intensity tropical rains. The flow depth varies from below to above critical depth. Shallow flow has a complex hydrology because most variables (run-on, throughflow, or direct fall, runoff, infiltration) occur simultaneously and are constantly changing with respect to time and space (Emmett, 1978). The average length of shallow overland flow, before it is channeled, varies and depends on watershed characteristics. Horton (1945) advanced this formula relating average length of sheet flow to drainage density over the watershed:

$$L_0 = \frac{1}{2D_d} \qquad (6.3)$$

$$D_d = \frac{\Sigma L_s}{A} \qquad (6.4)$$

Here L_0 is the average length of overland flow, D_d is the drainage density, ΣL_s is the sum of stream lengths for the watershed, and A is the watershed drainage area.

Runoff and hydrograph

Overland flow or surface runoff is water that travels over the ground to a channel, natural depression, or waterway. A channel carries small rivulets of water in turbulent flow during and after a rain. The distance that overland flow usually traverses to join a channel is computed by Eq. (6.3). Runoff in a channel comprises various types of flow (Table 6.1), including surface runoff, interflow or subsurface flow, and groundwater flow or base flow. The total flow is customarily divided into two parts: storm, or direct runoff, and base flow. The distinction between the two is based on when they arrived in the stream rather than on the path followed. Direct runoff consists of overland flow and a part of interflow, whereas base flow is mostly groundwater. Horton's model of overland flow is the basis for separating stream hydrographs into the two components of the hydrograph, i.e., surface runoff and base flow. In many regions, however, base flow may occur even when overland flow does not.

A hydrograph is a plot of time vs. flow. The hydrograph consists of a rising limb, crest segment, and falling limb, or recession. The shape of the rising limb is influenced by the rainfall characteristics. There are different types of hydrographs in relation to parent material, topography, and soil and vegetation (Gregory and Walling, 1973).

Horton's concept of overland flow is easily explained for a runoff hydrograph from a compacted or relatively impermeable soil surface. Some of the initial rainfall fills the surface detention. At any instant the amount of water in surface detention equals the difference between rainfall and overland flow. At equilibrium, rainfall equals outflow. When the rain ceases, there is no further inflow and both the outflow and detention volume decrease. The unit hydrograph is a typical hydrograph for a watershed where runoff volume under the hydrograph is commonly adjusted to 1 in.

Sediment detachment and entrainment by overland flow

Forces acting on a submerged particle in a shallow flow are complex and difficult to define precisely. A schematic of various forces acting on such a particle is presented in Fig. 5.1 (Kirkby, 1980). Under such conditions, detachment and deposition occur simultaneously. It is the velocity of shallow flow and its shearing force that detach soil particles. Particles are detached and transported by drag and lift due to moving water. A drag is exerted by the differential stress on the upstream and downstream faces of the particle. Drag is the component of the resultant force exerted by overland flow on soil particles lying parallel to the relative motion of the water:

$$\text{Drag} = \frac{C_D \rho A V^2}{2} \tag{6.5}$$

The total drag force consists of both friction drag and pressure drag. Lift is a component of the resultant force exerted by overland flow on soil particles lying perpendicular to the relative motion of the water:

$$\text{Lift} = \frac{C_L \rho A V^2}{2} \tag{6.6}$$

Here C_D and C_L are drag and lift coefficients, respectively, ρ is the density of water, A is the area of soil particles projected on a plane perpendicular to the relative motion of the water, and V is the relative velocity of overland flow with respect to the soil particles. The relative velocity of overland flow differs for soil particles already detached and in motion vis-à-vis soil aggregates yet to be detached.

Sheet flow generates only a small shearing stress. Nonetheless, detachment caused by shallow flow increases exponentially with the slope angle up to a critical angle, where the flow changes from sheet to rill flow. This critical angle for sheet flow is rather small and depends on soil properties, surface roughness, and land use. The relative quantity of sediments detached and transported by sheet flow is small compared with splash and rill flow.

Particles detached in the interrill area by raindrop impact are transported to rills by the combined action of the thin, interrill sheet flow and raindrop impact. Raindrop impact entrains the larger particles while sheet flow transports only small particles. With gentle slopes, sediments originating in interrill areas are not delivered to rills because of limited transport capacity. The reverse is true on steeper slopes (Foster and Meyer, 1975).

Rill Erosion

Even excessive interrill erosion can go unnoticed, but rill erosion is easily observed. Freshly plowed land is often susceptible to rill erosion (Plate 6.2). Variations in topography lead to concentrated sheet flow in small, closely spaced channels called rills. The rills carry both overland flow from the interrill area and direct flow. The depth and velocity of water in channeled flow are much greater than in prechannel flow. The depth of channeled flow may be 50 times that of overland flow, and the velocity 10 times greater (Young and Wiersma, 1973). Shear stress exerted by the concentrated flow causes soil detachment along channel sides and floor.

Sheet or interrill erosion occurs independent of erosion in the rills. Sediments originating in the interrill areas, however, move downslope mostly by

Plate 6.2 Rill erosion on freshly plowed land in Brazil.

flow in the rills. So the magnitude of erosion from a field or watershed depends on the origin of sediment in interrill areas of transport capacity of rills. The principle of erosion control essentially lies in controlling either detachment or transport capacity or both. Appropriate soil and crop management practices are adopted to control erosion by detachment- or transport-limited techniques.

The concentrated flow in rills causes shear stress, which is strongest at the bottom of the channel. Rills are initiated by a gullylike headcut developing along the slope. The rate at which headcuts advance depends on many factors, e.g., slope, soil properties, and flow. Rill depths keep increasing until a resistant layer restricts further deepening. The rill then widens by lateral development of headcuts and undercutting of rill walls. Deepening or broadening of rills (rill erosion) is more severe if the sediment provided by the interrill areas is more than the transport capacity of the rill system. Some important factors affecting rill erosion are the concentrated flow, topography, and soil properties.

Concentrated flow

Both depth and velocity of flow are important in determining rill erosion. Rill erosion is a combination of rill shear erosion and rill headcut erosion (Meyer et al., 1975)

$$A_R = A_s + A_H \tag{6.7}$$

where A_R is the rill erosion, A_s is the rill shear erosion between headcuts, and A_H is the rill headcut erosion due to overall effects. Foster and Meyer (1972) observed that A_s is generally linearly related to the flow rate and that A_H perhaps varies with the flow rate to a power of 1.0 to 1.5. There is also a critical discharge before rill erosion begins; this critical discharge depends on soil properties, topography, and surface conditions (Meyer, 1981). Rill erosion can be expressed as a function of critical discharge as

$$A_R = D_R(Q - Q_c) \tag{6.8}$$

where D_R is a characteristic indicative of soil's susceptibility to rill erosion, Q is the flow rate, and Q_c is the critical discharge below which rill erosion is negligible. Combining Eqs. (6.7) and (6.8) leads to this equation describing rill erosion:

$$A_R = a_1(Q - Q_s)^{a_2} + a_3(Q - Q_H)^{a_4} \tag{6.9}$$

where Q_s and Q_H are the flow rates below which shear erosion and headcut erosion, respectively, are negligible and a_1 to a_4 are constants. Critical flow also depends on the interaction with raindrop impact and is much lower when rills are exposed than when they are protected by mulch or canopy cover.

The energy of running water is related to both depth and velocity. For equal masses of water, however, runoff energy is less than raindrop energy by several orders of magnitude. The shear strength or erosive energy of water is increased by sediment concentration. For an equal volume, sediments have 2.65 times more energy than water and are thus highly abrasive. Similarly, turbulent flow has more erosive power than laminar flow. Laminar flow becomes turbulent only after attaining critical depth and velocity. The velocity of water in small channels is given by Manning's formula

$$V = \frac{R^{2/3} S^{1/2}}{\eta} \tag{6.10}$$

where V is the average velocity of flow (m/s), R is the hydraulic radius (m), S is the land slope (m/m), and η is the coefficient of surface roughness. The hydraulic radius R is related to the cross-sectional area A^* and the wetted perimeter P by

$$R = \frac{A}{P} \tag{6.11}$$

The transport of eroded particles is related to the power function of the flow velocity by

$$G = aV^b \tag{6.12}$$

where G is the quantity of transported material, V is the flow velocity, and a and b are constants. The value of b is about 4 (Laursen, 1958; Meyer and Monke, 1965). Thus sediment entrainment is highly sensitive to change in velocity of concentrated flow.

Topography

Various topographical factors influence runoff depth and its velocity. The amount, depth, and velocity of runoff differ for different slope lengths, steepness, shape, and aspect (see Chap. 4). Rill erosion is influenced more than splash erosion by slope steepness (Lettanzi, 1973; Lettanzi et al., 1974). The force of water moving downhill is related to slope steepness according to

$$F = mg \sin \theta \tag{6.13}$$

where m is the mass, g is the acceleration due to gravity, and θ is the gradient angle. The numerical value of $\sin \theta$ increases from 0 for flatland to 1 for a right angle. Numerically $\sin \theta$ is equal to the gradient, i.e., units of vertical fall per unit of distance along the land surface, expressed as a decimal fraction.

Soil properties

Similar to detachability, soil's susceptibility to rill erosion also depends on its physical, chemical, mineralogic, and biologic properties. Some soils are more susceptible to rill erosion than others. Susceptibility of a soil to erosion is called *erodibility*. The term *erodibility* refers to both detachability and transportability and is an inherent characteristic of soil. Soil erodibility is not, however, static but dynamic. It is always changing, even during a single rain event (Poesen, 1981). The magnitude and nature (increase or decrease) of change depend on the particle size distribution, aggregate stability, amount of dispersible clay, antecedent moisture content, and rainfall characteristics. Dry soil's erodibility normally changes more than that of moist soil during a rain event. Duration of rainfall is also an important factor, because once the soil surface has attained steady state, detachability changes only slightly. Soils containing expanding-lattice clay minerals swell on wetting, and their erodibility changes drastically during the rain event. In soils containing low-activity clays, change in erodibility during a rain event is caused by slaking of aggregates, closure of intraparticle pores by particle reorientation, changes in soil bulk density and water transmission characteristics. Soil erodibility depends on the particle size distribution, soil structure, and soil strength.

Particle size distribution. Particles of different sizes are affected to various degrees by detachment and transport by channeled flow. Cohesive soils (e.g., heavy-textured or clayey soils) have more cohesive force and require more energy to detach than noncohesive soils (Bubenzer and Jones, 1971). There is no consensus among researchers on the most detachable size fraction because it depends on many factors. For example, the most detachable size range was reported to be 60 to 110 µm by Terwindt et al. (1968), 105 to 210 µm by Mazurak and Mosher (1970), a highly variable range by Moeyersons and de Ploey (1976), and about 100 µm by Poesen (1981). Farmer (1973) observed that particles more easily detached by running water are in the range of 238 to 1041 µm. He reported that the detachability rate is highest in coarse- and medium-sand size material and decreases with either larger or smaller particles. The threshold flow velocity for particle transport is also greater for coarse than fine fractions. The flow depth and flow velocity required for rill and fluvial erosion are less for fine- than coarse-textured soils. Detachability increases with decreasing median grain size, but transportability decreases with increasing median grain size. An increase in the physical mass of the coarse fraction and an increase in the cohesion of the smaller fraction are responsible for the low detachability of these size fractions. Rather than size, the weight of hydrated or dehydrated grains is also a factor in detachment and subsequent transport (Yalin, 1972; Yariv, 1976).

Young and Onstad (1976, 1978) proposed the concept of "rillability" while relating particle size distribution to the soil's susceptibility to erosion. They observed that the particle size distribution of a soil susceptible to rill erosion differed from that of soil susceptible to splash or sheet wash. Different agents of erosion are involved in rill and sheet wash erosion, and different magnitudes of forces are involved. Whereas splash and sheet wash are caused mostly by raindrop impact with and without overflow, the channeled flow and its velocity are responsible for detachment and transport in rill erosion. Young and Onstad related soil's rillability to the particle size distribution, organic matter content, and soil moisture retention at −15 bar suction. They expressed rillability by

$$F_r = \text{Sa} \times \frac{\text{Cl}}{\text{OM}} \times 100 \qquad (6.14)$$

where F_r is the soil factor related to rillability, OM is the organic matter content, Sa is the percentage of sand, and Cl is the percentage of clay.

Soil structure and soil strength. Structurally stable soils have lower rillability than unstable soils. Soils of massive structure are also less easily rilled than those with loose, friable structure. Rill erosion depends on the resistance of aggregates to detachment and transport by running water. It was the importance of aggregate stability to running water that led to the development of wet-sieving techniques (Baver, 1935; Yoder, 1936; Bryan, 1969, 1974, 1976).

Gully Erosion

Rills are channels small enough to be obliterated by normal tillage operations; large erosion channels that cannot be removed by tillage are called *gullies* (Plate 6.3). A large gully is also called a *ravine*. A gully channel may be U or V shaped depending on the strength of the subsoil's resistance or its resistance to water's cutting action. U-shaped gullies are formed when the surface and subsoil material are uniformly weak. V-shaped gullies are formed when the subsoil is more resistant to erosion than the surface soil. Gully formation has been attributed to removal of vegetation cover, overgrazing, and other social factors.

An important factor of severe gully erosion is lack of provision for safe disposal of water from roads, construction sites, public buildings, and footpaths. Water runoff from a newly constructed road made up and down the slope often initiates severe gullying (Plates 6.4 and 6.5). A drainpipe installed with no provision to safeguard the soil against the cutting action of water leads to severe gullying (Plate 6.6). Footpaths made up and down a slope readily turn into gullies (Plate 6.7). Overgrazing of steep hills nearly devoid of protective vegetation cover is also a serious cause of gully erosion (Plate 6.8).

Processes involved in transition from rill to gully erosion are not well understood. Schumm (1977) advocated the "threshold concept" in relation to gully erosion. Gully erosion initiates where the longitudinal profile of an alluvial valley floor has become too steep from sediment deposition. There is a critical gra-

Plate 6.3 Channels that cannot be removed by tillage are called gullies. Gully head on Jos plateau in Nigeria.

dient for a given discharge that initiates a gully. For different parent materials and land uses, a geomorphic threshold exists for gully initiation. The gully advances by undercutting, e.g., "the removal of material by running water at the base of a steep slope or a cliff by water falling in a stream." Once initiated, a gully propagates farther because undercutting steepens the slope by producing an overhanging cliff.

Gully erosion is a serious problem in loess soils in the United States (Piest et al., 1975) and China (Kung and Chiang, 1977), in alluvial soils of India, and in soils derived from alluvial deposits in southeastern Nigeria (Plate 6.9).

Pipe or Tunnel Erosion

Pipe or tunnel erosion is caused by subsurface flow of two types: the low and diffuse matrix flow called *interflow* or the quick, pipe or tunnel flow. It is pipe flow that removes and transports soil particles. Pipe flow may be initiated by existing cracks, rills, or animal burrows. Piping provides a macropore network for quick transmission of throughflow. The process of removing soil materials through quick subsurface flow or pipes developed by seepage water is called *piping* (Plate 6.10). The subsurface flow is the part of water that has already infiltrated the soil surface and moves through preferential channels that eventually develop into pipes. Soft material of low resistance is gradually detached and carried away, widening the pipes. Adjacent pipes become interconnected, and the roof may cave in, transforming the pipes into gully erosion. Piping is associated with certain soil and site properties, e.g., drainage, depth to impermeable horizons, permeability, and slope angle (Gerrard, 1981). Soils prone to piping are those with coarse-textured surface horizons and impermeable layers

Erosion Processes over a Watershed   173

Plate 6.4 Water runoff from roads leads to gullying.

at shallow depths. Two important processes lead to piping in these soils, i.e., eluviation of clay with rapid subsurface flow and desiccation (Hughes, 1972; Rathjens, 1973). Although enlargement of desiccated cracks is the principal mechanism in cohesionless soils (Parker, 1964), topographic factors play an important role in pipe erosion (Burt and Anderson, 1980; Anderson and Burt,

Plate 6.5 Unplanned drain outlet on newly constructed roads often leads to severe gullying.

174   Basic Processes

Plate 6.6 Faulty drain outlet from highways causes gully erosion.

1982; Anderson and Kneede, 1980). Pipe flow occurs in all slopes including convex ones. The network of tunnels developed in areas even farther from the channel effectively increases the stream network and provides a quick response to rainfall and storm channel flow. Pipe flow occurs in fluvial landscapes in a wide range of rainfall regimes (Drew, 1982; Imeson et al., 1982). Pipe flow can

Plate 6.7 Footpaths made up and down a slope readily turn into gullies.

Plate 6.8 Overgrazing is a cause of gully erosion.

occur at velocities as high as that of the surface flow (Yair et al., 1980; Jones, 1981).

Mass Movements

Mass movement of soil occurs on steep slopes under the influence of gravity. The term *mass movement* (or soil movement en masse) refers to processes that involve transfer of slope-forming materials from higher to lower grounds under the influence of gravity and without the primary assistance of a fluid. Various types of mass movement are described on the basis of the type of materials or

Plate 6.9 Severe gullying occurs in soils derived from sandstone parent material in Nigeria.

the type of movement involved. Rapp (1963) distinguished three movement types: fall, slide, and flow. On the basis of the material involved, there are four types of mass movement; debris slide, debris avalanche, debris flows, and mud flows. While debris slide movement occurs as slide, the other three are flows. In comparison, Carson and Kirkby (1972) also described three main types of movements: slide, flow, and heave, and they classified different mass movement processes. Brundsden (1979) described four main types of mass movement: creep, flow, slide, and fall. Some mass movements involve a combination of two or all types of flows.

There are two major causes of mass movement: a decrease in shear stress and an increase in shear resistance. Both lead to hillslope failure on steep slopes (Terzaghi, 1960).

Brundsden (1979) described different modes of hillslope failure:

1. *Falls.* Fall implies free movement of material from a steep slope (Plate 6.10). The various causes of rock or soil fall include undercutting by erosion.

2. *Plane failures.* Plain failures occur along highly ordered structures; they may be circular, plane, wedge, or toppling failure.

3. *Slides.* Landslides are relatively rapid movements of slope-forming materials in which failure of one or more discrete surfaces limits and defines the failed mass. Slides may be single, multiple, or successive (Hutchinson, 1968). Mudslides are relatively slow-moving, lobate or elongate masses of softened, argillaceous debris that advance chiefly by sliding on discrete, boundary shear surfaces. Mudslides usually occur at slopes of 5° to 15°.

4. *Debris flows.* Such flows represent a transitional set of processes between stream flow (or mass transport) and the drier form of mass movement. De-

Plate 6.10 An example of piping leading to gullying in Brazil.

bris flow involves mass movement of a wet mixture of granular solids, clay minerals, water, and air under the influence of gravity. Their three main categories are catastrophic mud flows, simple hillside debris flows, and flows that originate in small mountain catchments.

5. *Creep.* Gravity-influenced slow movement of soil on hillslopes at a fairly constant rate is called *creep.*

Two important factors of mass movement on steeplands are the soil strength and groundwater conditions. Both factors are influenced by vegetation cover; both factors, in turn, influence soil strength.

Mass movement is triggered by perturbations caused by people. Downward movement of soil and debris is generally slow even on steep slopes if the land is protected by the natural vegetation cover. Mass movement occurs when the land-climate-vegetation equilibrium is disturbed by deforestation, fire, overgrazing, construction, farming, or mining practices.

Stream Bank Erosion

Erosion along the banks of streams and rivers is caused by the force of running water and by undercutting. Fertile soils in flood plains are damaged by stream bank erosion. Fluvial processes involved in stream bank erosion are discussed by Thornes (1980) and Derbyshire et al. (1981).

Coastal Erosion

Coastal erosion is caused by the energy of ocean waves striking against the coastline. Comprehensive treatises on this subject are provided by Clark (1979) and Derbyshire et al. (1981).

References

Anderson, D. M, and Linville, A. 1962. Temperative fluctuations at a welting front. I. Characteristic temperature-time curves. *Soil Sci. Soc. Am. Proc.* 26:14–18.
Anderson, M. G., and Burt, T. P. 1982. Throughflow and pipe monitoring in the humid temperate environment. In: *Badland Geomorphology and Piping* (R. B. Bryan and A. Yair, eds.), Geobooks, Norwich, England, pp. 337–353.
Anderson, M. G., and Kneede, P. E. 1980. Topography and hillslope soil water relationships in a catchment of low relief. *J. Hydrol.* 47:115–128.
Atkinson, T. C. 1978. Techniques for measuring surface flow in hillslopes. In: M. J. Kirkby (ed.), *Hillslope Hydrology,* J. Wiley and Sons, Chichester, U.K., pp. 73–120.
Baver, L. D. 1935. Factors contributing to the genesis of soil micro-structure. *Am. Soil Survey Assoc. Bull.* 16:55–56.
Bodman, G. B., and Coleman, L. A. 1944. Moisture and energy conditions during downward entry of water into soils. *Soil Sci. Soc. Am. Proc.* 8:116–122.
Bruce-Okine, E., and Lal, R. 1975. Soil erodibility as determined by raindrop technique. *Soil Sci.* 119:149–157.
Brundsden, D. 1979. Weathering. In: *Process in Geomorphology* (C. Embleton and J. Thomas, eds.), pp. 73–129, Edward Arnold, London.
Bryan, R. B. 1969. The relative erodibility of the soil developed in the peak district of Derbyshire. *Geograf. Ann.* 51A:145–149.

Bryan, R. B. 1974. Water erosion by splash and wash and the erodibility of Albertan soils. *Geograf. Ann.* 59A:159–181.

Bryan, R. B. 1976. Consideration on soil erodibility and sheetwash. *Catena* 3:99–112.

Bubenzer, G. D., and Jones, B. A. 1971. Drop size and impact velocity effects on the detachment of soils under simulated rainfall. *Trans. ASAE* 14:624–628.

Burt, T. P., and Anderson, M. G. 1980. Soil moisture conditions on an instrumented slope. In: *Atlas of Drought in Great Britain* (J. C. Doornkamp and K. J. Gregory, eds.), Geobooks, Norwich, U.K., vol. 44.

Carson, M. A., and Kirkby, M. J. 1972. Hillslope form and process. Cambridge University Press, Cambridge.

Charreau, C., and Nicou, R. 1971. L'amelioration du profil cultural dans les sols sableux et sablo-argileux de la zone tropicale sèche ouest-africaine et ses incidences agronomiques. *L'agron. Trop.* 26:209–255.

Chow, V. T. 1964. *Handbook of Applied Hydrology,* McGraw-Hill, New York.

Clark, M. 1979. Marine processes. In: *Process in Geomorphology* (C. Embleton and J. Thomas, eds.), pp. 352–377, Edward Arnold, London.

Collis-George, N., and Lal, R. 1970. Infiltration into columns of swelling soils as studied by high speed photography. *Aust. J. Soil Res.* 8:195–207.

Collis-George, N., and Lal, R. 1971. Infiltration and structural changes as influenced by initial moisture content. *Aust. J. Soil Res.* 9:107–116.

Collis-George, N., and Lal, R. 1973. The temperature profiles of soil column during infiltration. *Aust. J. Soil Res.* 11:93–105.

Cook, H. L. 1946. The nature and controlling variable of the water erosion process. *Soil Sci. Soc. Am. Proc.* 1:487–494.

Derbyshire, E., Gregory, K. J., and Hails, J. R. 1979. *Geomorphological Processes.* Butterworths, London.

Derbyshire, E., Gregory, K. J., and Hails, J. R. 1981. *Geomorphological Processes.* Butterworths, London.

Drew, D. P. 1982. Piping in the Big Muddy Badlands, southern Saskatchewan, Canada. In: *Badland Geomorphology and Piping* (R. B. Bryan and A. Yair, eds.), pp. 293–304, Geobooks, Norwich, England.

Dunne, T. 1978. Field study of hillslope flow processes. In: *Hillslope Hydrology* (M. J. Kirkby, ed.), pp. 227–293, Wiley, Chichester, U.K.

Emmett, W. W. 1978. Overland flow. In: *Hillslope Hydrology* (M. J. Kirkby, ed.), pp. 145–176, Wiley, Chichester, U.K.

Farmer, E. E. 1973. Relative detachability of soil particles by simulated rainfall. *Soil Sci. Soc. Am. Proc.* 37:629–633.

Foster, G. R. 1982. Modeling the erosion process. In: *Hydrologic Modeling of Small Watersheds* (C. T. Hahn, H. P. Johnson, and D. L. Brakensiek, eds.), pp. 297–380, ASAE Monographs, St. Joseph, Mich.

Foster, G. R., and Meyer, L. D. 1972. A closed-form soil erosion equation for upland areas. In: *Sedimentation* (H. W. Shen, ed.), pp. 1–13, Colorado State University, Fort Collins.

Foster, G. R., and Meyer, L. D. 1975. Mathematical simulation of upland erosion by fundamental erosion mechanics. In: *Present and Prospective Technology of Predicting Sediment Yield and Sources,* ARS-40, USDA, Washington, pp. 190–207.

Gerrard, A. J. 1981. *Soils and Land Forms: An Integration of Geomorphology and Pedology,* Allen and Unwin, London.

Gregory, K. J., and Walling, D. E. 1973. *Drainage Basin Form and Process: A Geomorphological Approach,* Edward Arnold, London.

Hack, J. T., and Goodlett, J. G. 1960. *Geomorphology and Forest Ecology of a Mountain Region in the Central Appalachians.* U.S. Geological Survey, Professional Paper, no. 347, Reston, Va.

Horton, R. E. 1945. Erosional development of streams and their drainage basins: Hydrophysical approach to quantitative morphology. *Bull. Geol. Soc. Am.* 56:275–370.

Hudson, N. W. 1981. Instrumentation for studies of the erosive power of rainfall. In: *Erosion and Sediment Transport Measurement Symposium,* pp. 383–390, IAHS Publ. 133, Washington.

Hughes, P. J. 1972. Slope, aspect, and tunnel erosion in the loess of Banks Peninsula, New Zealand. *NZ J. Hydrol.* 11:94–98.
Hutchinson, J. N. 1968. Mass movement. In: *Encyclopedia of Geomorphology* (R. W. Fairbridge, ed.), pp. 688–695, Reinhold, New York.
Imeson, A. C., Kwaad, F. J. P. M., and Verstrataen, J. M. 1982. The relationship of soil physical and chemical properties to the development of badlands in Morocco. In: *Badlands Geomorphology and Piping* (R. B. Bryan and A. Yair, eds.), pp. 47–70, Geobooks, Norwich, England.
Jones, J. A. A. 1981. *The Nature of Soil Piping: A Review of Research*, Geobooks, Norwich, England.
Jones, M. J., and Wild, A. 1975. *Soils of the West African Savanna*, Commonwealth Bureau of Soils, Harpenden, U.K.
Kirkby, M. J. 1980. Modelling water erosion processes. In: *Soil Erosion* (M. J. Kirkby and R. P. C. Morgan, eds.), pp. 183–216, Wiley, Chichester, England.
Kung, S., and Chiang, T. 1977. Soil erosion and its control in small gully watersheds in the rolling loess area on the middle reaches of the Yellow River, Beijing, China.
Lal, R. 1981. Analyses of different processes governing erosion by water in the tropics. In: *Erosion and Sediment Transport Measurement Symposium*, pp. 351–364, IAHS-AISH Publ. 133, Florence, Italy.
Laursen, E. M. 1958. The total sediment load of streams. *Proc. Am. Soc. Civil Eng.* 84 (HY1).
Lettanzi, A. 1973. Influence of straw mulch rate and slope steepness on inter-rill detachment and transport of soil. M.Sc. thesis, Purdue University, W. Lafayette, Ind.
Lettanzi, A. R., Meyer, L. D., and Baumgardner, M. F. 1974. Influence of mulch rate and slope steepness on interrill erosion. *Soil Sci. Soc. Am. Proc.* 38:946–950.
Mazurak, A. P., and Mosher, P. N. 1970. Detachment of soil aggregate by simulated rainfall. *Soil Sci. Soc. Am. Proc.* 34:789–800.
McDonald, R. C., Isbell, R. F., Speight, J. G., Walker, J., and Hopkins, M. S. 1985. Hardsetting soils. *Australian Soil and Land Survey Field Handbook*, CSIRO, Canberra, Australia.
Meyer, L. D. 1981. How rain intensity affects interrill erosion. *Trans. ASAE* 24:1472–1475.
Meyer, L. D., Foster, G. R., and Romkens, M. J. M. 1975. Source of soil eroded by water from upland slopes. In: *Present and Prospective Technology for Predicting Sediment Yield and Sources*, ARS-S-40, USDA, Washington, pp. 177–189.
Meyer, L. D., and Monke, E. J. 1965. Mechanics of soil erosion by rainfall and overland flow. *Trans. ASAE* 8:572–577, 580.
Moeyersons, J., and de Ploey, J. 1976. Quantitative data on splash erosion, simulated on unvegetated slopes. *Z. Geomorph. Suppl. Bd.* 25:120–131.
Nijhawan, S. D., and Olmstead, L. B. 1947. The effect of sample pre-treatment upon soil aggregation in wet sieve analysis. *Soil Sci. Soc. Am. Proc.* 12:50–53.
Northcote, K. H., Hubble, G. D., Isbell, R. F., Thompson, C. F., and Bettany, E. 1975. *A Description of Australian Soils*, CSIRO, Canberra, Australia.
Panabokke, C. R., and Quirk, J. P. 1957. Effect of initial water content on stability of soil aggregates in water. *Soil Sci.* 83:185–195.
Parker, G. G. 1964. Piping, a geomorphic agent in landform development of the drylands. *IAHS Bull.* 65:103–113.
Piest, R. F., Kramer, L. A., and Heinemann, H. G. 1975. Sediment movement from loessial watersheds. In: *Present and Prospective Technology for Predicting Sediment Yield and Sources*, ARS-S-40, USDA, Washington, pp. 130–141.
Poesen, J. 1981. Rainwash experiments on the erodibility of loose sediments. *Earth Surface Processes Landforms* 6:285–307.
Rapp, A. 1963. The debris slides at Ulradal, western Norway: An example of catastrophic slope processes in Scandinavia. *Nachr. Akad. Wiss.* 13:195–210.
Rathjens, C. 1973. Subterrane Abtrazeng (piping). *Z. Geomorph. Suppl.* 17:168–176.
Robinson, D. O., and Page, J. B. 1950. Soil aggregate stability. *Soil Sci. Soc. Am. Proc.* 15:25–29.

Schumm, S. A. 1977. *An Experimental Study of Geomorphic Thresholds*. Final report, Colorado State University, Fort Collins.

SCSA, 1982. *Resource Conservation Glossary,* Soil Conservation Society of America, Ankeny, Iowa.

Sinclair, J. 1985. Crusting, soil strength and seedling emergence in Botswana. Ph.D. thesis, Aberdeen University, U.K.

Terwindt, J. H. J., Brensers, H. N. C., and Svasek, J. N. 1968. Experimental investigations on the erosion sensitivity of a sand clay lamination sedimentology. *Sedimentology* 11:105–114.

Terzaghi, K. 1960. *From Theory to Practice in Soil Mechanics: Selections from the Writings of Karl Terzaghi,* Wiley, London.

Thompson, A. L., and James, L. G. 1965. Water droplet impact and its effect on infiltraton. *Trans. ASAE* 28:1506–1510.

Thornes, J. B. 1980. Erosional processes of running water and their spatial and temporal controls: A theoretical viewpoint. In: *Soil Erosion* (M. J. Kirkby and R. P. C. Morgan, eds.), J. Wiley and Sons, Chichester, U.K., pp. 129–182.

Whipkey, R. Z., and Kirkby, M. J. 1978. Flow within the soil. In: *Hillslope Hydrology* (M. J. Kirkby, ed.), J. Wiley and Sons, Chichester, U.K., pp. 121–144.

Yair, A., Bryan, R. B., Lavee, H., and Adar, E. 1980. Runoff and erosion processes and rates in the Zin Valley badlands, northern Negev, Israel. *Earth Surface Processes* 5:205–255.

Yalin, M. S. 1972. *Mechanisms of Sediment Transport,* Pergamon, Oxford, England.

Yariv, S. 1976. Comments on the mechanism of soil detachment by rainfall. *Geoderma* 15:393–399.

Yoder, R. E. 1936. A direct method of aggregate analysis of soils and study of the physical nature of erosion losses. *J. Am. Soc. Agron.* 28:337–351.

Young, R. A. 1984. A method of measuring aggregate stability under water drop impact. *Trans. ASAE* 27:1351–1354.

Young, R. A., and Onstad, C. A. 1976. Predicting particle-size composition of eroded soil. *Trans. ASAE* 19:1071–1075.

Young, R. A., and Onstad, C. A. 1978. Characterization of rill and inter-rill eroded soil. *Trans. ASAE* 21:1126–1130.

Young, R. A., and Wiersma, J. L. 1973. The role of rainfall impact in soil detachment and transport. *Water Resources Res.* 9:1629–1636.

Part

3

Measurement and Prediction

Chapter 7

Erosion Measurement and Evaluation

Introduction

The measurement of soil erosion rates is a relatively young science. Some of the earlier reported data are based on measurements initiated in the first and second decades of the twentieth century. Consequently, most of the techniques used still require standardization. Furthermore, new methods are rapidly being developed.

The technique used to evaluate the erosion rate depends on the type of erosion to be monitored, the scale of measurement, and the objectives. In some cases, the type of erosion and the scale of measurement are not necessarily independent. Because the objectives and the degree of precision required are different for different scales, the equipment used and analytical techniques also vary (Table 7.1).

Scales for Measuring Soil Erosion

There are three scales of measurement: macroscale, mesoscale, and microscale. The macroscale involves hundreds to thousands of square kilometers of area and deals with streams and river basins. Some of the geographic, ecological, and regional aspects of soil erosion are studied at macroscale. For example, measurements at macroscale are used to assess denudation rates of major river basins, mountain systems, continents, and ecological regions. These assessments are necessary to evaluate sediment transport in rivers and streams and to plan development strategies at the regional or national level.

TABLE 7.1 Objectives of Measurement for Different Scales

Macroscale (river basin)	Mesoscale (agricultural watersheds)	Microscale (small field plot)
1. Evaluate the effects of climate, parent rocks, lithology, and land use on sediment transport.	1. Evaluate the effects of agricultural practices, topography, and slope length on sediment origin.	1. Evaluate processes of soil erosion in relation to soil type, rainfall characteristics, and overland flow.
2. Assess erosion potential in different geographic regions.	2. Determine relative importance of rill vs. interrill erosion.	2. Study soil-raindrop and soil-rill flow interaction.
3. Plan regional or national conservation program.	3. Assess pollution of natural waters by movement of chemicals from agricultural watersheds.	3. Determine soil erodibility and factors affecting it.
4. Evaluate soil and water resources of a region and plan regional development activities.	4. Study soil erosion–crop productivity relationship.	4. Assess effects of slope type and aspect on sediment origin.
5. Develop and validate predictive models of sediment transport through river systems.	5. Analyze predominant factors and processes of soil degradation.	5. Evaluate relative effectiveness of cultural practices and cropping systems in controlling runoff and erosion.
6. Units of measurement: mm/(km$^2 \cdot$ yr), m^3/(km$^2 \cdot$ yr), t/(km$^2 \cdot$ yr)	6. Units of measurement: kg/(ha $\cdot$ yr) or t/(ha $\cdot$ yr)	6. Soil erosion is assessed in g/(m$^2 \cdot$ yr), kg/(m$^2 \cdot$ yr), t/(ha $\cdot$ yr)

The mesoscale involves evaluation of sediment sources at the scale of farm units e.g., a few hectares to a few hundred hectares. Measurements of erosion rates at this scale are needed to evaluate the effects of farming practices, land use systems, and topographic factors (e.g., slope gradient, length, shape, and aspect) on runoff and erosion. The effects of agricultural practices on pollution of environments and eutrophication of natural waters are also assessed at this scale.

The microscale involves study of hillslope erosion at a scale of a few square meters to a few hundred square meters. The basic process governing soil splash, detachability and transportability, and initiation of overland flow and of sediment transport by rill erosion is studied at microscale. Studies at the microscale also involve assessment of soil degradation hazard in relation to changes in soil properties.

Some of the processes operating at the microscale are also studies on soil samples under more controlled conditions in the laboratory. Soil detachability and the effects on it of soil properties are studied on soil samples in the laboratory. In addition, the study of processes at meso- and microscale can be facilitated by using artificial rain. A wide range of rainfall simulators have been designed to study soil erodibility, soil detachments, and effects of depth and/or velocity of overland flow by using rainfall simulators.

The choice of appropriate scale of measurement apparently depends on the objectives. Because of the emphasis of this book on agriculture, the discussion of methodology is confined to measurement techniques at mesoscale and microscale. Only a brief description is given of basic issues involved at macroscale.

Evaluation of Erosion at Macroscale

The world's average yield of sediment and solutes by rivers is equivalent to a lowering of the earth's surface by 3 cm every 1000 years or 42 t/(km^2 · yr). The denudation rates are 27, 35, 45, 63, 96, and 600 t/(km^2 · yr) for Africa, Europe, Australia, South America, North America, and Asia, respectively (Gregory and Walling, 1973). Global maps of erosion rates have been prepared by using this approach (Fournier, 1960; Strakhov, 1967; Walling, 1984). This type of global, or regional, rate of erosion is computed by using the techniques of measurements of water runoff and sediment transport in stream, rivers, and large drainage basins. The basic principle involves monitoring sediment transport rates past a point in the river channel at the watershed outlet. The data thus obtained indicate trends in erosion, but it is difficult to determine the sediment source in relation to soil properties, landforms, or land use. Different techniques used are discussed in books on hydrology and geomorphology. For detailed and comprehensive reviews, see Gregory and Walling (1973) and Chow (1964).

Measurement of runoff and sediment transport in rivers

Runoff measurements. The major parameters that are measured to compute runoff include stage or water level, velocity, discharge, and their variations through time. A range of continuous-stage recording equipment is available, and the procedures of installation and calibration are described in books such as the *Field Manual for Research in Agricultural Hydrology* (USDA Agriculture Handbook 224). Several types of equipment are commercially available to measure the velocity of channel flow, e.g., the current meter, floats, and pendulum current meter. Different dyes and other tracers are also used to measure the velocity of channel flow. To compute the velocity, it is common to use Manning's formula (Chap. 6).

The discharge can be computed by the velocity-area technique, in which discharge is the product of the velocity and area. The channel section selected for discharge measurement must meet the following requirements: a regular and stable stream bed, velocity lines parallel and normal to the stream cross-section, velocities exceeding 10 to 15 cm/s, a depth of flow preferably in excess of 30 cm, and minimum growth of aquatic plants.

Some control structures are specifically installed to measure the discharge. Weirs, flumes, etc., are constructed to control the cross-sectional area, and the flow velocity can be computed by measuring the changes in stage through time.

Methods of installation and calibration of these structures are also discussed in the references cited above.

Sediment load. Sediment transport in channels, streams, and rivers must be assessed to compute the erosion over the watershed. The concentrations of sediment vary at different locations in the stream cross section. A range of sampling devices are available to obtain samples of suspended sediments and bed load transport (Gregory and Walling, 1973; Nordin and Mielke, 1984; Walling, 1982). Photoelectric turbidity meters and neutron or gamma probe devices are available to directly record the sediment load in flowing streams and rivers.

Different types of sediment are transported by river: suspended load, bed load, and dissolved load. Consequently, different sampling techniques are used depending on the type of sediment to be sampled. The concentration of suspended and bed load sediments varies with the discharge. Calibration curves are obtained relating sediment load to discharge for the portion of the stream representing the watershed.

Sediment yield and reservoir sedimentation

Erosion rates over a delineated watershed can be calculated if the major stream draining it passes through a well-defined reservoir. The sediments accumulated in the reservoir over the known period are converted to the erosion rate over the entire watershed. This technique has been used in Tanzania by Rapp et al. (1972) and Christiansson (1981), in Lesotho by Chakela (1981), and elsewhere (Jolly, 1982; Lambert, 1982).

The data on sediment deposition from reservoirs and stock ponds are used to compute the denudation ratio over the watershed. The small reservoirs, however, may not trap all the sediments. If the information on trap efficiency is known, the volume of sediment transported out of the watershed over a design duration can be estimated. There are many examples where this technique has been used. A revelation is the computations of denudation rates over the watershed of River Linth, Switzerland, by Lambert. Another example is based on the erosion rates computed for Tanzania by Stromquist (1981), as shown in Table 7.2. The sedimentation rates for major reservoirs in India have been reported by Dandekar (1981). The reservoir life was computed on the basis of a sedimentation rate of 257 $m^3/(km^2 \cdot yr)$. This rate, however, was exceeded in all cases. For the Kotmale reservoir in Sri Lanka in part of the Mahaweli Development, Russell (1981) estimated the suspended sediment yield at 180 $m^3/(kg^2 \cdot yr)$. In Java, Indonesia, Hardjowitjitro (1981) evaluated erosion rates on the basis of sedimentation in different reservoirs. The data in Table 7.3 show extremely high erosion rates. The sedimentation rates of major reservoirs in India are shown in Table 7.4. The denudation rate can be calculated, given the watershed area and the trap efficiency. In central India, Shaxon et al. (1981) estimated the average sediment yield per unit area on the basis of the sedimentation rate in Yashwant Sagar Dam. The erosion rate was estimated to be 4.5 $t/(ha \cdot yr)$. Dearing et al. (1981, 1982) compared the results from this technique with that of the erosion rates computed for a small watershed on the basis of the hydrologic monitoring of a stream. The data of stream monitoring produced a suspended-

TABLE 7.2 Reservoir Data, Sedimentation, and Soil Denudation Rate for Five Catchments in Semiarid Areas of Tanzania

Catchment	Catchment area, km²	Relief ratio, m/km	Annual sediment yield, m³/km²		Soil denudation rate, mm/yr	Reservoir completed (sediment survey), year	Capacity, m³	Percentage of original volume	Annual loss of capacity through sedimentation, %[a]		Expected total life of reservoir, years
Ikowa	640	730/50	191	1957–74	0.1–0.36	1957	3,807,000	100.0	1957–74	2.8	30–40
			362	1957–60		(1960)	3,110,000	81.6	1957–69	6.13	
			193	1960–63		(1963)	2,740,000	71.9	1960–63	3.23	
			111	1963–69		(1969)	2,315,000	60.8	1963–69	1.85	
			99	1969–74		(1974)	2,000,000	52.5	1969–74	1.66	
Matumbulu	(18.1) effective 15.0	257/4.4	581	1962–74	0.44–0.63	1962 (–60)[b]	333,000	100.0	1962–74	2.6	35–45
			626	1962–71		(1971)	248,500	74.6	1962–71	2.8	
			445	1971–74		(1974)	228,500	68.6	1971–74	2.0	
Msalatu	8.7	183/4.1	556	1944–74	0.44–0.62	1944 (theor.)	421,000[c]	100.0	1944–74	1.15	80–90
			623	1944–50		(1950)	388,500	92.3	1944–50	1.3	
			443	1950–60		(1960)	358,000[d]	85.0	1950–60	0.9	
			622	1960–71		(1971)	298,000	70.8	1960–71	1.3	
			536	1971–74		(1974) (theor.)	284,000[c]	67.4	1971–74	1.1	
Imagi	2.2	122/1.6	610	1934–71	0.52–0.70	1934 (–29)[e]	171,500	100.0	1934–71	0.8	120–130
			521	1934–50		(1950)	152,000	88.6	1934–50	0.67	
			659	1950–60		(1960)	146,500[d]	85.4	1950–60	0.85	
			703	1960–71		(1971)	129,500	75.5	1960–71	0.90	
Kisongo[f]	0.3	225/5.7	481	1960–71	0.45–0.64	1960	121,000	100.0	1960–71	3.7	25–30[g]
			447	1960–69		(1969)	83,600	69.1	1960–69	3.3	
			640	1969–71		(1971)	71,700	59.3	1969–71	4.7	

[a]Changes in capacity due to raised spillway and excavations have not been accounted for in this column.
[b]Katumbula reservoir was completed in 1960. In February 1961 a part of the embankment was washed away. The embankment was repaired in 1962.
[c]The spillway of Msalatu reservoir was raised by 2 ft in 1950 and by 1 ft in 1972. The spillway of Imagi reservoir was raised by 4 ft in 1932–33 and by 2 ft in 1972.
[d]Both Msalatu and Imagi reservoirs have been subject to sediment excavations, Imagi in 1952 (9000 m³) and Msalatu in 1953 (8000 m³).
[e]Imagi reservoir was completed in 1929. However, the first proper survey of the total volume of the reservoir was not undertaken until 1934.
[f]Data from Rapp, Murray-Rust, and Christiansson (1972b).
[g]In early 1974 the embankment gave way. By the end of 1974 it had not yet been repaired.

SOURCE: Stromquist, 1981.

TABLE 7.3 Transported Sediment in Several Rivers in Java

River	Size of watershed, km²	Transported sediment, 10⁶ t/yr	Sediment, t/(km² · yr)	Years of observation
Cimanuk	3200	25	7,810	1958–1969
Cipeles	410	2	4,880	1948–1969
Cilutung	600	7.2	12,000	1948–1969
Citeruh	250	2.8	11,200	1948–1969
Citandug	2540	9.44	3,740	1973–1974
Cimuntur	578	1.75	3,030	1973–1974
Cijolang	382	0.73	1,910	1973–1974
Cikawung	550	1.90	3,450	1973–1974
Ciseel	190	0.28	1,470	1973–1974
Bengawan Solo	5782	13.77	2,280	1952–1971
Kali Madium	3907	8.20	2,100	1952–1971
Kali Brantas	8460	6.22	957	1951–1970

SOURCE: Hardjowitjitro, 1981.

sediment yield of about 17 kg/(ha · yr) that was estimated to be about 10 percent of the total output. The erosion rate calculated on the basis of sediments deposited in the lake was about 100 kg/(ha · yr). The sediment lake data, however, represented erosion over a time scale of 10 to 100 years whereas the stream monitoring represented a time scale of 1 to 10 years. It is important to know the land use history to be able to relate the sedimentation data to past land use.

One of the difficulties of converting sediment deposited in lakes and reservoirs to erosion rates is the lack of precise measurements of the bulk density of the sediments. The bulk density of reservoir sediments varies widely (Dunne, 1977), and erroneous results are obtained by using one density for all sediments.

Regression models. Various empirical models have been developed to relate sediment yield to the characteristics of rainfall, runoff, and watershed. Fournier (1960) related sediment yield to the climatic coefficient p^2/P, altitude, and slope by

$$\log E = 2.65 \log \frac{p^2}{P} + 0.46 \log H \tan \Phi - 1.56 \qquad (7.1)$$

where E is suspended sediment yield [t/(km² · yr)], H is the mean height, Φ is the mean slope in a drainage basin, p is the rainfall (mm) in the month with greatest precipitation, and P is the mean annual precipitation. Douglas (1967) developed a similar relationship relating suspended load to p^2/P for the watersheds in eastern Australia:

$$\log E = 0.737 \log \frac{p^2}{P} + 0.380 \qquad r = 0.69 \qquad (7.2)$$

Relief is an important factor in sediment transport. For example, in the catchment of Irrawaddy in Burma, the sediment load passing Mandalay is 75.5 m³/(km² · yr), but below Mandalay the sediment rate of 347.0 m³/(km² · yr) is derived from steep relief at Chindwin (Douglas, 1967). Fournier (1960) also developed two regression equations, one for low and one for high relief. The specific equations for humid tropical regions are

$$E_L = \frac{27.12 p^2}{P} - 475.4 \qquad (7.3)$$

$$E_H = \frac{52.49 p^2}{P} - 513.21 \qquad (7.4)$$

where E_L and E_H refer to low and high relief, respectively. On the basis of data from nine rivers in southeast Australia, Douglas (1967) related the suspended-sediment yield to the mean annual runoff as

$$\log E = 0.527 \log Qy + 0.0668 \qquad r = 0.498 \qquad (7.5)$$

where E is the suspended-sediment yield [m³/(km² · yr)] and Qy is the runoff (mm). Sediment yield is also related to rainfall because the sediment yield is a result of the erosive influence of rainfall:

TABLE 7.4 Sedimentation Rates for Major Reservoirs in India

Site no.	Name of reservoir	Catchment area, km²	Year of impounding	Observed average sedimentation rate, m³/(km² · yr)
1.	Ukai	62,225	1971	497
2.	Bhakra (Gobindsagar)	57,000	1959	595
3.	Tungbhadra	28,180	1953	601
4.	Gandhi Sagar	23,025	1960	964
5.	Nizam Sagar	21,694	1931	690
6.	Mata tila	20,729	1958	444
7.	Pancher Hill	10,690	1956	1,048
8.	Maithon	6,294	1956	1,310
9.	Lower Bhavani	4,134	1953	412
10.	Ramganga	3,120	1974	1,730
11.	Mayurakshi	1,860	1955	1,648
12.	Shivaji Sagar	892	1961	1,524
13.	Dheel Sagar	689	1911	517
14.	Sukhna Lake	44	1958	831

SOURCE: Dandekar, 1981.

$$E = \frac{1.631(0.03937P)^{2.3}}{1 + 0.0007(0.03937P)^{3.3}} \qquad (7.6)$$

where E is sediment load and P is the effective precipitation (mm), defined as the amount of precipitation required to produce a known amount of runoff under specified temperature conditions. Douglas (1968) developed an equation on the basis of a detailed investigation of 11 drainage basins in Queensland, Australia:

$$\log E = 8.41 + 2.074 \log \frac{P^2}{P} + 5.603R_b + 2.967 \log D \qquad r = 0.9 \qquad (7.7)$$

where R_b is the bifurcation ratio and D is the drainage density. Jansen and Painter (1974) proposed empirical equations relating sediment yield in tropical regions to runoff discharge, basin area, relief, and mean annual temperature:

$$\log E = 4.354 + 1.527 \log Q - 0.302 \log A + 0.296 \log R - 3.417 \log T \qquad r = 0.967$$
$$(7.8)$$

Here Q is runoff discharge (m³/km²), A is watershed area, R is relief/length ratio (m/km), and T is mean annual temperature (°C).

While using these empirical equations for the assessment of sediment yield from watersheds may give useful preliminary information, the data so obtained are merely an estimate. The equations developed in one watershed may be applicable to another only within the ecological range of the original watershed.

The problem of sediment delivery ratio. The ultimate aim is to assess mean soil erosion over the entire watershed. The sediment yield over the watershed is expressed as the erosion rate computed in t/(km² · yr). An attempt to estimate on-site erosion rates within a watershed on the basis of the sediment yield at the outlet requires knowledge of processes and sources of sediments, deposition within the watershed, and storage and transport in the channel. The amount of sediment delivered to the watershed outlet is only a fraction of the gross erosion that occurs within the watershed. The term *delivery ratio* is defined as the ratio of sediment delivered at the watershed outlet to gross erosion within the watershed. The delivery ratio depends on a wide range of geomorphologic and environmental factors including the size of the watershed, land use, vegetation, nature of flow, properties of sediment, and channel characteristics. There is generally an inverse relationship between the delivery ratio and the watershed size (Robinson, 1977). The sediment delivery ratio ranges from 5 percent for large watersheds of about 1000 km² to about 70 percent for small plots of about 0.2 ha. Trimble (1975, 1977) observed that in many large watersheds, sediment storage in the form of colluvium and alluvium may be much greater than sediment yield. Trimble reported that while upland erosion in the southeastern United States was proceeding at a rate of 95 mm per 100 years, sediment yield in rivers was only 5.3 mm per 100 years. In southeastern United States, the delivery ratio was found to vary inversely with the fifth root (0.2 power) of the watershed area (Roehl, 1962). These relationships are locale-specific and vary from one watershed to another (Rapp, 1977). For example, an apparently anal-

TABLE 7.5 Scheme of Mapping of Soil Erosion by Calculation of Visible Erosion Features

Degree of erosion	Description
0	No erosion
1	Visible erosion within the field from the higher to the lower parts; only slight erosion forms
2	Rills and furrows up to 10 cm wide and deep; erosion rills cross the limits of field
3	Rills and furrows more than 10 mm wide and deep; erosion rills cross the limits of the fields

SOURCE: Hempel, 1968.

ogous observation was made by Heusch (1980) who reported that sediment delivery ratio increased with increase in area of the watershed. These anomalies arise from the confounding effects of other geomorphologic and environmental factors (Walling, 1983). For example, the watershed size determines the time lag involved in the origin of sediments within a small plot and sediment transport out of a large watershed. The data in Fig. 7.1 show the problems of scale as highlighted by Rodriguez-Iturbe and Gupta (1983). It is apparent from the data in Fig. 7.1 that the sediment yield is a function of the runoff volume which increases with an increase in the size of the watershed. Similar problems in relation to some watersheds in Kenya were observed by Edwards (1979).

Soil surveys

Soil surveys have long been used to qualitatively estimate erosion hazard over large areas. The USDA soil classification system uses special designations to denote different types and degrees of soil erosion (Table 7.5). The USDA soil survey staff have also developed a special rating system. The descriptive classes are shown in Table 7.6. Some surveys also use the vegetation cover as an indicator of erosion rate. The possible indicators of erosion used are percentage of bare soil, canopy density, and density of ground cover. Erosion rates within the

TABLE 7.6 Classes of Erosion

Class	Description
1	No apparent, or slight, erosion
2	Moderate erosion: moderate loss of topsoil generally and/or some dissection by runoff channels or gullies.
3	Severe erosion: severe loss of topsoil generally and/or marked dissection by runoff channels or gullies
4	Very severe erosion: complete truncation of soil profile and exposure of subsoil (B horizon) and/or deep and intricate dissection by runoff channels or gullies

SOURCE: USDA Soil Survey Staff, 1951.

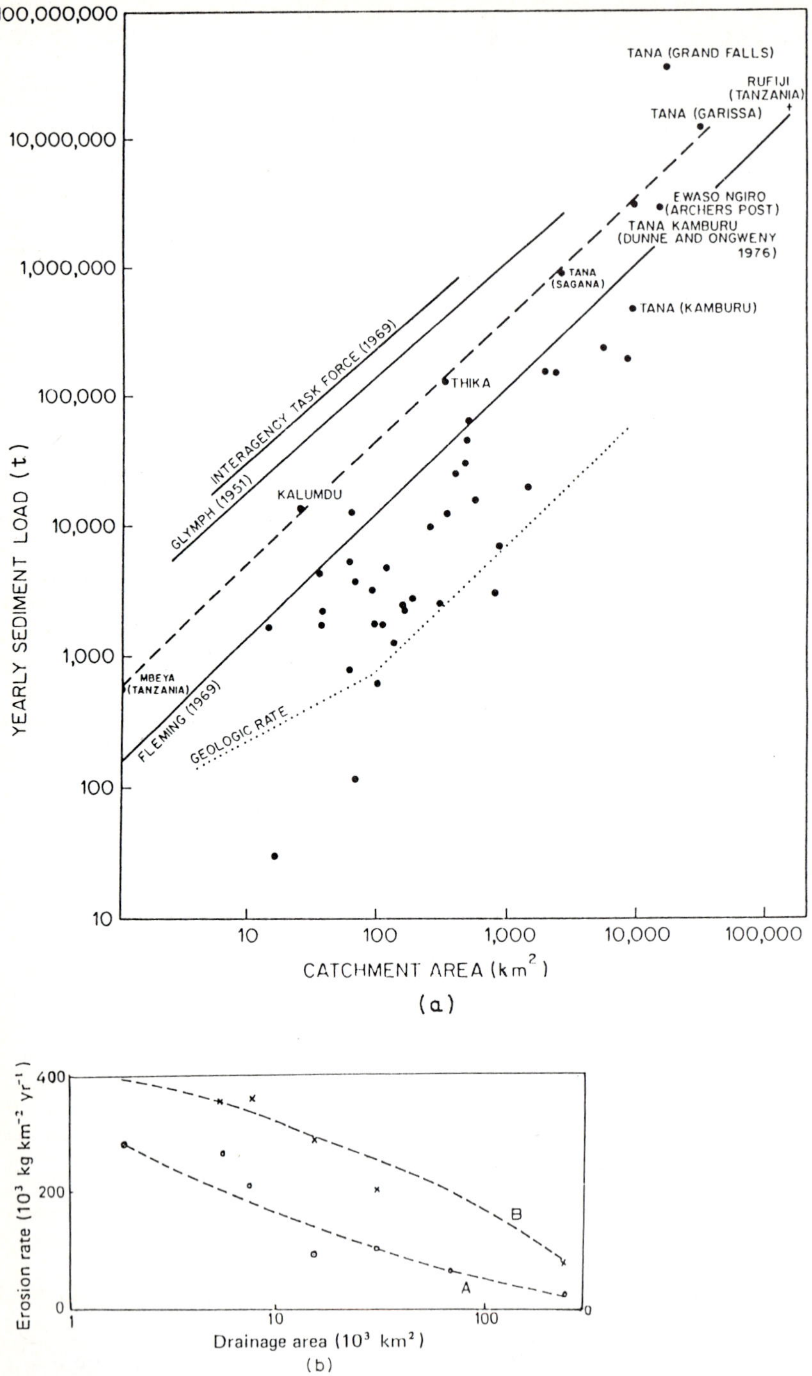

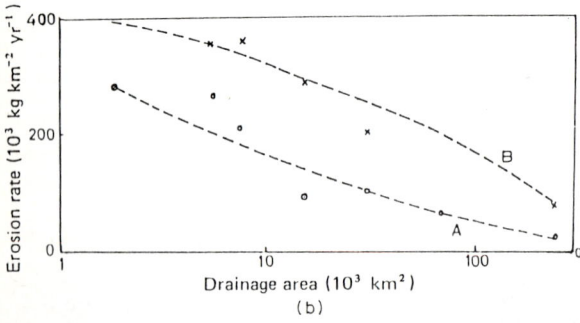

Figure 7.1 Effect of the area of drainage basin on (a) total sediment yield from some basins in Kenya (Edwards, 1979), and (b) erosion rate from the Krishna River Basin in India (curve A is the average for the 6-year period, curve B is values for 1975–1976) (Subramanian, 1982).

small plots reflect the upland erosion processes related to rainfall patterns while those from the large watershed rating system used for gully erosion are described in Chap. 15. These rating systems based on field observations are relatively crude techniques and provide only a qualitative description of the erosion hazard.

Richter (1980) used mapping techniques to assess actual and potential erosion in Germany and Czechoslovakia. He made an inventory of the soil damage by observing the pattern of sheet and rill erosion, by measuring soil deposits in closed depressions, and by investigating soil profile. Richter discussed different rating systems suitable for different types of erosion, different scales of mapping, or different size watersheds to be mapped. A comprehensive classification of erosion processes in Hungary has been developed by Salamin (1982). Within the two broad categories of erosion, static and dynamic, there are further subclassification systems (Table 7.7). On the basis of this survey, it has been estimated that 15.6 percent of the land surface of Hungary (93,000 km^2) is eroded and that a further 2.3 percent is represented by outcrops of bare rocks. A recent book by Siderius (1986) outlines in detail the procedures involved in land evaluation in relation to erosion hazard.

Such surveys and rating systems have been developed by many, and regional and national maps have been prepared. An erosion hazard map of India thus prepared is shown in Fig. 7.2. Knowles (1979) described techniques of aerial photographs and field surveys for assessment of erosion in Australia. Using this technique, a national survey of soil erosion and land degradation conducted from 1975 to 1977 estimated that 51 percent of the agricultural land in Australia required erosion control measures. In New Zealand, Cuff (1985) devel-

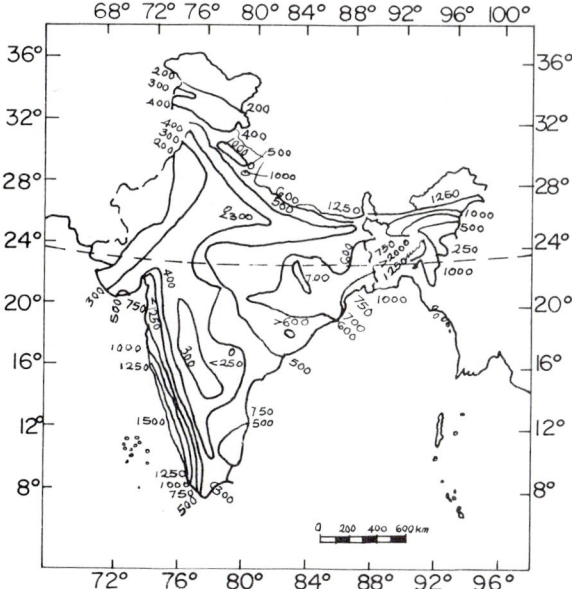

Figure 7.2 An erosion hazard map of India (Singh et al., 1985).

TABLE 7.7 Classes of Erosion Used in Hungary

1. Static state: E_S
    a. Biological weathering: E_{SB}
    b. Chemical weathering: E_{SCh} (dissolution hydration, dehydration, etc.)
    c. Physical weathering: E_{SPh}
2. Dynamic state: E_D
    Destructive (negative) processes: E_{-D}
    a. Processes localized at a point: E_p (rockfalls, slides, flows, subsistence, creep)
    b. Areal latent processes: E_{al}
        Raindrop impact or splash erosion: E_{als}
        Erosion by wash and dissolution of the soil: E_{ald}
        Erosion due to silting (puddle erosion): E_{alp}
        Erosion within the soil: E_{ali} (e.g., tunnel erosion, pothole erosion, piping, subcrustaneous erosion, and soil liquefaction)
        Attrition erosion: E_{ala}
        Ground frost erosion: E_{alf}
        Effects of ice needle: E_{SPh}
    c. Areal open processes: E_{ao}
        Surface sheet erosion: E_{ao1}
        Solifluction-gelisolifluction: E_{ao2}
        Erosion by dissolution: E_{ao3}
    d. Linear erosion: E_L
        Microrill erosion: E_{L1}
        Rill erosion (erosion in rills or shallow troughlike depressions): E_{L2}
        Ditch erosion: E_{L3}
        Gully erosion gullies, gorges, ravines, canyons, etc.: E_{L4}
        Erosional valleys: E_{L5}
        Linear ground frost erosion: E_{L6}
    e. Depositional (positive) processes associated with erosion: E_{+D}
        Areal depositional processes: E_{+a}
        Linear depositional processes: E_{+L}
        Microforms of the river bed, surfaces with flat and oblique dunes and antidunes, etc.)
        Mesoforms of the river bed: E_{+L2} (sand shoals, high beds, bottom banks, point bars, fords, deltas, alluvial cone accumulations or gravel terraces, erosion accumulation terraces, etc.)
    f. Particular erosion processes: E_p
        Shore erosion (E_p and E_L): E_{PB}
        Vertical (well-like) erosion: E_{PV} (karst and loess wells)
        Complex erosion and deflation: E_{PED}
        Particular erosion processes in lowlands: E_{PP} (E_p, E_{als}, E_{ald}, E_{alf}, E_{ao}, E_{PB}, E_{PED})
        Particular erosion processes in hilly and mountainous districts: E_{PHM}
        Human effects: E_{PA} (accelerated erosion)
        Deforestation: E_{PAL}
        Inadequate land cultivation: E_{PA2}
        Overgrazing: E_{PA3}
        Mining activity: E_{PA4} (e.g., mainly open-cast mining)
3. Great complex surface destruction: E

SOURCE: Salamin, 1982.

oped a system to map erosion on a watershed. Because mass movement is a serious problem in this region, the parameters considered include slope angle, altitude, mean annual rainfall, type and amount of bare ground surface, horizontal slope position, aspect, and geology. In the U.S.S.R., Shikula et al. (1974) proposed six intensity groups for mapping a territory according to erosion processes: group 1, no erosion; group 2, less than 0.5 mm/yr; group 3, 0.5 to 1.0 mm/yr; group 4, 1.0 to 2.0 mm/yr; group 5, 2.0 to 5.0 mm/yr; and group 6, greater than 5.0 mm/yr.

Aerial photography is increasingly being used to assess regions most susceptible to erosion (Arnoldus, 1974). In Brazil, Shaxson (1981) conducted rapid surveys to assess erosion hazard and land use incapability. In Sao Paulo municipality area, Politano et al. (1980) presented the results of an aerial survey showing the extent of various types of erosion. They estimated that erosion limits agricultural production in about 25 percent of the area. In Kenya, Muchena (1983) described the role of soil surveys and land evaluation in assessing soil erosion hazard. Heusch (1980) prepared a soil map of the Sahara in the massif region covering an area of about 1000 km^2. He showed that at least 25 percent of the soils are subject to sheet erosion and that unstable sectors of the hydrographic network suffer from gully erosion in excess of 500 m/km^2 of catchment area. Millington et al. (1982) used a flexible system of computer grid-square mapping of soil erosion in Sierra Leone. They prepared maps of potential and actual soil erosion risks, using this technique. The authors claim that this technique can be easily used to construct maps at different scales for different-size areas.

Imagery produced by Earth Resources Technology Satellites (ERTS-1 and LANDSAT-2) is also used to assess erosion hazard (Mathews et al., 1973; Weismiller and Kaminsky, 1978; Westin and Freeze, 1976; Baumgardner et al., 1978). Detailed information is available on computer-compatible tapes.

The satellite data are used for both small-scale and detailed surveys. It is possible to obtain new imagery of a specified area on the earth's surface every ninth day as a satellite passes overhead. Weismiller et al. (1985) were able to distinguish different degrees of erosion by using the LANDSAT imagery. They argued that this technique permits detection and delineation of eroding agricultural land due to water erosion. The photographs may be difficult to obtain, however, for humid regions with continuous cloud cover.

Evaluation of Erosion at Mesoscale

The mesoscale refers to farmland and agricultural watersheds. The soil survey and aerial photographic techniques can be used for this scale as well. However, the information required for the effective conservation planning needed for farmlands should be quantitative and more precise than that obtained at macroscale. Some of the commonly used techniques are described now.

Hydrologic measurements

Hydrologic control structures are used to measure surface runoff at the watershed outlet. The most commonly used structures are sharp-crested weirs, broad-crested weirs, and 90° V notch (Figs. 7.3 and 7.4, Plates 7.1 and 7.2). The

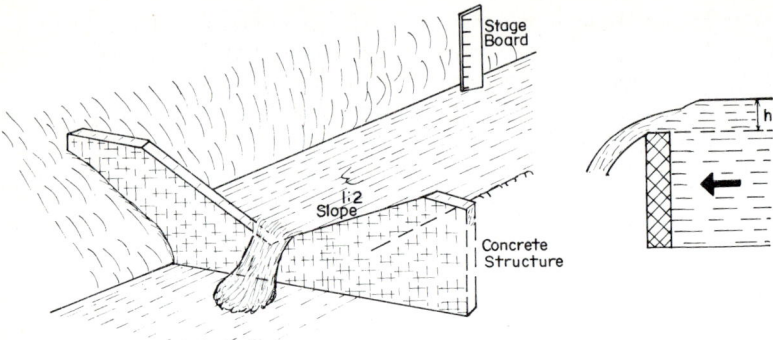

Figure 7.3 A broad-crested, triangular weir (Gregory and Walling, 1973).

large H-flume can be used for watersheds of about 5 ha (Fig. 7.5, Plate 7.3). Details of these structures are outlined in USDA Agriculture Handbook 242. Obtaining a representative sediment sample is a problem, indeed. The Coshocton wheel sampler is one of various techniques available. An example of the use of an H-flume and Coshocton wheel sampler to estimate runoff and soil erosion from a 4-ha watershed is shown in Fig. 7.5 and Table 7.8.

Radioisotopes and other tracers

Radioisotope tracer techniques have been used to identify the source of sediments over the watershed (Ritchie et al., 1974, 1975). Some researchers have proposed the use of magnetic properties to identify the areas of the watershed contributing the sediments (Walling et al., 1979). Schwertmann and Schmidt (1980) used copper as a tracer to estimate soil erosion in the hilly hop-growing area of lower Bavaria, Germany. The areas were chosen where hop has been grown for 13 to 46 years and copper has been used as an aqueous $CuSO_4$ solution as a fungicide. In undisturbed soil, most of the copper remains in the soil.

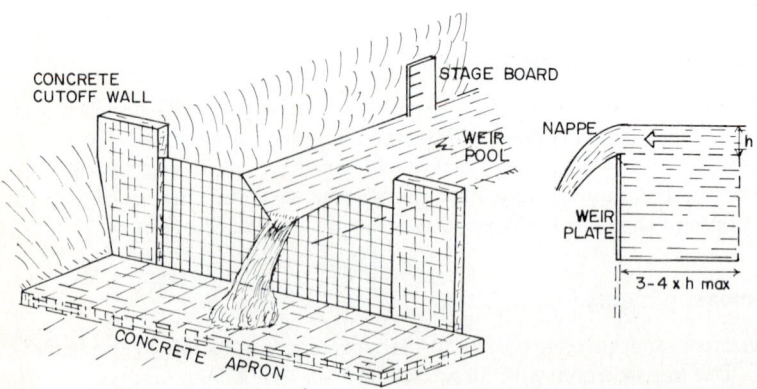

Figure 7.4 A sharp-crested weir with a 90° V notch (Gregory and Walling, 1973).

Erosion Measurement and Evaluation 197

Plate 7.1 A broad-crested weir.

The redistribution of copper along the slope was used as an indication of erosion and deposition.

The environmental tracer cesium 137 (^{137}Cs) has a half-life of 30 years and can be used to estimate erosion or deposition rates over small delineated watersheds. This isotope is produced by nuclear explosions and therefore has been

Plate 7.2 A partial flume.

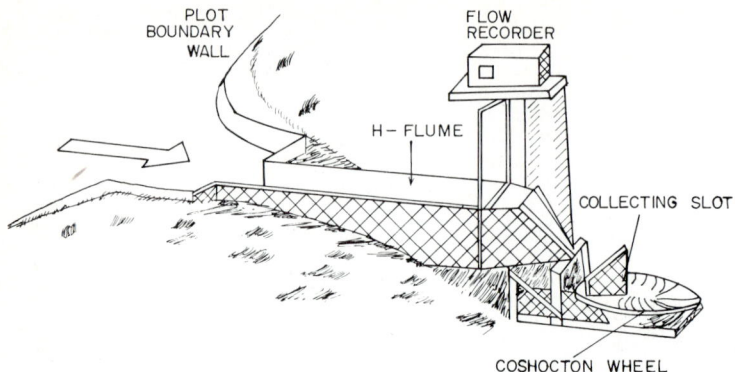

Figure 7.5 An outline of an H-flume and a Coshocton wheel sampler.

present in the environment only since the mid-1940s in the vicinity of the detonation and since 1952 globally. The ^{137}Cs is strongly absorbed by soil particles and is thus concentrated in the surface layers of undisturbed soil. If the soil particles move, the absorbed ^{137}Cs is redistributed in the surface layers. The eroded soils thus have lower concentration of ^{137}Cs than uneroded soils. Similarly, areas of sedimentation and deposition have proportionately higher concentrations. This behavior has been confirmed by the studies conducted by McHenry et al. (1973), Ritchie et al. (1974, 1975), and Mitchell et al. (1980) in the northern hemisphere and by McCallan and Rose (1977), McCallan et al. (1980), and Campbell et al. (1982) in the southern hemisphere. Longmore et al. (1983) estimated the accumulation of ^{137}Cs fallout in Brisbane from 1952 to 1978 to be 23.2 mCi/km^2. The fallout input patterns, however, vary among geographic re-

Plate 7.3 An H-flume with water-stage recorder.

TABLE 7.8 **Examples of Runoff and Erosion on Tropical Watersheds of about 4 ha Each**

Method of land clearing	Erosion, kg/ha	Total runoff, mm	Flow duration, min
Traditional farming	298.2	2.68	88
Manual—no-till (unterraced)	34.2	2.74	100
Manual—conventional till (terraced)	170.1	10.64	260
Shear blade—no-till (unterraced)	620.8	11.20	108
Tree pusher—root rake (conventional till)	3132.8	17.10	278
Tree pusher—root rake (no-till; unterraced)	1536.8	19.26	120

SOURCE: Lal, 1986.

gions due to differences in the number, nature, and location of exploded devices. The input value of fallout must therefore be estimated for each watershed of interest. The input value must also be corrected for decay to the year of sampling. The areal concentration is calculated by integrating the vertical ^{137}Cs activity profile at the site. Longmore et al. (1983) have outlined the procedure for using this technique. This method is, however, suitable for a qualitative and survey-type erosion and deposition. In contrast to the universal soil loss equation (USLE), this method takes into account both erosion and deposition.

Measuring changes in soil level

Remnants of vegetation and tree root exposure are used as indications of surface lowering by erosion. Remnants of the original soil surface, pedestals formed by stone covers, and exposed tree roots are used to estimate the depth of soil eroded (Plate 7.4, Fig. 7.6). Extreme caution is needed to use such information in estimating soil erosion. Some tree species grow with their roots exposed even on undisturbed soil while other trees with dominant taproots may show no erosion even if 20 cm of the surrounding soil was washed away. The methodology and limitations of using tree root exposure to estimate erosion rates have been discussed by Lamarche (1968).

Measurement and Evaluation at Microscale

Detailed erosion processes are studied at microscale involving an area of a few square meters to a few hundred square meters. A range of field and laboratory techniques used for these studies is briefly described now.

Field studies

Splash measurement. Measurement of soil splash due to raindrop impact has been done since the pioneering work of Ellison in the early 1940s. The design requirements for equipment have been explained by Moeyersons and de

Plate 7.4 Exposed tree roots.

Ploey (1976). The design depends on the objectives and involves a range of simple equipment such as splash boards (Ellison, 1944; Kwaad, 1977), receptacles buried in the soil (Sreenivas et al., 1947; Bollinne, 1975; Gorchichko, 1977), evaluation of the movement of painted soil particles (Kirkby and Kirkby, 1974; Lewis, 1976, 1981), radioactive tracers (Coutts et al., 1968; de Ploey, 1969), and field splash cups (Morgan, 1978, 1981). The detailed designs of the splash cup used by Bolline (1980) and Morgan (1981) are explained in Figs. 7.7 and 7.8. One of the major obstacles in measuring soil splash is the separation of the

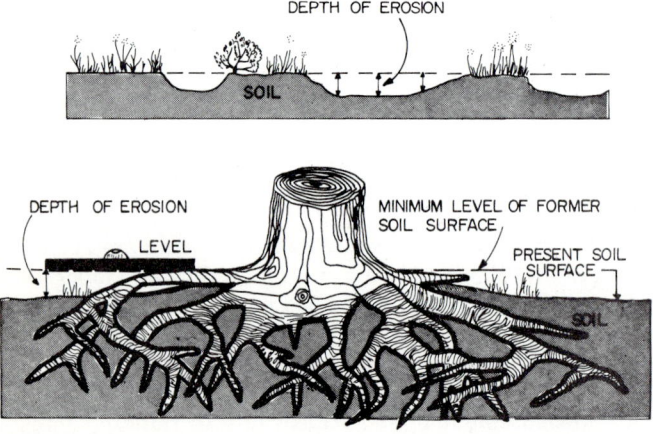

Figure 7.6 Exposed tree roots are a possible indication of the past erosion.

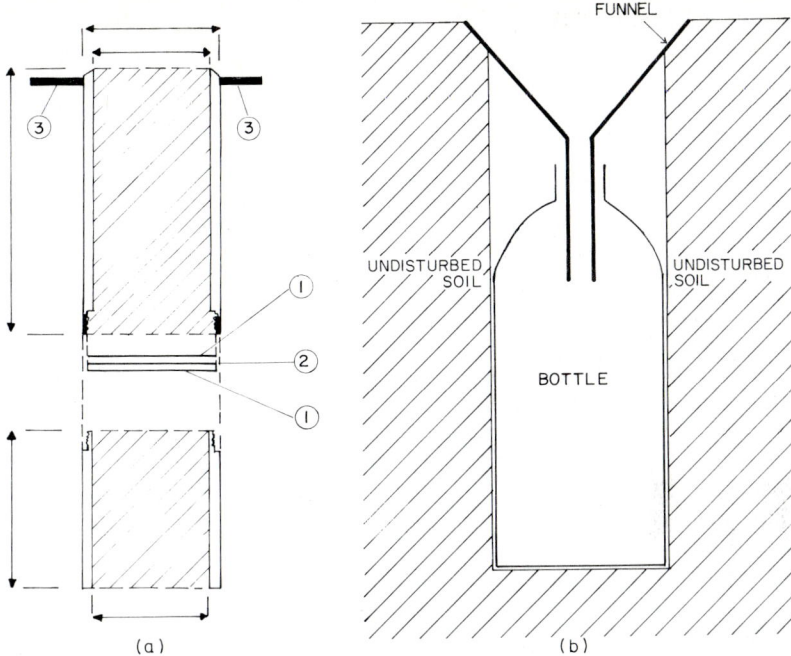

Figure 7.7 (*a*) A splash cup design: 1, stainless-steel filter; 2, paper filter; 3, stop pin. The inner cylinder is 50 mm in diameter. (*b*) An alternate splash cup design. (Bollinne, 1980.)

overland flow from the soil splash. The equipment designed must isolate the area of soil from which the splash is to be measured and must also identify the distance through which the splashed particles are transferred under the raindrop impact.

Some techniques monitor both splash and soil transport in overland flow over short distances. Young (1963) described one such technique. The technique of tracing movement of soil particles by using paints and dyes also measures sediment movement by splash and the overland flow (Lewis, 1981).

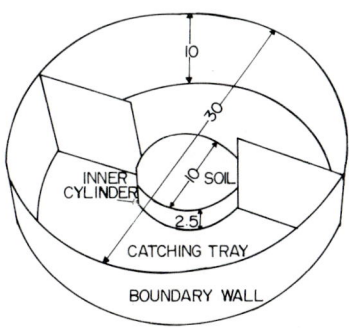

Figure 7.8 Another splash cup design. Dimensions in centimeters. (Morgan, 1981.)

202 Measurement and Prediction

Some scientists in the U.S.S.R. have attempted to monitor the height and distance of splashed particles. The height of the receptacle is carefully designed to catch all spattered particles. Gorchichko (1976) proposed a device to study the characteristics of splashing soil by raindrops (Fig. 7.9). The receptacles are installed at different levels to determine the quantity of soil splashed at the desired height.

Morgan (1981) outlined the merits and limitations of different techniques (Table 7.9). Considerable improvements are needed in most of the methods used. One indication of the improvements needed is the high variability in data obtained. The sample size is often so small that results are highly variable. For example, in Hong Kong, Lam (1977) observed a large spread of data with a coefficient of variation of 37 percent. A large sample size may reduce the variability but introduce other problems, e.g., knowing the exact site from which the particles are splashed. Furthermore, splash measured by different techniques is so different that it is hard to determine what may be the valid results. The discrepancy between results obtained by two techniques was pointed out by the data of Reeve (1982). He compared soil splash measured by two techniques. He used a splash board consisting of a vertical 30- by 60-cm sheet-metal surface to intercept splashed particles collected in a 2-cm-wide trough. The results from the splash board technique were compared with those obtained with a modified Bolline splash trap. The splash erosion measured by the splash board technique was consistently more than that measured by the Bolline trap.

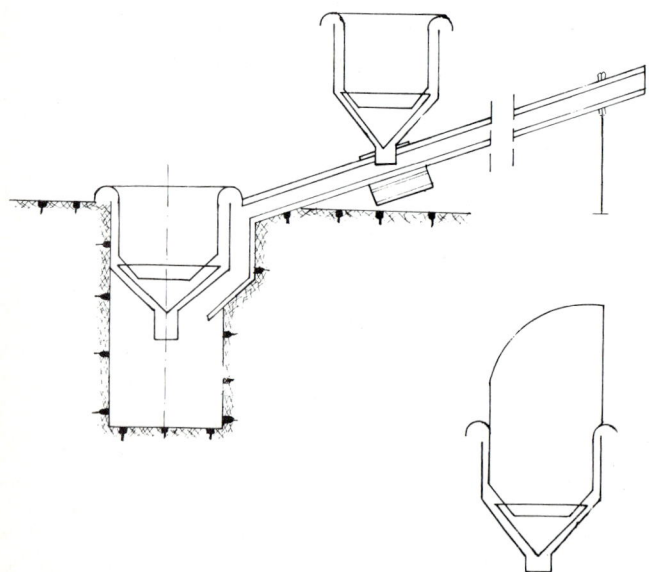

Figure 7.9 A device to evaluate the characteristics of splashing soil particles by raindrop impact (Gorchichko, 1976).

TABLE 7.9 Design Requirements for a Field Measuring Device for Splash Erosion

Method	Measurement constraints				Measurement requirements		
	Isolation of splash	Rim effect	Rainfall interference	Environmental acceptability	Total splash	Direction of splash	Height/distance of splash
Splash boards	−	0	−	+	+	+	−
Bottles/funnels	−	−	+	+	+	−	−*
Marked stones†	−	0	+	+	+	+	+
Tracers	−	0	+	−‡	+	+	+
Field splash cups	+	0	0	+	+	+	−

+ Points favoring technique
0 Points of no strong influence
− Points against technique
* Gorchichko (1977) provides data on the height of the splashed particles.
† Not feasible for soil particles.
‡ Radioactive tracers.
SOURCE: Morgan, 1981.

Measurements of interrill and rill erosion

Paints and dyes. Many researchers have estimated soil movements over slopes by using painted soil particles as tracers. These techniques are useful to evaluate the relative importance of different mechanisms involved in soil movement over short distances. Lewis (1976) measured soil creep in Puerto Rico, using this technique. He observed that the antecedent moisture, texture, temperature, throughflow, and density of plant roots play an important role in soil movement. The same technique was used to evaluate soil movement on different slopes in western Nigeria (Lewis, 1981). He observed that although a majority of rains caused soil particle movement by rain splash, only rains with intensities exceeding 2.5 cm/h caused substantial soil movement. The maximum distance of the movement of soil particles by splash downslope was about 4 m. This technique was also used by Mtakwa et al. (1987) and was found to be completely unsuitable. The painted soil particles get artificially cemented together into aggregates, and their movement is drastically different from that of the unpainted soil.

Buried nails and stakes. Repeated measurements of the ground height at prefixed stakes and pins are taken over different time intervals to estimate the erosion rates along a hillslope (Fig. 7.10). The methodology and limitations of this technique are described by Hudson (1964), Leopold et al. (1964), Temple et al. (1972), and Ciesiolka (1984). The use of washers as a reference mark may improve the precision of measurements of the change in ground height (Schumm, 1954, 1956), but may sometimes hinder the erosion process in a gentle rainstorm. Nails are usually painted bright red to improve their visibility. Millington (1981) used the erosion pin technique in Sierra Leone. Soil losses obtained

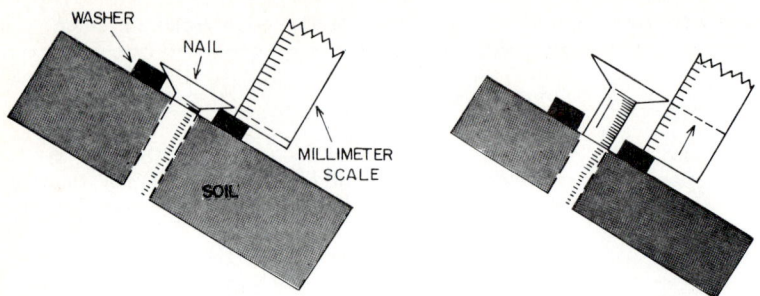

Figure 7.10 Buried-nail technique to estimate erosion along a hillslope.

from the pin measurements tended to be much higher than those obtained by plot measurement.

For alfisols in southwestern Nigeria, Mtakwa et al. (1987) compared soil erosion measured on field runoff plots with that estimated by the buried-nail technique and the rill volume method. The data in Fig. 7.11 show that the rill method was the least satisfactory and that it underestimated soil erosion. The buried-nail technique of measuring soil erosion, however, was considered satisfactory for estimating erosion on small plots. The buried-nail technique is apparently not satisfactory for measuring erosion on large watersheds or even on plots of about 0.5 ha. On small plots of 5 to 100 m^2, however, the buried-nail technique may be adequate. Mtakwa et al. (1987) observed that on small plots (20 m^2 or less) at least 20 nails are required to estimate erosion or deposition after every storm. In comparison with the runoff plot, the buried-nail technique is economical, and nails are readily available; they are easy to install and mea-

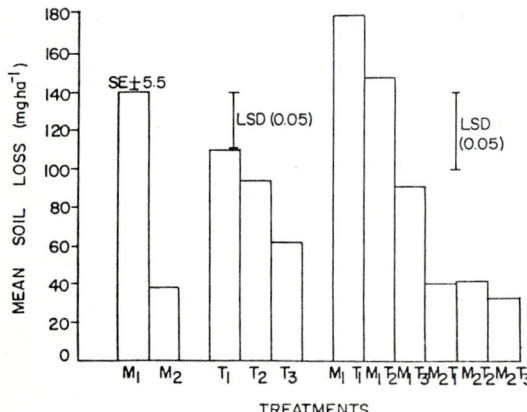

Figure 7.11 Comparison of estimate of soil erosion by rill method with that measured by other techniques. M_1 = bare-fallow plot, M_2 = maize plot, T_1 = nail technique, T_2 = conventional technique, T_3 = rill technique. SE = standard error, LSD = least significant difference. (Mtakwa et al., 1987.)

sure, do not require much maintenance, suffer less from vandalism, are easily portable, and can be used even in remote areas. There are some severe limitations of the buried-nail technique, however: the nails themselves can obstruct soil movement; cultivation, weeding, and other soil management operations can artificially displace soil around the pin or nail; measurement of ground height upslope, downslope, or on sides can introduce bias; and the effect of erosion can be easily masked by expansion in swelling soils.

Measuring volume of rills. The volume of soil eroded can be estimated by computing the volume of rills made after an erosive storm. This technique is applicable for relatively small plots 10 to 20 m^2 in area. In the U.S.S.R. Gerasimenko (1980) evaluated the accuracy of determining soil erosion by the rill method. He observed that the accuracy depends on the type of microchannels and on agricultural practices. The accuracy of determination increases with an increase in the number of measurements. The minimum number of determinations required is 30 measurement sites along the length of each small field and 5 to 10 depth measurements of each microchannel. This technique has been used in

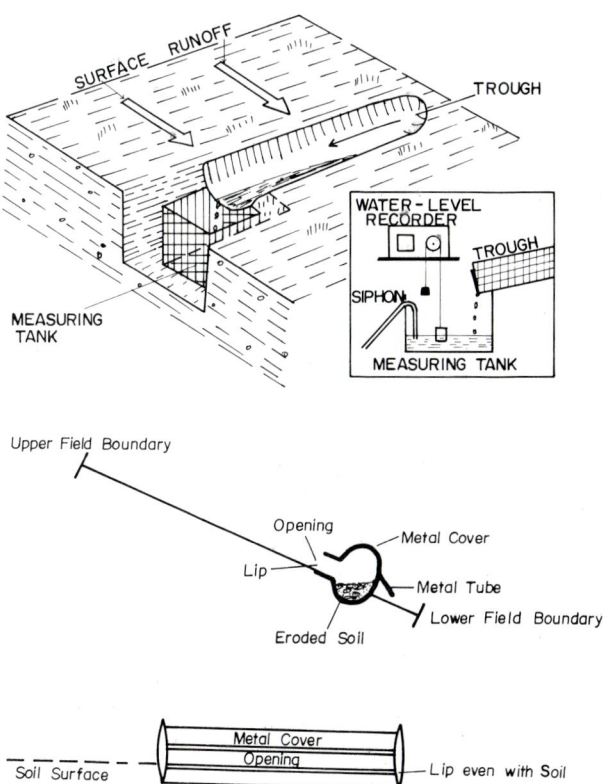

Figure 7.12 Design of troughs usually installed along a hillslope to measure erosion (Gregory and Walling, 1973; Lewis, 1984).

western Nigeria by Franzen (1986) and Mtakwa et al. (1987) to measure erosion rates after every storm. Mtakwa et al. observed that the rill method underestimated the erosion even on small plots. First, the sheet erosion or soil splash is not measured by the rill method because the overland flow does not cause rill until the critical flow is attained. Second, the slope must be long. The rill effect cannot be measured on plots of less than 10-m length. The method may be satisfactory for a large slope length, however. For example, in the U.S.S.R., Gerasimenko (1980) concluded that the method was very objective and practical for estimating erosion from plots 100 to 400 m long. For greater accuracy, he cautioned, one must have many measurement sites along the slope as well as many depth measurements within the rill cross section. The technique estimates erosion better from long than short slopes. The use of commercially available rill meters can facilitate field measurements (Curtis and Cole, 1972; McCool et al., 1976, 1981). A rill meter or a microtopographic profile gauge can be either a simple micrometer scale or an elaborate electronic device.

Trapping the soil removed. Trapping and measuring the amount of soil removed from a hillslope is a widely used technique to measure combined rill and interrill erosion. Some of the devices used for this purpose are briefly described here.

Collecting troughs. Sediments are trapped in a collecting trough installed along the contour. The trough is connected to a trap in which eroded sediments and runoff are measured. The longer the trough, the better the sample size. The boundaries of large plots are defined by topographic surveys. Pairs of traps,

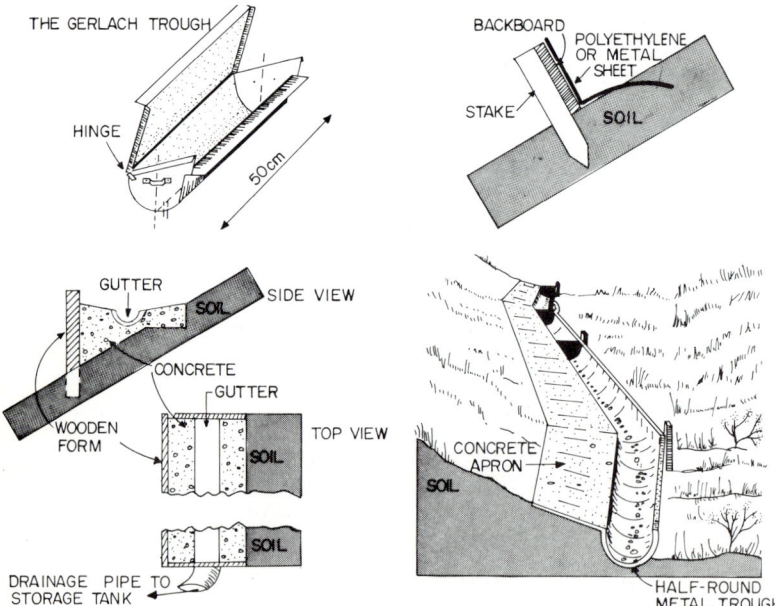

Figure 7.13 More schematics of troughs installed along a hillslope (Gregory and Walling, 1973; Lewis, 1984).

Erosion Measurement and Evaluation   207

Plate 7.5   Soil trough installed along a hillslope (U.K.).

placed a fixed (downslope) distance apart, are used to measure net soil loss or gain over the intervening portion of the slope (Townshend, 1970). A wide range of troughs are used for erosion measurement (*Revue de Geomorphologie Dynamique,* 1967; Figs. 7.12 and 7.13, Plate 7.5). The most commonly used trap is the "Gerlach trough" (Gerlach, 1967). This trough is 50 cm long, is made of sheet metal, and is installed in a shallow trench. The trough has a hinge lid to prevent rain falling into it. Anderson et al. (1959) used half-round steel troughs to measure soil movement along steep slopes. Young (1960) described another type of simple trap (Fig. 7.14) that he successfully used in Mato Grasso, Brazil,

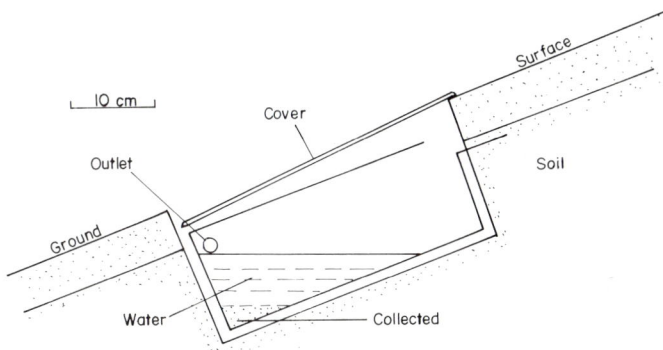

Figure 7.14   A commonly installed sampling trap along a hillslope (Young, 1960).

and elsewhere in the tropics. Best results are obtained if traps are used in conjunction with pins and stakes.

Field plots. Rather than the use of traps, runoff plots are established by defining a specified area of a hillslope which is surrounded by a cutoff wall. The downslope boundary is specially constructed to convey the surface runoff and entrained sediments through an outlet where they can be measured (Fig. 7.15). The runoff plots may vary from 5 m^2 to 0.1 ha (Van Doren et al., 1950). Small-sized plots are particularly useful to study the infiltration rate. When the plot size is big and the runoff volume too large to collect in a storage tank, a sampling device is used to collect a small portion of the aliquot sample in the storage tank. A commonly used fraction divider was developed by Geib (1933) and is called the *Geib multislot divisor* (Fig. 7.16). The runoff is usually fractioned into an odd number (3, 5, 7, 11, etc.), and the central pipe is connected to the

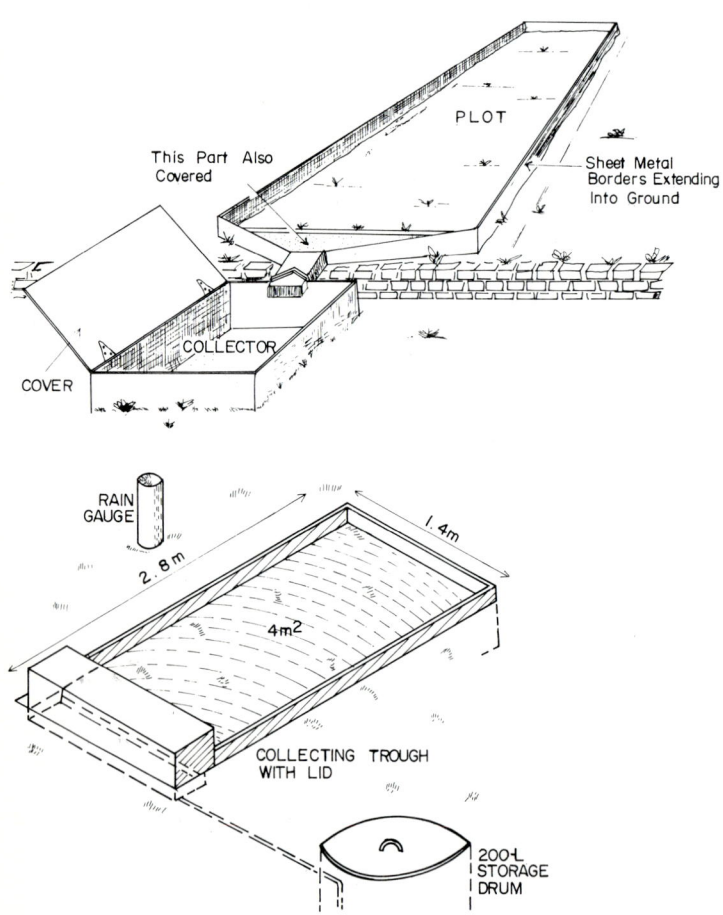

Figure 7.15 Field runoff plots of different designs (Gregory and Walling, 1973).

sample storage tank. When the plot size is large, a rate-measuring flume is used with a water stage recorder, and an aliquot sample is collected for determining suspended sediments and dissolved nutrients (Plate 7.6). The design requirements of runoff plots are described in detail by Mutchler et al. (1987).

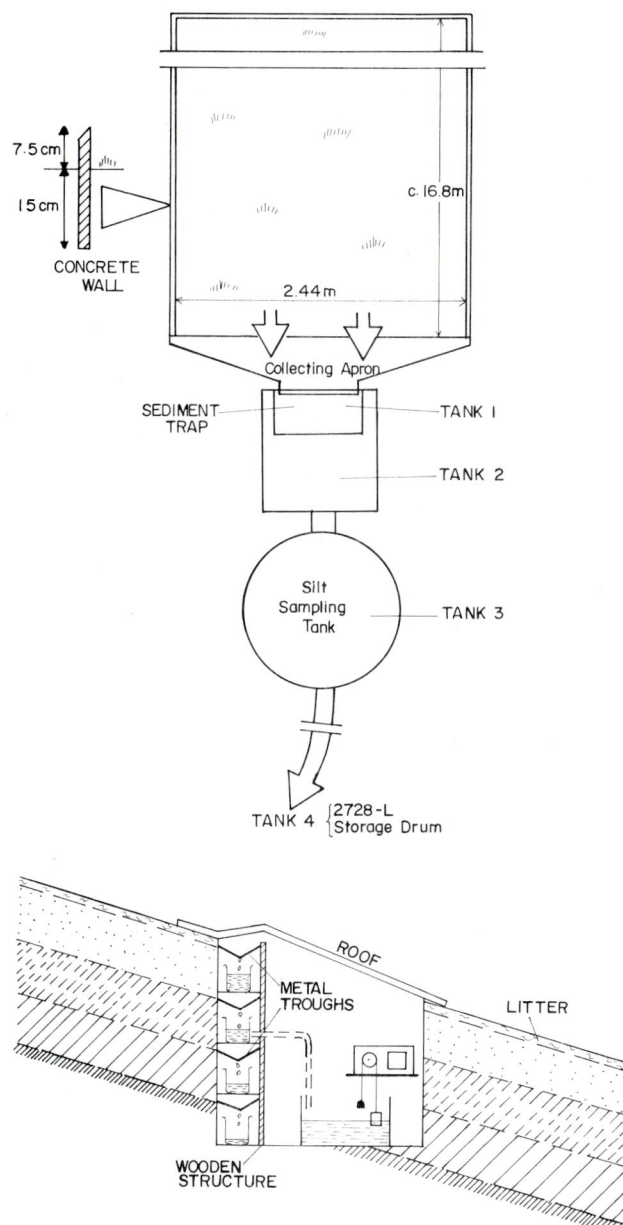

Figure 7.15 (*Continued*)

The size of runoff plots used depends on the objectives. The plot size is often less than 5 m^2 if the objective is to compare simple treatments, e.g., the effects of residue mulch on water runoff and soil loss. The size of the runoff plot may be as large as 0.1 ha if the objective is to compare the effects of agricultural practices, e.g., tillage systems, crop rotation, or an agroforest technique. Different sizes of runoff plots commonly used are shown in Tables 7.10 and 7.11. Some of the commonly used collection systems for different size plots are shown in Plates 7.6 to 7.11. Designs of field runoff plots and collection systems are described by Lal (1976) in Nigeria, Felipe-Morales et al. (1977) in Peru, Fearnside (1980) in the Transamazonia Highway colonization area of Brazil, Roose and Lelong (1976) in Ivory Coast, and Jinze (1981) in China, among others.

The dominant processes governing soil movement along the hillslope can be studied with runoff plots by using either natural or simulated rainfall. Portable rainfall simulators are designed for more intensive studies involving rainfall of different drop size distribution, intensity, amount, or frequency. The use of a rainfall simulator on runoff plots for measuring soil erosion in the tropics is described for Brazil by Mondardo and Vieira (1975) and Biscaia (1982); for Australia by Grierson and Oades (1977), Loch (1982), and Loch and Donnollan (1983); for Kenya by Barber et al. (1979); for Nigeria by Olayemi and Yadav (1983); and for Francophone countries in west Africa by Roose and Asseline (1978) and Collinet and Valentin (1985). Design criteria of simulators for use of field runoff plots have been described by Meyer (1987).

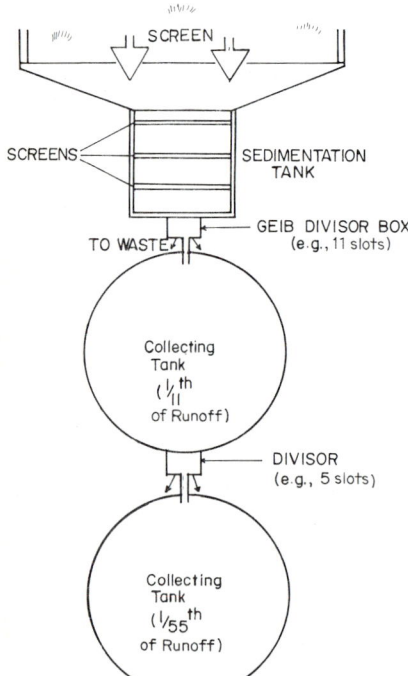

Figure 7.16 Geib multislot divisor.

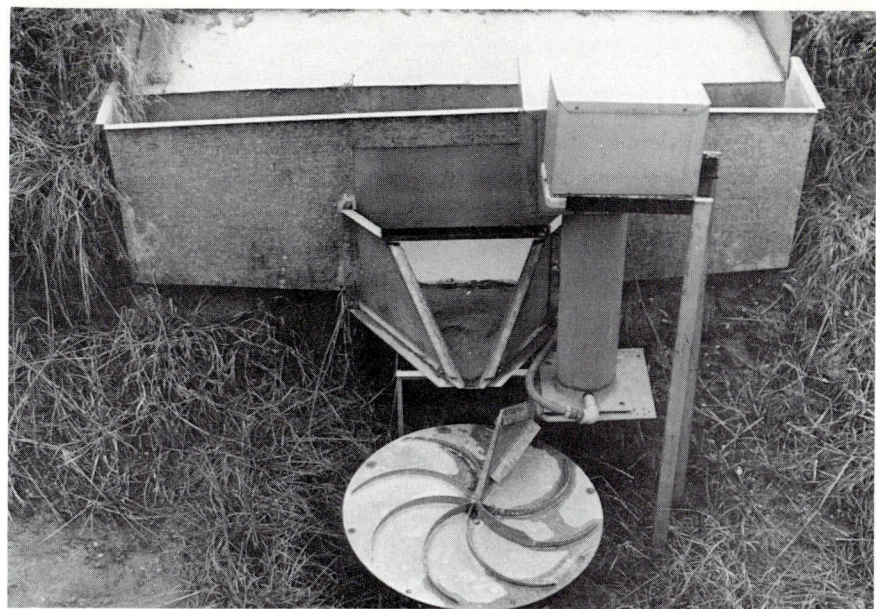

Plate 7.6 Coshocton wheel sampler.

Tracer techniques are also used in conjunction with runoff plots to study soil splash. De Ploey (1967) used scandium 46 (^{46}Sc) to study splash erosion. Hills (1971) described the use of a fluorescent dye to indicate the time required for flow to reach a known distance. Use of other tracers with low energy radiation is also possible (Pilgrim, 1966).

TABLE 7.10 Size of Bounded Runoff Plots and the Type of Experiment

Type of plot	Size	Experiment
Microplots	1–2 m²	Relative erodibility of soil
		Comparison of ground covers
		Comparison of other "no space" dependent treatments
Standard plots	40 m² (22.6 m × 1.8 m)	Evaluation of standard erodibility on a 9% slope
		Cropping and rotation experiments
Macroplots	>200 m²	Field scale operation treatments
		Tillage treatments
		Tree covers
		Mechanical protection measures

SOURCE: Chisci, 1981.

TABLE 7.11 Classification of Common Devices for Measuring Runoff and Soil Loss from Plots of Different Size

Type of plot	Plot size	For runoff		For sediment rate	
		Continuous measurement	Total for a storm or longer period	Continuous measurement	Chosen time intervals during a storm	Total for a storm or longer period
Microplots	1–2 m²	Calibrated weir, triangular or logarithm, with hydrometer or hydrograph devices	Collecting tanks	Turbidimeters	Manual sampling Pumping sampler	Collecting tanks
Standard plots	40–100 m²	Calibrated weirs with hydrograph device: HS flumes, Venturi-type flumes (i.e., Parshall) Weighing tanks with recording devices (Linsalata-Cavazza type)	Collecting tanks with divisors: Geib-type or Coshocton-type	Turbidimeters Weighing tanks with recording device	Manual sampling Pumping sampler	Collecting tanks
Macroplots	>200 m²	Calibrated weirs with hydrograph device: H and HL flumes, Venturi-type flumes (i.e., Parshall) Weighing tank with recording device (Linsalata-Cavazza type)	Collecting tanks with divisors: Coshocton-type	Turbidimeters Weighing tanks with recording device	Manual sampling Pumping sampler	Collecting tanks

SOURCE: Chisci, 1981.

Plate 7.7 Plastic tanks used in Passo Fundo, Brazil.

The Scale Problem

The hydrologic processes should be measured and evaluated at a scale at which these processes occur in nature. Making measurements at one scale and extrapolating the results to another are bound to create problems because the processes involved are different at different scales. If the erosion processes are

Plate 7.8 Storage system in Bondung, Indonesia.

Plate 7.9 Runoff collection system in Peru used by C. Felipe-Morales.

monitored at the level of splash involving a few square meters, it is illogical to extrapolate the data to even a scale of the runoff plot. The processes occurring at different scales, though interrelated, are different in magnitude, trends, and forces involved. Conceptual problems of scale involved in hydrologic studies have been discussed at length by Klemes (1983). A few attempts have been

Plate 7.10 Storage cum multidiver tank in Nigeria designed by the author.

Erosion Measurement and Evaluation   215

Plate 7.11 Cement tanks in Dominican Republic designed by T. J. Logan.

made to extrapolate the results from one scale to another (Millington et al., 1982). It is not surprising that results obtained at one scale differ from those of another scale by several orders of magnitude. In the United States, Rogowski et al. (1985) concluded that estimation of erosion on a 1-ha basis is an appropriate scale for predicting erosion hazard on large scale in mined and reclaimed lands in Appalachia. Ciesiolka and Freebairn (1982) evaluated runoff and soil loss data from Toowoomba, Queensland, Australia, at three different scales. The scales of measurements were a 0.2-ha plot within a contour discharging through a rill, a 1-ha contour catchment discharging from a graded channel into a waterway, and a 250-ha catchment leading to a stream. All three scales studied are, in fact, within the mesoscale. Nevertheless, authors observed that peak runoff rates declined rapidly with increasing size of the plot. There also occurred a large reduction in sediment concentration between the 0.2-ha rill outlet and the 1.0-ha contour catchment, with a lesser reduction in sediment con-

Plate 7.12 A multidivisor system designed by the author.

centration at 250 ha. Both the origin and the entrainment of sediment depend on not only the predominant processes involved but also the physiographic and environmental factors.

The problem of scale has been adequately highlighted by Blandford (1981) in connection with soil erosion on rangelands. He indicated that sediment yield from a watershed of a few thousand square kilometers does not tell anything about erosion from a small patch, a sideslope, or a watershed draining into a first- or second-order stream. The measurements made on large areas indicate nothing about the soil relocation within the catchment. Similarly, measurements made on small plots have little, if any, bearing on the sediments leaving a large watershed. The problem of scale poses serious questions in defining strategies for erosion control. If the results of erosion measurement are so scale-dependent, one wonders whether we understand the magnitude of the erosion problem. If not, how can we decide on the appropriate solutions to solve this problem?

Laboratory Measurements

The relative magnitude and trends in some processes at microscale can be evaluated under controlled conditions in the laboratory.

Splash by raindrop impact has been extensively studied under laboratory conditions. Ghadiri and Payne (1980) used cinephotography to study the raindrop impact. Perrens and Reeve (1982) studied the dynamics of splash transport from photographs taken by a twin flash stroboscope. In Hungary, Kerenyi (1981) conducted raindrop experiments on sand by using a Kazo-type rainfall simulator. He defined the threshold kinetic energy required to initiate splash for different grain sizes and slope angles.

Soil erodibility and factors affecting it have been extensively studied under laboratory conditions. The single-drop technique has been widely used to determine the kinetic energy required to break a structural aggregate (Bruce-Okine and Lal, 1975). Laboratory-based rainfall simulators are used to determine soil detachability and soil splash (Aina et al., 1979; Gabriels et al., 1973; Van Elslande et al., 1976). A schematic of one such simulator is shown in Fig. 7.17. The degree of erosion caused by the previous land use has been evaluated under laboratory conditions by the application of infrared spectroscopy (Chandra and De, 1974).

In the U.S.S.R., Kuznetsov (1977) and Grigorev and Kuznetsov (1976) have developed a laboratory technique to evaluate the erosion resistance of soils. From the weighted mean diameter of the aggregates and soil adhesion, the authors calculated the critical flow velocity required to detach particles. Soil adhesion, however, depended on the antecedent moisture content. This technique has an application to assess erosion potential under furrow irrigation (Grigorev et al., 1979).

Conclusions

A wide range of techniques are available to measure soil erosion. Most of these techniques have not been standardized yet, and the results are greatly influ-

Erosion Measurement and Evaluation   217

enced by the technique used. Standardization of measuring techniques is an important consideration in reliability and precision of the results obtained.

There are also serious questions regarding the scale of measurement. The results obtained are scale-dependent. Assessment of soil erosion should be done at the same scale to which the results are to be applied for erosion control. Extrapolating results from one scale to another can lead to erroneous conclusions.

The choice of technique and scale of measurement depends on the objectives. Some erosion processes (e.g., soil detachment) can be studied at microscale or even in the laboratory. Others can be studied only at the scale of an agricultural field (e.g., rill erosion as influenced by agricultural practices). The danger lies

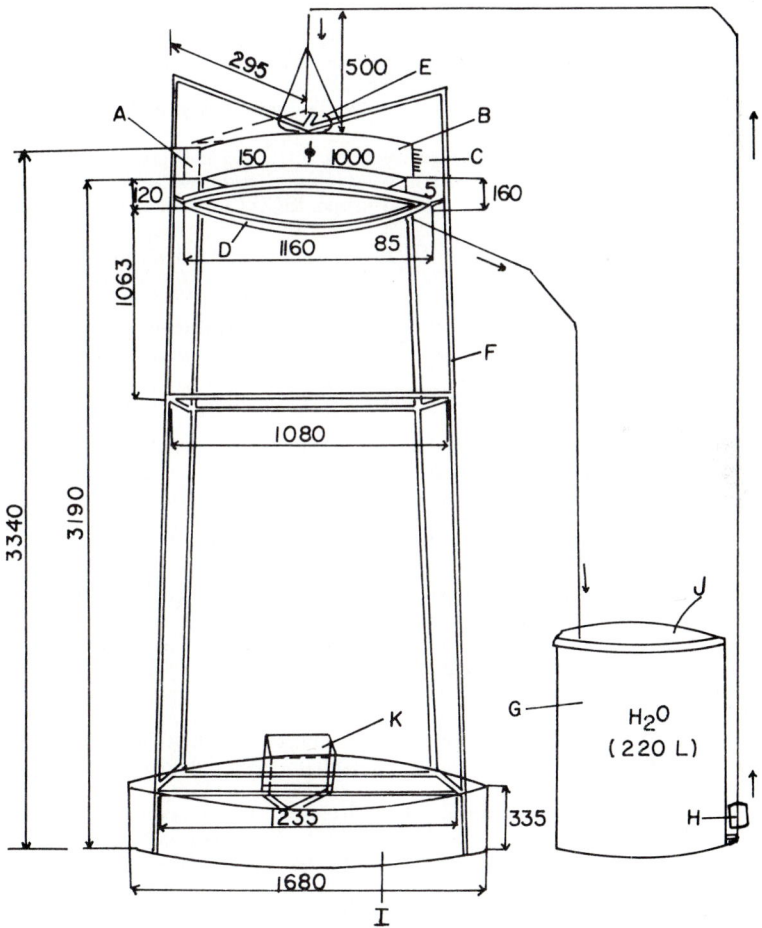

Figure 7.17 Schematics of a laboratory rainfall simulator. Measurements are in centimeters. A = motor, B = distribution tank, C = overflow system, D = overflow drain, E = chain and cog wheel, F = supporting legs, G = water supply tank, H = pump, I = catchment basin, J = water inflow, K = soil pan. (Gabriels et al., 1976.)

in measuring erosion at the level of the microscale and extrapolating the results to that of the watershed.

Reliable results can be obtained from simple, direct, and inexpensive techniques as long as they are designed with due consideration of the processes involved. The technique used must take into consideration the processes involved in runoff initiation and in sediment origin and entrainment.

References

Aina, P. O., Lal, R., and Taylor, G. S. 1979. Effects of vegetal cover on soil erosion on an alfisol. In: *Soil Physical Properties and Crop Production in the Tropics* (R. Lal and D. J. Greenland, eds.), pp. 501–508, Wiley, Chichester, England.

Anderson, H. W., Coleman, G. B., and Zinke, P. J. 1959. *Summer Slides and Winter Scour; Dry Wet Erosion in Southern California Mountains.* U.S. Forest Service, Pacific Southwest Forest and Ranger Experiment Station, Berkeley, Calif., Technical Paper 36.

Arnoldus, H. M. J. 1974. *Soil Erosion: A Review of Processes, Assessment Techniques, Expert Consultation on Soil Degradation,* FAO/UNEP, Rome.

Barber, R. G., Moore, T. R., and Thomas, D. B. 1979. The erodibility of two soils from Kenya. *J. Soil Sci.* 30(3):579–591.

Baumgardner, M. F., Weismiller, R. A., and Kirschner, F. R. 1978. Digital Analysis of LANDSAT Data for Soil Survey and Erosion Assessment. Presented at the Workshop on Assessment of Soil Erosion in the USA and Europe. ISCO Conference Proceedings, Ghent, Belgium.

Biscaia, R. C. M. 1982. Soil and water loss with the rotating-boom rainfall simulator, for the succession wheat-soybeans. In: *Proc. of the 9th Conference of the Soil Tillage Research Organization,* Ponta Gross, Parana, Brazil, pp. 569–573.

Blandford, D. C. 1981. Rangelands and soil erosion research: A question of scale. In: *Soil Conservation: Problems and Prospects* (R. P. C. Morgan, ed.), pp. 105–122, Wiley, Chichester, U.K.

Bollinne, A. 1975. La mésure de l'intensité du splash sur sol limoneux. *Pedologie* 25:199–210.

Bollinne, A. 1978. Study of the importance of splash and wash on cultivated loamy soils of Hesbaye Belgium. *Earth Surface Processes* 3(1):71–74.

Bollinne, A. 1980. Splash measurements in the field. In: *Assessment of Erosion* (M. De Boodt and D. Gabriels, eds.), pp. 441–454, Wiley, Chichester, U.K.

Bruce-Okine, E., and Lal, R. 1975. Soil erodibility as determined by raindrop technique. *Soil Sci.* 119:149–157.

Bubenzer, G. D., and Sons, B. A. 1971. Drop size and impact velocity effects on the detachment of soil under simulated rainfall. *Trans. ASAE* 14(4):625–628.

Campbell, B. L., Loughran, R. J., and Elliot, G. L. 1982. Caesium-137 as an indicator of geomorphic processes in a drainage basin system. *Aust. Geogr. Stud.* 20:49–64.

Chakela, Q. K. 1981. *Soil Erosion and Reservoir Sedimentation in Lesotho.* Scondinarin Inst. of African Studies, Uppsala, Sweden.

Chandra, S., and De, S. K. 1974. Application of infrared spectroscopy to soil erodibility studies. *Agrokemia es Talajtan* (Dept. of Chemistry, Allahabad Univ., India). 23(3/4):461–470.

Chisci, G. 1980. Physical soil degradation due to hydrological phenomena in relation to change in agricultural systems in Italy. In: *Proc. Series on Soil Degradation, EEC Land Use and Rural Resources Management Committee,* Soil Degradation Group, Wageningen, The Netherlands.

Chisci, G. 1981. *Upland Erosion: Evaluation and Measurement.* IAHS Publication 133, Washington, pp. 331–349.

Chow, V. T. (ed.) 1964. *Handbook of Applied Hydrology; A Compendium of Water-Resources Technology,* McGraw-Hill, New York.

Christiansson, C. 1981. Soil Erosion and Sedimentation in Semi-Arid Tanzania, Second Inst. of Af. Studies/Dept. Phys. Geog., Univ. Stockholm, Uppsala.

Ciesiolka, C. A. A. 1984. Problems of erosion measurement in Fairbairn Dam Catchment. *Proc. National Soils Conference,* 13–18 May 1984, Brisbane, Australia.

Ciesiolka, C. A. A., and Freebairn, D. M. 1982. The influence of scale on runoff and erosion. In: *Agricultural Engineering Conference, Resources—Efficient Use and Conservation,* Barton, ACT, Australia, Inst. of Engrs. pp. 203–206.

Collinet, J., and Valentin, C. 1985. Evaluation of factors influencing water erosion in West Africa using rainfall simulation, pp. 451–461. In: *Challenges in African Hydrology and Water Resources,* IAHS Publication 144, Washington.

Coutts, J. R. H., Kandil, M. F., Nowland, J. L., and Tinsley, J. 1968. Use of radioactive ^{59}Fe for tracing soil particle movement. I. Field studies of splash erosion. *J. Soil Sci.* 19(2):311–324.

Cuff, J. R. I. 1985. Quantifying erosion-causing parameters in a New Zealand watershed. In: *Soil Erosion and Conservation* (S. A. El-Swaify, W. C. Moldenhauer, and A. Lo, eds.), pp. 99–112, SCSA, Ankeny, Iowa.

Curtis, W. R., and Cole, W. D. 1972. Micro-topographic profile gauge. *Agric. Eng.* 53:17.

Dandekar, M. M. 1981. Indian experience of reservoir sedimentation. In: *Problems of Soil Erosion and Sedimentation* (T. Tingsanchali and H. Eggers, eds.), pp. 395–404, AIT, Bangkok, Thailand.

De, S. K., and Varma, A. 1968. A simple apparatus of studying soil erosion in the laboratory. *Indian J. Agric. Chem.* 1:53–55.

Dearing, J. A., Elner, J. K., and Happy-wood, C. M. 1981. Recent sediment flux and erosional processes in a Welsh upland lake-catchment based on magnetic susceptibility measurements. *Quatern. Res.* 16.

Dearing, J. A., Foster, I. D. L., and Simpson, A. D. 1982. *Timescales of Denudation: The Lake-Drainage Basin Approach.* IAHS Publication 137, Wallingford, U.K., pp. 351–360.

De Ploey, J. 1967. Erosion pluvial au Congo Occidentale. *Isotopes in Hydrology,* IAEA, Vienna, pp. 291–300.

De Ploey, J. 1969. L'erosion pluviale: experimences à l'aide de stables traceurs et bilans morphogeniques. *Acta Geogr. Lovan* 7:1–28.

De Ploey, J., and Savat, J. 1968. Contribution à l'étude de l'erosion par le splash. *Z. Geomorph.* 12:174–193.

Douglas, I. 1967. National and man-made erosion in the humid tropics of Australia, Malaysia and Singapore. *Int. Assoc. Sci. Hydrol.* 75:17–30.

Douglas, I. 1968. Erosion in the Sungei Gombak catchment, Selangor, Maloyeia. *J. Trop. Geogr.* 26:1–16.

Dunne, T. 1979. Sediment yield and landuse in tropical catchments. *J. Hydrol.* 42:281–300.

Edwards, K. A. 1979. Regional contrasts in rates of soil erosion and their significance with respect to agricultural development in Kenya. In: *Soil Physical Properties and Crop Production in the Tropics* (R. Lal and D. J. Greenland, eds.), pp. 441–454, Wiley, Chichester, U.K.

Ellison, W. D. 1944. Two devices for measuring soil erosion. *Agric. Eng.* 25:53–55.

Farquharson, F. A. K., Mackney, D., Newson, M. D., and Thomasson, A. J. 1978. *Estimation of Runoff Potential of River Catchments from Soil Surveys.* Special Soil Survey Report, No. 11, Harpenden, U.K.

Fearnside, P. M. 1980. Prediction of soil erosion losses under various land uses in the Transamazonian Highway colonization area of Brazil. *Acta Amazonica* 10(3):505–511.

Felipe-Morales, C., Meyer, R., Alegre, C., and Vittorelli, C. 1977. Determination of erosion and runoff under various cultivation systems in the Santa Amahuancayo Region. I. Preliminary results of the 1974–75 and 1975–76 seasons. *An. Cient. Univ. Nac.* 15(1–4):75–84.

Fournier, F. 1960. Climat et erosion: La relation entre l'erosion du sol par l'eau et les precipitations atmosphèriques, University of Paris, France.

Franzen, H. 1986. Physikalische Eigenschaften und Ertragsleistung eines alfisols in Sud-West-Nigeria in Abhangigkeit von Bodenbearbeitung und Mulchbedeckung. Ph.D. dissertation, Univ. Gottingen.

Free, G. R. 1960. Erosion characteristics of rainfall. *Agric. Eng.* 41:447–449, 455.

Gabriels, D., De Boodt, M., and Minjanw, W. 1973. Description of a rainfall simulation for soil erosion studies. *Model Fak. Landbouw, R.U. Gent* 38:294–303.

Geib, H. V. 1933. A new type of installation for measuring soil and water losses from control plots. *J. Am. Soc. Agron.* 24:429–440.

Gerasimenko, V. P. 1980. Evaluation of the accuracy of determination of soil erosion by the method of water rills. *Sov. Soil Sci.* 12:318–323.

Gerlach, T. 1967. Evolutions actuelles des versants dans les carpathes, d'apres l'exemple d'observations fixes. *Slopes Comm. Rep.* 5:129–138.

Ghadiri, H., and Payne, D. 1980. A study of soil splash using cine-photography. In: *Soil Erosion Assessment* (M. De Boodt and D. Gabriels, eds.), pp. 185–192, Wiley, Chichester, U.K.

Gorchichko, G. K. 1976. Device for determining the amount of soil splashed by raindrops. *Prochvovedeniye* 9:121–124.

Gorchichko, G. K. 1977. Device for determining the amount of soil splashed by raindrops. *Sov. Soil Sci.* 8:610–613.

Gregory, K. J., and Walling, D. E. 1973. *Drainage Basin Form and Process: A Geomorphological Approach*, Edward Arnold, London.

Grierson, I. T., and Oades, J. M. 1977. A rainfall simulator for field studies of runoff and soil erosion. *J. Agric. Eng. Res.* 22(1):37–44 (Univ. of Adelaide, Australia).

Grigorev, V. Ya, and Kuznetsov, M. S. 1976. Experimental equipment for studying the erosion resistance of soils and certain results of its use. *Moscow. Univ. Soil Sci. Bull.* 31:38–41.

Grigorev, V. Ya, Kuznetsov, M. S., and Kim, D. A. 1979. Potential danger of irrigation erosion for the soils of Dzhizakskaya steppe and measures for its prevention. *Moscow Univ. Soil Sci. Bull.* 34:49–53.

Hardjowitjitro, H. 1981. Soil erosion as a result of upland traditional cultivation in Java Island. In: *Problems of Soil Erosion and Sedimentation* (T. Tingsanchali and H. Eggers, eds.), pp. 173–179, AIT, Bangkok, Thailand.

Hempel, L. 1951. Uber kartierungsmethoden von Bodenerosion durch Wasser. *Neue Arch. Niedersachsen H.* 26:590–598.

Hempel, L. 1954. Beobachfungen uber die Empfindlicht keit von Ackerboden gegenuber der Bodenerosion. *Z. Pflanzenernahrung, Dungung, Bodenkunde*, Bd. 64:42–54.

Hempel, L. 1968. Bodenerosion in Suddentschland. *Forschungen Z. dt. Landeskunde*, Bd. 1979, Bad Godesberg.

Heusch, B. 1980. Erosion in the Ader Dutchi massif (Niger): An example of mapping applied to water and soil conservation. In: *Assessment of Erosion* (M. De Boodt and D. Gabriels, eds.), pp. 521–529, Wiley, Chichester, U.K.

Hills, R. C. 1971a. The influence of land management and soil characteristics on infiltration and the occurrence of overland flow. *J. Hydrol.* 13:163–181.

Hills, R. C. 1971b. Lateral flow under cylinder infiltrometers—A graphical correction procedure. *J. Hydrol.* 13:153–162.

Holeman, J. N. 1968. The sediment yield of major rivers of the world. *Water Resources Res.* 4:737–747.

Hudson, N. W. 1964. Field measurement of accelerated soil erosion in localised areas. *Rhod. Agric. J.* 61:46–47, 60.

Hudson, N. W. 1965. The influence of rainfall on the mechanics of soil erosion with particular reference to Southern Rhodesia. M.Sc. thesis, University of Cape Town.

Jansen, J. M. L., and Painter, R. B. 1974. Predicting sediment yield from climate and topography. *J. Hydrol.* 21:371–380.

Jinze, M. 1981. *The Establishment of Experimental Plots for Studying Runoff and Soil Loss in the Rolling Loess Regions of China*, pp. 467–477, IAHS-AISH Publication 133, Wallingford, U.K.

Jolly, J. P. 1982. *A Proposed Method for Accurately Calculating Sediment Yields from Reservoir Deposition Volumes*. IAHS Publication 137, Wallingford, U.K., pp. 153–161.

Kerenyi, A. 1981. *A Study of the Dynamics of Drop Erosion under Laboratory Conditions*. IAHS Publication 133, Wallingford, U.K., pp. 365–372.

Kirkby, A. V. T., and Kirkby, M. J. 1974. Surface wash at the semi-arid break in slope. *Z. Geomorph. Suppl. Bd.* 21:151–176.

Klemes, V. 1983. Conceptualization and scale in hydrology. *J. Hydrol.* 65:1–23.
Knowles, G. H. 1979. Erosion assessment and control techniques in Australia. *J. Soil Conserv. Service New South Wales* 35(2):70–81.
Kuznetsov, M. S. 1977. A new method for evaluation of the influence of soil moisture content on erosion resistance. *Moscow Univ. Soil Sci. Bull.* 32:20–23.
Kwaad, F. J. P. M. 1977. Measurements of rainsplash erosion and the formation of colluvium beneath deciduous woodland in the Luxembourg Ardennes. *Earth Surface Processes* 2(2–3):161–173.
Lal, R. 1976. Soil erosion problems on Alfisols in Western Nigeria, Parts I–IV. *Geoderma* 16:366–431.
Lam, Kin-Che. 1977. Patterns of and rates of slopewash on the badlands of Hong Kong. *Earth Surface Processes* 2:319–332.
Lamarche, V. C. 1968. *Rates of Slope Degradation as Determined from Botanical Evidence, White Mountains, California.* Professional Paper 352, U.S. Geological Survey, Reston, Va.
Lambert, A. M. 1982. *Estimation of Erosion and Sediment Yield by Volume Measurements on a Lacustrine River Data.* IAHS Publication 137, Wallingford, U.K., pp. 171–176.
Leopold, L. B., Wolman, M. G., and Miller, J. P. 1964. *Fluvial Processes in Geomorphology,* W. H. Freeman, San Francisco.
Lewis, L. 1976. Soil movement in the tropics—A general model. *Geomorph. N.F.* 25:132–144.
Lewis, L. A. 1981. The movement of soil materials during a rainy season in western Nigeria. *Geoderma* 25(1/2):13–25.
Loch, R. J. 1982. Rainfall simulator methodology: Concepts for realistic research. In: *Agricultural Engineering Conference 1982, Resources—Efficient Use and Conservation.* Barton ACT, Australia, pp. 99–103.
Loch, R. J., and Donnollan, T. E. 1983. Field rainfall simulator studies on two clay soils of the Darling Downs, Queensland. I. The effect of plot length and tillage orientation on erosion processes and runoff and erosion rates. *Aust. J. Soil Res.* 21(1):33–46.
Longmore, M. E., O'Leary, B. M., Rose, C. W., and Chandica, A. L. 1983. Mapping soil erosion and accumulation with the fallout isotope caesium-137. *Aust. J. Soil Res.* 21(4):373–385.
Loughran, R. J., Campbell, B. L., and Elliott, G. L. 1982. *The Identification and Quantification of Sediment Sources Using* ^{137}Cs. IAHS Publication 137, Wallingford, U.K., pp. 361–369.
Mathews, H. L., Cunningham, R. L., Cipra, J. E., and West, T. R. 1973. Application of multispectral remote sensing to soil survey research in southeastern Pennsylvania. *Soil Sci. Soc. Am. Proc.* 37:88–93.
McCallan, M. E., O'Leary, B. M., and Rose, C. W. 1980. Redistribution of caesium-137 by erosion and deposition on an Australian soil. *Aust. J. Soil Res.* 18:119–128.
McCallan, M. E., and Rose, C. W. 1977. The construction of a geochronology for alluvial deposits on the Condamine Plain using ^{132}Cs and ^{210}Pb and the estimation of aerial variation in erosion intensity in the Upper Condamine drainage basin. *Aust. Environ. Stud. Tech. Rep.* February.
McCool, D. J., Dossett, M. G., and Yecha, S. J. 1976. *A Portable Rill Meter for Measuring Soil Loess.* ASAE Paper 76-2054, ASA, St. Joseph, Mo.
McCool, D. J., Dossett, M. G., and Yecha, S. J. 1981. *A Portable Rill Meter for Field Measurement of Soil Loss.* IAHS Publication 133, Wallingford, U.K., pp. 479–484.
McHenry, J. R., Ritchie, J. C., and Gill, A. C. 1973. Accumulation of fallout caesium-137 in soils and sediments in selected watersheds. *Water Resources Res.* 9:676–686.
Meyer, L. D. 1987. Rainfall simulators for soil conservation research. In: *Soil Erosion Research Methodology* (R. Lal, ed.), Soil Conservation Society of America, Ankeny, Iowa, pp. 75–96.
Millington, A. C. 1981. *Relationship between Three Scales of Erosion Measurement on Two Small Basins in Sierra Leone.* IAHS Publication 133, Wallingford, U.K., pp. 485–492.
Millington, A. C., Robinson, D. A., and Browne, T. J. 1982. *Establishing Soil Loss and Erosion Hazard Maps in a Developing Country: A West African Example* IAHS Publication 137, Wallingford, U.K., pp. 283–292.

Mitchell, J. K., Bubenzer, G. D., McHenry, J. R., and Ritchie, J. C. 1980. Soil loss estimation from fallout cesium-137 measurements. In: *Soil Erosion Assessment* (M. De Boodt and D. Gabriels, eds.), pp. 393–401, Wiley, Chichester, U.K.

Moeyersons, J., and de Ploey, J. 1976. Quantitative data on splash erosion simulated on unvegetated slopes. *Z. Geomorph. Suppl. Bd.* 25:120–131.

Mondardo, A., and Vieira, M. J. (eds.). 1975. First National Congress on Erosion Studies with Rainfall Simulators. Londrina, Brazil, EMBAPA/IAPAR (Empresa Brasileira de Pesquisa Agropecuria, and Fundacao Instituto Agronomico do Parana).

Morgan, R. P. C. 1978. Field studies of rainsplash erosion. *Earth Surface Processes* 3:295–299.

Morgan, R. P. C. 1981. *Field Measurement of Splash Erosion*. IAHS Publication 133, Wallingford, U.K., pp. 373–382.

Mtakwa, P. W., Lal, R., and Shama, R. B. 1987. An evaluation of the Universal Soil Loss Equation and field techniques for assessing soil erosion on a tropical alfisol in Western Nigera. *Hydrol. Processes* 1:199–209.

Muchena, F. N. 1983. The role of soil surveys and land evaluation in assessing soil erosion hazard. In: *Soil and Water Conservation in Kenya* (W. M. Senga and D. B. Thomas, eds.), pp. 79–86, Nairobi, Kenya, Institute for Dev. Studies & Fac. of Agric., University of Nairobi.

Munn, J. R., and Huntington, G. L. 1976. A portable rainfall simulator for erodibility and infiltration measurements on rugged terrain. *Soil Sci. Soc. Am. J.* 40(4):622–624.

Mutchler, C. K., Murphree, C. E., and McGregor, K. C. 1987. Laboratory and field plots for soil erosion studies. In: *Soil Erosion Research Methodology* (R. Lal, ed.), Soil Conservation Society of America, Ankeny, Iowa, pp. 9–38.

Nordin, C. F. Jr., and Mielke, P. W. Jr. 1984. *Confidence Limits for Determining Concentration of Tract Particles in Sediment Samples*, pp. 43–46. Water Supply Paper 2262, U.S. Geological Survey, Colorado State University, Fort Collins.

Olayemi, F. F., and Yadav, R. C. 1983. Rainfall simulator for tillage research in the tropics. *Soil Tillage Res.* 3(4):397–405.

Perrens, S. J., and Reeve, I. J. 1982. High speed photography of raindrop splash using a twin flash stroboscope. In: *Agricultural Engineering Conference. Resources—Efficient Use and Conservation*, pp. 22–24, Armidale, New South Wales, Australia.

Pilgrim, D. H. 1966. Radioactive tracing of storm runoff on a small catchment. I. Experimental technique. *J. Hydrol.* 4:289–305.

Politano, W., Corsini, P. C., and Douzelli, J. L. 1980. Characterization and mapping of surface anthropogenic erosion in the municipality of Monte Alto, Sao Paulo. Faculdade de Ciencias Agrerias e veterinarias, UNESP, 14870 Jaboticabal, S.P., Brazil.

Rapp, A. 1977. Methods of soil erosion monitoring for improved watershed management in Tanzania, pp. 85–98. *Guidelines for Watershed Management,* FAO Conservation Guide 1, FAO, Rome.

Rapp, A., Murray-Rust, D. H., Christiansson, C., and Beny, L. 1972. Soil erosion and sedimentation in four catchments near Dodoma, Tanzania, *Geogr. Ann.* 54A:255–318.

Reeve, I. J. 1982. A comparison of two methods of measuring soil erosion due to raindrop splash. In: *Agricultural Engineering Conference. Resources—Efficient Use and Conservation,* pp. 104–107, Armidale, New South Wales, Australia.

Revue De Geomorphologie Dynamique. 1967. Field methods for the study of slope and fluvial processes. *Rev. Geomorph. Dyn.* 17:145–158.

Richter, G. 1980. Soil erosion mapping in Germany and in Czechoslovakia. In: *Assessment of Erosion* (M. De Boodt and D. Gabriels, eds.), pp. 29–54, Wiley, Chichester, U.K.

Ritchie, J. C., Hawks, P. H., and McHenry, J. R. 1974. Estimating soil erosion from the redistribution of fallout ^{137}Cs. *Soil Sci. Soc. Am. Proc.* 38:137–139.

Ritchie, J. C., Hawks, P. H., and McHenry, J. R. 1975. Deposition rates in valleys determined using fallout caesium-137. *Geol. Soc. Am. Bull.* 86:1128–1130.

Robinson, A. R. 1977. Relationships between soil erosion and sediment delivery. *Int. Assoc. Hydrol. Sci. Publ.* 122:159–167.

Rodriguez-Iturbe, I., and Gupta, V. K. 1983. Scale problems in hydrology. *J. Hydrol.* 65 (special issue):1–257.

Roehl, J. W. 1962. *Sediment Source Area Delivery Ratios and Influencing Morphological Factors,* IAHS Publication 59, Washington, pp. 202–213.

Rogowski, A. S., Khanbilvardi, R. M., and De Angelis, R. J. 1985. Estimating erosion on plot, field, and watershed scales. In: *Soil Erosion and Conservation* (S. A. El-Swaify, W. C. Moldenhauer, and A. Lo, eds.), pp. 149–166, SCSA, Ankeny, Iowa.

Roose, E. J., and Asseline, J. 1978. Measurement of erosion phenomena under simulated rainfall at Adiopodonme. Solid and soluble loads in runoff water from bare soil and pineapple plantations. *Cah. ORSTOM, Pedologie.* 16:43–72.

Roose, E. J., and Lelong, F. 1976. Factors in water erosion in tropical Africa. Studies using small experimental soil plots. *Rev. Geogr. Physique Geol. Dynamique* 18(4):365–374.

Russell, J. R. 1981. Sedimentation in the proposed Kotmale Reservoir, Sri Lanka. In: *Problems of Soil Erosion and Sedimentation* (T. Tingsanchali and H. Eggers, eds.), pp. 405–417, AIT, Bangkok, Thailand.

Salamin, P. 1982. *Results of Recent Research on Erosion Processes in Hungary.* IAHS Publication 137, Wallingford, U.K., pp. 67–71.

Schumm, S. A. 1954. *The Relation of Drainage Basin Relief to Sediment Loss,* IAHS Publication 36(1), Washington, pp. 216–219.

Schumm, S. A. 1956. Evolution of drainage systems and slopes in badlands at Perth Amboy, New Jersey. *Bull. Geol. Soc. Am.* 67:597–646.

Schwertmann, U., and Schmidt, F. 1980. Estimation of long term soil loss using copper as a tracer. In: *Assessment of Erosion* (M. De Boodt and D. Gabriels, eds.), pp. 403–406, Wiley, Chichester, U.K.

Shaxson, T. F. 1981. Determining erosion hazard and land use incapability: A rapid subtractive survey method. *Soil Survey Land Eval.* 1(3):44–50.

Shaxson, T. F., Brabben, T. E., and Stenenson, E. C. 1981. Rain-runoff measurements on basalt soils in Central India. In: *Problems of Soil Erosion and Sedimentation* (T. Tingsanchali and H. Eggers, eds.), pp. 76–84, AIT, Bangkok, Thailand.

Shikula, N. K., Rozhkov, A. G., and Tregubov, P. S. 1974. Mapping of a territory according to erosion processes intensity. *Trans. 10th Int. Cong. Soil Sci.* XI:32–37.

Siderius, W. (ed.). 1986. *Proceedings of the International Workshop on Land Evaluation for Land Use Planning and Conservation in Sloping Areas,* PUDOC, Wageningen.

Sreenivas, L., Johnston, J. R., and Hill, H. O. 1947. Some relationships of vegetation and soil detachment in the erosion process. *Soil Sci. Soc. Am. Proc.* 11:471–474.

Strakhov, N. M. 1967. *Principles of Lithogenesis* (trans. J. P. Fitzsimmons, S. I. Tomkieff, and J. E. Hemingway), Oliver and Boyd, Edinburgh, Scotland.

Stromquist, L. 1981. Recent studies on soil erosion, sediment transport, and reservoir sedimentation in semi-arid Central Tanzania. In: R. Lal and E. W. Russell (eds.), *Tropical Agricultural Hydrology,* J. Wiley and Sons, Chichester, U.K., pp. 189–225.

Subramanian, V. 1979. Chemical and suspended-sediment characteristics of rivers of India. *J. Hydrol.* 44:37–55.

Subramanian, V. 1982. *Sediment Yield of the River Krishna.* IAHS Publication 137, pp. 183–188.

Temple, P. H., and Mworay-Rust, D. H. 1972. Sheet wash measurements on erosion plots at Mfumbwe, Eastern Uluguru Mountains, Tanzania. In: *Studies of Soil Erosion and Sedimentation in Tanzania* (A. Rapp, L. Berry, and P. Temple, eds.), *Geogr. Ann.* 54:195–202.

Townshend, J. R. G. 1970. Geology, slope form and process and their relation to occurrence of laterite in the Mato Grosso. *Geogr. J.* 136:392–399.

Trimble, S. W. 1975. Denudation studies: Can we assume steady state? *Science* 188:1207–1208.

Trimble, S. W. 1977. The fallacy of stream equilibrium in contemporary denudation studies. *Am. J. Sci.* 277:887.

USDA Soil Survey Staff. 1951. *Soil Survey Manual,* Agricultural Handbook 18, GPO, Washington.

Van Doren, C. A., Stauffer, R. S., and Kidder, E. H. 1950. Effect of contour farming on soil loss and runoff. *Soil Sci. Soc. Am. Proc.* 15:413–417.

Van Elslande, A., Lal, R., and Gabriels, D. 1976. The erodibility of some Nigerian soils: A

comparison of rainfall simulator results with estimates obtained from the Wischmeier nomogram. *Hydrol. Process* 1.
Walling, D. E. (ed.). 1982. *Recent Developments in the Explanation and Prediction of Erosion and Sediment Yield.* IAHS Publication 137, Washington.
Walling, D. E. 1983. The sediment delivery problem, *J. Hydrol.* 65:209–237.
Walling, D. E. 1984. The sediment yield of African rivers. *IAHS Publication 144,* Wallingford, U.K., pp. 265–283.
Walling, D. E., Peart, M. R., Oldfield, F., and Thompson, R. 1979. Suspended sediment sources identified by magnetic measurements. *Nature (Lond.)* 281(5727):110–113.
Weismiller, R. A., and Kaminsky, S. A. 1978. Application of remote sensing technology to soil survey research. *J. Soil Water Conserv.* 33:287–289.
Weismiller, R. A., Van Scoyoc, G. E., Pazar, S. E., Latz, K., and Baumgardner, M. F. 1985. Use of soil spectral properties for monitoring soil erosion. In: *Soil Erosion and Conservation* (S. A. El-Swaify, W. C. Moldenhauer, and A. Lo, eds.), pp. 119–127, SCSA, Ankeny, Iowa.
Westin, F. C., and Freeze, C. J. 1976. LANDSAT data, its use in a soil survey program. *Soil Sci. Soc. Am. J.* 40:81–89.
Young, A. 1960. Soil movement by denudational processes on slopes. *Nature (Lond.)* 188:120–122.
Young, A. 1963. Soil movement on slopes. *Nature* 200:129–130.

Chapter

8

Predicting the Erosion Potential

Measuring soil erosion losses for different soils, crops, management alternatives, and other conditions is an expensive and time-consuming undertaking. Field measurements are highly variable in time and space. The minimum time required to obtain reliable data from field plots is 2 or 3 years. Adequately equipped field plots are expensive and must be replicated enough to obtain the site-specific data. Results from field experiments can be extrapolated to other regions via empirical or physical models, but no model, regardless how valid, is a substitute for reliable measurements from well-designed, properly equipped field experiments. Still, models help us understand the cause-effect relationships involving major soil erosion agents and factors. Models also are useful for indicating knowledge gaps that must be filled through research. More comprehensive and sophisticated, physical-based models are now being developed by using computer facilities.

There are wide varieties of erosion prediction models, each developed to provide an answer to a specific question. No single model meets all requirements. There are different models for predicting erosion on long-term or event-based time scale, from a hillslope or watershed, from croplands or rangelands, in humid regions or semiarid environments, by sheet erosion or mass movement, etc. Data input also differs for different models.

Attempts have been made since the early 1940s to predict soil erosion from basic factors. Some of the early models were empirical and simple and involved only a few of the most important variables. Some of these early models are reviewed by Gerrard (1981), among others. In the 1970s and 1980s, the emphasis has shifted to developing conceptual or physical models that consider the basic principles of major processes involved. Mitchell and Bubenzer (1980) and Haan et al. (1982) have presented detailed reviews on such conceptual or physical models.

Interrill and Rill Erosion

Major processes governing soil erosion from arable lands concern interrill and rill erosion. Interrill erosion is the principal cause of sediment detachment. The concentration of overland flow, caused by surface irregularities and variations in topography, leads to rill formation. Rills erode downward until they reach the soil layer of low susceptibility to erosion. After that, the rills widen by lateral erosion. If the subsoil is also susceptible to erosion, rills eventually transform to gullies. A series of empirical models have been developed to predict the combined interrill and rill erosion from agricultural lands. Some of the most widely used ones are described below.

Universal Soil Loss Equation

This parametric equation was developed in the late 1950s (Wischmeier and Smith, 1958). The universal soil loss equation (USLE) has been the most widely used forecasting tool for two decades ending in mid-1980. Although developed mainly as a forecasting cum planning tool for agricultural lands, USLE has been modified and adapted to predict the erosion potential from watersheds and nonagricultural sites.

Attempts to develop a predictive model to estimate soil erosion were begun in the early 1930s. Baver (1933) proposed the empirical equation for estimating soil erosion

$$E = \left(\frac{KD}{Ap}\right) P^{-1} \tag{8.1}$$

where E is soil erosion, K is constant of proportionality, D is ease of dispersion, A is absorption, p is soil permeability, and P is size of soil particles. Chronologically, Baver's model was followed by A. W. Zingg's 1940 equation relating the effect of slope length and steepness on erosion. Subsequently, sheet erosion equations were proposed by Horton (1945) and Ellison (1947). A major forerunner of USLE was Musgrave's (1947) parametric equation. Musgrave proposed this factorial equation involving most variables affecting soil erosion:

$$A = 0.00527 KCS^{1.35} L^{0.35} I_{30}^{1.75} \tag{8.2}$$

Here A is soil erosion (mm/yr), K is inherent soil erodibility at 10 percent slope on a slope 22 m long (mm/yr), C is vegetation cover factor, S is slope percentage, L is slope length (m), and I_{30} is maximum 30-min rainfall with a 2-year frequency.

On the basis of the extensive field experiments conducted in the midwestern United States, Wischmeier and Smith (1958) revised Eq. (8.2) and developed another parametric equation that became known as the *universal soil loss equation*:

$$A = RKLSCP \tag{8.3}$$

Here A is average annual soil loss, R is erosivity of rainfall, K is soil erodibility, L is slope length, S is slope gradient, C is cropping management, and P is conservation practice. USLE defines the controlling factors more precisely than Musgrave's equation (8.2). The USLE is based on more than 25 years of research covering 10,000 plot years of data from natural-runoff plots and the equivalent of 1000 plot years of data from field plots using rainfall simulators. The aim in developing the USLE model was to collect field data on all variables that might affect soil erosion and to use statistical analyses for the multivariate regression equation.

The ecological limits of the data base are rather narrow, because the experiments were limited to the midwestern United States. Most data came from soil types and environments of the region east of the Rocky Mountains. Even though the USLE has greatly enhanced the usefulness of the original data base, it is very limited by its narrow ecological range. Since its development in 1958, however, all variables involved in USLE have been standardized and their potential and limitations assessed (Wischmeier and Smith, 1978; Mitchell and Bubenzer, 1980).

Erosivity index R

Climatic erosivity refers to the aggressiveness of the climate (rain, wind, snow) toward erosion. Rainfall, runoff, or wind is called erosive when its energy causes erosion. The factor R in Eq. (8.3), an index of rainfall erosivity, is an important variable. Wischmeier and Smith (1958) used the data of Laws and Parsons (1943) to develop the regression equation relating kinetic energy to rainfall intensity

$$E = 1.213 + 0.890 \log I \tag{8.4}$$

where E is the kinetic energy [kg · m/(m^2 · mm)] and I is rainfall intensity (mm/h). They imposed a limit of 76.2 mm/h intensity for Eq. (8.4). Wischmeier (1959) observed that soil erosion from bare fallow plots correlated highly with the cross-product of total kinetic energy and the maximum 30-min rainfall intensity. The product EI_{30} is a measure of rainfall's aggressiveness toward erosion. The factor R is obtained from

$$R = \frac{\sum_{i=1}^{n} EI_{30}}{100} \tag{8.5}$$

where n is the number of rainstorms per year. A plot of the average annual value of R for a region is called the *isoerodent map*. Lines representing the same annual erosion index are called *isoerodents*. Wischmeier and Smith (1978) compiled such a map for the continental United States (Fig. 8.1). Similar maps have since been compiled for other regions (see pp. 241–249).

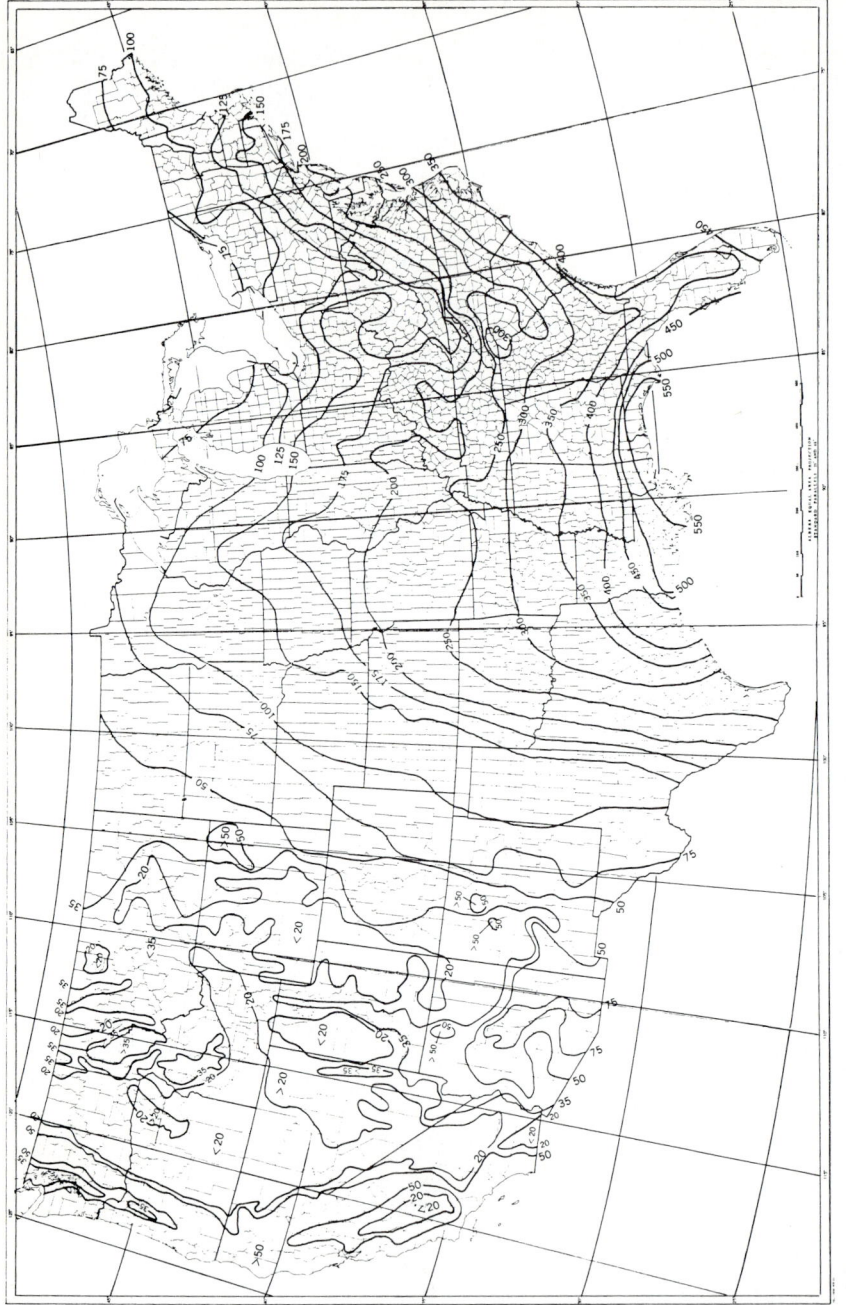

Figure 8.1 Isoerodent map of the continental United States (Wischmeier and Smith, 1978).

Soil erodibility factor K

Soils are inherently erodible; their physical, chemical, mineralogic, and biologic characteristics make some soils more susceptible to erosion than others. The soil erodibility factor K of the USLE was determined experimentally and was defined as the rate of erosion per unit of erosion index R from a unit plot on that soil. The unit plot is the reference plot. It uses an area 22.2 m long with a uniform slope of 9 percent in continuous bare fallow (no vegetation) and with tillage operations performed up and down the slope. The dimensions of the unit plot were representative of most of the field plots that generated the data used in developing the USLE. The values of factor K for major soil types in the United States are shown in Table 8.1. Wischmeier et al. (1969) developed a nomogram to estimate a soil erodibility factor K (Fig. 8.2). It has not been widely validated for soils outside the United States.

TABLE 8.1 Computed K Values for Soils on Erosion Research Stations

Soil	Source of data	Computed K
Dunkirk silt loam	Geneva, NY	0.69*
Keene silt loam	Zanesville, OH	0.48
Shelby loam	Bethany, MO	0.41
Lodi loam	Blacksburg, VA	0.39
Fayette silt loam	LaCrosse, WI	0.38*
Cecil sandy clay loam	Watkinsville, GA	0.36
Marshall silt loam	Clarinda, IA	0.33
Ida silt loam	Castana, IA	0.33
Mansic clay loam	Hays, KS	0.32
Hagerstown silty clay loam	State College, PA	0.31*
Austin clay	Temple, TX	0.29
Mexico silt loam	McCredie, MA	0.28
Honeoye silt loam	Marcellus, NY	0.28*
Cecil sandy loam	Clemson, SC	0.28*
Ontario loam	Geneva, NY	0.27*
Cecil clay loam	Watkinsville, GA	0.26
Boswell fine sandy loam	Tyler, TX	0.25
Cecil sandy loam	Watkinsville, GA	0.23
Zaneis fine sandy loam	Guthrie, OK	0.22
Tifton loamy sand	Tifton, GA	0.10
Freehold loamy sand	Marlboro, NJ	0.08
Bath flaggy silt loam with surface stones > 2 in removed	Arnot, NY	0.05*
Albia gravelly loam	Beemerville, NJ	0.03

*Evaluated from continuous fallow. All others were computed from rowcrop data.
SOURCE: Wischmeier and Smith, 1978.

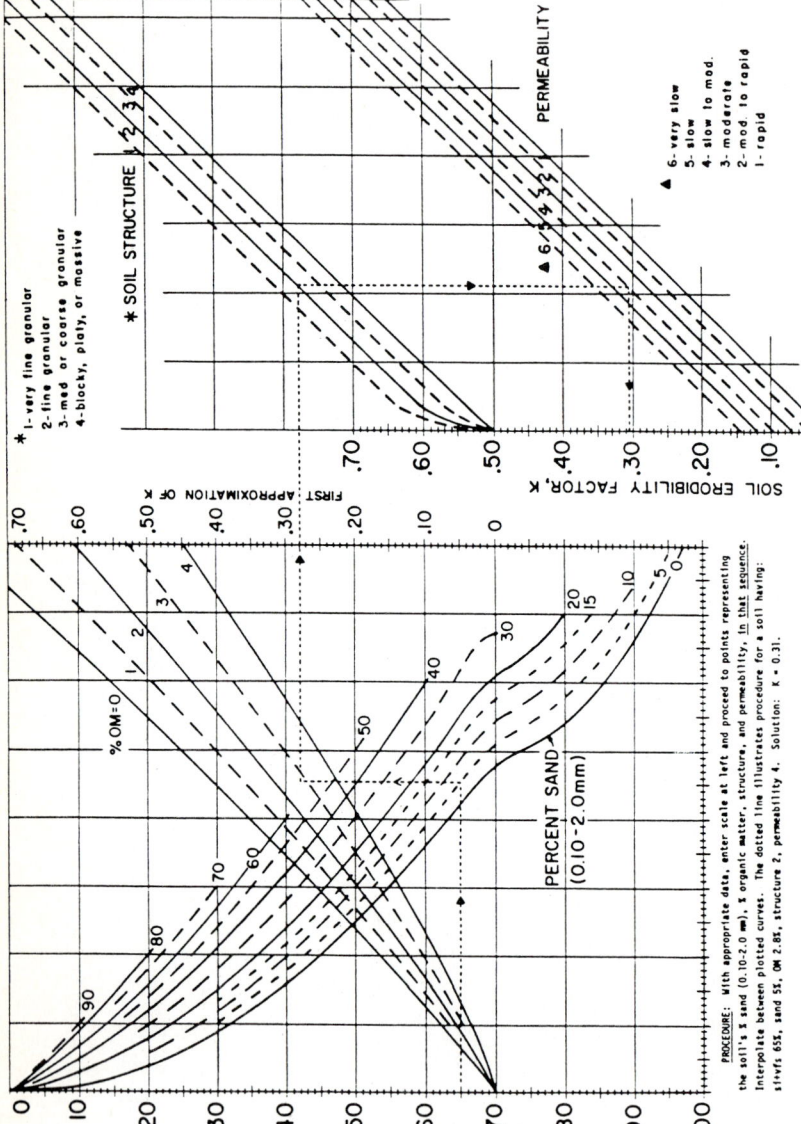

Figure 8.2 Nomogram to estimate soil erodibility factor K from soil properties. Where the silt fraction does not exceed 70%, the equation is $100K = 2.1M^{1.14}(10^{-4})(12 - a) + 3.25(b - 2) + 2.5(c - 3)$, where $M = (\%\text{ silt} + \text{very fine sand})(100 - \%c)$, $a = \%$ organic matter, $b =$ structure code, $c =$ profile permeability class. (Wischmeier et al., 1969.)

Topography factor *LS*

The factor *LS*, a dimensionless variable, is a ratio of soil erosion from a unit area of a field slope percentage to that from a 22.2-m-long area with a 9 percent slope with other conditions identical. This ratio also is obtainable from the slope-effect chart (Fig. 8.3 and Table 8.2). The *LS* factor represented in Table 8.2 and Fig. 8.3 is the average value for the entire slope. But erosion is not evenly distributed over the entire length. Foster and Wischmeier (1974) and Wischmeier (1974) therefore developed a procedure to estimate relative amounts of soil loss from successive segments of a slope where there is no deposition by overland flow. Wischmeier and Smith (1978) later developed a procedure to estimate the *LS* factor from irregular slopes.

Cover and management factor *C*

Factor *C* in the USLE is the ratio of soil loss from land cropped under specified conditions to the corresponding loss from clean-tilled, continuous fallow land. Factor *C* is the integrated effect of all variables, e.g., crop canopy, residue management, cropping systems, mixed cropping, tillage systems, and mulch rate. Wischmeier and Smith (1978) compiled elaborate lists of factor *C* for a range of crops and management practices in the United States. In addition to tables for

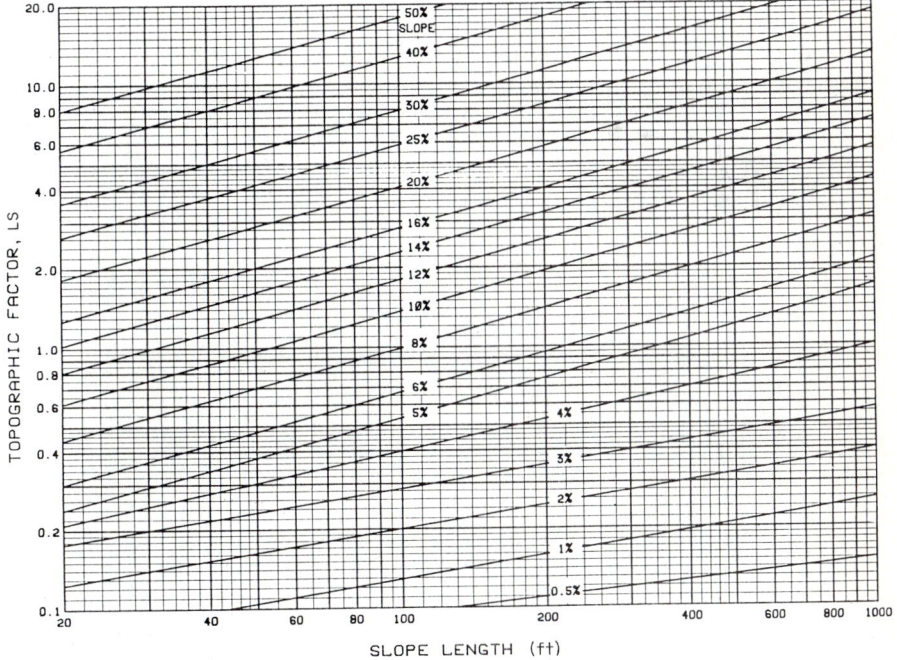

Figure 8.3 Slope-effect chart. $LS = (\lambda/72.6)^m (65.41 \sin^2 \theta + 4.56 \sin \theta + 0.065)$ where λ = slope length in feet; θ = angle of slope; and m = 0.2 for gradients less than 1%, 0.3 for 1–3% slopes, 0.4 for 3.5–4.5% slopes, and 0.5 for slopes of 5% or steeper. (Wischmeier and Smith, 1978.)

TABLE 8.2 Values of Topographic Factor LS for Specific Combinations of Slope Length and Steepness*

Percent slope	Slope length, ft											
	25	50	75	100	150	200	300	400	500	600	800	1,000
0.2	0.060	0.069	0.075	0.080	0.086	0.092	0.099	0.105	0.110	0.114	0.121	0.126
0.5	0.073	0.083	0.090	0.096	0.104	0.110	0.119	0.126	0.132	0.137	0.145	0.152
0.8	0.086	0.098	0.107	0.113	0.123	0.130	0.141	0.149	0.156	0.162	0.171	0.179
2	0.133	0.163	0.185	0.201	0.227	0.248	0.280	0.305	0.326	0.344	0.376	0.402
3	0.190	0.233	0.264	0.287	0.325	0.354	0.400	0.437	0.466	0.492	0.536	0.573
4	0.230	0.303	0.357	0.400	0.471	0.528	0.621	0.697	0.762	0.820	0.920	1.01
5	0.268	0.379	0.464	0.536	0.656	0.758	0.928	1.07	1.20	1.31	1.52	1.69
6	0.336	0.476	0.583	0.673	0.824	0.952	1.17	1.35	1.50	1.65	1.90	2.13
8	0.496	0.701	0.859	0.992	1.21	1.41	1.72	1.98	2.22	2.43	2.81	3.14
10	0.685	0.968	1.19	1.37	1.68	1.94	2.37	2.74	3.06	3.36	3.87	4.33
12	0.903	1.28	1.56	1.80	2.21	2.55	3.13	3.61	4.04	4.42	5.11	5.71
14	1.15	1.62	1.99	2.30	2.81	3.25	3.98	4.59	5.13	5.62	6.49	7.26
16	1.42	2.01	2.46	2.84	3.48	4.01	4.92	5.68	6.35	6.95	8.03	8.98
18	1.72	2.43	2.97	3.43	4.21	3.86	5.95	6.87	7.68	8.41	9.71	10.9
20	2.04	2.88	3.53	4.08	5.00	5.77	7.07	8.16	9.12	10.0	11.5	12.9

*$LS = (\lambda/72.6)^m (65.41 \sin^2 \theta + 4.56 \sin \theta + 0.065)$, where λ = slope length in feet, m = 0.2 for gradients < 1 percent, 0.3 for 1 to 3 percent slopes, 0.4 for 3.5 to 4.5 percent slopes, 0.5 for 5 percent slopes and steeper; and θ = angle of slope. (For other combinations of length and gradient, interpolate between adjacent values.)
SOURCE: Wischmeier and Smith, 1978.

croplands relevant to North America, USDA Agricultural Handbook 282 contains C values for construction areas, mulch factors, pastures, rangelands, woodlands, and idle lands. The lowest value of factor C is zero, an important planning consideration.

Support practice factor P

The factor P in the USLE is the ratio of soil loss with a specific support practice to the corresponding loss with upslope and downslope cultivation. The P factor is indicative of the effectiveness of such management practices as contouring, ridge furrow system, contour strip cropping, terracing, and other engineering devices. The value of P is rarely less than 0.25. Tabular values of P for different practices used in North America are published in USDA Agricultural Handbook 282.

As a planning tool, USLE is useful in selecting erosion control practices for agriculture and other land uses. Within the ecological limits it was developed for, it can also be used to predict the erosion potential of any land use.

Modifications in USLE

USLE is no longer universal, even within the ecological limits of the region for which it was developed. Values of empirically derived parameters are always restricted in validity to a few geographic regions. USLE is an empirical regression model that lacks conceptual basis. Because it is a statistical equation, it is severely limited in balancing the units. It contains nonhomogeneous factors (rainfall, soil, and vegetation) that some researchers argue simply cannot be multiplied (Kirkby, 1980). The model is based on correlations that are specific to the original data base derived from a limited range of soil types. USLE as a soil loss equation does not estimate deposition. It lumps together interrill and rill erosion, but does not consider gully erosion, stream bank erosion, or erosion from steeplands exceeding 20 percent slope. Morgan (1983) argues that some factors in USLE, such as erosivity R and erodibility K, are not independent.

USLE estimates the average soil loss over an extended period, i.e., average annual soil loss. The equation, as initially developed, does not predict storm-by-storm soil losses. Above all, USLE has a narrow data base. Only by coincidence could its control variables be valid for ecological regions that differ drastically from those of the midwestern United States.

USLE was specifically designed to do the following (Wischmeier, 1976):

- Predict average annual soil movement from a given field slope under specified land use and management
- Guide the selection of conservation practices for specific sites
- Estimate the reduction in soil loss possible by adopting conservation practices
- Determine acceptable cropping intensity with alternative conservation measures, e.g., contouring, terracing, or strip cropping

- Determine the maximum length of slopes that would tolerate given cropping and management practices
- Estimate soil losses from construction sites, rangelands, woodlands, and recreational areas

Used for other purposes, USLE gives erroneous results. Misuses of the equation include the following (Wischmeier, 1976):

- Applying USLE in geographic regions where basic information on various factors (R, K, C, and P) is not available or factor values cannot be accurately derived from existing data
- Computing soil erosion from complex watersheds by taking average slope length and making other adjustments
- Estimating soil erosion from specific rain events

Those limitations have led to several proposed modifications of USLE.

Changes in erosion index

We now discuss some of the most commonly used new erosion indices.

KE > 1. Experiments conducted in Zimbabwe by Hudson (1971) indicated that the accumulative kinetic energy of storms with intensity exceeding 2.5 cm/h (1 in/h) correlated better with soil loss than the EI_{30} index does.

p^2/P. Fournier (1956) and Fournier and Henin (1959) proposed the rainfall coefficient

$$\text{Rainfall coefficient} = \frac{p^2}{P} \tag{8.6}$$

where p is mean monthly rainfall in wettest month of the year and P is mean annual rainfall. Fournier observed that the specific degradation sediment yield, expressed in tonnes per square kilometer of catchment area, was linearly related to the rainfall coefficient. He also observed that the climatic index correlated sediment yield more significantly in the tropics than with sediment yield in rivers of the temperate-zone climate. Furthermore, sediment in watersheds with steep slopes was better correlated with the rainfall coefficient than with gentle slope. Application of the rainfall coefficient, however, is limited to agricultural lands. Low (1967) used the rainfall coefficient to estimate the erosion potential of the Andean region of Peru.

Modified p_m^2/P. FAO (1979) modified Fournier's rainfall coefficient for computing the erosion index of Africa north of the equator and in the Middle East. A regression equation was developed between the modified erosion index and the R factor of USLE (Arnoldus, 1977a,b, 1980). This regression equation was used to estimate R from the modified p_m^2/P:

$$\text{Modified Fournier's index} = \sum_{i=1}^{12} \frac{p_m^2}{P} \qquad (8.7)$$

Here p_m is average rainfall of the month with highest rainfall, and P is average annual rainfall. The rainfall erosivity map of Africa and the Middle East based on this index is shown in Fig. 8.4.

AI_m The AI_m index is the product of the rainfall amount (cm) and the maximum short-time (5- to 10-min) intensity. This index correlates better with soil erosion on a storm basis than the R index of USLE does (Lal, 1976). The AI_m index was used to prepare an isoerodent map of southeastern Nigeria (Armon, 1984). Lal (1976) also considered incorporating the wind factor in the AI_m index. Ahmad and Brechner (1974) also reported the importance of the wind factor in determining the erosivity of rains in Trinidad and Tobago.

Using rainfall amounts. When the long-term records of a recording rain gauge are not available, index R is sometimes computed from annual rainfall. Wischmeier (1962) related EI_{30} with other rainfall parameters by

$$EI_{30} = f(P \times I_1^{2\,yr} \times I_{24}^{2\,yr}) \qquad (8.8)$$

in which P is annual precipitation, $I_1^{2\,yr}$ is 1-h rainfall with a return period of 2 yr, and $I_{24}^{2\,yr}$ is 24-h rainfall with a return period of 2 yr. Similarly, Ateshian (1974) developed two empirical equations for estimating the average R factor:

$$R = 27.00 P^{2.2} \qquad (8.9)$$

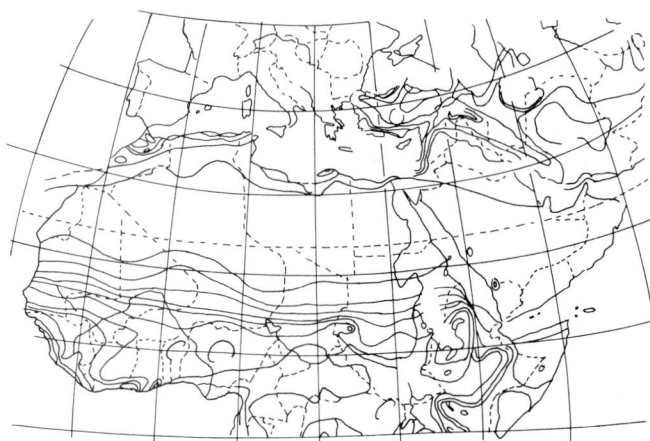

Figure 8.4 Rainfall erosivity map of Africa north of the equator and Middle East (FAO, 1978).

$$R = 16.55P^{2.2} \tag{8.10}$$

Here P is 6-h rainfall (in) with a return period of 2 yr. Equation (8.9) applies to the mainland United States except for the coastal sides of the Sierra Nevada and Cascade mountains. Equation (8.10) was developed for the islands of Hawaii and Alaska. In Belgium, Bollinne et al. (1980) observed a highly significant correlation between EI_{30} and the mean annual precipitation P (mm):

$$EI_{30} = 159.56 + 0.27P \quad r = 0.99 \tag{8.11}$$

Bollinne et al. also reported significant correlations between EI_{30} and Fournier's index F:

$$EI_{30} = -168.42 + 3.27F \quad r = 0.99 \tag{8.12}$$

For Africa north of the equator and for the Middle East, Arnoldus (1980) estimated factor R from Fournier's index. In the Netherlands, Bergsma (1980) estimated R from the average annual rainfall by using

$$R = K_1 abc + K_2 \tag{8.13}$$

where K_1 and K_2 are constants, a is average annual rainfall, b is 24-h rainfall recurring once in 2 yr, and c is 1-h rainfall recurring once in 2 yrs, and K_1 and K_2 were estimated to be, respectively, 0.75 to 1.05 and 20 to 38.

In semiarid west Africa, several researchers (Charreau, 1969; C.T.F.T., 1974; Galabert and Millogo, 1973; and Roose, 1977a,b) observed a linear relationship between kinetic energy and amount of rain. The equation proposed to compute R from the rainfall amount and 30-min intensity for Burkina Faso and Nigeria is

$$R = 0.0158HI_{30} - 1.2 \tag{8.14}$$

where H is the amount of rainfall. For the humid coastal region of Ivory Coast, Roose (1977a,b) observed a linear relation between R and the rainfall amount H for the rainy season, June to September:

$$R = 0.577H - 5.766 \tag{8.15}$$

However, a curvilinear relation exists for other months. Roose (1977a,b) also reported that a single empirical relationship exists between the average annual erosivity index R_{an} and the corresponding average annual rainfall H_{an}

$$\frac{R_{an}}{H_{an}} = 0.50 \pm 0.05 \tag{8.16}$$

In Venezuela, Guartsma et al. (1981) developed regression models to relate erosivity indices for individual rainstorm events to the rainfall amount. And in Thailand, Srikhajon et al. (1984) observed that the annual value of the erosivity index R correlated significantly with the annual rainfall amount. The regression equation, however, differed for various ecological regions, but the correlation coefficient was high for tropical regions and higher for the tropical rain forest than for the savanna (Fig. 8.5). A similar type of correlation coefficient was observed between the annual rainfall amount and the KE > 1 index. In east Java, Utomo and Mahmud (1984) reported significant correlation between R and weekly (W), monthly, (M), and annual (P) rainfall (cm):

$$R = 237.4 + 2.61P \qquad r = 0.84$$

$$R = 489.2 + 4.04 \sum_{i=1}^{n} \frac{M^2}{P} \qquad r = 0.78 \qquad (8.17)$$

for calculating annual index R, and

$$R = 2.80 + 4.15M \qquad r = 0.89 \qquad (8.18)$$

$$R = 25.48 + 8.11 \sum_{i=1}^{n} \frac{W^2}{M} \qquad r = 0.72 \qquad (8.19)$$

for calculating monthly values of R (r = correlation coefficient). In Malaysia, Jamal et al. (1984) also used the rainfall amount as an erosivity index to estimate soil erosion.

In Zimbabwe, contrary to observations in Venezuela, Thailand, and Ivory Coast, Stocking and Elwell (1976) observed no relationship on a yearly basis between erosivity index EI_{30} and annual total rainfall. In Hawaii, El-Swaify et al. (1985) reported poor correlation coefficients between the EI_{30} index and storm rainfall amounts, but reported a significant correlation ($R^2 = 0.89$) between average annual rainfall and the R index.

In southern Africa, Elwell and Stocking (1973, 1975) and Stocking (1977), after comparing different parameters of erosivity to best describe erosion, reported that parameters based on kinetic energy were the most nearly accurate predictors of soil loss from bare soil, explaining 96.4 percent of the variation in

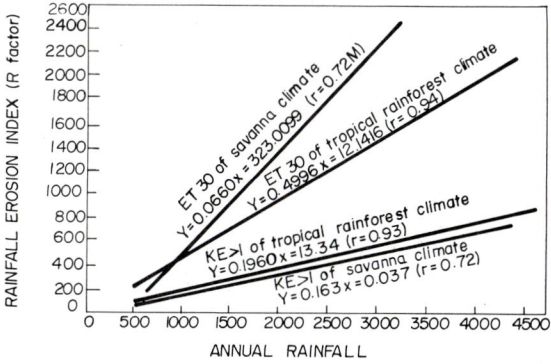

Figure 8.5 Relationship between erosivity index R and annual rainfall amount for different regions in Thailand. (Adapted from Srikhajon et al., 1984.)

results from a clay loam and 80.0 percent from a sandy soil. Little accuracy was lost, however, by using the more practical rainfall amount as an erosivity parameter. In fact, Elwell and Stocking reported little difference among momentum, energy, and rainfall amount. In Malaysia, Jamal et al. (1984) observed that soil loss from bare plots correlated significantly with rainfall amount, with no difference in correlation between soil loss and different erosivity indices, e.g., rainfall amount, EI_{30}, KE > 1, AI_m, etc. They proposed using the rainfall amount as an erosivity index.

Improving the energy-intensity relationship. Because the Laws and Parsons (1943) equation relating rainfall intensity and kinetic energy has a narrow data base, Mutchler and Murphree (1985) proposed some experimentally derived modifications in the energy equation:

$$E = 0.273 + 0.217 \exp(-0.048I) - 0.413 \exp(-0.072I) \qquad (8.20)$$

Here E is the kinetic energy per 1 mm of rain for a corresponding intensity I/(mm/h) for short time increments. This equation supposedly is applicable for intensities exceeding 100 mm/h. Mutchler and Murphree also proposed appropriate modifications in factors K, L, S, and C.

Universal index. Onchev (1985) proposed yet another index for east European countries, which he called the *universal index*:

$$R' = \frac{P}{St} \qquad (8.21)$$

Here R' is the universal index (UI), P is the quantity of rainfall greater than or equal to 9.5 mm with rainfall intensity i greater than or equal to 0.180 mm/min, and St is storm duration with i greater than or equal to 0.180 mm/min. The average annual value of UI is computed by summation of UI indices for all rain events in a year. To apply the EI_{30} index in an oceanic temperate climate such as in western Europe, Bolline (1985) thought the threshold rainfall of 12.7 mm was too high. He proposed a 1-mm threshold. He also questioned the application of both E and I_{30} parameters under the ecological conditions of western Europe.

Improvements in USLE

Many changes have been proposed in the USLE to achieve objectives it was not initially designed to attain:

For use on single-storm events. With the objective to improve the erosion prediction capability of USLE from a single-storm event, Foster et al. (1977a,b) proposed an improved erosivity factor R_m:

$$R_m = 0.5 R_{st} + 0.35 V_u P_u^{1/3} \qquad (8.22)$$

Here R_m is the modified erosivity factor, R_{st} is EI_{30} (N/h) for the storm, V_u is runoff volume (mm), and P_u is the peak runoff rate (mm/h) from a unit plot of the same soil. Slope length also varies from storm to storm, being higher with more severe rill erosion. The value of the exponent n of the function P_u^n should be increased by 0.1 when rill erosion exceeds the normal level and decreased by 0.1 when rill erosion is less than normal (Foster, 1982). Similar modifications were proposed by Chisci et al. (1985) for possible improvements in predicting erosion for individual storm events.

For use on watershed. Williams (1975) and Williams and Berndt (1972) proposed modifying USLE to estimate sediment yield from individual runoff events from a watershed by replacing the R factor of USLE with the R factor

TABLE 8.3 Calculated and Observed Soil Loss by USLE and MUSLE

Site	Crop code	C factor	USLE Observed	USLE Calculated	MUSLE Observed	MUSLE Calculated
Laupahoehoe	1	0.06	4.1	4.4	3.6	3.6
	2	—	—	—	—	—
	3	—	—	—	—	—
	4	0.01	1.9	2.3	0.3	0.2
	Total		6.0	6.7	4.0	3.8
Honokaa	1	0.08	1.2	3.9	2.3	1.5
	2	0.06	0.4	6.9	0.1	0.2
	3	0.05	0.0	0.7	—	—
	4	0.04	13.4	17.5	13.7	5.2
	Total		15.0	29.0	16.0	6.9
Waialua	1	0.15	11.5	15.9	11.5	8.4
	2	0.12	0.1	0.2	0.4	0.4
	3	0.08	6.9	8.7	9.7	5.7
	4	0.04	1.2	4.9	1.3	0.6
	Total		19.6	29.7	22.8	14.5
Mililani	1	0.16	13.8	23.5	29.2	11.9
	2	0.14	0.3	8.6	0.3	2.6
	3	0.12	0.0	0.8	0.0	0.1
	4	0.10	9.2	18.2	9.3	3.0
	Total		23.3	51.1	38.8	17.5
Kunia	1	0.55	5.8	2.6	5.8	3.2
	2	0.44	6.0	3.8	6.0	0.6
	3	0.33	0.8	2.4	3.7	1.3
	4	0.22	17.0	51.1	28.3	8.9
	Total		29.5	59.9	43.8	14.0

SOURCE: Cooley and Williams, 1985.

$$R_w = 9.05(VQ_p)^{0.56} \qquad (8.23)$$

where V is the volume of runoff (m^3) and Q_p is the peak discharge rate (m^3/s). Modified USLE with the R_w factor is called *MUSLE*. Further modifications have been proposed to account for channel and gully erosion or deposition due to impoundments. Cooley and Williams (1985), after testing MUSLE in Hawaii, observed that both USLE and MUSLE provide good estimates of soil erosion when the value of cover factor C is chosen carefully (Table 8.3).

To use USLE in estimating sediment yield from watersheds, the American Society of Civil Engineers (ASCE, 1975) proposed multiplying the gross erosion estimated via USLE by the appropriate delivery ratio factor. In Colombia, South America, Van Vuuren (1982) modified USLE to predict sediment yield from mountainous watersheds of Rio Canca and Rio Nima. A constraint that then had to be overcome in evaluating soil loss from the large watersheds is the many individual landscape elements that a large watershed contains. Using USLE for each element is obviously laborious and inefficient, so Van Vuuren used watershed average characteristics (slope gradient, length, etc.), although Wischmeier (1976) had advised against that. Still Van Vuuren modified the method of computing factors *LS* and *C*. The *C* factor for tropical rain forest was taken to be as low as 0.00022. The *R* factor was computed through a regression equation with Fournier's index. The proposed extrapolated formula that supposedly fits between 0 and 50 percent slope gradient is

$$\sin(\theta) = 48.9(\tfrac{1}{2}\sin 2\theta)^{1.61} \simeq \{16[\sin 2 \arctan(\sin/100\%)]\}^{1.6} \qquad (8.24)$$

where θ is the slope angle in degrees.

For use on rangelands. Renard et al. (1974) proposed modifications in USLE to estimate sediment yield from rangelands in semiarid regions of the southeastern United States. The modified USLE included the additional factor E_c for channel erosion:

$$A = (RKLSCP)E_c \qquad (8.25)$$

Factor E_c is similar to the delivery ratio. However, E_c may exceed unity, whereas the delivery ratio is always less than 1. Renard et al. observed that E_c for some rangelands exceeded unity because of sediment contributed from the channel, e.g., erosion of channel bed and banks.

For use on forest lands. Dissmeyer and Foster (1985) proposed yet another modification in USLE to estimate soil erosion from forest lands. Specific subfactors considered under forest conditions are surface area of bare (without cover) soil, canopy cover, soil reconsolidation, organic matter content, amount of fine roots, residual binding effect, on-site storage, and steps and contour tillage.

For use on flatlands. Mutchler and Murphree (1981) indicated the need to change USLE for use on flatlands with slopes of less than 3 percent. Because most of the data for USLE were gathered on slopes of 3 to 18 percent, Mutchler and Murphree thought adjustments were required for the equation to be used on flatland slopes. They suggested decreasing the R factor of 300 by a coefficient varying from 100 percent (for slopes of 1 percent) to 90 percent (for slopes of 0.1 percent). For an R factor of 600, the suggested decrease ranged from 90 percent (for slopes of 1 percent) to 70 percent (for slopes of 0.1 percent). They also suggested that some modifications in the K factor should be higher for flatlands, by modifying the slope exponent m [$L = (\lambda/22.13)^m$, where L is slope length] for flatlands as follows:

$$m = 1.2(\sin \theta)^{1/3} \qquad (8.26)$$

Here $\sin \theta$ is equivalent to the slope percentage divided by 100 for low slopes, and λ is the actual slope length.

Applications of USLE in the Tropics

USLE has been used widely throughout the tropics, although it was not designed for such ecological regions, it has been used widely because it seemed to meet the need better than any other tool available. The enthusiasm of young researchers in the tropics may have been counterproductive, however. Rather than developing original concepts and evaluating methods to control erosion under field conditions, they may have misdirected a lot of time and energy in applying USLE. Some relevant examples of using USLE in the tropics follow:

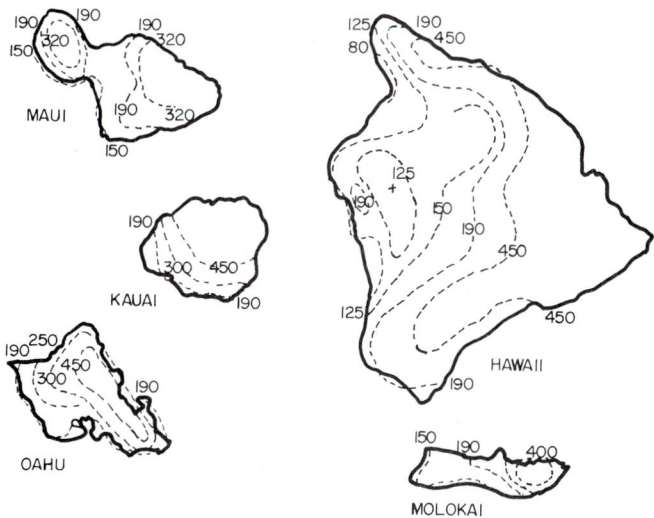

Figure 8.6 Isoerodent maps for Hawaiian islands (Wischmeier and Smith, 1978).

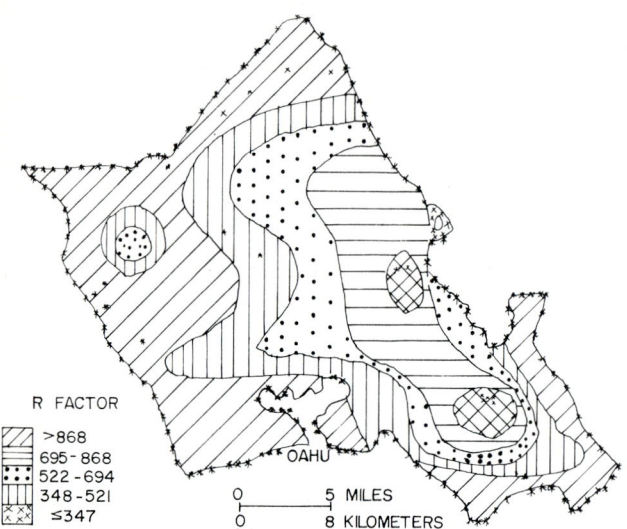

Figure 8.7 Isoerodent map for Oahu, Hawaii, based on hourly rainfall intensity (Lo et al., 1985).

Erosion index R

Although few, if any, attempts have been made to evaluate the validity of the R index in the tropics, it has been computed for a wide range of ecological regions. USLE's erosion index R has been computed even for those tropical regions where records of daily rainfall amount are hard to obtain (Arnoldus, 1977a,b).

In Hawaii, Brooks (1977) and El-Swaify and Dangler (1977) compiled isoerodent maps for different islands (Fig. 8.6). El-Swaify et al. (1985) computed the R index from the annual rainfall amount. Isoerodent maps thus computed are shown in Figs. 8.7 and 8.8. Erosion index R has also been computed for different

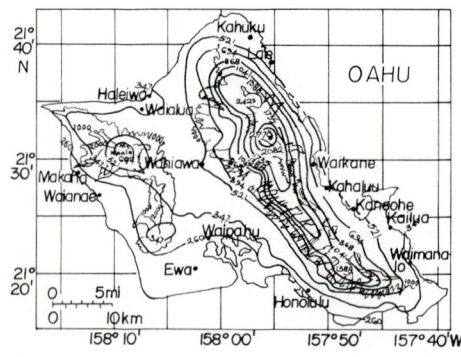

Figure 8.8 Isoerodent map for Oahu, Hawaii, based on average annual rainfall amount (Lo et al., 1985).

regions of Brazil (Pereira et al., 1978; Cogo et al., 1978; Dedecek, 1974; Castro Filho et al., 1978, 1982; and Brito Chaves and Freire, 1978). In Piracicaba, Freire and Castro Filho (1977) calculated the erosion index from the 10-year rainfall records for the period from 1957 to 1966. The erosion index varied from year to year, as one would expect. Cataneo et al. (1982) developed a computer program for calculating the erosion index. The erosion potentials of rainfall also have been computed for different locations in Uruguay (Koolhaas, 1979) and Chile (Brito and Pena MacCaskill, 1980).

Roose (1977a, 1977b, 1980) compiled the isoerodent map of west Africa, using the R index of USLE (Fig. 8.9). The R index in relation to rainfall characteristics of different regions in west Africa was studied by Wilkinson (1975a,b), Lal (1976), Lal et al. (1980), Aina et al. (1977), Kowal and Kassam (1977), and Kowal (1970). Similar estimates were made for Kenya by Barber et al. (1979). An isoerodent map for South Africa was compiled by Smithen and Schulze (1982).

In south Asia, the erosion index R has been computed for Sri Lanka (Fig. 8.10) by Joshua (1977) and for India (Fig. 8.11) by Babu et al. (1978), Ramaiah and Sreenivas (1973), and Das and Babu (1979). Similar maps have been compiled, directly or indirectly, for Zimbabwe (Stocking and Elwell, 1976), Morocco (Arnoldus, 1977a), Kenya (Barber et al., 1981), and Malaysia (Maene et al., 1977) (Fig. 8.12).

Soil erodibility factor *K*

Like the erosion index, estimates of soil erodibility by the nomogram method of Wischmeier et al. (1969) have been made for many tropical soils. The erodibility factor K for some tropical soils of Hawaii was determined with a rainfall simulator (El-Swaify, 1977; Dangler et al., 1976; Dangler and El-Swaify, 1976). The data in Table 8.4 show that the erodibility of 10 soils, representing 5 soil orders, ranged from 0 to 0.60 m · t/(ha · R), where R is the rainfall erosivity factor. The data for Hawaiian soils indicated that the soil erodibility factor was generally greater for soils developed under low annual rainfall than for soils developed under high rainfall.

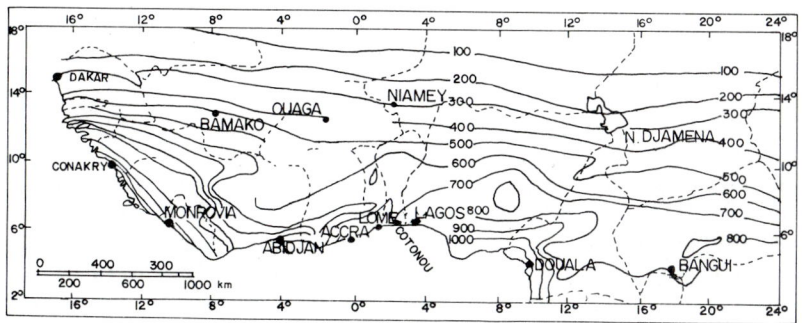

Figure 8.9 Isoerodent map of west Africa (Roose, 1977a,b).

Measurement and Prediction

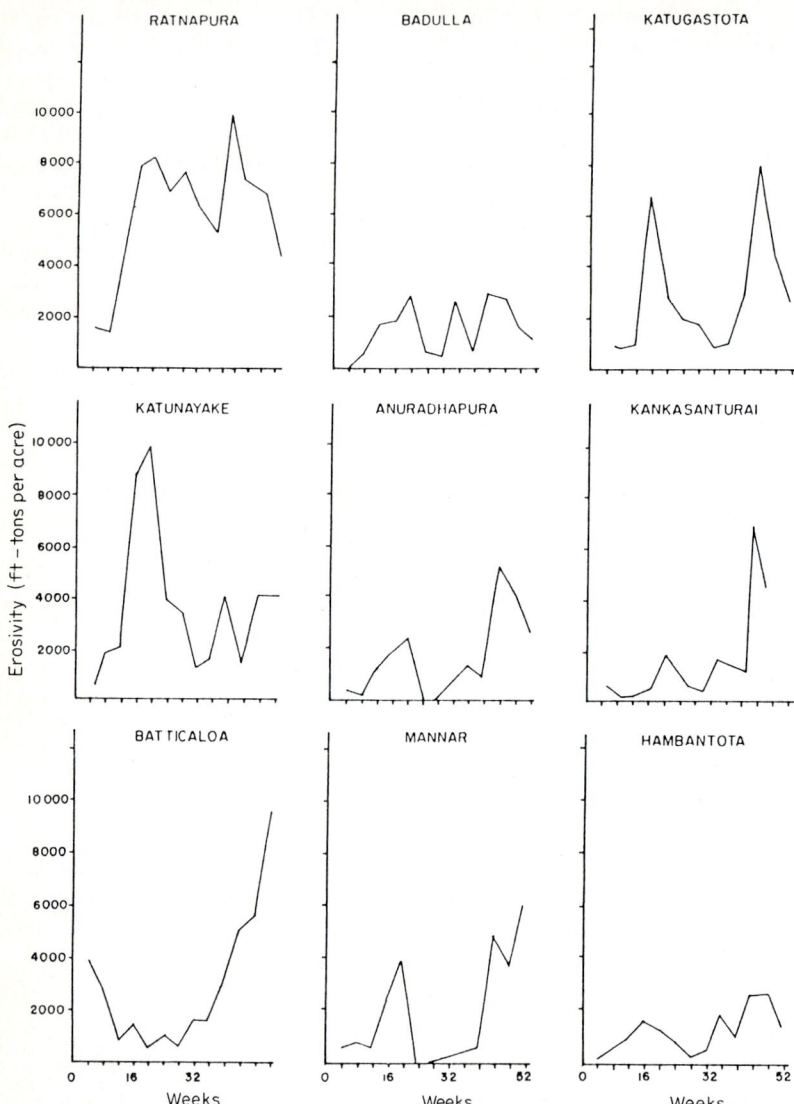

Figure 8.10 Isoerodent map for Sri Lanka (Joshua, 1977).

Considerable efforts have been made to determine the erodibility factor K for soils in Brazil. Mota and Lima (1976) measured the erodibility of seven soil series (representing aridisol, ultisol, mollisol, and entisol orders) from the São Paulo region of Brazil. In Rio Grande, Do sul, Dedecek and Cabeda (1977) concluded that the erodibility factor K of oxisols was generally low. Unless erodibility and erosivity are independent variables, their conclusion is erroneous. The severe erosion in these soils resulted from poor soil management and harsh cli-

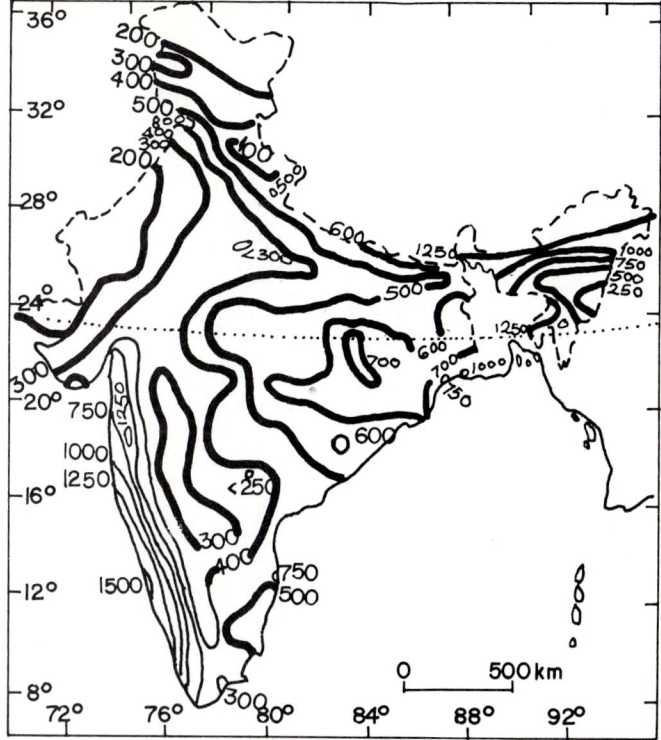

Figure 8.11 Isoerodent map for India (Babu et al., 1978).

matic and topographic factors. Freire and Pessotti (1978) estimated the erodibility factor K for soils of Piracicaba, using the nomogram of Wischmeier et al. (1969). The factor K ranged from 0.06 to 0.57 (English units). In another study, Freire and Vasques (1982) estimated the K value to be as low as 0.02 and 0.03. Resck et al. (1978) used the rainfall simulator to measure K for soils of Mato Grosso. In metric units, the factor K ranged from 0.03 to 0.15.

Measurements of soil erodibility K have also been made for soils in Parana (Mondardo et al., 1978). The erodibility of oxisols with rainfall simulator and that with natural rains were similar and ranged from 0.09 to 0.37. In Parana, Biscaia et al. (1981) reported K values for two latosols to be 0.238 and 0.268. In Paraiba state, Brazil, the K factor was in the following order: vertic noncalcic brown soil and noncalcic brown soil, 0.29; red-yellow podzolic soil, 0.26; and lithosol, 0.18.

Machado (1978) estimated K for some soils from Colombia on the basis of soil texture. In Puerto Rico K was determined for major soil orders. In Trinidad, Lindsay and Gumbs (1982) concluded that the nomogram of Wischmeier et al. (1969) correctly predicted erodibility for four soil orders.

Soil erodibility has also been determined for soils of tropical Africa. In Machakos district, Kenya, Barber et al. (1981) measured the K factor by using

246 Measurement and Prediction

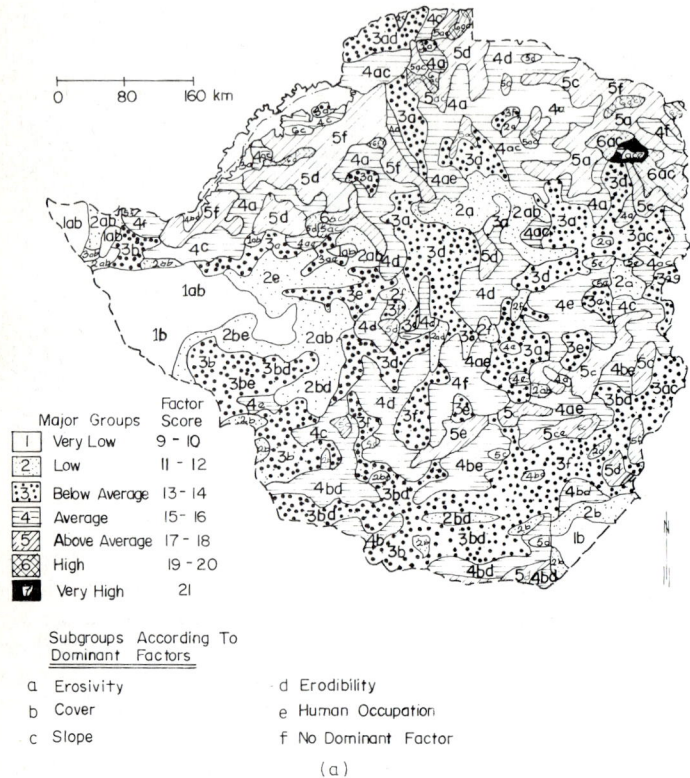

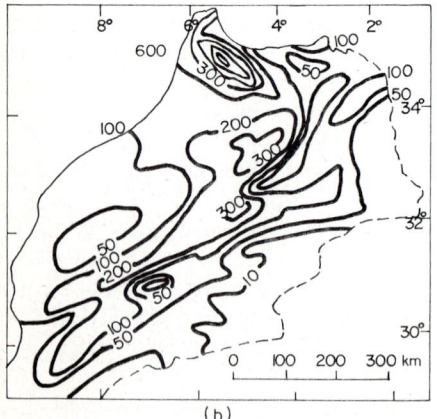

Figure 8.12 Isoerodent map for (*a*) Zimbabwe, (*b*) Morocco.

a portable rainfall simulator. The K values of those soils ranged from 0.05 for humic nitosol to 0.52 for ferral-chromic luvisol. Roose (1974, 1977a, 1977b) measured K for many soils in Francophone west Africa. The data in Table 8.5 show that K ranged from 0.02 to 0.32. Lal (1984) observed that the erodibility factor K of some soils in western Nigeria was low and that erodibility changed with time, apparently because soil properties changed (Fig. 8.13). Vanelslande et al. (1985) measured K for some Nigerian soils and observed that it ranged from 0 to 0.535. They also compared the measured value of K with that estimated by the nomogram. The data in Table 8.6 show that the measured erodibility was generally greater than that estimated by the nomogram. Vanel-

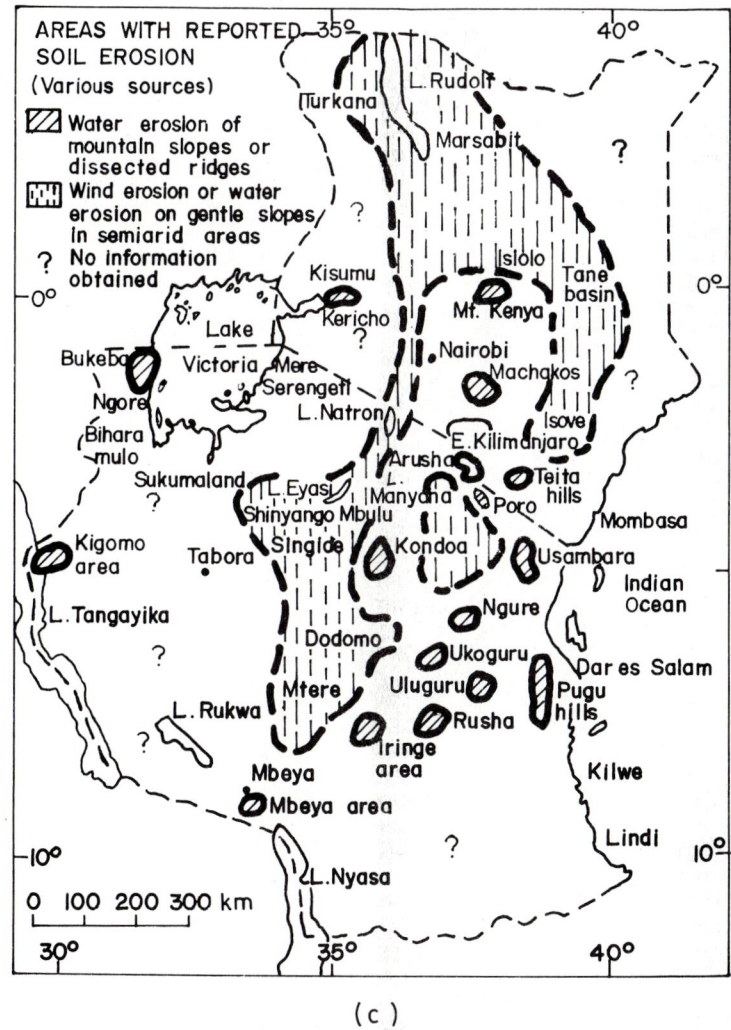

(c)

Figure 8.12 (*Continued*) Isoerodent map for (c) Kenya.

slande et al. (1984) measured the erodibility of three soils of Nigeria with contrasting physical and chemical properties as well as the erodibility of the same soil by using the nomogram of Wischmeier et al. (1969). The data in Table 8.7 show considerable differences between measured and estimated values. Mtakwa et al. (1987) compared measured soil erodibility with that estimated from the nomogram for alfisols in western Nigeria. Their data (Table 8.8) show that the nomogram underestimated erodibility. In contrast, Wilkinson (1975b) observed that the nomogram overestimated the soil erodibility factor K. In Tanzania, Ngatunga et al. (1984) compared measured and estimated soil erodibilities for three soils at Mlingano. The soils were classified as rhodic ferralsols (FAO, 1979). The measured K value was 0.160, 0.153, and 0.121, respectively, for soils of 10, 19, and 22 percent slopes. The K factors estimated from the nomogram for the same soils were 0.09, 0.085, and 0.110, respectively. In this case, the nomogram underestimated soil erodibility. Thomas and Barber (1978) mea-

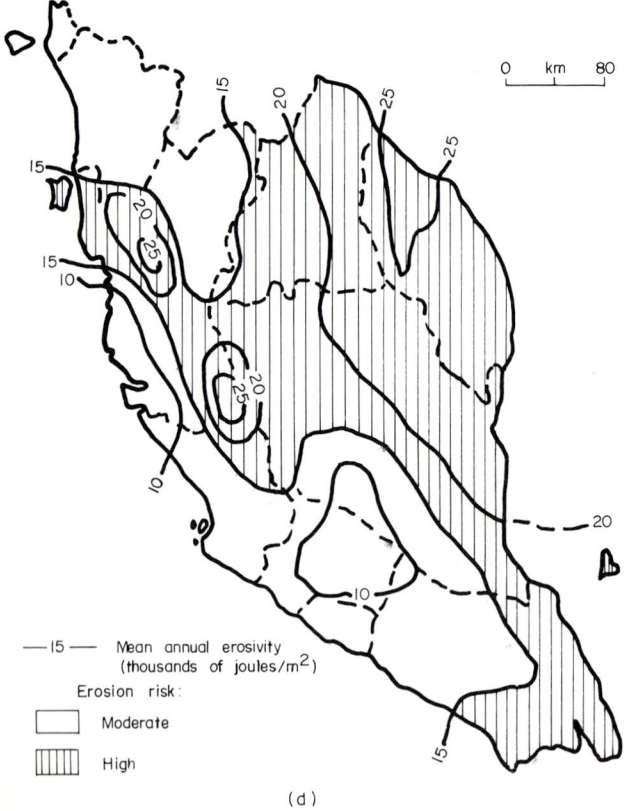

Figure 8.12 (*Continued*) Isoerodent map for (*d*) Malaysia.

sured soil erodibility in Kenya. Their data (Table 8.9) also show differences between measured and estimated factor K for similar reported soils of west Africa. The differences in measured and estimated values indicate the need for appropriate modifications to adapt the nomogram to soils of tropical Africa.

In contrast to soils of tropical Africa, measured and estimated values of K for soils of subtropical and temperate-zone climates in South Africa closely agree. For example, Platford (1982) used a rotating-boom rainfall simulator to measure the erodibility of soils ranging from weak, structureless clays to well-structured clays. Measured erodibility correlated closely with erodibility determined by the nomogram for the less structured soils but not for the well-structured soils. McPhee et al. (1983) also used a rotating-boom rainfall simulator for soils in the pineapple-growing areas. Field measurements, in general, agreed with the nomogram estimates of K.

Measurements of soil erodibility have also been made for a wide range of soils in south and southeast Asia. In Malaysia, Maene et al. (1977) compared measured values with those estimated by nomogram. Their data show that estimated and measured values are closely related. In east Java, Utomo and Mahmud (1984) made the same comparisons for some humus-rich andosols (Table 8.10). They observed that the nomogram-estimated values were always less than the field-determined ones. The high organic matter content, 5 to 10 percent, handicapped the nomogram. Erodibility values K have also been measured for many soils in the Indonesian islands. Sudjadi (1984) and Sukmana (1987) measured the factor K for some soils of Sumatra and other outer islands (Tables 8.11 and 8.12). Using the nomogram, Srikhajon et al. (1984) computed

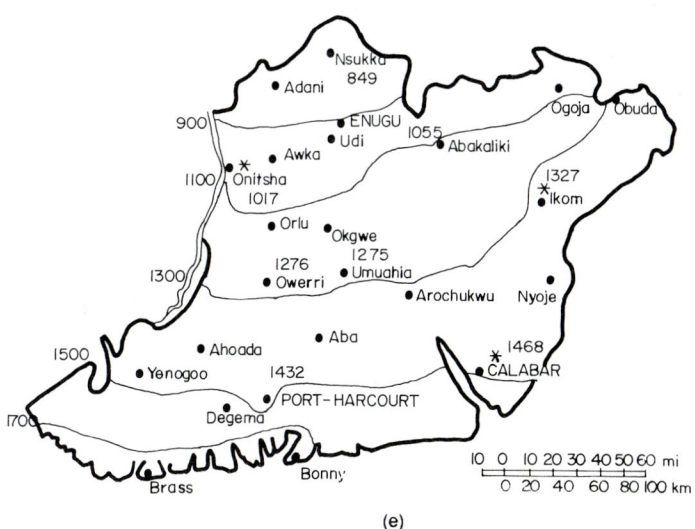

(e)

Figure 8.12 (*Continued*) Isoerodent map for (*e*) eastern Nigeria.

TABLE 8.4 Approximate Values of Soil Erodibility Factor K for 10 Benchmark Soils in Hawaii

Order	Suborder	Great group	Subgroup	Family	Series	K
Ultisols	Humults	Tropohumults	Humoxic Tropohumults	Clayey, kaolinitic, isohyperthermic	Waikane	0.10
Oxisols	Torrox	Torrox	Typic Torrox	Clayey, kaolinitic, isohyperthermic	Molokai	0.24
Oxisols	Ustox	Eutrustox	Tropeptic Eutrustox	Clayey, kaolinitic, isohyperthermic	Wahiawa	0.17
Vertisols	Usterts	Chromusterts	Typic Chromusterts	Very fine, montmorillonitic, isohyperthermic	Lualualei	0.28
					Kawaihae	0.32
Aridisols	Orthids	Comborthids	Ustoltic Comborthids	Medial, isohyperthermic	(Extremely stony phase)	
Inceptisols	Andepts	Dystrandepts	Hydric Dystrandepts	Thixotropic, isothermic	Kukaiao	0.17
Inceptisols	Andepts	Eutrandepts	Typic Eutrandepts	Medial, isohyperthermic	Noolehu (variant)	0.20
Inceptisols	Andepts	Eutrandepts	Entic Eutrandepts	Medial, isohyperthermic	Pakini	0.49
Inceptisols	Andepts	Hydrandepts	Typic Hydrandepts	Thixotropic, isohyperthermic	Hilo	0.10
Inceptisols	Tropepts	Ustropepts	Vertic Ustropepts	Very fine, kaolinitic, isohyperthermic	Waipahu	0.20

SOURCE: El-Swaify and Dangler, 1977.

TABLE 8.5 Measured Index of Erodibility *K*

Location	Soil types	Measured *K* Max.	Measured *K* Min.	Value used	No. measures
Adiopodoume	Low-base, saturated ferralitic on argillic-sandy tertiary material	0.17	0.05	0.10	24
Agonkamey	Medium-based, saturated ferralitic on argillic-sandy tertiary material	0.11	0.03	0.10	4
Bouake	Eroded, reworked ferralitic on granite	0.16	0.02	0.12	4
Korhogo	Impoverished, reworked ferralitic on granite	0.02	0.01	0.02	6
Gampela	Tropical ferruginous on lateritic pan at 20 cm	0.32	0.05	0.25	5
Saria	Tropical ferruginous on lateritic pan at 50 cm	0.28	0.06	0.25	3
Sefa	Leached tropical ferruginous with stains and concretions	0.17	0.05	0.25	2

SOURCE: Roose, 1977a,b.

K for soils of different ecological regions in Thailand. Their data (Table 8.13) show the *K* factor ranging form 0.04 to 0.57. In west Java, Ambar and Wiersum (1980), after testing seven erodibility indices of different soils and land use types, concluded that the *K* factor was the most reliable index. But they cautioned that it should be used only for soils with characteristics similar to the U.S. soils for which the nomogram was developed. Also in Java, Kurnia and Suwardjo (1985) estimated the *K* factor of some oxisols, ultisols, alfisols, and inceptisols. The value ranged from a low of 0.03 for an oxisol to a high of 0.22 for an alfisol.

Nomogram-estimated factor *K* also differed slightly from measured *K* in some soils of western Europe. Bollinne (1985) reported a *K* factor of 0.62 estimated by nomogram compared with the measured 0.53 (with EI_{30} of rainfalls greater than or equal to 12.7 mm), a difference of 17 percent. Differences in tropical soils are often much greater.

The literature presented above indicates that the estimates of *K* may agree with the measured values for structureless or weakly structured soils. The nomogram may also be satisfactorily applied where soils have properties similar to those it was developed to measure, but not where soils contain high clay contents or high-activity clays (vertisols), soils with high proportions of skeletal materials (gravelly and concretionary soils), or soils high in iron or aluminum oxides.

The cover and management factor

Phenology and canopy characteristics of staple food crops of the tropics have not been studied intensively, especially in relation to soil erosion. Important crops

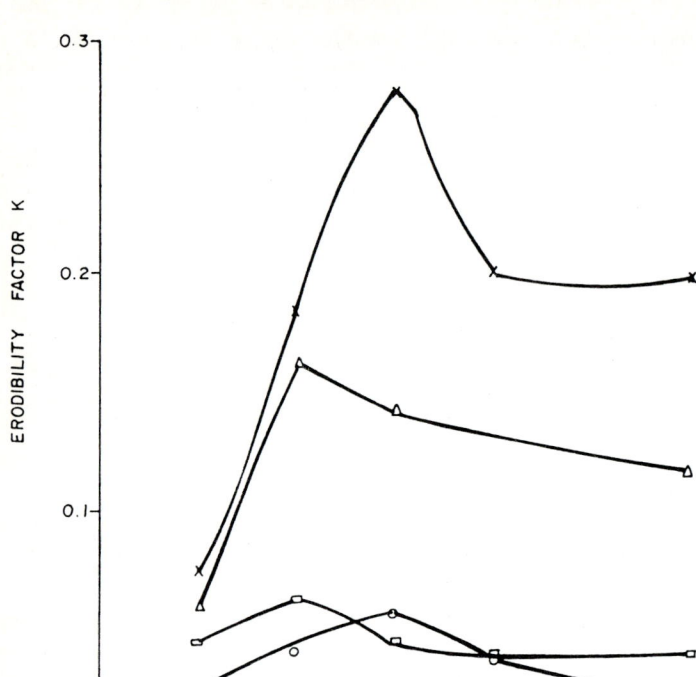

Figure 8.13 Change in soil erodibility factor K with time (Lal, 1984).

are cassava (*Manihot esculenta*), yam (*Dioscorea* spp.) and plantation crops, e.g., oil palm (*Elaeis guineensis*), rubber (*Hevea brasiliensis*), cocoa (*Theobroma cacao*), tea (*Camellia thea*), and coffee (*Coffea arabica*). Nor have the effects on erosion by some of the cultural practices exclusively used in traditional tropical farming systems been systematically assessed, e.g., canopy, tied ridges, ridges up and down the slope, and mulch farming. Relevant examples of some of the available data are discussed below.

Roose (1977a,b) measured the cover and management factor C for a range of crops and management systems in Francophone west Africa. The data in Table 8.14 show that the C factor for some conservation-effective systems may be as low as 0.01, for example, surface residue mulch, a quick-growing cover crop. Similar measurements were reported for mulch farming and no-till systems for soils of western Nigeria (Lal, 1976). The data in Table 8.15 show that the C factor for soil loss from maize-maize and maize-cowpea rotation for mulch and no-till farming approached zero even for steep slopes up to 15 percent. The C factor for water runoff was also low for the mulch and no-till systems (Table 8.16). For the mulch system, however, the C factor varied with the slope and mulch rate

TABLE 8.6 Erodibility according to Textural Class: Minimum, Maximum, Mean, and Standard Deviation

Measured K

Textural class	No. samples	Measured erodibility K			
		Minimum	Maximum	Mean	Standard deviation
Sandy loam	24	0.013	0.378	0.160	0.104
Sandy clay loam	16	0.000	0.237	0.126	0.078
Loam, silt loam	4	0.116	0.342	0.230	0.063
Clay loam	5	0.123	0.355	0.237	0.082
Clay	6	0.000	0.535	0.316	0.232

Nomogram-computed K_c

Textural class	No. samples	Nomogram-computed erodibility K_c			
		Minimum	Maximum	Mean	Standard deviation
Sandy loam	24	0.03	0.42	0.085	0.079
Sandy clay loam	16	0.02	0.13	0.052	0.035
Loam, silt loam	4	0.23	0.37	0.322	0.063
Clay loam	5	0.08	0.29	0.224	0.082
Clay	6	0.01	0.27	0.152	0.111

(Figs. 8.14 and 8.15). The C factor for soil loss approached 0.001 for a mulch rate of 6 t/ha from slopes of less than 10 percent. The C factor is apparently related to the proportion of rainfall energy intercepted by the canopy. The greater the canopy cover, the greater the energy interception (Fig. 8.16). Ngatunga et al. (1984) computed C factor for both runoff and soil loss for three management systems and three slopes at Mlingano, Tanzania. The data in Table 8.17 show that the C factor for soil loss was as low as 0.002 for mulch and grass fallow treatments.

TABLE 8.7 Difference between Measured and Estimated Erodibility Factor for Three Soils of Nigeria

Location	Soil	Parent material	Erodibility factor* K	
			Measured	Estimated
Ikom	Rhodustalf	Basalt	0.015	0.039
Heipang	Paleustalf	Basement complex	0.04	0.18
Onne	Paleudult	Coastal deposits	0.04	0.025

*SI units.
SOURCE: Vanelslande et al., 1984.

TABLE 8.8 Estimated and Measured Erodibility Factor *K* in Bare Fallow Plots

Slope, %	Season	K factor, mg/ha	
		Estimated	Measured
11	1	0.148	0.193
	2	0.160	0.168
11	1	0.148	0.162
	2	0.139	0.147
8	1	0.154	0.206
	2	0.136	0.245

SOURCE: Mtakwa et al., 1987.

In India, Singh et al. (1981) determined the cover factor *C* for different tropical crops. The *C* factor ranged from 0.1 to 0.54 (Table 8.18). Compared with cultivated crops, the *C* factor for tropical rain forest and perennial covers is usually low, as evidenced by the data of Srikhajon et al. (1984) from Thailand shown in Table 8.19. The *C* factor for tropical rain forest ranged from 0.0003 to 0.017 and that of human-created forest from 0.019 to 0.09. In comparison, the *C* factor for disturbed forest was 0.400. The *C* factors for different cropping systems in Nigeria and for a number of annual crops are show in Tables 8.20 and 8.21, respectively. Additional measurement of the *C* and *P* factors for different cropping systems in Indonesia are reported by Abdurachman et al. (1984). Similar to the data from elsewhere, the *C* factor is low for cultural practices that maintain an effective soil cover. Such conservation-effective measures as no-till and cover crops are applicable throughout the African, Asian, and South American tropics.

Specific examples of using USLE in the tropics

The application of USLE to geographic regions outside the United States can be improved by using locally derived data of the various factors. With the basic

TABLE 8.9 Calculated Soil Erodibility *K* Values for the Kabete and Katumani Soils

Rainfall intensity, mm/h	Moisture state			
	Kabete soil		Katumani soil	
	Dry	Wet	Dry	Wet
25	—	—	0.03	0.07
50	0.02	0.06	0.12	0.60
100	0.02	0.05	0.11	0.49
150	0.04	0.15	—	—

SOURCE: Barber et al., 1979.

TABLE 8.10 Comparison of Nomogram-Estimated and Field-Determined Values of Factor K for Humus-Rich Andosols in East Java

Soil	Depth	Texture	Factor K Nomogram	Factor K Field
Association of brown regosol and humid gleic	D	M	0.26	0.21
	D	M	0.16 (<0.10)	0.07
	D	M	0.22	0.18
Association of yellowish brown andosol and brown regosol	D	C	0.22	—
	D	M	0.18 (<0.10)	0.08
	D	C	0.18 (<0.10)	0.09
Complex of brown andosol and lithosol	D	M	0.10	—
	D	C	0.26	0.22

D = deep, M = medium texture, C = clay texture.
SOURCE: Utomo and Mahmud, 1984.

data made available, USLE has been used to choose between appropriate soil conservation practices and to estimate the erosion potential of a specific ecology. USLE has been widely used in Brazil, especially in the subtropical regions where wheat-soybean rotation is commonly practiced. In Chile, Pena Mac-Caskill (1983) computed local values of R, K, and C of the USLE for an area of the Andean foothills. The K factor for a silty loam soil (Dystrandept) was 0.09, and the value of C for various cropping systems ranged from 0.004 to 0.36.

In Israel, Degani et al. (1979), using USLE, developed an interactive computer program, SOILCART, to predict the impact of change in land use on soil erosion. In Australia, Edwards and Charman (1980) adapted USLE for specific problems using experimentally determined K, R, and C values.

The USLE has been used extensively in southeast and east Asia. In Philippines, Bruce (1985) considered using USLE for land planning. O'Sullivan (1985) observed USLE to be a useful tool on a comparative rather than an absolute basis. In Thailand, Srikhajon et al. (1984) used USLE to assess the erosion potential for different ecological regions. They computed factor K from the nomogram and values of T, LS, and P from USDA Agricultural Handbook 224 for the United States. In Java, Utomo and Mahmud (1984) explored the possibility of using USLE in a mountainous area of east Java with humus-rich andosols and concluded that, with adjustments, it could provide useful guidelines for erosion control. In India, Singh et al. (1985) stated that the availability of specific values for some USLE factors permits one to use USLE to estimate soil erosion under different management conditions. In China, Shi and Yang (1983) used locally derived data on erodibility, slope, intensity of land use, and damage from unsuitable land use to predict erosion by water in a loessial, hilly region. In Taiwan, Cook (1970) used the R and C values from Iowa to predict soil losses under location conditions. Chan (1981), however, used locally derived data to select appropriate soil conservation measures on steeplands. In Korea, Shin et

TABLE 8.11 Measured K Factor (USLE) and Nomograph of Soils at Eight Locations of Erosion Experimental Plots

Soil	Textural class	Location	Slope degrees, %	EI_{30}	Years of observation	Soil loss, t/(ha · yr)	Measured K Ranges	Measured K Average	K nomograph
Haplorthox	Heavy clay	Damoga (West Java)	15	1476–3539	4	81.9–385.8	0.02–0.05	0.04	0.05
			18	1476–3539	4	121.0–391.4	0.02–0.04	0.03	0.05
			22	1476–3539	4	113.9–376.6	0.01–0.03	0.03	0.05
Haplorthox	Heavy clay	Citayam (West Java)	14	2573–3548	3	440.7–532.0	0.06–0.09	0.09	0.09
Troporthents	Clay	Taojungharjo (Yogyakarta)	10	1092–1410	3	133.1–249.3	0.11–0.16	0.14	0.16
Chromuderts	Clay	Jegu (East Java)	7	1106–1683	2	152.8–285.0	0.24–0.30	0.27	0.27
Tropudults	Clay	Pekaiongan (Lampung)	3.5	1194–2922	3	97.7–144.5	0.14–0.27	0.16	0.19
Tropohumults	Clay	Citaman (West Java)	14	811–1892	7	101.5–421.2	0.09–0.11	0.10	0.12
Tropudalfs	Heavy clay	Putat (Yogyakarta)	9	1017–2627	7	259.1–607.2	0.16–0.29	0.23	0.21
Tropuqualfs	Clay	Punung (East Java)	10	830–2159	7	220.2–459.2	0.18–0.25	0.22	0.22

SOURCE: Kurnia and Suwardjo, 1985.

TABLE 8.12 Soil Physical Characteristics and K Factor (Nomograph) of 13 Soils in West Sumatra and Northern Caram

Soil	Textural class	Silt + fine sand, %	Sand, %	Organic matter, %	Codes		Predicted K
					Structure	Permeability	
West Sumatra							
Tropohumults	Heavy clay	32	6	3.55	4	3	0.16
Tropohumults	Silty clay	47	4	6.00	3	3	0.15
Tropudults	Heavy clay	27	10	4.50	3	5	0.19
Haplorthox	Silty clay	25	39	5.71	4	3	0.16
Haplorthox	Clay	32	26	3.84	4	3	0.21
Tropudults	Clay	30	9	4.29	4	3	0.17
Tropudults	Silty clay	44	7	4.71	4	3	0.21
Tropudults	Sandy clay loam	33	39	2.88	4	3	0.27
Tropudults	Sandy loam	25	47	2.74	4	3	0.21
Northern Caram							
Tropudalfs	Loam	65	11	3.40	4	5	0.40
Tropudalfs	Sandy loam	31	52	1.86	2	2	0.19
Tropudalfs	Sandy clay loam	42	38	3.52	4	3	0.33
Tropudults	Silty clay loam	54	7	6.00	4	5	0.31

SOURCE: Kurnia and Suwardjo, 1985.

TABLE 8.13 Soil Erodibility Factor K for Soils from Different Ecological Regions in Thailand

Soil texture	Soil		North		Average K value Northeast		East		Central and west	
	Upland	Lowland	Upland	Lowland	Upland	Lowland	Upland	Lowland	Upland	Lowland
Sand	0.04	0.04	—	—	—	—	0.05	0.05	—	—
Loamy sand	0.07	0.04	0.05	0.06	0.04	0.05	0.07	0.08	0.08	0.07
Sandy loam	0.20	0.30	0.27	0.30	0.24	0.26	0.19	0.34	0.30	0.26
Loam	0.33	0.34	0.33	0.35	0.29	0.35	0.30	0.33	0.33	0.43
Silt loam	0.40	0.34	0.49	0.34	0.37	0.34	0.21	0.44	0.56	0.47
Silt	—	0.37	—	—	—	—	—	—	—	—
Sandy clay loam	0.19	0.21	0.21	0.22	0.24	0.20	0.25	0.23	0.20	0.21
Clay loam	0.29	0.31	0.24	0.27	0.25	0.36	0.30	0.25	0.28	0.29
Silty clay loam	0.31	0.21	0.35	0.42	0.46	0.43	0.37	0.38	0.38	0.29
Sandy clay	—	0.31	—	0.17	—	—	—	0.18	0.15	0.17
Silty clay	0.22	0.29	0.21	0.27	0.23	0.27	0.19	0.29	0.25	0.23
Clay	0.11	0.14	0.15	0.18	0.13	0.15	0.12	0.14	0.14	0.18

Remarks: Upland means the undulating to hilly terrain while lowland is the flat or nearly flat terrain. Some textural classes have no K value because in that region and/or landform appropriate information does not exist. Those K values are the average values of the soils which have the same surface textural class and are situated in the same landform of the same region.

SOURCE: Srikhajon et al., 1984.

TABLE 8.14 Vegetal Cover Factor and Cultural Techniques in West Africa

Cultural techniques	Annual average C factor
Bare continuously fallowed	1
Forest or dense shrub, high mulch crops	0.001
Savanna, prairie in good condition	0.01
Overgrazed savanna or prairie	0.1
Crop cover of slow development or late planting—first year	0.3–0.8
Crop cover of rapid development or early planting—first year	0.01–0.1
Crop cover of slow development or late planting—second year	0.01–0.1
Corn, sorghum, millet (as a function of yield)	0.4–0.9
Rice (intensive fertilization)	0.1–0.2
Cotton, tobacco (second cycle)	0.5–0.7
Peanuts (as a function of yield and date of planting)	0.4–0.8
First year cassava and yam (as a function of date of planting)	0.2–0.8
Palm tree, coffee, cocoa with crop cover	0.1–0.3
Pineapple on contour (as a function of slope)	
(burned residue)	0.2–0.5
(buried residue)	0.1–0.3
(surface residue)	0.01
Pineapple and tied ridging (slope 7%)	0.1

SOURCE: Roose, 1977a,b.

al. (1976) used locally derived data to compute C for different soil and crop management systems. In Nepal, Fetzer (1977) used USLE to predict soil erosion from steeplands. USLE has also been extensively used in Africa. In Zimbabwe, Wandelaar (1978) reported significant differences between measured and estimated soil erosion. In general, predicted erosion was more than that measured by field plot techniques. USLE has been used in west Africa by Roose (1977b).

TABLE 8.15 Mean Cover Factor C for Indicated Slope Percentage Cropping Systems

Slope, %	First season 1973				Second season 1973			
	A	B	C	D	A	B	C	D
1	0.00	0.20	0.00	0.06	0.00	0.11	0.00	0.19
5	0.00	0.10	0.00	0.06	0.00	0.04	0.00	0.08
10	0.00	0.08	0.00	0.04	0.00	0.04	0.00	0.06
15	0.00	0.014	0.00	0.04	0.01	0.16	0.03	0.39

A = maize-maize (mulch at 6/ha), B = maize-maize (plowed), C = maize-cowpea (no-till), D = cowpea-maize (plowed).
SOURCE: Lal, 1976.

TABLE 8.16 Mean C Factor for Water Runoff for Indicated Soil and Crop Management Systems

Crop/soil management	C factor for water runoff	
	First season	Second season
Bare fallow	1.00	1.00
Maize-maize (mulch)	0.04	0.05
Maize-maize (plowed)	0.35	0.29
Maize-cowpea (no-till)	0.05	0.10
Cowpea-maize (plowed)	0.16	0.40

SOURCE: Lal, 1976.

In western Nigeria, Mtakwa et al. (1987) compared soil erosion estimated by USLE with that measured on field runoff plots. The estimates of soil erosion by the USLE were lower than those determined by the runoff plot technique (Table 8.22). The lower estimate results from using underestimated values of both R and K in USLE. Contrary to the results from tropical Africa, Banasik (1985) re-

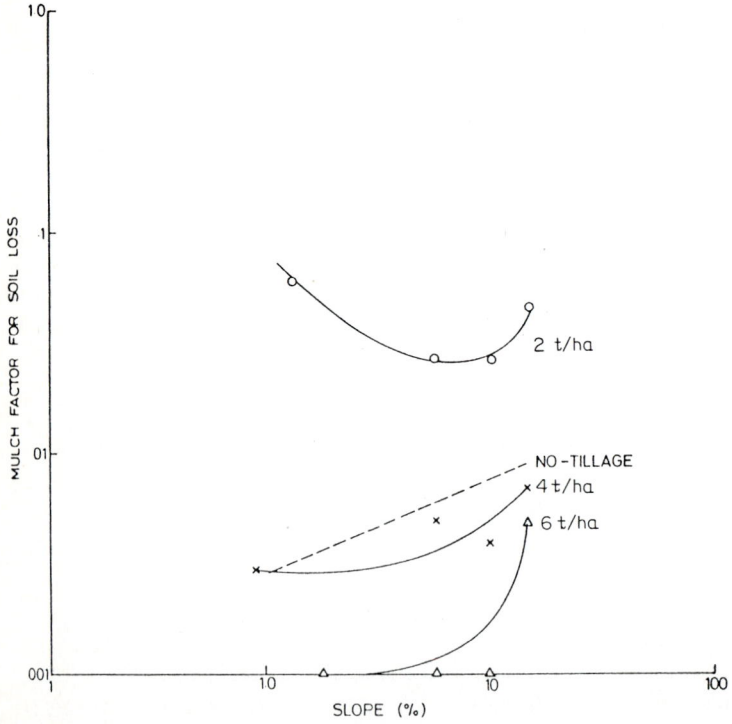

Figure 8.14 Effects of slope on the C factor for a tropical alfisol for different mulch rates (Lal, 1976).

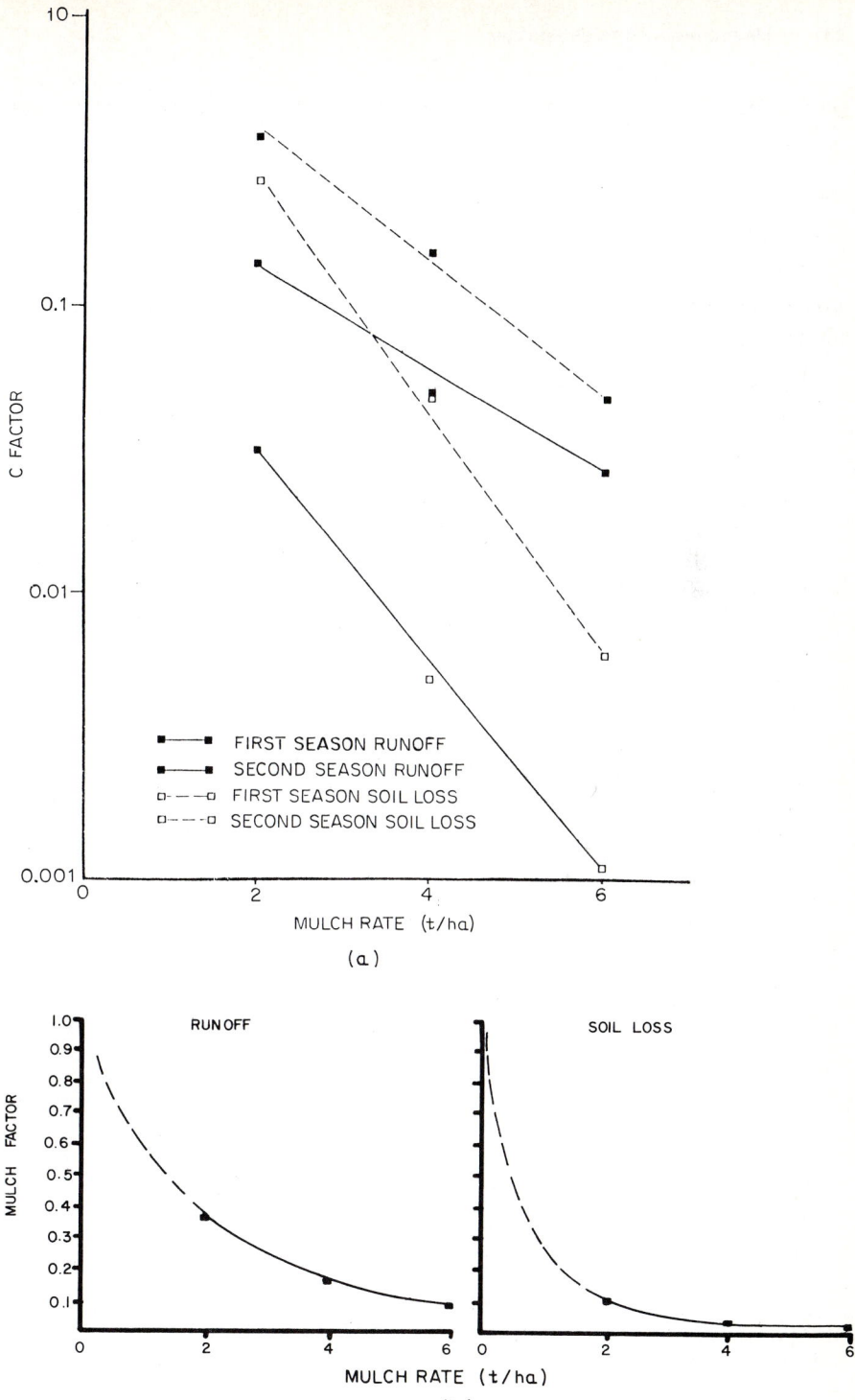

Figure 8.15 Effects of the mulch rate on (*a*) C factor for a tropical alfisol and (*b*) mulch factor for runoff and soil loss (Lal, 1976).

ported from Poland that soil erosion computed by USLE was greater than that measured by runoff plots.

The discrepancy between measured and estimated results is a problem typical of the difficulties that arise from the use of USLE without validating its applicability. When its applicability is not validated, any good that its estimates can do in planning conservation measures or choosing a cultural practice or cropping system is questionable. Applying USLE to do what it was designed to do may be acceptable for some geographic and ecological conditions, provided that locally obtained data are used for the different variables involved. If variables developed in the United States are used for regions in Asia, Africa, or elsewhere, it is doubtful that interpretations based on such data are useful. It is precisely because of these problems that attempts are now under way to develop predictive models based on conceptual principles. The USDA Water Erosion Prediction Project (WEPP) is one such attempt.

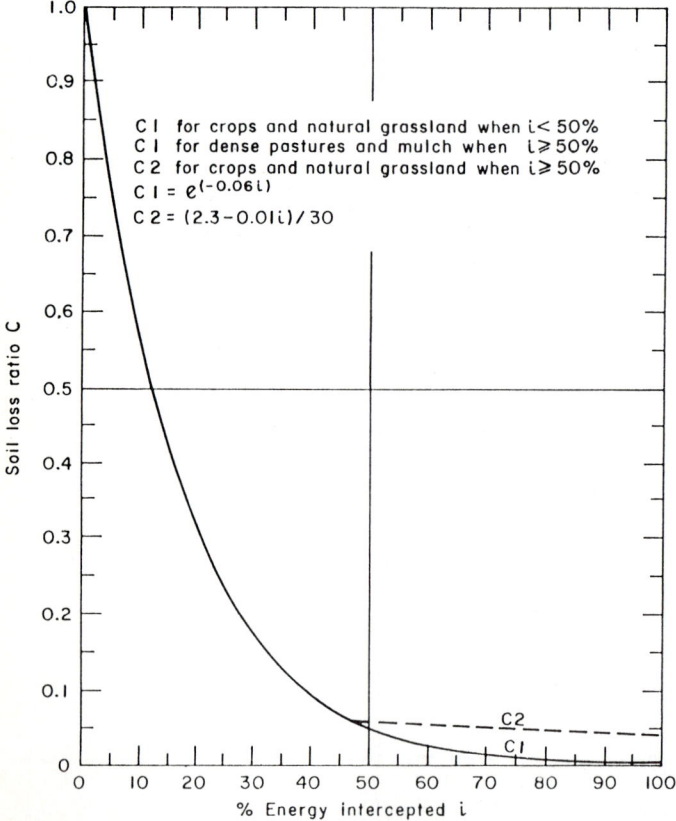

Figure 8.16 The C factor in relation to the percentage of energy intercepted for grassland in Zimbabwe (Elwell, 1984).

TABLE 8.17 The C Factor for Runoff and Soil Loss for Three Soils at Mlingano, Tanzania

Season	Slope, %	Mulch		Plowing		Grass fallow	
		Soil loss	Runoff	Soil loss	Runoff	Soil loss	Runoff
Vuli*	10	0.004	0.068	0.389	0.324	0.003	0.072
	19	0.002	0.074	0.319	0.223	0.002	0.046
	22	0.004	0.053	0.369	0.239	0.002	0.043
Masika†	10	0.003	0.052	0.348	0.492	0.002	0.043
	19	0.002	0.036	0.443	0.570	0.001	0.014
	22	0.002	0.028	0.249	0.449	0.001	0.013
Annual	10	0.003	0.055	0.359	0.460	0.002	0.019
	19	0.002	0.045	0.408	0.481	0.002	0.023
	22	0.002	0.35	0.266	0.392	0.001	0.021

*Vuli is short rainy season from September to December.
†Masika is long rainy season from March to June.
SOURCE: Ngatunga et al., 1984.

Other Predictive Models

Because the erosion potential of rains in the tropics is assumed to be greater than in the temperate latitudes, attempts have been made to develop erosion forecasting models specifically for the tropical environments.

Erosion prediction models for southern Africa

With a long history of both basic and applied erosion research in Zimbabwe (formerly Rhodesia), development of erosion prediction techniques was a natural outcome of the voluminous data collated. In the early 1960s Hudson (1961) proposed a parametric prediction equation similar to USLE

$$A = TSLPMR \tag{8.27}$$

where A is erosion and remaining variables are soil type T, slope gradient S, slope length L, agricultural practice P, mechanical protection M, and rainfall R. Naturally, Hudson used the KE > 1 erosion index and Eq. (8.3) to compute the kinetic energy of rainfall.

Also in Zimbabwe, Elwell (1977) and Elwell and Stocking (1982) developed another empirical model called SLEMSA (for soil loss estimation system for southern Africa). This model is shown in Eq. (8.28):

$$A = KCX \tag{8.28}$$

where A is the predicted mean annual soil loss, K is the mean annual soil loss from a standard field plot 30 m by 10 m on a 4.5 percent slope with soil of known

TABLE 8.18 Crop Cover Factor C for Some Tropical Crops Grown in India

Crop	C
Green gram (*Vigna radiata*)	0.392
Black gram (*V. mungo*)	0.538
Groundnut (*Arachis hypogeer*)	0.406
Soybean (*Glycine max*)	0.421
Cowpea (*V. unguiculata*)	0.386
Maize (*Zea mays*)	0.502
Paddy (*A. sativa*)	0.28
Pigeon pea (*Cajanus cajan*)	0.38
Pineapple (weed control)	0.38
Pineapple (weed)	0.10

SOURCE: Singh et al., 1981.

erodibility under fallow, C is the ratio of soil loss from a cropped plot to loss from the standard plot, and X is the ratio of soil loss from a plot of length L and slope S to that lost from the standard plot. The input data consist of energy intercepted by the crop, rainfall energy, soil erodibility, slope steepness, and slope length.

Elwell (1978) reported that Eq. (8.28) predicts mean annual soil losses to within 1.7, 2.6, and 4.3 t/(ha · yr) in 50, 70, and 90 percent of cases, respectively, for the normal range of soil losses from agricultural lands. Considerable field and laboratory work was done to refine and standardize the variables in Eq. (8.28) (Elwell, 1979). Because the factors of soil erosion by water are the same in temperate and tropical climates, the apparent similarity between Eqs. (8.27) and (8.28) and USLE [Eq. (8.3)] is neither surprising nor coincidental.

TABLE 8.19 Crop Cover Factor C for Tropical Vegetation

Forest type	C
Tropical rain forest	0.001
Hill evergreen forest	0.0003
Mixed deciduous with teak	0.014
Monsoon forest + savanna forest ecotone	0.017
Dry evergreen forest	0.019
Mixed deciduous forest	0.048
Dry dipterocarp forest	0.064
Teak plantation	0.088
Dipterocarp savanna forest	0.09
Disturbed hill evergreen with early state of pine	0.400

SOURCE: Srikhajon et al., 1984.

TABLE 8.20 C Factors for Different Cropping Systems in Nigeria

Cropping system	C factor	
	Runoff	Erosion
Uncropped control*	1.0	1.0
Maize	0.79	0.99
Yam (mounds)	0.74	0.50
Cassava (ridges)	—	—
Melon	—	—
Sweet potato	0.89	0.90
Maize + sweet potato	0.61	0.51
Cassava + melon	—	0.92
Yam + melon	—	0.99
Cassava + maize	—	—

*Runoff and erosion under uncropped control were 250.9 mm and 4.4 t/ha, respectively.
SOURCE: Unpublished data of Lal, 1980.

Erosion prediction on a watershed scale for eastern Europe

One of the USLE limitations is that it does not predict erosion from large watersheds. So geomorphologists and geographers interested in mapping the soil erosion have developed simple models to predict erosion from large areas. For example, Stehlik (1975) proposed an empirical model to estimate potential soil erosion at a scale of 1:500,000. He defined the potential as the combined effect of stable natural factors which influence soil erosion

$$A_1 = DGPS \qquad (8.29)$$

where A_1 is the potential water erosion rate (mm/yr), D is the coefficient of climatic conditions, G is the coefficient of geologic conditions, and S is the coefficient of inclination. Stehlik used a map thus developed to prepare the erosion potential map of central Europe.

TABLE 8.21 C Factors for a Number of Annual Crops

Crop	C factor
Maize	0.39
Maize with lalang mulch	0.02
Cowpea	0.27
Groundnut	0.28
Mung bean	0.04
Chilli	0.39

SOURCE: Mokhtaruddin et al., 1985.

TABLE 8.22 Paired t Test for Soil Loss (mg/ha) Determined by USLE and Runoff Plot Technique, Bare Fallow Plots, Season 1*

	Technique		Difference	t_H
	USLE	Runoff plot		
Sum	1077.0	1380.1	−303.1	
Mean	119.7	153.4	−33.7	4.88†
			SE +13.3	

SE = standard error.
*Average of nine replications.
†Significant at 1% level.
SOURCE: Mtakwa et al., 1987.

In the U.S.S.R., Golubev (1982) proposed another model to predict erosion from large watersheds. His model considered the land use to be an important determinant of soil erosion:

$$EG = ERN(AT - AC) + K \cdot ERN \cdot AC \tag{8.30}$$

Here EG is the total soil erosion in a given type of landscape [t/(km$^2 \cdot$ yr)], ERN is the soil erosion rate under natural vegetation in the given type of landscape [t/(km$^2 \cdot$ yr)], AT and AC are the total area of the given type of landscape and the area of the cropland (km^2), respectively, and K is ratio of soil erosion yield from cropland to that from natural areas. The first component in the equation represents soil erosion from areas with natural vegetation, and the second relates to cropland areas. But the schematics presented do not suggest how to predict the components.

Model to predict the development of level-bench terraces

In Kenya, grass strips are effective as a soil conservation measure (see Chap. 11), which led Barber et al. (1981) to develop a model to predict how much level-bench terraces developed on steep backslope terraces would control erosion and to compare erosion from level between terraces with that from normal tillage practices. The model is based on USLE.

Empirical models are site-specific, so it is inappropriate to use them to predict erosion outside the ecological limits they were developed for. Information obtained by inappropriate use of USLE in geographic regions where its application has not been validated is questionable. The same is true where variables have not been obtained or standardized. Careless and inappropriate use of USLE, especially in the tropics, may have been counterproductive, by leading to erroneous conclusions, faulty planning, and improper conservation measures. It is a mere coincidence if empirical models happen to apply in regions or situations other than those they were developed for. One reason for the numerous modifications in USLE is its regional specificity. Theoretically, each research site in a different ecological environment would require a USLE modification. That is one of the reasons for the proliferation of empirical models (Table

8.23). It is the time now for researchers around the world, and especially those in the tropics, to be sure they have a conceptual understanding of erosion processes and that they establish strong data bases with technological alternatives for effective erosion control. Enough resources have been diverted away from basic research in proving or disproving empirical models. Although site-specific empirical models can be useful, there is no limit to their development. Theoretically, one can have almost as many empirical models as the researchers involved in generating them.

Erosion Prediction for Gully and Channel Erosion

It is more difficult to predict gully and channel erosion than rill-interrill erosion. Traditionally, gully erosion has been estimated from cross-sectional surveys of channel sections taken at suitable time intervals. Aerial photographs taken at regular time intervals have also been used to estimate gully erosion. Many researchers have proposed using bed-load transport equations for rough estimates of bed-load transport in gulleys.

The rate at which gully growth advances in loess soils in western Iowa was estimated by Beer and Johnson (1963) who used the equation

$$G_A = 81.41 R_I^{0.0982} A_t^{-0.0440} L_g^{0.7954} L_w^{-0.2473} e^{-0.0014 P} \qquad (8.31)$$

where G_A is gully surface growth (m^2), R_I is the index of surface runoff (mm), A_t is the terraced area of watershed (m^2), L_g is the beginning length of gully (m), L_w is the distance from the end of the gully to watershed divide (m), e is the base of natural logarithm, and P is the deviation of precipitation from normal (mm). Thompson (1964) developed a similar equation for estimating gully advance in Minnesota:

$$G_1 = (7.13 \times 10^{-5}) A^{0.49} S^{0.14} R^{0.74} Cl \qquad (8.32)$$

Here G_1 is the gully head advancement for the period considered (m), A is the drainage area above gully head (m^2), S is the slope of the approach channel

TABLE 8.23 Empirical Models to Predict Erosion from Plots and Watersheds

Plots		Watersheds	
Number	Authors	Number	Authors
1	Baver (1933)	1	Horton (1945)
2	Zingg (1940)	2	Dragoun and Miller (1966)
3	Ellison (1947)	3	Stehlik (1975)
4	Musgrave (1947)	4	Golubev (1982)
5	Wischmeier and Smith (1958)		
6	Hudson (1961)		
7	Elwell and Stocking (1982)		

above the gully head (%), R is the sum of total rainfall from 24-h rains equal to or exceeding 12.7 mm for the period considered, and Cl is the clay content of the eroding soil profile. The average gully head advance is also estimated by using Eq. (8.33), developed by USDA (1966):

$$G_1 = (4.25 \times 10^{-3})A^{0.46}R^{0.20} \tag{8.33}$$

Physical or Conceptual Models

Some progress has been made in developing deterministic models based on fundamental erosion processes. To date, the few physical or conceptual models developed have not been widely tested. Some of the basic climatic and hydrologic functions used in them are the same as those described in Chaps. 2 and 3.

Major factors governing the upland erosion process, i.e., interrill and rill erosion, were outlined by Meyer and Wischmeier (1969) (Fig. 8.17). Basic factors involved in erosion are similar to those outlined in empirical models, erosivity, erodibility, management, etc. Processes affected by these factors are defined in precise physical or mathematical forms, rather than by regression equations. Once these processes are defined, two laws are used to assess sediment balance: the law of conservation of energy or momentum and the equation of continuity.

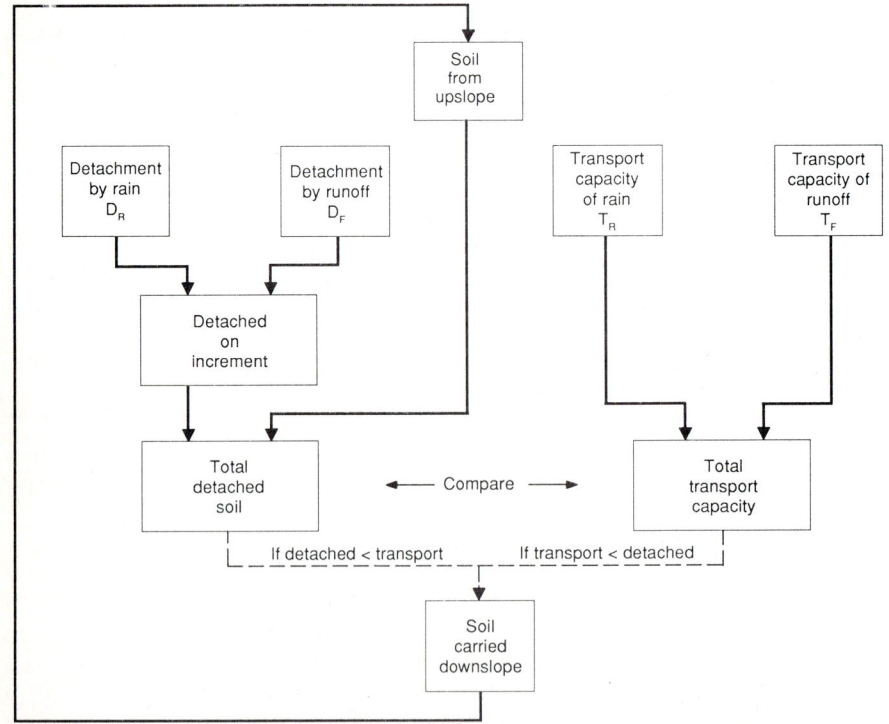

Figure 8.17 Factors governing upland erosion (Meyer and Wischmeier, 1969).

TABLE 8.24   Deterministic or Conceptual Models in Estimating Soil Erosion and Sedimentation

Erosion		Sedimentation	
Number	Authors	Number	Authors
1	Meyer and Wischmeier (1969)	1	Negev (1967)
2	Foster and Meyer (1975)	2	Fleming and Fahmy (1973)
3	David and Beer (1975)	3	Fleming and Walker (1976)
4	Martinez et al. (1980)	4	Kirkby (1976)
5	Brenneman and Laflen (1980)	5	Standford model
6	Khanbilvardi et al. (1982)	6	Leite et al. (1982)
7	Rose and Freebairn (1985)	7	Piest et al. (1975)
8	Boon and Savat (1981)		

Examples are given by David and Beer (1975), Foster (1982), Kirkby (1980), Rose and Freebairn (1985), and Rose (1987). There is a separate model for each major process involved, i.e., soil detachment and transport model for upland areas (Martinez et al., 1980), modeling erosion and sediment transport in rills (Khanbilvardi et al., 1982), effect of crop residue on rill erosion (Brenneman and Laflen, 1980), sediment transport models (Piest et al., 1980), model for estimating on-site soil erosion (Li et al., 1977), and many others (Table 8.24).

Upland erosion model

This model is proposed by Foster and Meyer (1975) and Foster (1982). It distinguishes erosion processes involved in interrill and rill erosion. The model adds the sediment contribution from interrill erosion. It assumes a uniform downslope overland flow and that flow and sediment are uniformly distributed across the slope; therefore, variables are expressed on a total-width or total-area basis. The basic relationship is the continuity equation

$$\frac{q_s}{x} + \frac{\rho_s(cy)}{t} = D_r + D_i \qquad (8.34)$$

where q_s is the sediment load (mass per unit width per unit time), x is the distance along the slope, ρ_s is the mass density of the sediment particles, c is the concentration of sediment in the flow (volume of sediment per volume flow), y is the flow depth, t is time, D_r is the delivery rate of sediment from interrill areas (mass per unit area per unit time). The term q_s/x is the rate of change (buildup or loss) of sediment with distance, $\rho_s(cy)/t$ is the storage rate of sediment within the flow depth, and D_r and D_i are contributions of sediment from lateral inflow. By assuming quasi-steady sediment movement, Eq. (8.34) is reduced to

$$\frac{dq_s}{dx} = D_r + D_i \tag{8.35}$$

Equation (8.35) simply adds interrill and rill sediment contributions.

Water erosion prediction project (WEPP)

The upland erosion model developed by Foster and Meyer (1975) and Foster (1982) is now revised and updated and is called WEPP (Foster and Lane, 1987). This model was developed primarily by the United States Department of Agriculture (USDA) Agricultural Research Service (ARS). This model will replace USLE as the primary means of predicting soil erosion by the Soil Conservation Service (SCS) (Foster and Lane, 1987; Foster, 1990).

Basically the model separately describes rill and interrill erosion as two major components of upland erosion. The interrill detachment is modeled as follows (Elliott et al., 1989):

$$D_i = K_i I^2 \tag{8.36}$$

where D_i is the interrill detachment rate [kg/(m²/s)], K_i is the interrill erodibility [(kg · s)/m⁴)], and I is rainfall intensity (m/s). Because interrill erosion is related to slope gradient (S), the comprehensive interrill erosion equation involving all factors is as follows:

$$D_i = K_i I^2 S_f \tag{8.37}$$

$$S_f = 1.05 - 0.85 \exp[-4 \sin(S)] \tag{8.38}$$

where S_f is a slope factor. The S_f varies from 0.2 for a flat slope to 1.0 for a slope 45°, to 1.05 for a slope of 90°.

The rill erosion is modeled as follows:

$$D_c = K_r(\pi - \pi_c) \tag{8.39}$$

where D_c is the detachment capacity of the clear water [kg/(m²/s)], K_r is the rill erodibility of soil due to hydraulic shear (s/m), and π_c is shear below which there is no detachment (P_a). π is hydraulic shear of flowing water (P_a) and is computed as follows:

$$\pi = \gamma r_h s \tag{8.40}$$

where γ is density of water (N/m³), r_h is hydraulic radius (m), and s is hydraulic gradient or slope of the rill bottom.

The mechanics of the use of this model are described in detail by Nearing and Lane (1990).

The WEPP model is in the development stages and has not yet been validated and adapted to major soils and environments within the United States. Apparently, the model has not been tested anywhere in the tropics. However, it is important that researchers in the tropics obtain the data needed for validation and adaptation of this and other conceptual models. The basic input

Griffith University model

Rose and associates developed another process-oriented model at Griffith University, Australia (Rose et al., 1983a,b,c; Rose, 1985a). Like the upland erosion model, their model assumes that landslope and soil strength relationships are such that landslides, mud flows, and gully erosion do not occur and upland erosion is mainly associated with the shallow overland flow and concentrated rill flow. Their model estimates sheet and rill erosion from

$$\text{Total erosion} = \text{Sediment flux} + \text{water flux} \times C \qquad (8.41)$$

where C is the concentration of sediment in runoff water expressed as mass of sediment per unit volume. The term *flux* denotes the rate of flow across a unit width of slope. Sediment concentration is the net sediment entrained in the flow—gross sediments minus deposition.

The source of gross sediment is rainfall detachment and runoff entrainment. The model uses known mathematically derived relations for these three processes:

$$C(L, t) = \frac{aC_e P}{QI} \sum_{i=1}^{n} \frac{1}{\gamma_i} + \rho g S K C_r \left(\frac{1 - x_*}{L}\right) \quad \text{kg/m}^3 \qquad (8.42)$$

The first term on the right-hand side of Eq. (8.42) is rainfall detachment, and the second term is due to runoff entrainment; both are net values over disposition. The term a is the detachability of the soil by rainfall, C is the sediment concentration, $C(L, t)$ is the sediment concentration at x_*/L at time t, C_e is the fraction of soil surface unprotected from raindrop detachment, C_r is the fraction of soil surface unprotected from entrainment by overland flow, g is acceleration due to gravity, i is the subscript referring to particular sediment size range, n is the number of sediment size ranges, I is the infiltration rate, K is 0.276, L is the length of plane, P is the rainfall rate, Q is the runoff rate per unit plane area, S is the slope of the plane (the sine of the angle of the land surface inclination), x is distance downslope from the top of the plane, x_* is the value of x beyond which entrainment commences (and L is greater than x_*), and ρ is the density of water. The term j_i is given by

$$j_i = 1 + \frac{V_i}{Q} \qquad (8.43)$$

where V_i is the settling or fall velocity of sedimentary particles of size range i, and Q is the efficiency of net entrainment and transport by overland flow. The stream power Ω at $x = L$ (the bottom of the plane) is given by

$$\Omega = \rho g S Q L \qquad (8.44)$$

The term x_*/L is equal to

$$\frac{x_*}{L} = \frac{\Omega_0}{\rho g S Q} \tag{8.45}$$

where Ω_0 is the threshold value of stream power. According to this model, soil erodibility is determined by the three parameters a, ρ, and Ω_0. The deposition rate is determined by the distribution of the fall or settling velocity V_i; detachment and transport include hydrology (rainfall and runoff), topography (slope gradient, length, shape, and aspect), soil erodibility (soil characteristics), soil transportability (size, aggregation, and charge properties), soil cover, residue management, land use history, soil surface conditions, and tillage systems. Once the sediments have been carried to the channel, a range of factors affect the channel process (Foster, 1982).

Chisci (1981) outlined an even more comprehensive scheme of the processes involved in upland erosion. An unknown and not well understood factor is rill erosivity. It includes inflow, soil erodibility, soil transportability, tillage, presence of nonerodible layer, cover, and channel characteristics, e.g., control, sidewall stability, and alignment.

Two additional models are being developed. A comprehensive model was developed to assess chemical runoff and erosion from agricultural management systems (CREAMS) (Knisell, 1980). It evaluates non-point-source pollution from field-size areas. CREAMS consists of three submodels, one each for hydrology, erosion sedimentation, and chemistry. It has been used to predict water, soil, and chemical losses from sugarcane fields in South Africa (Platford, 1983).

At present, the state of development modeling of soil erosion processes is more art than science. Emphasis must be put on the "white-box approach" in which the physical laws and nature of the system are well understood and can be synthesized into a system operating without recourse to observation of inputs and outputs. We still have a long way to go.

References

Abdurachman, A., Abujamin, S., and Kurnia, U. 1984. Soil and crop management practices for erosion control. *Pemberitaan Peelitian Tanah Dan Pupuk* 3:7–19.

Ahmad, N., and Brechner, E. 1974. Soil erosion on three Tobago soils. *Trop. Agric.* 51:313–324.

Aina, P. O., Lal, R., and Taylor, G. S. 1977. Soil and crop management in relation to soil erosion in the rainforest of Western Nigeria, pp. 75–82. *Soil Erosion: Prediction and Control*. SCSA Spec. Publ. 21, Soil Conservaton Society of America, Ankeny, Iowa.

Ambar, S., and Wiersum, K. F. 1980. Comparison of different erodibility indices under various soil and land use conditions in West Java. *Indonesian J. Geogr.* 1:1–15.

American Society of Civil Engineers (ASCE). 1975. *Sedimentation Engineering*, ASCE, New York.

Armon, M. N. 1984. Soil erosion and degradation in southeastern Nigeria in relation to biophysical and socio-economic factors. Ph.D. thesis, University of Ibadan, Ibadan, Nigeria.

Arnoldus, H. M. J. 1977a. Methodology used to determine the maximum potential average annual soil loss due to sheet and rill erosion in Morocco. *FAO Soils Bull.* 34:39–51.

Arnoldus, H. M. J. 1977b. Predicting soil losses due to sheet and rill erosion. *FAO Conserv. Guide* 1:99–124.

Arnoldus, H. M. J. 1980. An approximation of the rainfall factor in the USLE. In: *Assessment of Erosion* (M. De Boodt and D. Gabriels, eds.), pp. 127–132, Wiley, Chichester, U.K.

Ateshian, J. K. H. 1974. Estimation of rainfall erosion index. *J. Irr. Drainage Div.* 100:293–307.

Babu, Ram, Tejwani, K. G., Agarwal, M. C., and Bhushan, L. S. 1978. Distribution of erosion index and iso-erodent map of India. *Indian J. Soil Conserv.* 6(1):1–12.

Banasik, K. 1985. Applicability of the USLE for predicting sediment yield from small watershed in Poland, pp. 85–95. *Proc. Int. Symp. on Erosion, Debris Flow and Disaster Prevention,* Erosion Control Engineering Society, Tsukuba, Japan.

Barber, R. G., and Van Eijusbergen, A. C. 1981. A proposed model to predict the development of level bench terraces from steep backslope terraces, *J. Agric. Eng. Res.* 26(3):271–276.

Barber, R. G., Moore, T. R., and Thomas, D. B. 1979. The erodibility of two soils from Kenya. *J. Soil Sci.* 30:579–591.

Barber, R. G., Thomas, D. B., and Moore, T. R. 1981. Studies on soil erosion and runoff and proposed design procedure for terraces in the cultivated, semiarid areas of Machakos District, Kenya. In: *Soil Conservation* (R. P. C. Morgan, ed.), pp. 219–238, Wiley, Chichester, U.K.

Barn, R., Tejwani, K. G., Agarwal, M. C., and Bhushan, L. S. 1978. Distribution of erosion index and iso-eroded map of India. *Indian J. Soil Conserv.* 6(1):1–12.

Baver, L. D. 1933. Some factors affecting erosion. *Agric. Eng.* 14:51–52.

Beer, C. E., and Johnson, H. P. 1963. Factors in gully growth in the deep loess area of Western Iowa. *Trans. ASAE* 6:237–240.

Bergsma, E. 1980. Provisional rain-erosivity map of the Netherlands. In: *Assessment of Erosion* (M. De Boodt and D. Gabriels, eds.), pp. 121–125, Wiley, Chichester, U.K.

Betson, R. P., and Ardis, A. P. Jr. 1979. Implications for modelling surface-water hydrology. In: *Hillslope Hydrology* (M. J. Kirkby, ed.), pp. 295–324, Wiley, Chichester, U.K.

Biscaia, R. C., Rufino, R. L., and Henklain, J. C. 1981. Estimation of soil erodibility factor (K factor) in two soils of Parana State. *Rev. Bras. Cienc. Solo* 5(3):183–186.

Bollinne, A. 1985. Adjusting the universal soil loss equation for use in western Europe. In: *Soil Erosion and Conservation* (S. A. El-Swaify, W. C. Moldenhauer, and A. Lo, eds.), pp. 206–213, SCSA, Ankeny, Iowa.

Bollinne, A., Laurant, A., Rosseau, P., Panwels, J. M., Gabriels, D., and Aelterman, J. 1980. Provisional rain erosivity map of Belgium. In: *Assessment of Erosion* (M. De Boodt and D. Gabriels, eds.), pp. 111–120, Wiley, Chichester, U.K.

Boon, W., and Savat, J. 1981. A nomogram for the prediction of rill erosion. In: *Soil Conservation* (R. P. C. Morgan, ed.), pp. 303–320, Wiley, Chichester, U.K.

Brenneman, L. G., and Laflen, J. M. 1980. *Modeling the Effect of Corn Residue on Rill Erosion and Deposition.* ASAE Paper 80-2032, St. Joseph, Mich.

Brito Chaves, I. de, and Freire, O. 1978. Erosividade das chuvas na microrregiao homegenea Brasileira no 98 Estado da Paraibo, pp. 175–180, Encontro Nacional de Pesquisa sobre Conservacao do Solo, 2, Anais, EMBRAPA, Centro Nacional de Pesquisa de Trigo, Passo Fundo.

Brito, O. J., and Pena MacCaskill, L. 1980. Determination of the *"R"* factor of the universal soil loss equation in Nuble Province, Chile. *Agric. Tecn.* 40(4):152–156.

Brooks, F. L. 1977. Use of the universal soil loss equation in Hawaii. In: *Soil Erosion—Prediction and Control* (G. R. Foster, ed.), pp. 22–30. SCSA Special Publ. 21, Ankeny, Iowa.

Bruce, R. C. 1985. Assessment of some soil erosion prediction models for application to the Philippines. In: *Soil Erosion Management* (E. T. Craswell, J. V. Remenyi, and L. G. Nallana, eds.), pp. 42–49, ACIAR Proc. Series 6, Canberra, Australia.

Cassol, E. A., Eltz, F. L. F., and Guerra, M. 1978. Erodibilidate do solo "Sao Jeronimo" (Lateritico Bruno Avermelhado Distro fico) determinada com simulador de chuvas—Resultados Preliminares: 203–208.

Castro Filho, C. de, Cataneo, A., and Biscaia, R. C. M. 1982. Use of Wilkinson method for calculating rainfall erosivity potential in five localities of Parana, Brazil. *Rev. Bras. Cienc. Solo* 6(3):140–241.

Castro Filho, C. de, Mondardo, A., Bigarella, L. P., Farais, G. S. de, Vieira, M. J., Rufino, R. L., Henklain, J. C., Derpsch, R., and Kemper, B. 1978. Erosivdade das chuvas para alguns locains do Estada processa prento eletronico de dados, pp. 167–174.

Cataneo, A., Castro Filho, C., and Acquarole, R. M. 1982. Computer program for calculating rainfall erosivity index. *Rev. Bras. Cienc. Solo* 6(3):236–239.
Chan, C. C. 1981. Evaluation of soil loss factors on cultivated slopelands of Taiwan. *Tech. Bull.* 55, ASPAC Food and Fertilizer Technology Center, Taipei, Taiwan.
Charreau, C. 1969. *Influence des techniques culturales sur le developpement du ruissellement et de l'erosion en Casamance.* VIIe Congress International du Genie Rural, C.N.R.A., Bambey, Senegal.
Chisci, G. 1981. Empirical studies of soil erosion and conservation. In: *Soil Conservation: Problems and Prospects* (R. P. C. Morgan, ed.), pp. 155–174, Wiley, Chichester, England.
Chisci, G., Sfalanga, M., and Torri, D. 1985. An experimental model for evaluating soil erosion on a single rainstorm basis. In: *Soil Erosion and Conservation* (S. A. El-Swaify, W. C. Moldenhauer, and A. Lo, eds.), pp. 558–565, Soil Conservation Society of America, Ankeny, Iowa.
Cogo, N. P. 1978. Erodibilidade de alguns solos do Rio Grande do sul avaliada pelo metodo do nomograma, pp. 215–218.
Cogo, N. P., Drews, C. R., and Gianello, C. 1978. Indiceda erosividade das chuvas dos municipios de Guaiba, Ijui e Passo fundo, No Estado do Rio Grande do Sul, pp. 145–152.
Collinet, J., and Valentin, C. 1979. Analysis of the different factors affecting surface hydrodynamics. New prospects. Agronomic applications. *Cah. ORSTOM, Pedologie* 17(4):283–328.
Cook, R. L. 1970. Predicting rainfall-erosion soil losses in Taiwan. *Soils Fertil. Taiwan.* 1969:11–17.
Cooley, K. R., and Williams, J. R. 1985. Applicability of the universal soil loss equation (USLE) and modified USLE to Hawaii. In: *Soil Erosion and Conservation* (S. A. El-Swaify, W. C. Moldenhauer, and A. Lo, eds.), pp. 509–522, Soil Conservation Society of America, Ankeny, Iowa.
C.T.F.T./Burkina Faso. 1974. *Report de synthèse 1973.* C.T.F.T.I. Ministere de l'Agriculture de Burkina Faso, Ouagadougou.
Dangler, E. W., and El-Swaify, S. A. 1976. Erosion of selected Hawaii soils by simulated rainfall. *Soil Sci. Soc. Am. J.* 40(5):769–773.
Dangler, E. W., El-Swaify, S. A., Ahuja, L. R., and Barnett, A. P. 1976. *Erodibility of Selected Hawaiian Soils by Rainfall Simulation.* ARS-W, USDA Publication 35.
Das, D. C., and Babu, R. 1979. Computation and estimation of rainfall energy and erosion indices in metric units. *Ind. Forester* 105(3):217–222.
David, W. P., and Beer, C. E. 1975. Simulation of soil erosion. *Trans. ASAE* 18:126–137.
Dedecek, R. A. 1974. Caracteristica fisica e fator de erodibilidade de Oxissolos do Rio Grande do Sul. In: Unidades Erexim, Passo Fundo e Santo Angelo, Porto Alegre, Faculdade de Agronomia da UFRGS, Tese M.S.
Dedecek, R. A. 1978. Capacidade erosiva das chuvas de Brasilia-DF. II Encontro Nacional de Pesquisa Sobre conservacao do solo Anais, pp. 157–166.
Dedecek, R. A., and Cabeda, M. S. V. 1977. Fator de erodibilidade de oxissolos no Rio Grande do Sul. *Pesquisa Agropec. Bras.* 12:91–95.
Degani, A., Lewis, L. A., and Downing, B. B. 1979. Interactive computer simulation of spatial process of soil erosion. *Professional Geogr.* 31(2):184–190.
Dernadin, J. E., Ramos, P. D. de C., and Wunscher, W. A. 1978. Comprimento de rampa de um Latossolo ver melho Escuro Alico (Unidade de Mapeamento Passo fundo). II Encontro Nacional de Pesquisa sobre conservacao do solo, Anais, pp. 219–230.
Dissmeyer, G. E., and Foster, G. R. 1985. Modifying the universal soil loss equation for forest lands. In: *Soil Erosion and Conservation* (S. A. El-Swaify, W. C. Moldenhauer, and A. Lo, eds.), pp. 480–495, Soil Conservation Society of America, Ankeny, Iowa.
Dragoun, F. J., and Miller, C. R. 1966. Sediment characteristics and two small agricultural watersheds. *Trans. ASAE* 9(4):66–70.
Edwards, K., and Charman, P. E. V. 1980. The future of soil loss prediction in Australia. *J. Soil Conserv. Service N.S.W.* 36(2):211–218.
Elliot, W. J., Liebenow, A. M., Laflen, J. M., and Kohl, K. D. 1989. *A Compendium of Soil Erodibility Data from WEPP Cropland Soil Field Erodibility Experiments, 1987 and 1988.* National Soil Erosion Research Laboratory Report No. 3, West Lafayette, Ind.

Ellison, W. D. 1947. Soil erosion studies I and II. *Agric. Eng.* 28:145–146, 197–201.
El-Swaify, S. A. 1977. Susceptibility of certain tropical soils to erosion by water. In: *Soil Conservation and Management in the Humid Tropics* (D. J. Greenland and R. Lal, eds.), pp. 71–77, Wiley, Chichester, U.K.
El-Swaify, S. A., and Dangler, E. W. 1977. Erodibilities of selected tropical soils in relation to structural and hydrological parameters, pp. 105–114. *Soil Erosion: Prediction and Control,* Soil Conservation Society of America, Ankeny, Iowa.
El-Swaify, S. A., Dangler, D. W., and Armstrong, C. L. 1982. Soil erosion by water in the tropics. *Res. Ext. Series 24,* College of Tropical Agriculture and Human Resources, Univ. Hawaii, Honolulu.
El-Swaify, S. A., Pathak, P., Rego, T. J., and Singh, S. 1985. Soil management for optimized productivity under rainfed condition in the semi-arid tropics. *Adv. Soil Sci.* 1:1–64.
Elwell, H. A. 1977. Soil loss estimation for southern Africa, Dept. of Conservation and Extension. *Res. Bull.* No. 22, Harare, Zimbabwe.
Elwell, H. A. 1978. Modelling soil losses in Southern Africa. *J. Agric. Eng. Res.* 23(2):117–127.
Elwell, H. A. 1979. *Modelling Soil Losses in Zimbabwe, Rhodesia.* ASAE Paper 79-2051, Washington.
Elwell, H. A., and Stocking, M. A. 1973. Rainfall parameters for soil loss estimation in a subtropical climate. *J. Agric. Eng. Res.* 18(3):167–177.
Elwell, H. A., and Stocking, M. A. 1975. Parameters for estimating annual runoff and soil loss from agricultural lands in Rhodesia. *Water Resources Res.* 11(4):601–605.
Elwell, H. A., and Stocking, M. A. 1982. Developing a simple yet practical method of soil-loss estimation. *Trop. Agric.* 59(1):43–48.
FAO/UNEP/UNESCO, 1979. *A Provisional Methodology for Soil Degradation Assessment,* FAO, Rome, Italy.
Fetzer, K. D. 1977. The influence of soil erosion on soils of the northwestern mountain ridge of the Kathmandu valley, Nepal. *Giesseuer Beitrage Zur Entwicklungsforschung* 1(3):111–118.
Fleming, G., and Fahmy, M. 1973. *Some mathematical concepts for simulating the Water and Sediment Systems of Natural Watershed Areas.* Dept. of Civil Engineering, Univ. of Strathclyde, Glasgow, Scotland.
Fleming, G., and Walker, R. 1976. *A Runoff-Erosion Model for Land Use Assessment and Management.* Technical Report, Dept. of Civil Engineering, Univ. of Strathclyde, Glasgow, Scotland.
Foster, G. R. 1982. Modeling the erosion process. In: *Hydrologic Modeling of Small Watersheds* (C. T. Haan, H. P. Johnson, and D. L. Brakensiek, eds.), pp. 297–380, ASAE, St. Joseph, Mich.
Foster, G. R. 1990. Major developments in prediction of soil erosion by water. In: *Soil Management for Sustainability* (R. Lal and F. J. Pierce, eds.), Soil and Water Conservation Society, Ames, Iowa.
Foster, G. R., and Lane, L. J. 1987. *User Requirements: USDA-WEPP (Draft 6.3).* National Soil Erosion Research Laboratory, West Lafayette, Ind.
Foster, G. R., and Meyer, L. D. 1975. Mathematical simulation of upland erosion by fundamental erosion mechanics, pp. 190–207. *Present and Prospective Technology for Predicting Sediment Yields and Sources,* ARS-S-40, USDA, Washington.
Foster, G. R., Meyer, L. D., and Onstad, C. A. 1977a. An erosion equation derived from basic erosion principles. *Trans. ASAE* 20:678–682.
Foster, G. R., Meyer, L. D., and Onstad, C. A. 1977b. A runoff erosivity factor and variable slope length exponents for soil loss estimates. *Trans. ASAE* 20:683–687.
Foster, G. R., and Wischmeier, W. H. 1974. Evaluating irregular slopes for soil loss prediction. *Trans. ASAE* 17:305–309.
Fournier, F. 1956. The effect of climatic factors on soil erosion: Estimates of solids transported in suspension in runoff. *Assoc. Int. Hydrol. Publ.* 38:6.
Fournier, F. 1960. *Climat et erosion,* Presses Universitaries de France, Paris.
Fournier, F., and Henin, S. 1959. A new climatic formula for evaluating the specific degradation of soil. *C.R. Acad. Sci.* 248:1694–1699.

Freire, O., and Castro Filho, C. 1977. Evaluation of rainfall erosion potential in Piracicaba. *Rev. Agric. Piracicaba* 52(2/3):105–111.

Freire, O., and Castro Filho, C. 1978. Erosividade de chuva em Pira cicabo-SP II Encontro Nacional de Pesquisa Sobre conservacao do solo, Anais, Passo Fundo-RS, pp. 153–156.

Freire, O., and Pessotti, J. E. 1978. II Encontro Nacional De Pesquisa sobre Conservacao do solo, Anais, EMBRAPA-Centro Nacional de Pesquisa De Trigo Servico Nacional De Levantamento E Conservacao do solo, Passo Fundo-RS, pp. 185–192.

Freire, O., and Vasques, J. 1982. Erodibility of the Bora watershed soils, Brazil. *Rev. Agric. Bras.* 57(1/2):77–91.

Galabert, J., and Millogo, E. 1973. *Indice d'erosion de la pluie en Haute Volta* C.T.F.T., Ouagadougou.

Gerrard, A. J. 1981. *Soils and Land Forms,* George Allen & Unwin, London.

Golubev, G. N. 1982. *Soil Erosion and Agriculture in the World: An Assessment and Hydrological Implications.* IAHS Publication 137, Washington, pp. 261–268.

Guartsma, R., Paez, M. L., and Rodriguez, O. 1981. *Rainfall Erosion Indexes Estimated from Daily Precipitation Amount.* ASAE Paper 81-2530, Central Univ. of Venezuela.

Haan, C. T., Johnson, H. P., and Brakensiek, D. L. (eds.). 1982. *Hydrologic Modeling of Small Watersheds,* ASAE, St. Joseph, Mich.

Horton, R. E. 1945. Erosional development of streams and their drainage basins: Hydrological approach to quantitative morphology. *Geol. Soc. Am. Bull.* 56:275–370.

Hudson, N. W. 1961. An introduction to the mechanics of soil erosion under conditions of subtropical rainfall. *Rhodesia Sci. Assoc. Proc.* 49:14–25.

Hudson, N. W. 1971. *Soil Conservation,* Cornell University Press, Ithaca, New York.

Jamal, T., Mokhtaruddin, A. M., and Wan Sulaiman, M. 1984. Use of rain height as an erosivity index to estimate soil erosion, pp. E7.1–E7.11. *Fifth ASEAN Soil Conference, 10–23 June 1984,* DLD, Bangkok, Thailand.

Jansson, M. B. 1982. *Land Erosion by Water in Different Climates.* UNGI Rapport 57, Dept. of Physical Geography, Uppsala University, Sweden.

Joshua, W. D. 1977. *Soil Erosive Power of Rainfall in the Different Climatic Zones of Sri Lanka.* IAHS Publication 122, Washington, pp. 51–61.

Khanbilvardi, R. M., Rogowski, A. S., and Miller, A. C. 1982. *Modelling Erosion and Sediment Transport in Rills.* Paper 82-2610, American Society of Agricultural Engineers, St. Joseph, Mich.

Kinnell, P. I. A. 1985. Runoff effects on the efficiency of raindrop kinetic energy in sheet erosion. In: *Soil Erosion and Conservation* (S. A. El-Swaify, W. C. Moldenhauer, and A. Lo, eds.), pp. 399–405, Soil Conservation Society of America, Ankeny, Iowa.

Kirkby, M. J. 1976. Hydrological slope models: The influence of climate. In: *Geomorphology and Climate* (E. Derbyshire et al., eds.), pp. 247–267, Wiley, Chichester, U.K.

Kirkby, M. J. 1980. Modelling water erosion processes. In: *Soil Erosion* (M. J. Kirkby and R. P. C. Morgan, eds.), pp. 183–216, Wiley, Chichester, U.K.

Knisell, W. G. (ed.). 1980. CREAMS: A field-scale model for chemicals, runoff and erosion from agricultural management systems. *USDA Conserv. Res. Rep.* 26.

Koolhaas, M. H. 1979. Erosive potential of rainfall in Uruguay (El potencial erosivo de la Iluvia en el Uruguay). *Turrialba* 29(1):3–9.

Kowal, J. M. 1970. Some physical properties of soil at Samaru, Zaria, Nigeria: Storage of water and its use by crops. I: Physical status of soils. *Samaru Res. Bull.* 111.

Kowal, J. M., and Kassam, A. H. 1977. Energy load and instantaneous intensity of rainstorm at Samaru, Northern Nigeria. In: D. J. Greenland and R. Lal (eds.), *Soil Conservation and Management in the Humid Tropics,* pp. 57–70, Wiley, Chichester, U.K.

Kurnia, U., and Suwardjo, H. 1985. Effect of mechanical conservation methods on soil erosion in Tropudalf and Tropothent in Yogyakarta. *Pembr. Pen. Tanah Dan Pupuk* 4:46–50.

Lal, R. 1976. Soil erosion on alfisols in Western Nigeria. III. Effects of rainfall characteristics. *Geoderma* 16(15):389–401.

Lal, R. 1984. Soil erosion from tropical arable lands and its control. *Adv. Agron.* 37:183–248.

Lal, R., Lawson, T. L., and Anastase, A. H. 1980. Erosivity of tropical rains. In: *Assessment of Erosion* (M. de Boodt, and D. Gabriels, eds.), pp. 143–151, Wiley, Chichester, U.K.

Laursen, E. M. 1958. Sediment transport mechanics in stable-channel design. *ASCE Trans.* 123:195–206.
Laws, J. O., and Parsons, D. A. 1943. The relation of raindrop size to intensity. *Trans. Am. Geophys. Union* 24:452.
Leite, J. A., Cavalcante, L. F., Medina, B. F., and Montenegro, J. O. 1982. Erodibility factor of four soils from the municipality of Sao Mamede, Paraila State, Brazil. *Revista Ceres* 17(2):319–321.
Li, R. M., Simons, R. K., and Shiao, L. Y. 1977. Mathematical modeling of on-site soil erosion. In: *Proceedings, International Symposium on Urban Hydrology, Hydraulics and Sediment Control, July 18–21, 1977,* ASCE, New York.
Lindsay, J. I., and Gumbs, F. A. 1982. Erodibility indices compared to measured values of selected Trinidad soils. *Soil Sci. Soc. Am. J.* 46(2):393–396.
Low, K. F. 1967. Estimating potential erosion hazard in developing countries. *J. Soil Water Conserv.* 22:147–148.
Lugo Lopez, M. A. 1982. K values of the soils of Puerto Rico. *J. Agric. Univ. Puerto Rico* 66(4):311–312.
Machado, S. A. 1978. First approximation of values of the erodibility factor (K), in some Colombian soils. *Rev. Facult. Nac. Agron. Medellin, Colombia* 31(1):1–22.
Maene, L. C., Maesschalk, G. G., Huan, L. H., and Manan, M. 1977. *Soil Physics Annual Report October 1976–September 1977,* Faculty Agric., Univ. Pertanian, Malaysia.
Martinez, M. R., Fogel, M. M., and Lane, L. J. 1980. *Erosion Model for Upland Areas.* Paper 80-2505, ASAE, St. Joseph, Mich.
McPhee, P. J., Hartman, M. O., and Kieck, N. F. 1983. *Soil Erodibility and Crop Management Factors of Soils under Pineapple Production.* Paper 83-2073, American Society of Agricultural Engineers, St. Joseph, Mich.
Meyer, L. D., and Monke, E. J. 1965. Mechanics of soil erosion by rainfall and overland flow. *Trans. ASAE* 8:572–577.
Meyer, L. D., and Wischmeier, W. H. 1969. Mathematical simulation of the process of soil erosion by water. *Trans. ASAE* 12:754–758, 762.
Mitchell, J. K., and Bubenzer, G. D. 1980. In: *Soil Erosion* (M. J. Kirkby and R. P. C. Morgan, eds.), pp. 17–62, Wiley, Chichester, U.K.
Mondardo, A., de Forias, G. S., Henklain, J. C., Vieira, M. J., Castro, C. de Filho, Rufino, R. L., Kemper, B., and Derpsch, R. 1978. Indices de erodibilidade de alguns solos do Estado do pa rana. II Encontro Nacional De pesquisa sobre conservacao do solo, Anais, EMBRAPA-Centro Nacional de pesquisa de trigo servico National de Levantamento e conservacao do solo, pp. 199–202.
Moore, T. R. 1979. Rainfall erosivity in East Africa. *Geogr. Ann.* 61A(3–4):147–156.
Morgan, C. 1983. The non-independence of rainfall erosivity and soil erodibility. *Earth Surface Processes Landforms* 8(4):323–338.
Morgan, R. P. C. 1979. *Soil Erosion,* Longmans, London.
Mota, F. O. B., and Lima, F. A. M. 1976. Erodibility of the soils of FEVC (Farende Experimental Vale do curn). *Erodibilibade dos solos das FEVC. Solo* 68(2):60–62.
Mtakwa, P. W., Lal, R., and Sharma, R. B. 1987. An evaluation of the Universal Soil Loss Equation and field techniques for assessing soil erosion on a tropical alfisol in Western Nigeria. *Hydrol. Processes* 1:199–209.
Musgrave, G. W. 1947. Quantitative evaluation of factors in water erosion: A first approximation. *J. Soil Water Conserv.* 2:133–138.
Mutchler, C. K., and Murphree, C. E. 1981. Prediction of erosion on flatlands. In: *Soil Conservation* (R. P. C. Morgan, ed.), pp. 321–327, Wiley, Chichester, U.K.
Mutchler, C. K., and Murphree, C. E. 1985. Experimentally derived modification of the USLE. In: *Soil Erosion and Conservation* (S. A. El-Swaify, W. C. Moldenhauer, and A. Lo, eds.), pp. 523–527, Soil Conservation Society of America, Ankeny, Iowa.
Nearing, M., and Lane, L. J. 1990. Modeling fundamental soil erosion and sedimentation processes on field and watershed scales. In: *Soil Erosion Research Methods* (R. Lal, ed.), 2d ed., Soil and Water Conservation Society, Ankeny, Iowa.
Negev, M. 1967. *A Sediment Model on a Digital Computer.* Technical Report 76, Department of Civil Engineering, Stanford University, Stanford, Calif.

Ngatunga, E. L. N., Lal, R., and Uriyo, A. 1984. Effects of surface management on runoff and soil erosion from some plots at Mlingano, Tanzania. *Geoderma* 33:1–12.

Nowak, M. D. 1985. Soil loss and time to equilibrium for rill and channel erosion. *Trans. ASAE* 28:1790–1793.

Onchev, N. G. 1985. Universal index for calculating rainfall erosivity. In: *Soil Erosion and Conservation* (S. A. El-Swaify, W. C. Moldenhauer, and A. Lo, eds.), pp. 424–431, Soil Conservation Society of America, Ankeny, Iowa.

O'Sullivan, T. E. 1985. Farming systems and soil management: The Philippines/Australian Development Assistance Program experience. In: *Soil Erosion Management* (E. T. Craswell, J. V. Remeniyi, and L. G. Nallana, eds.), pp. 77–81, ACIAR Proceedings, Series 6, Canberra, Australia.

Pena MacCaskill, L. 1983. Determination of the R, K, and C factors of the universal soil loss equation for the Andean foothills area. VIII region. Preliminary study. *Agric. Jeimica* 43(2):151–158.

Pereira, W., da Silva, T. C. A., and Gomes, F. R. 1978b. Evaluation of rainfall erosivity in various parts of Minas Gerais. *Rev. Ceres* 25(142):506–548.

Pereira, W., da Silva, T. C. A, and Gomes, F. R. 1978a. Avaliacao da erosividade das chuvas em diferentes locais do Estado de Minas Gerais. II Encontro Nacional de Pesquisa Sobre conservacao do solo, Anais, EMBRAPA-Centro Nacional de Trigo Servico Nacional De Levantamento e conservacao do solo, pp. 141–144.

Piest, R. F., Kramer, L. A., and Heinemann, H. G. 1975. *Sediment Movement from Loessial Watersheds: Present and Prospective Technology for Prediction of Sediment Yields and Sources*, pp. 130–141, ARS-S-40, USDA, Washington.

Piest, R. F., Laflen, J. M., and Soiner, R. G. 1980. *Integrated Erosion Studies for Improved Sediment Transport Models*. ASAE Paper 80-2029, St. Joseph, Mich.

Platford, G. G. 1982. The determination of some soil erodibility factors using a rainfall simulator. *Proceedings of the 56th Annual Congress, South African Sugar Technologists Association* (1982), Johannesburg, South Africa, pp. 130–133.

Platford, G. G. 1983. The use of the CREAMS computer model to predict water, soil and chemical losses from sugarcane fields and to improve recommendations for soil protection. *Proc. of the 57th Annual Congress, South African Sugar Technologists Association*, Johannesburg, South Africa, pp. 144–150.

Ramaiah, R., and Sreenivas, G. N. 1973. Rainfall erosion index of Bangalore agro-climatic area. *Mysore J. Agric. Sci.* 9(3):448–453.

Renard, K. G., Simanton, J. R., and Osborn, H. B. 1974. Applicability of the Universal Soil Loss Equation to semi-arid rangeland conditions in the southwest. In: *Hydrology and Water Resources in Arizona and the Southwest*, vol. 4, American Water Resources Association, pp. 18–31.

Resck, D. V. S., Figueiredo, M. de S., Fernandes, B., Resen, M. de T. C. A. da Silva. 1978. Determinancao da erodibilidade de um Podzolico Vermelho Amarelo Cambico Distrofico fase ter raco, localizado na zona da mata (MG), utilizando o simu lador de chuva, pp. 193–198. INTO II Encontro Nacional De Pesquisa Sobre conservacao do solo, Anais, EMBRAPA-Centro Nacional de Pesquisa de Trigo Servico Nacional De Levantamento E conservacao Do solo, Passo Fundo-RS.

Roose, E. J. 1974. Investigation of the stability of some tropical soils to erosion. *Trans. 10th Int. Congr. Soil Sci.* 11:54–61.

Roose, E. J. 1977a. Application of the Universal Soil Loss Equation of Wischmeier and Smith in West Africa. In: *Conservation and Soil Management in the Humid Tropics* (D. J. Greenland and R. Lal, eds.), pp. 177–187, Wiley, Chichester, U.K.

Roose, E. J. 1977b. Use of the Universal Soil Loss Equation to predict erosion in West Africa. In: *Soil Erosion—Prediction and Control* (G. R. Foster, ed.), pp. 60–74, SCSA Special Publication 21, Ankeny, Iowa.

Roose, E. J. 1980. Approach to the definition of rain erosivity and soil erodibility in West Africa. In: *Assessment of Erosion* (M. de Boodt and D. Gabriels, eds.), Wiley, Chichester, U.K.

Rose, C. W. 1985a. Developments in soil erosion and desposition models. *Adv. Soil Sci.* 2:1–63.

Rose, C. W. 1985b. Progress in research on soil erosion processes and a basis for soil conservation practices. In: *Soil Erosion Management* (E. T. Craswell, J. V. Remenyi, and L. G. Nallana, eds.), pp. 32–41, ACIAR Proc. Series 6, Canberra, Australia.

Rose, C. W. 1985c. Soil erosion and conservation: From processes to practices to policies. In: *Soil Erosion and Conservation* (S. A. El-Swaify, W. C. Moldenhauer, and A. Lo, eds.), pp. 775–784, Soil Conservation Society of America, Ankeny, Iowa.

Rose, C. W. 1987. Modelling erosion. In: *Soil Erosion Research Methods* (R. Lal, ed.), Soil and Water Conservation Society, Ames, Iowa, pp. 119–140.

Rose, C. W., and Freebairn, D. M. 1985. A new mathematical model of soil erosion and deposition processes with applications to field data. In: *Soil Erosion and Conservation* (S. A. El-Swaify, W. C. Moldenhauer, and A. Lo, eds.), pp. 549–557, SCSA, Ankeny, Iowa.

Rose, C. W., Parlange, J. Y., Sander, G. C., Campbell, S. Y., and Barry, D. A. 1983a. Kinematic flow approximation to runoff on a plane: An approximate analytical solution. *J. Hydrol.* 62:363–369.

Rose, C. W., Williams, J. R., Sander, G. C., and Barry, D. A. 1983b. A mathematical model of soil erosion and deposition processes. I. Theory for a plane land element. *Soil Sci. Soc. Am. J.* 47:991–995.

Rose, C. W., Williams, J. R., Sander, G. C., and Barry, D. A. 1983c. A mathematical model of soil erosion and deposition processes. II. Application to data from an arid zone catchment. *Soil Sci. Soc. Am. J.* 47:996–1000.

Shi, D. M., and Yang, Y. S. 1983. Discussion on the development of eroded soils in the loessial hilly region by means of numerical analysis. *Acta Pedologica Sinica* 20(2):167–175.

Shin, J. S., Jung, Y. S., and Shin, Y. H. 1976. *Soil Loss Prediction for Uplands*. Research Reports of the Office of Rural Development, Soil Science, Fertilizer, Plant Protection and Microbiology, vol. 18, pp. 1–8.

Singh, G., Babu, R., and Chandra, S. 1981. Research work done on K factor in soil loss prediction research in India. *ICAR Bull.* T12/D-9, Dehradun, India.

Singh, G., Babu, R., and Chandra, S. 1985. Research on the Universal Soil Loss Equation In India. In: *Soil Erosion and Conservation* (S. A. El-Swaify, W. C. Moldenhauer, and A. Lo, eds.), pp. 496–508, SCSA, Ankeny, Iowa.

Smith, D. D. 1941. Interpretation of soil conservation data for field use. *Agric. Eng.* 22:173–175.

Smithen, A. A., and Schulze, R. E. 1982. The spatial distribution in Southern Africa of rainfall erosivity for use in the Universal Soil Loss Equation. *Water SA* 8(2):74–78.

Srikhajon, M., Somrang, A., Pramojanee, P., Pradabwit, S., and Anecksamphant, C. 1984. Application of the Universal Soil Loss Equation for Thailand. In: *Proc. 5th ASEAN Soil Conf.*, pp. E8.1–E8.15, DLD, Bangkok, Thailand.

Stehlik, O. 1970. Geograficka rajonizace eroze pudy V CSR. *Studia Geographica* No. 13, Geograficky Ustav CSAV, Brno.

Stehlick, O. 1975. Potencialni eroze pudy proudici vodou na uzemi CSR. *Studia Geographica* 42, Brno.

Stocking, M. A. 1973. Towards a model of soil erosion—An example from Rhodesia. *South African Geogr.* 4(3):253–258.

Stocking, M. A. 1977. *Rainfall Energy in Erosion: Some Problems and Applications*. Research discussion paper 13, Dept. of Geology, Univ. of Edinburgh.

Stocking, M. A., and Elwell, H. A. 1976. Rainfall erosivity over Rhodesia. *Trans. Instit. Brit. Geogr.* 1(2):231–245.

Sudjadi, M. 1984. *Problem Soils in Indonesia and Their Management*, FFTC Book Series, Centre for Soil Research, Ministry of Agriculture, Bogor.

Sudjadi, M., Widjaja-Adhi, I. P. G., McIntosh, J. L., and Palmer, B. 1985. Wetland soil tests for crop production, pp. 347–358. *Wetland Soils: Characterization, Classification and Utilization*, Proceedings of a Workshop held 26 March to April 1984, IRRI, Los Banos.

Sukmana, S. 1987. Alang-alang land in Indonesia: Problems and prospects, pp. 397–411. *Soil Management under Humid Conditions in Asia and Pacific*, IBSRAM Proceedings No. 5, Bangkok.

Thomas, D. B., and Barber, R. G. 1978. Report on rainfall simulator trials at Iiuni, Machakos, Kenya. Dept. Agricultural Engineering, University of Nairobi, Nairobi.

Thompson, J. R. 1964. Quantitative effect of watershed variables on rate of gully head advancement. *Trans. ASAE* 7:54–55.

Undang, K., and Suwardjo, A. 1984. Erodibility of selected soils in Jara as measured by USLE method. *Pembr. Pen. Tanah Dan Pupuk* 3:17–20.

United States Soil Conservation Service (USDA). 1966. *Procedure for Determining Rates of Land Damage, Land Depreciation and Volume of Sediment Produced by Gully Erosion.* USDA Technical Release 32, Washington.

Utomo, W. H., and Mahmud, N. 1984. The possibility of using the Universal Soil Loss Equation in mountainous areas of east Java with humus-rich andosols. In: *Proc. 5th ASEAN Soil Conference,* pp. E5.1–E5.13. DLD, Bangkok, Thailand.

Vaneslande, A., Lal, R., and Gabriels, D. 1985. Erodibility of some Nigerian soils, *Proc. Int. Symp. on Erosion, Debris Flow and Disaster Prevention,* Tsukuba, Japan. Erosion Control Engineering Society, pp. 51–56.

Vaneslande, A., Rousseau, P., Lal, R., Gabriels, D., and Ghuman, B. S. 1984. *Testing the Applicability of a Soil Erodibility Nomogram for Some Tropical Soils.* IAHS Publication 144, Washington, pp. 463–473.

Van Vuuren, W. E. 1982. *Prediction of Sediment Yield for Mountainous Basins in Colombia, South America.* IAHS Publication 137, Washington, pp. 313–325.

Wandelaar, F. E. 1978. *Applying the Universal Soil Loss Equation in Rhodesia.* Soil and Water Engineering Section, Institute of Agricultural Engineering, Salisbury, Zimbabwe.

Wilkinson, G. E. 1975a. Effect of grass fallow rotations on the infiltration of water into a savanna zone soil of Northern Nigeria. *Trop. Agric. (Trinidad)* 52:97–103.

Wilkinson, G. E. 1975b. Rainfall characteristics and soil erosion in the rain forest area of Western Nigeria. *Exp. Agric.* 11:247–255.

Williams, J. R. 1975. Sediment yield prediction with Universal Equation using runoff energy factor. In: *Present and Prospective Technology for Predicting Sediment Yields and Sources.* ARS-S-40, USDA, Washington, pp. 244–252.

Williams, J. R. 1978. A sediment yield routing model. *Proc. of the Special Conference on Verification of Mathematical and Physical Models in Hydraulic Engineering,* ASCE, New York, pp. 602–670.

Williams, J. R., and Berndt, H. D. 1972. *Sediment Yield Prediction Based on Watershed Hydrology.* Paper 76-2535, ASAE, St. Joseph, Mich.

Williams, M. A. J. 1969. Prediction of rain splash erosion in the seasonally wet tropics. *Nature (Lond.)* 222:763–765.

Wischmeier, W. H. 1959. A rainfall erosion index for a Universal Soil Loss Equation. *Soil Sci. Soc. Am. Proc.* 23:246–249.

Wischmeier, W. H. 1962. Rainfall erosion potential. *Agric. Eng.* 43:212–214, 225.

Wischmeier, W. H. 1974. New development in estimating water erosion, pp. 179–186. *Proc. 29th Annual Meeting,* Soil Conservation Society of America, Ankeny, Iowa.

Wischmeier, W. H. 1976. Use and misuse of the Universal Soil Loss Equation. *J. Soil Water Conserv.* 31:5–9.

Wischmeier, W. H., Johnson, C. B., and Cross, B. C. 1969. A soil erodibility nomogram for farmland and construction sites. *J. Soil Water Conserv.* 26(5):189–193.

Wischmeier, W. H., and Smith, D. D. 1958. Rainfall energy and its relationship to soil loss. *Trans. Am. Geophys. Union* 39:285–291.

Wischmeier, W. H., and Smith, D. D. 1978. *Predicting Rainfall Erosion Losses.* Agriculture Handbook 537, USDA, Washington.

Wischmeier, W. H., Smith, D. D., and Uhland, R. B. 1958. Evaluation of factors in soil loss equation. *Agric. Eng.* 39:458.

Wunsche, W. A., and Denardin, J. E. 1978. Erodibilidade de Latosso 10 Vermelho Escuro alico (Unidade de Mapeamento Passo Fun do) 1[a] Aproximacao. IN TO II Encontro Nacional De Pesquisa Sobre conservacao Do Solo, *ANAIS,* EMBRAPA-Centro Nacional De Pesquisa De Trigo Servico Nacional De Levantamento E conservacao Do Solo.

Yoder, R. E. 1936. A direct method of aggregate analysis of soils and a study of the physical nature of erosion losses. *J. Am. Soc. Agric.* 28:337–351.

Zingg, A. W. 1940. Degree and length of land slope as it affects soil loss in runoff. *Agric. Eng.* 21:59–64.

Part 4

Erosion Control

Chapter 9

Controlling Upland Erosion

Upland or interrill erosion and rill erosion are caused by the impact of raindrops, shear strength of overland and channelized flow, and interaction between the two factors. Effective erosion control, therefore, lies in reducing the impact of the raindrop, increasing the shear resistance of the soil, and decreasing the shear strength of erosive fluids (Fig. 9.1). Techniques of erosion control based on the principle of reducing raindrop impact are mostly vegetative or biologic. Increasing the shearing resistance of soil improves the stability of soil structure. Both agronomic and engineering techniques are used to decrease the velocity and shear strength of overland flow. The same general principles outlined in Fig. 9.1 apply to both wind and water erosion.

This chapter discusses the role of mulch in increasing the shear resistance of the soil. The next three chapters describe the role of land use, crop management, and soil management practices in reducing the raindrop impact and decreasing the shear strength of water runoff. Chapters 14 and 15 are devoted to erosion control on steeplands and wind erosion control.

Reducing Raindrop Impact

The practical way to reduce raindrop impact on soil is to maintain a ground cover. The ground cover on agricultural lands can be of different types (Fig. 9.2). For a cover to be effective, it must be as complete and as close to the ground surface as possible. In general, erosion is proportional to the ground cover according to a power function:

$$A = aC^b \tag{9.1}$$

284   Erosion Control

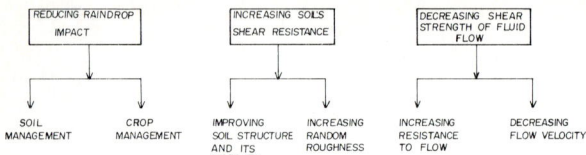

Figure 9.1   Principles of erosion control.

where A is soil erosion per unit area per unit time, C is ground cover in percent, and a and b are empirical constants. A complete cover within 50 cm of the soil surface is extremely effective in minimizing raindrop impact. The effectiveness of the cover, however, decreases with increasing height above the ground surface. The tree canopy 10 to 20 m above the ground surface loses its effectiveness because the coalescing drops falling from the canopy are generally big, attain terminal velocity, and have high impact energy.

A widely used, ancient practice to provide ground cover is mulching. Mulching has traditionally been practiced to enhance soil fertility through addition to soil of plant nutrients and organic matter, regulation of soil temperature and moisture regimes, and improvement of soil tilth by enhancing activity of soil fauna, especially the termites and earthworms. Mulch acts as a buffer because it dampens the influence on soil of the environments. The magnitude of the buffering effects of mulch on soil, however, depends on its quantity, quality, and durability. An important factor affecting mulch quality in relation to erosion is its height above the ground surface. A better-quality mulch for erosion control is the one placed close to the ground surface.

Mulch is a uniform layer of a dissimilar material placed between the soil surface and the atmosphere. Two important criteria in this definition are found in the terms *dissimilar* and *between the soil and the atmosphere*. The criterion of dissimilarity implies that a layer of soil placed on top of the soil is not mulch. The second criterion suggests that any material, organic or inorganic, arranged even as a uniform layer but placed below the soil surface is also not a mulch.

There are different types of mulch materials, depending on the method of procurement and soil and crop management practices (Fig. 9.3). Cultural practices associated with mulch material are briefly outlined now.

1. *Brought-in mulch.* The mulch material, usually a high-yielding grass or a perennial shrub, may be grown in a contiguous area, cut, and brought in as needed. Brining in the mulch material is a labor-intensive, expensive practice. The practice can be economical for some high-value cash crops. A wide range of mulch materials are brought in, e.g., farmyard manure and compost. Mulch applied at the rate of 4 to 6 t/ha is an effective erosion control measure in a wide range of ecologies (Lal, 1976). In the Ivory Coast, Roose

Height	Density	Canopy structure	Stories	Continuity	Nature
• High: trees	• High canopy cover	• Close canopy	• Multistory (tropical rain forest)	• Continuous over time	• Organic (live and dead)
• Intermediate: shrubs	• Low canopy cover	• Open canopy	• Single story	• Discontinuous	• Inorganic
• Low: crops					

Figure 9.2   Factors affecting type of ground cover.

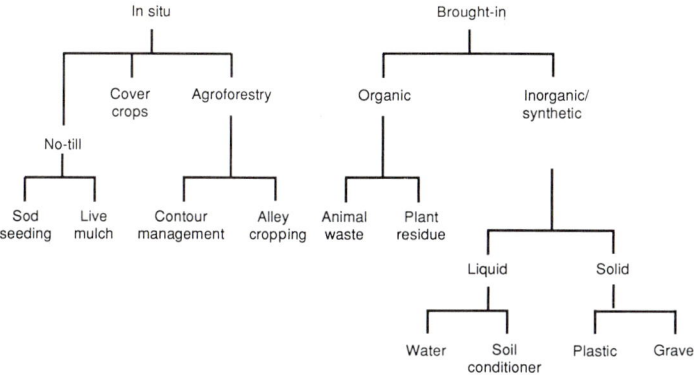

Figure 9.3 Types of mulch materials and the associated cultural practices to procure them.

(1975) observed that mulching a 7 percent slope provided as good a control for runoff and erosion as did undisturbed rain forest (Table 9.1). Similarly, in Tanzania, Lundgren (1980) observed that soil erosion from arable land with good management (mulching without burning) was even less than from forested lands (Table 9.2).

A wide range of by-products of agro-based industries are also used as mulch. Notable among these are rice husk and sawdust. Both these mulches, though of organic origin, may cause nutritional disorders in some crops. Their benefits in soil water conservation are significant, however.

2. *Crop residue mulch.* The residue from the previous crop is used as a source of mulch for the following crop. This is often done in association with no-tillage or minimum tillage practices. In regions with a high cattle population, however, crop residue is used as fodder. The residue is also used as a source of fuel and for fencing and thus is not available.

3. *In situ mulch.* Mulch is produced in place by growing a cover crop, a strip of grass, or a hedge of perennial shrubs and trees that can be regularly pruned. Prunings are used as mulch.

4. *Special mulch.* A wide range of inorganic materials and synthetic products are also being used as mulch. The most common is the polythene mulch (Plate 9.1) used for orchards, vegetable crops, and seed materials. Polythene

TABLE 9.1 Erosion and Runoff under a Straw Mulch (Slope 7%) and under Secondary Rain Forest Vegetation (Slope 23%) at Adiopodoume, Ivory Coast

		Erosion, kg/ha		Runoff, %	
Year	Rainfall, mm	Mulch	Rain forest	Mulch	Rain forest
1960	1897	5	13	0.47	0.58
1961	2289	11	15	0.53	0.34

SOURCE: Roose, 1975.

TABLE 9.2 Comparative Effects of Good Crop Management (Mulching and No Burning) on Runoff and Soil Erosion from Gentle and Steep Slopes in Tanzania

Land use	Rainfall, mm	Runoff, % of rainfall		Soil erosion, g/m^2	
		A	B	A	B
Lowland rain forest	1115	0.4	1.0	4.2	10.1
Highland rain forest	820	0.8	1.3	3.7	7.5
Cropland	635	0.1	0.1	1.0	0.9

A = gentle slope, B = steep slope.
SOURCE: Lundgren, 1980.

mulches eliminate raindrop impact, regulate soil temperature and moisture regimes, control weed growth, and have beneficial effects on mineralization of soil nutrients. Another variant of plastic mulch is the use of water-filled plastic bags to regulate the soil temperature. Soil conditioners and chemical amendments are sprayed on the surface to improve soil structure and reduce losses through runoff and erosion. A brief description of the usefulness of soil conditioners is given in another section.

Though not a common practice, gravels have been used as mulch in some special circumstances. Since gravel is inorganic and inert, the benefits of gravel mulch are restricted to decreasing the raindrop impact and evaporation losses. There are no drastic improvements in soil properties from using the gravel mulch.

Effects of mulches on evaporation

The rate of soil evaporation depends on the vapor pressure gradient between soil and the atmosphere, the resistance provided by the nonturbulent layer of

Plate 9.1 Polythene mulch used on some crops to conserve water and control weeds.

air adjacent to the soil surface, and the presence or absence of a barrier layer (Fig. 9.4). The barrier layer may be a mulch or a vegetation cover. In case of a porous barrier such as straw mulch, the evaporation rate declines. The rate of decline, however, depends on the soil moisture content. If the barrier layer is impermeable, such as plastic sheet or aluminum foil, the evaporation rate is reduced drastically.

The effect of a porous barrier, such as straw mulch, on the rate of soil evaporation depends on the degree of wetness or soil moisture content. The loss of soil moisture during evaporation occurs in three arbitrary stages (Fig. 9.4). A rapid loss of water takes place when the soil is very wet. During the first stage, the evaporation rate is controlled by the evaporative demand of the atmosphere. Soil water is not limiting the evaporation rate during this stage. The rate of evaporation from bare soil during the first stage is similar to that from a free water surface, if both surfaces are at the same temperature. In field conditions, however, bare soil surface is warmer than open water. The water movement essentially occurs as liquid flow under saturated conditions. A rapid decline in the evaporation rate occurs, once the soil moisture content goes below a certain critical value. During this second stage of declining evaporation rate, the soil moisture content determines the evaporation rate—the greater the soil moisture content, the greater the evaporation. As the soil becomes progressively drier, the liquid flow is replaced by vapor flow, and the rate of water loss is governed

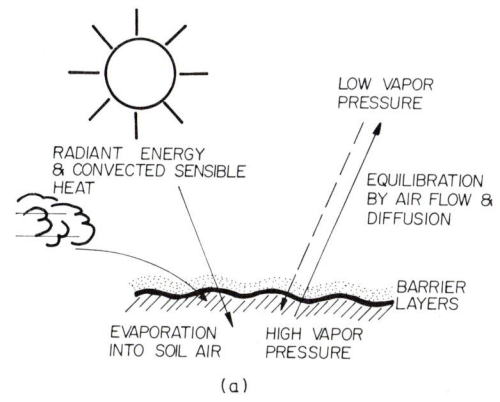

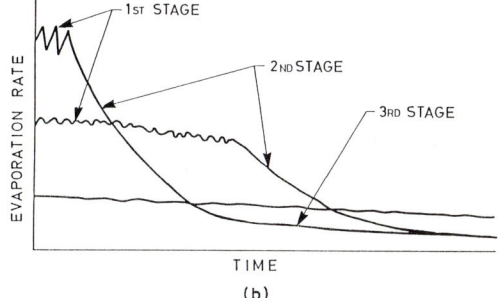

Figure 9.4 Soil water evaporation: (*a*) the process; (*b*) different stages of evaporation. (Adapted from Heinonen, 1985.)

by the vapor diffusion within the soil and from soil into the air through the barrier layer. In the third evaporation stage, the evaporation rate is extremely slow and constant. At this stage the evaporation rate is independent of both soil moisture content and evaporative demand. The soil moisture is held strongly by forces of cohesion and adhesion in the dry top layer.

Increasing Shearing Resistance of Soil

Using mulches for improving soil structure

Organic materials have long been used to improve soil structure. The effects are, however, transient and are easily eliminated by an intensive land use. Furthermore, the magnitude of the effect of mulch material on soil's physical and nutritional properties varies according to soils, mulch material, and crop management practices.

Straw mulch limits the transfer of heat to soil and of vapor from soil. It conserves water when there is water available in the soil to be conserved—during the first and second evaporation stages. The following is merely an example of the effects of a range of mulch materials on physical properties of a tropical alfisol. The data in Table 9.3 show that, with intensive cropping based on the maize-cassava rotation, soil physical properties are better maintained with any mulch in comparison with the unmulched control. Notable differences are observed in physical properties that influence soil erosion, such as bulk density (Table 9.3), earthworm activity (Table 9.4), saturated hydraulic conductivity (Table 9.5), mean weight diameter of aggregates (Table 9.6), moisture retention characteristics (Table 9.7), and penetrometer resistance (Table 9.8). Furthermore, soil physical properties are maintained better by organic than inorganic mulch materials. Some organic mulch materials, however, may adversely influence soil physical properties. These materials inhibit or restrict the biomass production by causing nutrient imbalance. For example, crop growth is adversely affected by the use of the mulch of typha, a plant with a high manganese content that grows in swamps. In contrast, some organic mulches drastically improve soil physical properties, especially those that enhance the activity of earthworms (Plate 9.2).

In comparison with the unmulched control, the data from IITA show that application of different mulch materials also reduced water runoff and soil erosion (Tables 9.9 and 9.10). The reduction in water runoff and soil erosion, however,

Plate 9.2 Earthworm casting of the *Hyperiodrilus africanus* and worm holes (biopores) through soil clods.

TABLE 9.3 Effect of Different Mulch Materials on Bulk Density of 0- to 10-cm Layer of a Tropical Alfisol (1978 Data)

	\multicolumn{8}{c}{Bulk density, g/cm³}							
	Cassava		Cowpea		Maize		Soybean	
Mulch treatment	A	B	A	B	A	B	A	B
Bare (without mulch)	1.47 ± 0.05	1.48 ± 0.12	1.25 ± 0.02	1.39 ± 0.04	1.36 ± 0.07	1.47 ± 0.07	1.32 ± 0.03	1.43 ± 0.03
Panicum maximum	1.17 ± 0.90	1.308 ± 0.76	1.32 ± 0.47	1.49 ± 0.93	1.28 ± 0.12	1.4 ± 0.1	1.27 ± 0.15	1.43 ± 0.14
Elephant grass	1.35 ± 0.15	1.42 ± 0.14	1.37 ± 0.10	1.46 ± 0.12	1.35 ± 0.12	1.42 ± 0.11	1.43 ± 0.13	1.5 ± 0.13
Black plastic	1.33 ± 0.19	1.39 ± 0.19	1.39 ± 0.16	1.47 ± 0.15	1.50 ± 0.10	1.53 ± 0.10	1.51 ± 0.09	1.55 ± 0.10
Translucent plastic	1.34 ± 0.16	1.41 ± 0.15	1.27 ± 0.17	1.3 ± 0.17	1.21 ± 0.17	1.33 ± 0.09	1.35 ± 0.17	1.43 ± 0.18
Sand dust	1.28 ± 0.07	1.39 ± 0.08	1.25 ± 0.07	1.33 ± 0.07	1.45 ± 0.06	1.48 ± 0.07	1.37 ± 0.10	1.41 ± 0.11
Rice husk	1.29 ± 0.09	1.44 ± 0.05	1.47 ± 0.10	1.54 ± 0.12	1.41 ± 0.03	1.46 ± 0.04	1.3 ± 0.07	1.42 ± 0.08
Pigeon pea tops	1.24 ± 0.16	1.37 ± 0.08	1.3 ± 0.17	1.4 ± 0.15	1.18 ± 0.12	1.3 ± 0.10	1.24 ± 0.13	1.39 ± 0.14
Maize stover	1.23 ± 0.08	1.36 ± 0.08	1.24 ± 0.17	1.42 ± 0.12	1.7 ± 0.05	1.39 ± 0.05	1.24 ± 0.17	1.41 ± 0.16
Pennisetum straw	1.23 ± 0.10	1.41 ± 0.09	1.15 ± 0.15	1.34 ± 0.13	1.28 ± 0.04	1.45 ± 0.07	1.27 ± 0.06	1.44 ± 0.08
Andropogon straw	1.19 ± 0.10	1.36 ± 0.07	1.36 ± 0.07	1.51 ± 0.05	1.23 ± 0.06	1.24 ± 0.60	1.34 ± 0.03	1.45 ± 0.06
Pigeon pea stems	1.33 ± 0.08	1.47 ± 0.08	1.45 ± 0.09	1.54 ± 0.09	1.3 ± 0.14	1.38 ± 0.16	1.38 ± 0.11	1.41 ± 0.10
Oil palm leaves	1.37 ± 0.17	1.48 ± 0.10	1.5 ± 0.13	1.59 ± 0.09	1.35 ± 0.17	1.44 ± 0.14	1.37 ± 0.10	1.43 ± 0.08
Chipped soybean tops	1.51 ± 0.07	1.58 ± 0.07	1.42 ± 0.12	1.51 ± 0.10	1.27 ± 0.16	1.39 ± 0.14	1.20 ± 0.08	1.36 ± 0.11
Rice straw	1.47 ± 0.08	1.56 ± 0.08	1.54 ± 0.04	1.58 ± 0.03	1.34 ± 0.10	1.46 ± 0.05	1.14 ± 0.15	1.38 ± 0.12
Typha straw	1.46 ± 0.11	1.49 ± 0.11	1.55 ± 0.07	1.6 ± 0.08	1.35 ± 0.06	1.46 ± 0.08	1.23 ± 0.08	1.44 ± 0.07
Cassava stems	1.46 ± 0.15	1.49 ± 0.14	1.45 ± 0.11	1.50 ± 0.11	1.27 ± 0.07	1.35 ± 0.16	1.30 ± 0.03	1.39 ± 0.05
Pigeon pea husks	1.5 ± 0.03	1.90 ± 0.03	1.51 ± 0.08	1.59 ± 0.05	1.24 ± 0.10	1.38 ± 0.05	1.34 ± 0.06	1.44 ± 0.04
Chipped mixed twigs	1.32 ± 0.11	1.4 ± 0.12	1.38 ± 0.05	1.47 ± 0.04	1.25 ± 0.13	1.39 ± 0.07	1.21 ± 0.11	1.33 ± 0.12
Chipped maize cobs	1.38 ± 0.14	1.44 ± 0.13	1.34 ± 0.08	1.42 ± 0.09	1.28 ± 0.12	1.38 ± 0.14	1.32 ± 0.12	1.41 ± 0.14
Eupatorium tops	1.47 ± 0.06	1.53 ± 0.03	1.33 ± 0.23	1.47 ± 0.23	1.23 ± 0.09	1.43 ± 0.08	1.11 ± 0.07	1.37 ± 0.06
Fine gravel	1.52 ± 0.07	1.66 ± 0.04	1.47 ± 0.02	1.65 ± 0.03	1.43 ± 0.07	1.58 ± 0.07	1.17 ± 0.07	1.47 ± 0.12

A = fine earth, B = overall bulk density.
SOURCE: Unpublished data of R. Lal.

TABLE 9.4 Effects of Mulch Material on the Surface Casting Rate of Earthworm *Hyperiodrilus africanus*, no./(day · m^2)

Mulch treatment	Crop				Mean
	Cassava	Cowpea	Maize	Soybean	
Bare (without mulch)	0	5	2	1	1 ± 2
Panicum maximum	2	25	5	38	18 ± 17
Elephant grass	11	23	26	30	23 ± 8
Black plastic	1	1	4	1	2 ± 2
White plastic	1	1	2	1	1 ± 1
Sawdust	5	14	23	43	21 ± 16
Rice husk	33	41	51	72	49 ± 17
Pigeon pea tops	23	27	16	33	25 ± 7
Pigeon pea stems	16	16	8	22	16 ± 6
Maize stover	34	22	16	11	21 ± 10
Pennisetum straw	13	33	29	29	26 ± 9
Andropogon straw	24	25	10	30	22 ± 9
Oil palm leaves	9	26	19	42	24 ± 14
Soybean husk	8	21	25	28	21 ± 9
Rice straw	12	10	25	26	18 ± 8
Typha straw	4	11	21	15	13 ± 7
Cassava stems (chipped)	7	14	28	29	20 ± 11
Cowpea husk	7	3	20	14	11 ± 8
Mixed twigs (chipped)	17	17	17	28	20 ± 6
Maize cobs (chipped)	19	15	39	13	22 ± 12
Eupatorium stover	6	13	29	20	17 ± 10
Fine gravels	3	1	0	1	1 ± 1

SOURCE: Unpublished data of R. Lal.

is in accord with the soil physical properties. Those mulches that have ameliorative effects on soil physical properties also control runoff and soil erosion more effectively than those that cause moderate or slight improvements in soil physical properties. The runoff losses with the gravel mulch are often greater than those under mulch of organic origin. The plastic mulch has to be applied very carefully. The runoff losses can be substantial with plastic mulch unless seep holes are provided through the plastic sheet (Plate 9.3). Even the infiltration rate under the plastic mulch may be less than that in soil receiving organic mulches. The reduction in infiltration rate is particularly noticed if the plastic mulch is consecutively applied for two to three growing seasons. Because of the lack of food material, the activity of earthworms is drastically less under plastic mulch than under organic mulch (Table 9.4). As a result of low earthworm activity, the infiltration rate and saturated hydraulic conductivity are also low

TABLE 9.5 Effects of Different Mulch Materials on Saturated Hydraulic Conductivity of 0- to 10-cm Layer (Data Obtained in 1978 Growing Season)

Kind of mulch	Hydraulic conductivity, cm/h			
	Cassava	Cowpea	Maize	Soybean
Bare (without mulch)	22.97 ± 9.36	89.2 ± 20.25	204.36 ± 152.61	4.61 ± 2.78
Panicum maximum	316.5 ± 250.28	295.64 ± 269.52	520.3 ± 170.53	274.18 ± 184.91
Elephant grass	370.55 ± 103.08	313.41 ± 87.67	324.58 ± 152.85	27.44 ± 17.2
Black plastic	104.56 ± 110.77	44.64 ± 16.83	2.27 ± 2.94	25.75 ± 27.65
Translucent plastic	456.61 ± 177.42	369.51 ± 218.52	376.75 ± 312.54	1022.48 ± 1003.04
Sawdust	81.28 ± 67.08	118.25 ± 69.37	265.97 ± 291.06	57.73 ± 7.81
Rice husk	302.32 ± 71.94	58.24 ± 32.17	89.59 ± 150.28	10.3 ± 4.42
Pigeon pea tops	103.33 ± 65.27	117.68 ± 120.99	233.43 ± 142.81	95.24 ± 129.74
Maize stover	439.56 ± 91.92	488.36 ± 132.28	1121.11 ± 734.25	176.32 ± 299.2
Pennisetum straw	367.82 ± 131.18	1186.46 ± 548.3	725.33 ± 283.06	125.61 ± 70.35
Andropogon straw	650.43 ± 152.47	858.01 ± 472.22	681.97 ± 109.63	155.31 ± 116.31
Pigeon pea stems	320.58 ± 183.40	256.29 ± 31.37	388.91 ± 386.26	35.56 ± 37.86
Oil palm leaves	188.50 ± 42.08	134.50 ± 90.24	372.55 ± 228.31	11.69 ± 11.78
Chipped soybean tops	159.41 ± 65.05	191.59 ± 132.37	518.77 ± 179.14	565.00 ± 185.87
Rice straw	237.18 ± 59.00	72.97 ± 41.57	325.02 ± 216.79	435.54 ± 291.30
Typha straw	27.89 ± 7.78	69.95 ± 43.44	571.69 ± 324.65	570.40 ± 248.98
Cassava stems	27.67 ± 12.87	32.23 ± 23.55	322.65 ± 143.69	310.32 ± 163.00
Pigeon pea husks	34.65 ± 6.10	606.79 ± 424.18	540.93 ± 342.88	116.78 ± 62.86
Chipped mixed twigs	109.56 ± 30.04	253.9 ± 256.54	606.38 ± 337.59	328.70 ± 126.96
Chipped maize cobs	88.63 ± 26.99	203.55 ± 195.73	888.27 ± 391.76	342.18 ± 244.41
Eupatorium tops	93.33 ± 27.37	382.74 ± 281.56	783.74 ± 225.64	741.11 ± 564.66
Fine gravel	213.46 ± 45.10	280.45 ± 209.89	800.91 ± 454.94	1135.24 ± 950.20
Crop mean	214.41 ± 169.5	292.02 ± 284.58	484.79 ± 273.54	1022.48 ± 1003.04

SOURCE: Unpublished data of R. Lal.

TABLE 9.6 Percentage of Aggregates Greater than 0.5 mm in Relation to the Nature of Mulch Material

Mulch material	Percentage of aggregates (>0.5 mm)
Unmulched control	52.7
Panicum maximum	77.0
Pennisetum purpureum	77.1
Black plastic film	83.9
Transparent plastic film	64.5
Sawdust	90.1
Rice husk	59.9
Maize stover	74.2
Andropogon stover	68.6
Pigeon pea straw	74.9
Soybean straw	77.5
Rice straw	76.7
Typha straw	48.1
Gravel	77.3

SOURCE: Unpublished data of R. Lal.

Plate 9.3 Seep holes through the plastic mulch are essential in facilitating water percolation and decreasing runoff losses.

TABLE 9.7 Effect of Different Mulch Materials on Soil Moisture Retention (pF) Curves of 0- to 10-cm Layer

Mulch	\multicolumn{6}{c}{Soil moisture suction, bar}					
	0	0.1	0.3	1.0	3	5
Bare (without mulch)	22.93 ± 2.37	15.40 ± 0.28	13.7	n.d.	9.63 ± 0.92	9.38 ± 1.91
Panicum maximum	26.95 ± 3.67	15.65 ± 1.34	15.1	12.6	9.50 + 2.35	10.23 + 0.92
Elephant grass	26.03 ± 2.51	19.15 ± 0.07	18.1	16.2	11.33 ± 2.44	11.38 ± 3.12
Black plastic	28.75 ± 3.76	16.50 ± 0.85	15.2	12.4	11.03 ± 2.22	10.48 ± 1.54
Translucent plastic	29.93 ± 3.65	15.80 ± 0.70	n.d.	13.0	11.05 ± 1.20	9.10 ± 2.41
Sand dust	30.28 ± 2.50	22.50 ± 0.90	20.1	17.9	14.13 ± 3.00	14.03 ± 2.81
Rice husk	29.40 ± 4.03	22.25 ± 1.20	21.8	19.8	11.65 ± 4.22	14.33 ± 1.88
Pigeon pea tops	26.43 ± 1.68	17.65 ± 0.78	15.8	14.5	12.17 ± 1.00	12.15 ± 0.96
Maize stover	27.33 ± 4.08	14.9 ± 1.56	13.8	12.4	10.20 ± 0.36	10.28 ± 1.36
Pennisetum straw	24.48 ± 1.90	12.70 ± 1.13	12.5	11.0	8.93 ± 0.84	8.37 ± 0.25
Andropogon straw	25.75 ± 1.77	13.35 ± 0.49	4.50	9.80	8.83 ± 0.76	8.47 ± 0.38
Pigeon pea stems	25.38 ± 3.97	17.65 ± 1.63	15.20	12.70	10.77 ± 1.65	10.5 ± 1.79
Oil palm leaves	26.65 ± 5.80	13.35 ± 0.64	13.30	9.60	8.97 ± 0.49	8.9 ± 1.14
Chipped soybean tops	25.18 ± 1.99	11.35 ± 0.07	9.30	8.60	7.73 ± 0.40	7.30 ± 1.20
Rice straw	27.98 ± 4.26	11.20 ± 0.71	10.80	9.20	8.23 ± 0.59	7.30 ± 1.47
Typha straw	25.98 ± 2.52	10.80 ± 1.56	n.d.	7.80	9.77 ± 0.67	7.63 ± 1.57
Cassava stems	28.53 ± 3.85	13.95 ± 1.34	n.d.	9.90	10.00 ± 0.56	8.80 ± 1.06
Pigeon peak husks	26.48 ± 2.28	14.35 ± 0.35	n.d.	10.6	11.20 ± 2.52	8.63 ± 1.07
Chipped mixed twigs	25.48 ± 0.79	13.35 ± 0.78	n.d.	9.30	8.97 ± 1.08	7.99 ± 0.83
Chipped maize cobs	30.4 ± 5.97	17.20 ± 1.56	n.d.	13.60	11.97 ± 3.84	9.80 ± 1.91
Eupatorium tops	26.53 ± 5.42	10.65 ± 1.20	n.d.	8.20	11.20 ± 4.59	6.63 ± 0.40
Fine gravel	20.98 ± 1.75	8.50	n.d.	0	7.60	6.90 ± 0.99

n.d. = not determined.
SOURCE: Unpublished data of R. Lal.

TABLE 9.8 Effect of Different Mulch Materials on Penetrometer Resistance of Surface Layer of a Tropical Alfisol (1978 Data)

Kind of mulch	Penetrometer resistance, kg/cm²			
	Cassava	Cowpea	Maize	Soybean
Bare (without mulch)	1.98 ± 0.19	2.24 ± 0.61	3.48 ± 0.85	2.61 ± 0.54
Panicum maximum	0.80 ± 0.23	3.69 ± 0.21	2.55 ± 0.52	1.86 ± 0.29
Elephant grass	0.95 ± 0.21	3.14 ± 0.70	3.05 ± 0.71	2.74 ± 0.39
Translucent plastic	1.61 ± 0.34	2.19 ± 0.56	1.98 ± 0.76	1.53 ± 0.76
Sawdust	3.28 ± 0.38	3.51 ± 0.46	4.11 ± 0.29	3.76 ± 0.29
Rice husk	2.93 ± 0.47	4.24 ± 0.32	3.34 ± 0.20	3.71 ± 0.34
Pigeon pea tops	2.49 ± 0.45	4.03 ± 0.38	3.81 ± 0.24	2.53 ± 0.49
Maize stover	2.01 ± 0.56	2.96 ± 0.35	2.91 ± 0.58	3.45 ± 0.70
Pennisetum straw	1.59 ± 0.13	3.53 ± 0.26	3.04 ± 0.53	3.08 ± 0.16
Andropogon straw	1.28 ± 0.03	1.20 ± 0.30	1.66 ± 0.33	2.74 ± 0.38
Pigeon pea stems	1.58 ± 0.12	1.86 ± 0.53	1.88 ± 0.19	4.16 ± 0.39
Oil palm leaves	1.51 ± 0.13	1.63 ± 0.05	2.00 ± 0.23	3.63 ± 0.29
Chipped soybean tops	1.55 ± 0.26	1.41 ± 0.19	1.58 ± 0.17	2.36 ± 0.35
Rice straw	1.74 ± 0.28	1.76 ± 0.17	1.53 ± 0.16	3.36 ± 0.41
Typha straw	1.45 ± 0.23	1.79 ± 0.19	1.60 ± 0.11	2.75 ± 0.60
Cassava stems	1.98 ± 0.23	1.79 ± 0.19	1.60 ± 0.11	2.75 ± 0.60
Pigeon pea husks	1.79 ± 0.26	1.93 ± 0.31	1.94 ± 0.48	2.45 ± 0.59
Chipped mixed twigs	1.50 ± 0.31	1.75 ± 0.13	1.85 ± 0.15	1.34 ± 0.17
Chipped maize cobs	1.55 ± 0.30	1.45 ± 0.17	1.85 ± 0.32	1.66 ± 0.32
Eupatorium tops	1.60 ± 0.18	1.61 ± 0.26	1.61 ± 0.18	1.56 ± 0.15
Fine gravel	2.21 ± 0.12	2.00 ± 0.39	2.56 ± 0.21	1.91 ± 0.32

SOURCE: Unpublished data of R. Lal.

TABLE 9.9 Effects of Mulch Material on Runoff and Soil Erosion under Maize for a Tropical Alfisol in Western Nigeria

Type of mulch	Runoff, mm/yr	Erosion, t/(ha · yr)
Unmulched control	52.9	2.5
Plastic sheet	23.6	0.65
Gravel	29.9	0.12
Grass straw	23.7	0.05
Leguminous straw	22.9	0.08

SOURCE: Unpublished data of R. Lal, 1978.

under plastic mulch in comparison with organic mulches. Regardless of the type of mulch, soil erosion is practically negligible in all mulches—organic, gravel, or plastic—in comparison with the unmulched control (Table 9.9).

Considering all pros and cons, crop residue mulch is the only viable method of using mulch for soil and water conservation on arable lands. Even then, bringing in an adequate amount of mulch material can be a constraint for managing large-scale agriculture. The availability of the large amount of mulch material required is in itself a severe constraint for many arid and semiarid regions.

TABLE 9.10 Effect of Rate of Mulch on Runoff and Soil Loss*

Mulch rate, t/ha	Slope, %				Mean	Percentage of rainfall
	1	5	10	15		
	Runoff, mm					
0	12.0	14.8	10.4	14.8	13.0	20
2	1.3	6.2	6.0	5.7	4.8	8
4	0.4	1.5	3.6	3.3	2.2	3
6	0.0	0.7	1.9	1.8	1.1	2
	Soil Loss, t/ha					
0	0.48	12.19	27.06	12.25	13.00	
2	0.01	3.49	0.82	0.64	1.24	
4	0.00	0.67	0.11	0.31	0.27	
6	0.00	0.16	0.03	0.08	0.07	

*Total rainfall = 64 mm.
SOURCE: Lal, 1976.

Soil conditioners

The shear resistance of the soil is increased by improving its structural stability. Soil organic matter content plays an important role in improving soil structure and its ability to withstand the disruptive effect of raindrop and overland flow: Splash erosion is decreased by improving the structural stability of aggregates through the application of soil conditioners and other amendments. Synthetic polymers cause the clay platelets to stick together by altering the interparticle forces. The changes in interparticle forces originate either by the adsorption of polymer over the surface of a single particle or by the attachment of one polymer molecule to several soil particles. The formation of these bonds aids flocculation and stabilizes the existing particle-to-particle arrangement

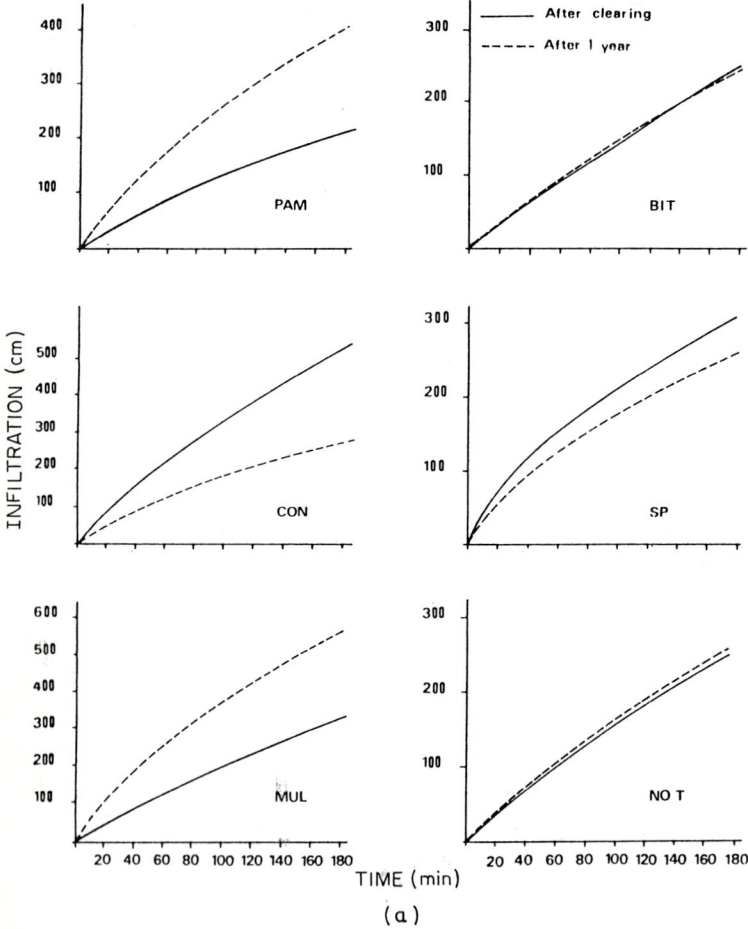

Figure 9.5 Effects of plastic mulches on water infiltration into a tropical alfisol: (*a*) field studies (PAM = polyacrylamide, BIT = bitumen, CON = control, SP = soil penetrant, MUL = mulch, NO T = no-till); (*b*), (*c*) laboratory studies. (De Vleeschauwer et al., 1978.)

(Overbeek, 1966; Greenland, 1972). Some of the most commonly used polymers are listed in Table 9.11. Another material of relevance under field conditions is the asphalt emulsion, which has been used, with varying degrees of success, to control both wind and water erosion. Experiments conducted on alfisols in Nigeria showed that application of asphalt improved water infiltration (Fig. 9.5). In Venezuela, Pla (1977) reported significant reductions in soil erosion by application of asphalt emulsion (Fig. 9.6).

A considerable amount of basic and applied research concerning soil conditioners has been done in temperate-zone countries (De Boodt, 1972; Gabriels, 1972). Some research experiments have also been done in the tropics (Alles, 1971; Pla, 1977; De Vleeschauwer et al., 1979; Lal and Greenland, 1978). Two relevant examples are the effects of PVA and of asphalt emulsion on runoff and erosion under maize at IITA, Ibadan, Nigeria. The data in Table 9.12 show that among treatments evaluated, residue mulch of rice straw was the most effective method of soil erosion control. Significant beneficial effects were also observed by the application of asphalt emulsion. The latter also improved crop growth and yield (Fig. 9.7). In the Ivory Coast, Roose (1975) evaluated the effect of Curasol and of grass mulch on runoff and soil erosion. Application of Curasol controlled soil erosion but was not as effective as grass mulch (Table 9.13). Similar beneficial results have been reported for some European soils (De Boodt, 1972a,b; Gabriels, 1972).

In Malaysia, the emulsion prepared from natural rubber has been tested for

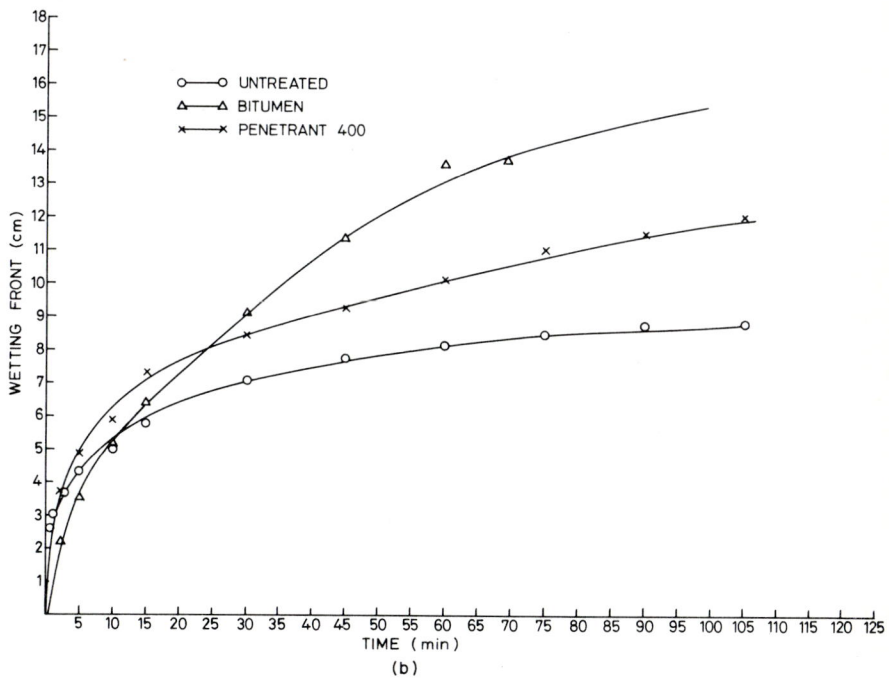

Figure 9.5 (*Continued*)

its effects in controlling erosion. Rubber and other natural elastomers impart hydrophobic properties to soils and stabilize soil aggregates against the dispersive action of impacting raindrops. The data of Soong and Yeoh (1975) show that an emulsion of 10 percent natural dry rubber prepared in 4 or 8 percent oil reduces soil erosion (Tables 9.14 and 9.15). The effectiveness of natural rubbers can be increased when they are used in combination with a grass or a leguminous cover (Table 9.15).

From the point of view of practical considerations for erosion control at field scale on arable lands, however, the best soil conditioner available is the crop residue mulch. Nonetheless, soil conditioners can be effectively used under special conditions, such as for stabilizing soils around individual trees planted on steep slopes, as is done for orchards in Asia. Soil conditioners and amendments are also used in stabilizing soil in the vicinity of engineering structures. In New South Wales, Australia, for example, failure of mechanical structures caused by tunnel erosion is prevented by using chemical amendments. The use of soil con-

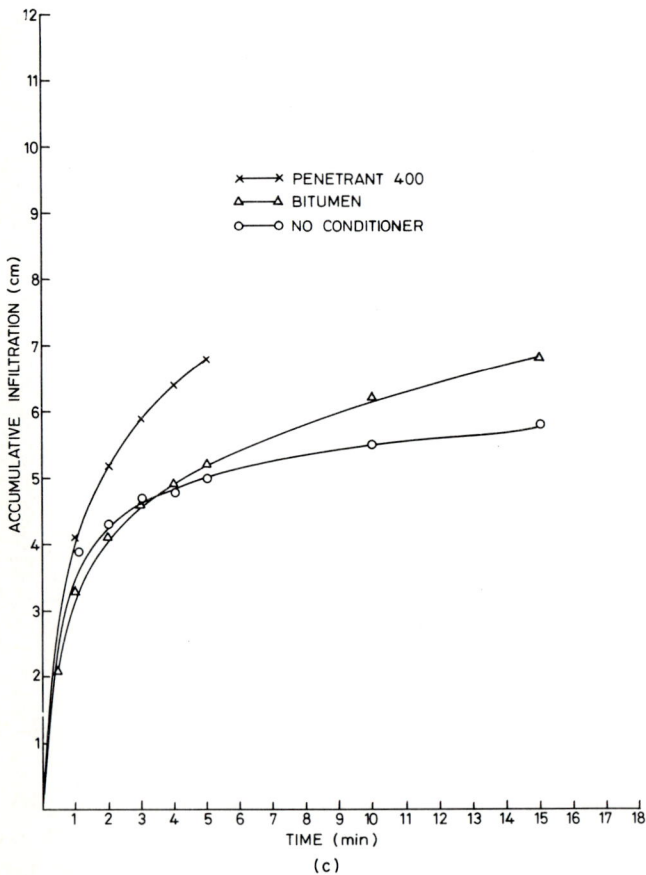

Figure 9.5 (*Continued*)

TABLE 9.11 Some Commonly Used Soil Conditioners

Trade name	Charge	Company	Form	Type
Elvanol 52-22	Nonionic	E.I. du Pont de Nemours	Powder	Polyvinyl alcohol
Elvanol 71-30	Nonionic	E.I. du Pont de Nemours	Powder	Polyvinyl alcohol
Humofina FB63	Anionic	Petrofina Belgium	Liquid	Hydrophobic asphalt emulsion
CRS 2	Cationic	Bitucoat, Des Moines, Iowa	Liquid	Hydrophobic asphalt emulsion
SS lh	Anionic	Bitucoat, Des Moines, Iowa	Liquid	Hydrophobic asphalt emulsion
Curasol AE	Nonionic	American Hoechst Corp.	Liquid	Polyvinyl acetate
Curasol AH	Nonionic	American Hoechst Corp.	Liquid	Polyvinyl acetate
DCA 70	Nonionic	Union Carbide	Liquid	Polyvinyl acetate
Petroset SB	Cationic	Phillips Petroleum Co.	Liquid	Rubber emulsion
Terratack	Nonionic	Grass Growers Inc.	Powder	Polysaccharide

source: De Boodt, 1972.

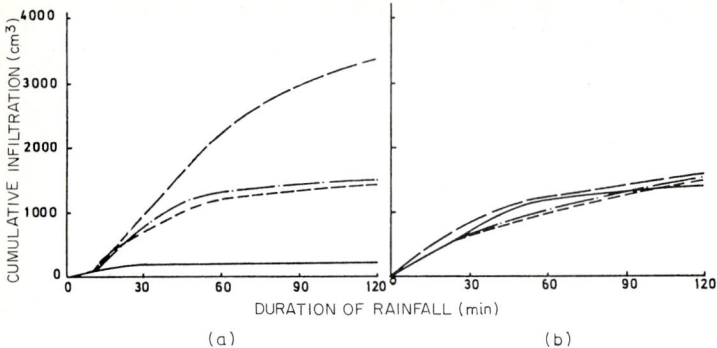

Figure 9.6 Effects of asphalt emulsion on erosion control in Venezuela. (*a*) Treated soil, (*b*) untreated soil. (Pla, 1977.)

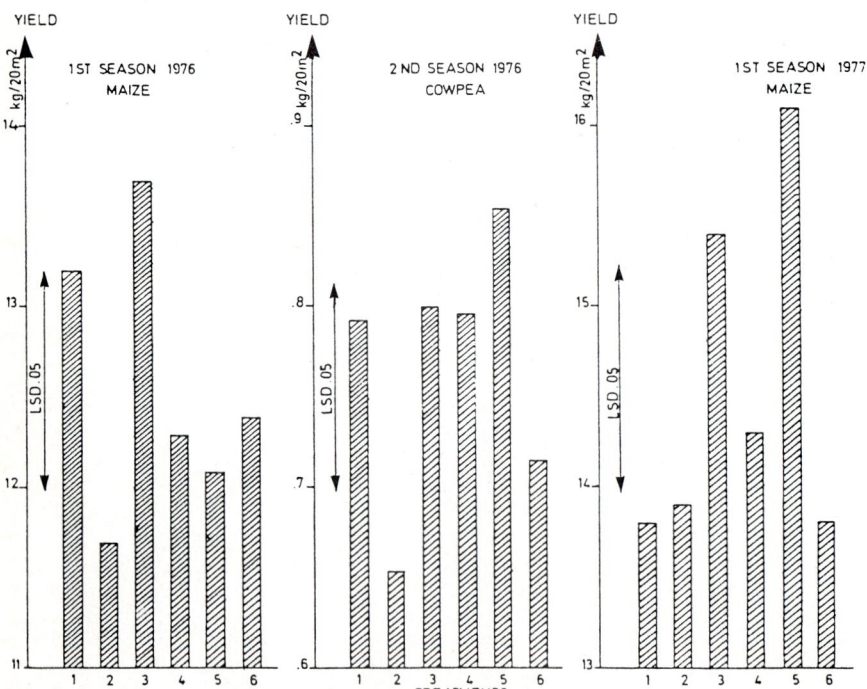

Figure 9.7 Effects of residue mulch and soil conditioners in yield of maize and cowpea on a tropical alfisol. 1 = polyacrylamide, 2 = control, 3 = bitumen, 4 = soil penetrant, 5 = mulch, 6 = no-till. LSD = least significant difference. (De Vleeschauwer et al., 1978.)

TABLE 9.12 Effects of Soil Conditioners and Crop Residue Mulch on Water Infiltration, Runoff, and Soil Erosion from a Tropical Alfisol

	Water infiltration		Runoff		
Treatment	Cumulative, cm/3 h	Rate, cm/min	mm	% of rainfall	Soil erosion, t/ha
Polyacrylamide	419.8	1.81	0.0	0.0	0.0
Control	277.6	1.18	35.0	8.1	4.8
Bitumen (asphalt)	247.0	1.15	1.4	0.3	0.0
Soil penetrant	246.1	0.99	26.1	6.0	1.9
Mulch	571.4	2.30	0.0	0.0	0.0
No-tillage	264.9	1.22	0.0	0.0	0.0

SOURCE: De Vleeschauwer et al., 1978.

ditioners is recommended for controlling tunnel formation. Elliott (1979) reported that hydrated lime is an effective flocculant to stop tunneling in an easily dispersed soil. Reddan (1980) also used hydrated lime to overcome failure of the drop structures due to tunneling in a highly dispersible subsoil.

Plate 9.4 Stone-wall contour terraces widely used in the Andean region of Peru. (Courtesy Prof. Felipe-Morales.)

Decreasing the Shear Strength of Overland Flow through Slope Management

A reduction in shear strength is achievable by reducing the amount and velocity of overland flow. The resistance to flow velocity can be increased by many practices, e.g., use of mulch, contour barriers, grass strip or a strip of perennial shrubs, rough seedbed, contour ridges, and engineering devices including graded-channel terraces and waterways.

Engineering devices are discussed in detail in many other texts (Richey et al., 1961; Hudson, 1971; Beasley, 1972; Blaisdell, 1981). Sheng (1981) described different types of terraces normally used for erosion control. These structures are described in Fig. 9.8 and include devices such as bench terraces, hillside

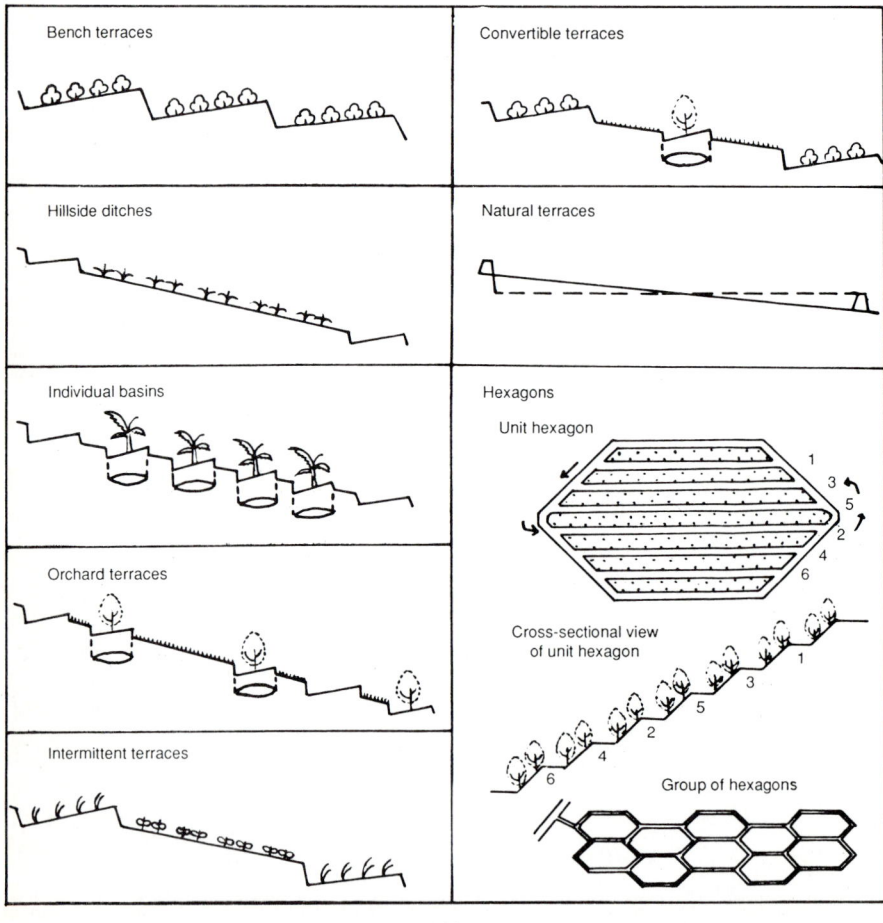

(a)

Figure 9.8 Different types of engineering structures used for erosion control. (*a*) Structures for land treatment. (*b*) Major types of waterways (sectional views; stilling basins not shown). (Sheng, 1981.)

ditches, individual basins, orchard terraces, miniconvertible terraces, and hexagonal terraces. Different terraces are used under different conditions. In New South Wales, Australia, Quilty (1973) proposed a scheme of water-retaining and water-spreading terraces on marginal cultivation lands. Special terraces are used for water conservation in arid and semiarid regions of unreliable rainfall. The contour seepage furrow is an effective measure for water conservation in arid regions (Hindson, 1981). Peasley (1975) recommended different types of engineering structures depending on the predominant erosion processes; for controlling sheet erosion, e.g., broad-based banks were recommended.

Stone-wall contour terraces are widely used in South America (Plate 9.4). In Venezuela, Comerma et al. (1973) suggested that the stone removed from the

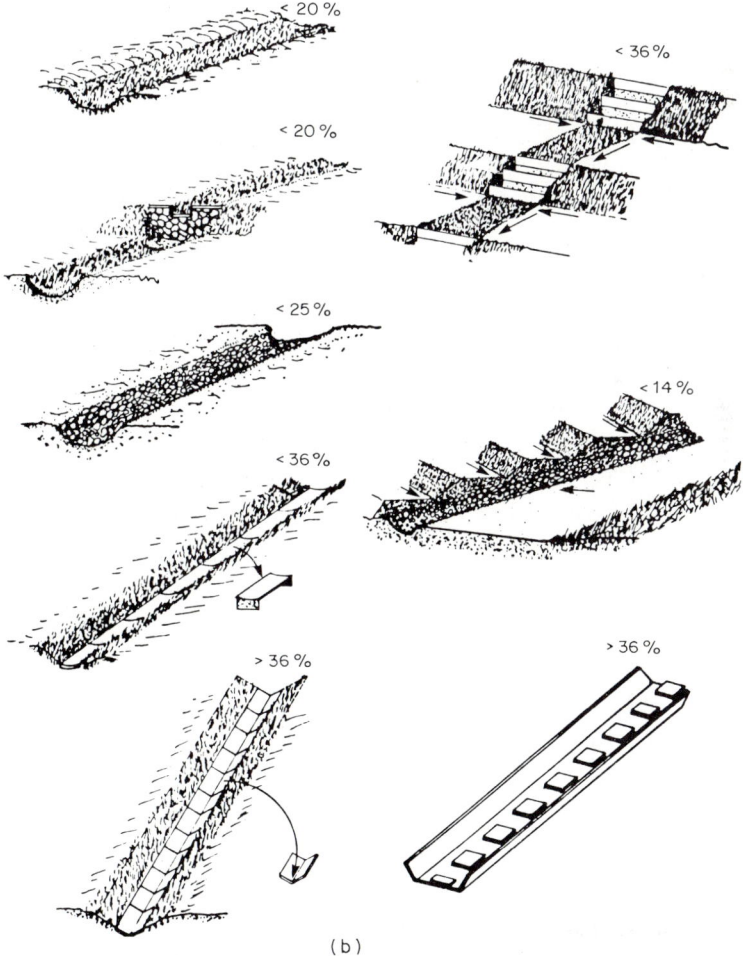

(b)

Figure 9.8 *(Continued)*

TABLE 9.13 Effects of Curasol Applications [60 g/(L · m^2)] on Runoff and Erosion Observed on Runoff Plots at Adiopodoume, Ivory Coast

Adiopodoume, 1970–1973	Rainfall height, mm	Erosion						Runoff					
		Panicum, $S = 7\%$		Bare fallow, $S = 7\%$		Bare fallow, $S = 23\%$		Panicum, $S = 7\%$		Bare fallow, $S = 7\%$		Bare fallow, $S = 23\%$	
Year		Check, t/ha	Curasol, % of check	Check, t/ha	Curasol, % of check	Check, t/ha	Curasol, % of check	Check, mm	Curasol, % of check	Check, mm	Curasol, % of check	Check, mm	Curasol, % of check
1970	1389.0	89.17	25	149.98	50	532.07	27	368.12	37	575.28	56	422.78	40
1971	1816.0	4.11	30	139.21	55	617.73	59	189.78	77	562.30	105	286.13	149
1972	1561.5	1.20	10	113.59	50	272.86	57	106.19	16	592.61	66	362.49	55
Average	1589	31.5	25	134.3	52	474.2	47	13.9	45	36.3	75	22.5	74

The plot without Curasol was tilled with a hoe to a 5-cm depth. This was enough to minimize erosion for 1 month following tillage.
SOURCE: Roose, 1975.

TABLE 9.14 Effect of Different Natural Rubber Formulations on Soil Erosion in Malaysia

Treatment	Soil erosion, t/ha
Control	288.6
5% dry rubber + 4% oil	49.6
5% dry rubber + 8% oil	96.6
10% dry rubber + 4% oil	36.7
10% dry rubber + 8% oil	45.2

SOURCE: Soong and Yeoh, 1975.

field be used for stabilizing terrace walls and installing ditch-drainage systems. Rocks, picked from within the field, thus become a resource rather than a liability. Williams and Walter (1987) observed that terrace construction in Venezuelan highlands provides the best advantages to the farmer. Sheng and Stennet (1975), on the basis of their work in Jamaica, presented a land classification scheme and recommended the type of engineering devices and conservation measures to be used for each land type.

Terraces have widely been used for rice culture on steeplands in Asia. In India, Gupta et al. (1975) described the use of different devices under different situations, e.g., bunding, bench terraces, diversion terraces or graded trenching, grossed waterways. A range of structures are used for steeplands in Taiwan (Chan, 1981). Terraces have been widely used in different regions of Africa, but with limited success (Lal, 1977). Different types of terraces used are described by Cormack (1972), Datiri (1974), and Millington (1983).

In Indonesia, Kurnia and Suwardjo (1981) compared soil erosion and water runoff losses from terraced land with those from land where soil and crop management techniques were used for erosion control. They observed that bench terraces even with mulch were less effective than intercropping with mulch.

In Nigeria, Lal measured erosion from terraced and unterraced watersheds. The data in Table 9.16 show that although installing graded-channel terraces decreased both runoff and soil erosion, tillage methods used within terraces had an overriding effect.

Engineering structures have been widely used throughout the world. These are capital-intensive measures, and their effectiveness depends on two factors:

TABLE 9.15 Comparative Effects of Natural Rubber Formulations and Grass Covers on Soil Erosion in Malaysia

Treatment	Soil erosion, t/ha
Control	338.5
Natural rubber formulation	62.0
Natural rubber formulation + grass	35.4

SOURCE: Soong and Yeoh, 1975.

TABLE 9.16 Runoff and Erosion Measurements with H-Flume for 1979 on about 5-ha Conventionally Tilled and No-Till Watersheds

Parameter	No-till (unterraced)			Conventional tillage (terraced)		
	First season	Second season	Total	First season	Second season	Total
	1979					
Runoff (mm)	3.60	17.93	21.53	173.64	51.42	225.06
Erosion (Mg/ha)	0.012	0.118	0.130	4.91	0.59	5.50
Rainfall (mm)	583.5	257.8	841.3	583.5	257.8	841.3
	1980					
Runoff (mm)	25.13	9.31	34.44	122.80	30.16	152.96
Erosion (Mg/ha)	0.307	0.025	0.33	1.61	0.286	1.89
Rainfall (mm)	621	279	900	621	279	900

SOURCE: Lal, 1984.

1. Adoption of a sound soil and crop management system within the terraces. Terraces are a backup system and can work only if the soil and crop management system is properly carried out.
2. Regular maintenance of terrace outlets and grass waterways. Terrace failure is a common phenomenon whenever the terraces are not properly installed or regularly maintained.

Conclusions

To be accepted by the farming community, the proposed soil conservation system must be effective, economical, and an integral part of the farming system and soil or crop management practices already used. Biologic measures of soil and water conservation, such as mulches and cover crops, are more effective and more economical than engineering measures. Soil erosion is caused by land misuse, and erosion can be controlled only by proper land use and appropriate soil and crop management practices.

References

Alles, W. S. 1971. Chemical sprays for controlling soil erosion. *Trop. Agric. Ceylon* 127(3–4):179–185.
Beasley, R. P. 1972. *Soil Erosion and Sediment Pollution Control.* Iowa State University Press, Ames.
Bennett, H. H. 1939. *Soil Conservation.* McGraw-Hill, New York.
Blaisdell, F. W. 1981. Engineering structures for erosion control. In: R. Lal and E. W. Russell (eds.), *Tropical Agricultural Hydrology,* Wiley, Chichester, U.K., pp. 325–356.
Chan, C. C. 1981. *Conservation Measures on the Cultivated Slopelands of Taiwan.* Extension Bull. (ASPAC/FFTC), 157.
Comerma, J. A. 1973. Productivity increase from soil conservation practices. *Agron. Trop. Venez.* 23(1):95–113.

Cormack, J. M. 1972. Efficient utilization of water through land management. *Rhod. Agric. J.* 69(1):11-6.

Datiri, B. T. 1974. *Soil and Water Conservation Practices, Particularly in Areas of Shifting Cultivation,* Soil Bull. (FAO), 24, pp. 237–241.

De Boodt, M. 1972a. Improvement of soil structure by chemical means. In: D. Hillel (ed.), *Optimizing the Soil Physical Environment Toward Greater Crop Yields,* Academic, New York, pp. 43–55.

De Boodt, M. (ed.). 1972b. *Fundamentals of Soil Conditioning.* Proc. Symp., 17–21 April 1972, University of Ghent, Ghent, Belgium.

De Vleeschauwer, D., Lal, R., and De Boodt, M. 1978. The comparative effects of surface application of organic mulch versus chemical soil conditioners on physical and chemical properties of the soil and on plant growth. *Catena* 5:337–349.

De Vleeschauwer, D., Lal, R., and De Boodt, M. 1979. Influence of soil conditioners on water movement through some tropical soils. In: *Soil Conditions and Crop Production in the Tropics* (R. Lal and D. J. Greenland, eds.), Wiley, Chichester, U.K., pp. 149–158.

Elliott, G. L. 1979. Preventing failure of earth-works in unstable soils. *J. Soil Conserv. Serv. N.S.W.* 35(2):101–107.

Elwell, H. A. 1972. Requirements of a modern soil conservation system. *Rhod. Agric. J.* 69(6):115–118.

Gabriels, D. 1972. Response of different soil conditioners to soils. In: M. De Boodt (ed.), *Fundamentals of Soil Conditioning,* Proc. Symp., 17–21 April 1972, University of Ghent, Ghent, Belgium, pp. 1014–1034.

Greenland, D. J. 1972. Interaction between organic polymers and inorganic soil particles. In: *Fundamentals of Soil Conditioning* (M. De Boodt, ed.), State University Ghent, Belgium, pp. 897–914.

Gupta, S. K., Das, D. C., Tejwani, K. G., Chittaranjan, S., and Srinivas, A. 1975. Mechanical measures of erosion control. In: *Soil and Water Conservation Research 1956–1971,* Indian Council of Agricultural Research, pp. 146–182.

Heinonen, R. 1985. *Soil Management and Crop Water Supply,* 4th ed. Dept. of Soil Science, Swedish Univ. of Agric. Sciences, Uppsala, Sweden.

Hindson, J. 1981. Proposed remedy for erosion: Contour seepage furrows. *Approp. Tech. UK* 8(1):10.

Kurnia, U., and Suwardjo, H. 1985. Effect of mechanical conservation methods on soil erosion on Tropudalf and a Tropothent in Yogyakarta. *Pembr. Pen. Tanah Dan Pupuk,* no. 4, pp. 46–50.

Lal, R., and Greenland, D. J. 1978. Effects of soil conditioners and initial water potential of a Vertisol on infiltration and heat of wetting. In: *Modification of Soil Structure* (W. Emerson Bond and A. Dexter, eds.), Wiley, Chichester, U.K., pp. 191–198.

Lundgren, L. 1980. *Comparison of Surface Runoff and Soil Loss from Runoff Plots in Forest and Small-Scale Agriculture in the Usambara Mts., Tanzania.* Department of Physical Geography, University of Stockholm, STOU-NG 38.

Mesquitela, J. C. 1973. Erosion control (Combate a erosao). Dionlgacao Agro-Pecuaria (Angola) no. 103.

Overbeek, J. T. H. G. 1966. Colloid stability in aqueous and non-aqueous media. *Disc. Faraday Soc.* 42:7–13.

Peasley, B. A. 1975. Soil conservation in the Narrabri area. *J. Soil Conserv. Serv. NSW* 31(1):9–18.

Pla, I. 1975. Effect of bitumen emulsion and polyacrylamide on some physical properties of Venezuelan soils. In: *Soil Conditioners,* Soil Science Society of America, Madison, Wisc., pp. 35–46.

Pla, I. 1977. Aggregate size and erosion control on sloping land treated with hydrophobic bitumen emulsion. In: *Soil Conservation and Management in the Humid Tropics* (D. J. Greenland and R. Lal, eds.), Wiley, Chichester, U.K.

Quilty, J. A. 1973. Soil conservation structures for marginal arable areas. A field study. *J. Soil Sci. Conserv. Serv. NSW* 29(3):119–129.

Reddan, B. 1980. Stabilization of a box drop inlet spillway structures in unconsolidated highly erodible soils. *J. Soil Conserv. Serv. NSW* 36(4):219–223.

Richey, C. B., Jacobson, P., and Hall, C. W. (eds.). 1961. *Agricultural Engineer's Handbook*. McGraw-Hill, New York.

Roose, E. J. 1975. Natural mulch or chemical conditioner for reducing soil erosion in humid tropical area. In: *Soil Conditioners,* Soil Science Society of America, Madison, Wisc., pp. 131–138.

Sheng, T. C. 1981. Engineering structures for erosion control. In: *Tropical Agricultural Hydrology* (R. Lal and E. W. Russell, eds.), Wiley, Chichester, U.K.

Sheng, T. C., and Stennet, R. 1975. *Forestry Development and Watershed Management in the Upland Regions, Jamaica.* FAO working document SF/JAM 505, Rome, Italy.

Soong, N. K., and Yeoh, C. S. 1975. Latex/oil emulsions for controlling soil erosion on exposed soil surfaces. *Proc. International Rubber Conference,* Rubber Research Institute of Malaysia, Kuala Lumpur, 1975.

Walker, G. T. 1978. Systems, soils and soil erosion. *Geogr. Educ.* 3(2):193–208.

Williams, L. S., and Walter, J. B. 1987. Controlled erosion terraces in Venezuela. In: *Proc. Soil and Water Conservation on Steeplands,* March 23–27, 1987, SCSA, San Juan, Puerto Rico.

Chapter 10

Land Use

Introduction

A relatively small proportion of the total arable land area of the world can be used intensively without causing severe soil erosion. Such lands are usually developed along flood plains and in river valleys. Most uplands are prone to erosion and require careful land use planning and appropriate managerial systems.

Choosing an appropriate land use is the foremost consideration in soil erosion management. Each land use has a different magnitude of soil erosion risk. Soil erosion becomes a serious problem whenever a land is used for whatever it is not capable of. The land use capability is assessed on the basis of a resource survey. The resource comprises soil, vegetation, water, climate, terrain, mineral deposits, accessibility, marketing, infrastructure, and logistics. It is on the basis of this survey that a rational decision can be made about whether land is suitable for agriculture, forestry, pastures, urban development, or mining. A land considered suitable for agricultural development may have a high potential for growing a wide range of crops or a moderate potential for growing a limited range of crops. Alternatively, a land may be suitable for growing artificial forests or for pastures. The potential for these uses may also be high or moderate. Land with moderate potential for pasture development can be grazed but at a low stocking rate.

The purpose of a resource survey is to rationally plan the development and use of the land according to its potential. Some lands are better left alone. Others, although capable of supporting agriculture, cannot be used for an intensive arable land use. If intensively used, such lands are easily degraded when accelerated erosion sets in. Periodic fallowing is necessary for these lands to regenerate soil structure and to restore ecological balance.

Land Use and Soil Erosion

An inappropriate land use accelerates the erosion rate beyond the tolerable level. The adverse effects of inappropriate land use are more severe in harsh than in moderate climates, e.g., regions with intense rains or prolonged dry periods with strong directional winds. Other factors remaining the same, erosion risks follow this trend: bare fallow > arable land > perennial crops > grass cover > natural vegetation (Fig. 10.1).

Forests

Forested lands have comparatively less erosion risk than arable lands or plantation crops. This is particularly true if the forested lands have undulating to steep terrains. For the same reason, the sparse natural vegetation cover in savanna regions causes more erosion than semideciduous rain forest in the subhumid regions. Deforestation for any purpose increases soil erosion. For example, in Cameron Highlands, Malaysia, Shallow (1956) observed that the soil erosion rate increased from 0.24 t/(ha · yr) under natural forest to 4.9 t/(ha · yr) under mature coffee and 7.32 t/(ha · yr) under vegetables. Many studies con-

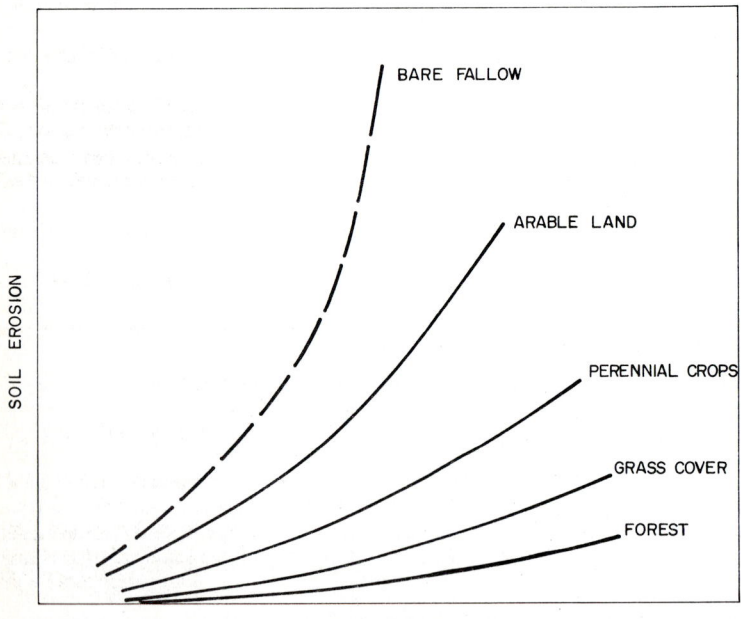

Figure 10.1 A hypothetic relation between erosion hazard, land use, and slope gradient.

ducted in Africa also support the conclusion regarding the protective effects of forest covers. In west Africa, Dabin (1959), Fournier (1967), Roose (1972), and Lal (1981) reported that erosion from cleared land was several thousand times greater than that from forested lands (Tables 10.1 and 10.2).

In Guinea, Dugain and Fauck (1959) observed significant but less drastic differences in soil erosion between natural vegetation and arable lands (Table 10.3). In Bamileke in the Bafoussam region of Cameroon, Rochette (1959) observed that erosion increased from 195 kg/ha for fallowed land with regenerated vegetation cover to 122,086 kg/ha for plowed and bare soil surfaces. In São Paulo, Brazil, Suarez de Castro (1979) also observed that erosion with arable land use was 36,000 times greater than that from forested lands (Table 10.4). In Barbados, Tam (1980) reported that large-scale deforestation for growing sugarcane has changed the hydrologic regime, causing accelerated erosion in

TABLE 10.1 Erosion and Runoff with Various Land Uses in the Ivory Coast and Burkina Faso

Stations	Slope	Erosion, t/(ha · yr)			Runoff, % of annual rains		
		Natural environment	Bare	Crop	Natural environment	Bare	Crop
ADIOPUDOUME ORSTOM (1956–1957)*	7%	0.03	138	0.1–90	0.14	38	0.5–30
Secondary evergreen forest	20%	0.1	570		0.7	24	
2100 mm: 4 seasons, 90- to 250-m² plots	65%	0.2–1			0.6–2.2		
BOUAKE IRAT-ORSTOM (1960–1970)†	4%	0.2	18–30	0.1–26	0.3	15–30	0.1–26
Dense, shrubby savanna, 1200 mm: 4 seasons, 200- to 250-m² plots†							
OUAGADOUGOU (CTFT-ORSTOM-IRAT) (1967–1971)‡	0.5%	0.15	10–20	0.6–8	10	40–60	2–32
Clear, wooded savanna, 850 mm: 200- to 5000-m² plots							

*Roose et al. (1971).
†Bertrand (1967).
‡CTFT (1971).
SOURCE: Roose, 1972.

TABLE 10.2 Soil Erosion and Water Runoff in the Ivory Coast under Different Land Use Systems

	Rainfall, mm	Runoff/erosion	Land use				
			Secondary forest	Natural fallow	Green-manure crop	Cassava	Bare soil
1956	1939	Runoff (%)	2.1	—	12.6	16.4	17.4
		Erosion (t/ha)	2.4	—	45.2	92.8	117.7
1957	2383	Erosion (t/ha)	0.03	—	0.1	28.7	104.5
1961	2289	Erosion (t/ha)	0.02	0.46	0.4	—	143.2

SOURCE: Fournier, 1967.

these areas. In Mexico's Texcoco River Basin, Solorio (1985) measured erosion losses under different land use systems for 10 years. Soil erosion from agricultural lands was 200 to 600 times greater than from forested lands (Table 10.5).

Similar differences in soil erosion have been observed in the United States (Table 10.6). On a 10 percent slope in southern Piedmont, Bennett et al. (1954) observed erosion losses of 0.005 t/(ha · yr) in forest, 0.76 t/(ha · yr) in grass, 37.26 t/(ha · yr) from agricultural rotation, 77.11 t/(ha · yr) under cotton, and 163.41 t/(ha · yr) from bare ground surface. Results of land uses in relation to erosion for soils in southern Piedmont, northeast Texas, and La Crosse, Wisc., shown in Table 10.6, show that erosion from cropped land was several hundred to thousand times greater than that from the forested land.

Plantation/tree crops

Soil erosion is generally less from well-managed plantations than from land grown to seasonal crops. This is evident from the data in Table 10.7 from Bujumbura, Burundi. Soil erosion from the bare fallow plots exceeded 400 t/(ha · yr), and that from cassava ranged from 5 to 87 t/(ha · yr). In comparison, there was no erosion from coffee or pinus plantations. Erosion can, however, be severe in mismanaged plantations. For example, Fournier (1967) observed erosion rates of about 25 t/(ha · yr) from Cinchona grown on 25 percent slope and 18

TABLE 10.3 Effects of Land Use on Runoff and Soil Erosion in Guinea

Land use	Runoff, % of rainfall	Erosion, t/(ha · yr)
Bare fallow	9.2	18.9
Pineapple	3.4	1.4
Fonio (cereals)	2.6	4.1
Cover crop	2.9	5.4
Natural vegetation	2.4	2.7

SOURCE: Dugain and Fauck, 1959.

TABLE 10.4 Effects of Different Vegetation Covers and Land Use Systems on Runoff and Erosion at Campinas, São Paulo, Brazil

Land use	Erosion, t/(ha · yr)	Runoff, % of rainfall
Forest	0.001	1.1
Pasture	1.0	1.6
Coffee	1.4	1.6
Cotton	36.0	8.2

SOURCE: Suarez de Castro, 1979.

t/(ha · yr) for citrus grown on 6 percent slope. These erosion rates are excessive and can cause severe soil degradation.

Pastures

As with plantation and tree crops, erosion is less from well-managed pastures with controlled grazing and optimum stocking rates. Research experiments conducted by the Indian Council of Agricultural Research (ICAR) have shown the least runoff and erosion from natural and planted grasses and leguminous pastures (ICAR, 1973, 1984). In South Africa, Menne (1959) reported the least erosion from protected veld or from moderately grazed but unburned veld. Erosion was also less from pastures sown to *Digitaria pentzii* and moderately grazed. Cultivation of maize, however, increased erosion from 20 to 80 t/ha. Less erosion from well-managed pastures than from croplands is also reported from Australia (Bligh, 1983) and Colombia (Suarez de Castro, 1979). In Madagascar, eastern Africa, Fournier (1967) observed that erosion from good grassland and forested lands was negligible (Table 10.8).

Accelerated erosion, however, occurs on mismanaged and overstocked pastures. There are many examples of severe soil erosion from overstocked rangelands, particularly from tropical Africa (Tables 10.8 and 10.9). In Kenya,

TABLE 10.5 Effects of Land Use Systems on Soil Erosion in Texcoco River Basin, Mexico

Land use	Average soil erosion, t/(ha · yr)	Relative erosion
Forest	0.016	1
Rangeland	0.028	1.75
Grass	0.280	17.58
Agriculture, lower slope	2.862	179.75
Agriculture, upper slope	3.065*	192.53
Bare ground (Tepetate)	9.176	576.41

*Based on 1974 data only.
SOURCE: C. A. Ortiz Solorio (personal communication, 1985).

TABLE 10.6 Effects of Land Use on Soil Erosion for Three Sites in the United States

Land use	Soil erosion, t/(ha · yr)		
	Southern Piedmont (10% slope)	Northeast Texas (9–16% slope)	La Crosse, Wisc. (16% slope)
Forest	0.005	0.124	—
Grass	0.76	0.012	0.224
Agricultural rotation	37.26	42.5	62.272
Maize	—	—	250.208
Cotton	77.11	57.33	
Cotton on subsoil	—	162.00	—
Bare ground	163.41	—	427.840

SOURCE: Bennett et al., 1954.

Reid (1983) reported that the distribution of gullied land in Machakos highlands was 45 percent on grazed land, 28 percent on lands adjacent to poorly constructed roads, 24 percent in the vicinity of stream banks, and only 3 percent on cultivated land. Barber (1983) observed that erosion from degraded grazing land was 50 times greater than that from good grazing land and about 3 times greater than that from cropped land (Table 10.9). Similar observations have been reported from Australia. Severe erosion and failure of conservation structures were reported in Fitzroy River Basin when a large expanse of Brigalow (*Acacia harpophylla*) and soft wood scrub was opened for pasture and agriculture. Uncontrolled grazing has caused severe erosion in the highlands of Nepal. Fleming (1983) observed that erosion rates were 8, 10, 15, and 35 t/(ha · yr) from forest land, well-managed and terraced scrubland, and grazing land, respectively.

Regeneration of degraded pastures decreased soil erosion risks. Appropriate

TABLE 10.7 Soil Erosion from Natural Forest, Artificial Forest, Coffee, and Cassava on Steeplands in Bujumbura, Burundi

Land use	Slope, %	Erosion, t/(ha · yr)	
		1981–1982	1982–1983
Forest	45–50	0	0
Coffee	45–50	0	0
Cassava (terraced)	49	5.2	11.1
Cassava (unterraced)	49	29.2	55.5
Cassava (unterraced)	49	87.4	71.6
Bare fallow (plowed)	40	441.4	428.2

SOURCE: Modified from Durand, 1984.

TABLE 10.8 Soil Erosion and Water Runoff Measured in 1959–1960 under Different Land Use Systems in Lake Alaotra, Madagascar*

		Land use		
Observations	Seasonal crops	Degraded grassland, 20% cover	Good grassland, 100% cover	Eucalyptus reafforestation (6 years old)
Slope, %	7	20	36	15
Runoff, %	15.4	29.0	6.9	3.7
Erosion, t/ha	59	12	0.026	0.025

*Total rainfall = 977.5 mm.
SOURCE: Fournier, 1967.

grasses and legumes are recommended for different soils and ecological regions for pasture establishment and renovation (Shankranarayan and Shankar, 1984; Winter et al., 1985).

Arable land use

All other factors remaining the same, potential erosion risks are higher in intensive arable land use than in forestry or pastures. The magnitude of erosion also depends on the cultural practices adopted.

Motorized farm operations usually cause more severe soil erosion than manual farm operations do. Erosion risks are also greater in monocropping than in rotation or mixed cropping, in open-row cereals than in close-canopy legumes, in clean cultivation than in mulch farming, and in plowed than in no-till lands.

Erosion Control in Different Land Use Systems

Choosing an appropriate land use in itself is no guarantee of automatic erosion control. An appropriate land use must also be accompanied by suitable cultural practices. A flowchart on the sequence of steps needed to choose an appropriate land use followed by a general guideline of suitable cultural practices for three major land uses is shown in Fig. 10.2. An appropriate land use system is chosen on the basis of a resource inventory. The latter involves the assessment of bio-

TABLE 10.9 Effects of Pasture Quality on Erosion Rate in Machakos, Kenya

Pasture (quality, land use)	Catchment, %	Erosion, t/(ha · yr)	Relative erosion
Good grazing land	20	1.10	1
Cropland	43	16.0	15
Degraded grazing land	37	53.3	50

SOURCE: Barber, 1983.

Erosion Control

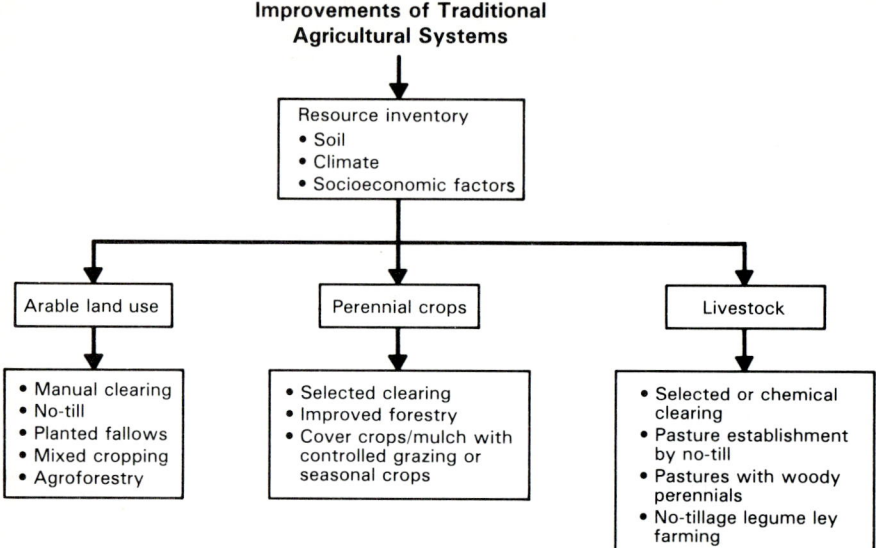

Figure 10.2 A useful guide of cultural practices for effective erosion control in different land use systems.

physical and socioeconomic factors. Suitable cultural practices listed for some major land uses are discussed at length in Chaps. 11, 12, and 13.

Soil Capability Classification and Erosion Hazard

High erosion risks on arable lands have prompted many researchers to assess soil suitability in relation to cultivation for different types of crops and cultural practices. Some soils, by virtue of their gentle slopes and deep topsoil layer, are suitable only for leguminous crops without mechanical seedbed preparation. A classification system has been developed to group soils according to their erodibility or erosion hazard:

1. *None.* No erosion risk for cultivation of any crops by any method or cultural practice.

2. *Slight.* Slight erosion risks, especially for cultivation of row crops with monocropping.

3. *Moderate.* Moderate erosion risks while cultivating row crops with plow-based system. Erosion can be curtailed by adopting suitable crop rotations and soil and crop management practices.

4. *High.* The soil should not be used for row crop production; crops can be grown with no-till and mulch farming techniques provided that the rotation adopted involves frequent use of planted fallows.

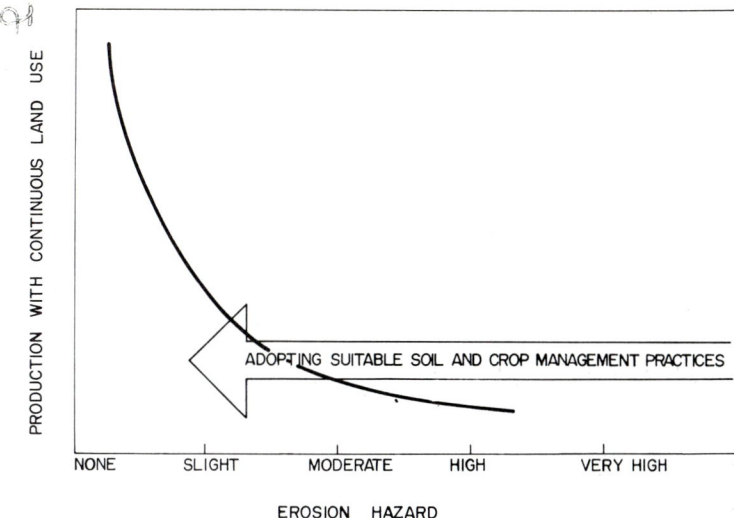

Figure 10.3 Effects of adopting suitable soil and crop management practices on the shift in the degree of erosion hazard.

5. *Very high.* This soil should not be used for production of row crops.

There is a specific set of cultural practices for each soil class. The adoption of these practices shifts the erosion hazard class from high to low (Fig. 10.3). One way to reduce erosion risks from arable lands is to create ecological diversity by combining different land uses, e.g., growing grain crops in association with trees and/or pastures. Although such a combination expectedly reduces net returns from crops proportional to the area or the time the land is out of crop production, erosion risks are drastically reduced.

Conclusions

Soil erosion is a function of the land use and its interaction with climate. Adopting an appropriate land use reduces erosion risks and preserves and sustains productivity. Rational land use planning is based on a resource survey. Soil capability classification is essential to define soil and crop management practices for each soil grouping. Adopting appropriate soil and crop management practices shifts the erosion hazard class from high to low. Erosion risks are high from a land use system that simplifies an ecosystem. Diverse and complex ecosystems, e.g., growing food crops in association with trees and pastures, have less erosion risks than monocultures.

References

Annual Report. 1985. Division of Arable Crops, Ministry of Agriculture, Agriculture Research Station, Sebele, Botswana.

Barber, R. G. 1983. The magnitude and sources of soil erosion in some humid and semi-arid parts of Kenya, and the significance of soil loss, tolerance values in soil conservation in Kenya. In: *Soil and Water Conservation in Kenya* (D. B. Thomas and W. M. Senga, eds.), Occasional Paper 42, University Nairobi, Kenya, pp. 20–46.

Beets, W. C. 1978. The agricultural environment of eastern and southern Africa and its use. *Agric. Environ.* 4:5–24.

Bennett, W. H., Pittman, D. W., and Tingey, D. C. 1954. Fifty years of dryland research at the Nephi Field Station. *Utah Agric. Exp. Sta. Bull. 371*.

Bligh, K. J. 1983. Microplots. *Proc. Soil Erosion Research Technique Conference,* 12–14 April 1983, Toowoomba, Old Department of Primary Industries, Australia, pp. 26–30.

CTFT (Centre Technique Forestier Tropical). 1979. *Conservation des sols au Sud du Sahara.* Ministère de la coopération, Paris, France.

Dabin, B. 1959. Bilan de trois anneés, d'erosion à la station d'Adiopodoume, Côte d'Ivoire. *Third Inter-African Soils Conference, Dalaba,* vol. 2, pp. 629–635.

Das, D. C. 1977. *Soil Conservation Practices and Erosion Control in India—A Case Study.* FAO Soils Bull. 33, Rome, Italy, pp. 11–50.

Dugain, F., and Fauck, R. 1959. Mésures d'erosion et de ruissellement en Moyenne-Guinea—relations avec certaines cultures. *Third Inter-African Soils Conference, Dalaba,* vol. 2.

Durand, P. 1984. Resultats des experimentations sur l'erosion des sols. Dept. Des Eaux et Forêts, Bujumbura, Burundi, Africa.

Fleming, W. M. 1983. Phewa Tal catchment management program: Benefits and costs of forestry and soil conservation in Nepal. In: *Forest and Watershed Development and Conservation in Asia and the Pacific* (L. S. Hamilton, ed.), Westview Press, Boulder, Colo., pp. 217–288.

Fournier, F. 1967. Research and soil erosion and soil conservation in Africa. *Afr. Soils* 12:53–96.

ICAR. 1973. *Soil and Water Conservation Research 1956–1971,* ICAR, New Delhi, India.

ICAR. 1984. *Soil Conservation, Annual Report,* Central Soil and Water Conservation Research and Training Institute, Dehra Dun, India.

Lawrence, P. A. 1984. The effects of landuse change on runoff, soil erosion and salinity on three Brigalow catchments. *Proc. National Soils Conference, 13–18 May 1984,* Brisbane, Australia, CSIRO.

Menne, T. C. 1959. A review of work done in the Union of South Africa on the measurement of runoff and erosion. *Third Inter-African Soils Conference,* Dalaba, Africa. vol. 2, pp. 615–627.

Nahal, I. 1975. *Principes de Conservation du Sol,* Masson et Cie, Editeurs, Paris.

Pereira, H. C., Dagg, M., and Hosegood, P. H., 1962. The water balance of bamboo thicket and of newly planted pines. *E. Afr. Agric. For. J.,* Special issue, pp. 95–103.

Rama Rao, M. S. V. 1960. *Soil Conservation in India,* ICAR, New Delhi, India.

Reid, L. M. 1983. A reconnaissance survey of erosion processes in Machakos Dist., Kenya. In: *Soil and Water Conservation in Kenya* (D. B. Thomas and W. M. Senga, eds.), Occasional Paper 42, University of Nairobi, Kenya, pp. 105–122.

Rensburg, H. J. Van. 1955. Runoff and soil erosion tests, Mpwapwa, Central Tanganyika. *E. Afr. Agric. J.* 20:228–231.

Rochette, C. 1959. Etude du ruissellement et de l'erosion sur les sols noirs de la region de Bafoussam, Cameroon Occidental. *Proc. Third Inter-African Soils Conference,* Dalaba, Africa. vol. 2, pp. 585–595.

Roose, E. J. 1972. *Quelques effets des pluies sur la mise en valeur des sols ferrallitiques et ferrugineaux tropicaux.* ORSOM, Abidjan.

Rubber Research Institute of Malaysia. 1980. *Soil Erosion and Conservation in Peninsular Malaysia,* Kuala Lumpur, Malaysia.

Shallow, P. G. 1956. *River Flow in the Cameroon Highlands Hydroelectric,* Tech. Memo 3, Central Electricity Board, Kuala Lumpur.

Shankranarayan, K. A., and Shankar, V. 1984. *Grasses and Legumes for Forage and Soil Conservation,* ICAR, New Delhi, India.

Solorio, C. A. 1985. Personal communication.

Staples, R. R. 1936. *Runoff and Soil Erosion Tests in Semi-arid Tanganyika Territory,* Annual Report, Department Veterinary Science & Animal Husbandry for 1935, pp. 131–141.

Suarez de Castro, F. 1979. *Conservacion de Suelos.* Instituto Inter-Americano de Ciencias Agricoles, San Jose, Costa Rica.

Tam, S. W. 1980. Causes of environmental deterioration in eastern Barbados since Colonization. *Agric. Environ.* 5:285–308.

Thomas, D. B., Edwards, K. A., Barber, R. G., and Hogg, I. 1981. Runoff, erosion and conservation in a representative catchment of Machakos District, Kenya. In: *Soil Conservation* (R. P. C. Morgan, ed.), Wiley, Chichester, U.K., pp. 219–239.

Tondeur, G. 1954. *Erosion du Sol: Specialement au Congo Belge,* Royaume De Belgique, Ministere des Colonies, Bruxella.

Winter, W. H., Cameron, A. G., Reid, R., Stockwell, T. G., and Page, M. C. 1985. Improved pasture plants. In: *Agro-research for the Semi-arid Tropics: North-West Australia* (R. C. Muchow, ed.), Old University Press, St. Lucia, Australia, pp. 225–269.

Chapter 11

Crop Management

Introduction

Soil conservation is not synonymous with erosion control. Erosion control merely implies decreasing physical displacement of soil from one site to another. In comparison, soil conservation means preserving and improving soil's life support systems for a high and sustained production. The latter involves reducing physical displacement of soil from one place to another plus preserving and improving its productivity. While erosion control may involve installation of mechanical or engineering structures, soil conservation is the judicious management of the natural resource—the soil.

There are two approaches to soil conservation—erosion prevention and erosion control. Erosion prevention measures involve use of good farming practices so that either soil erosion does not occur or the erosion rate does not exceed the tolerable level. Erosion prevention measures are integrated within the suitable farming system used. Erosion control measures, in comparison, refer to a narrow-based engineering approach of runoff management. The latter may be necessary under some conditions, e.g., steeplands controlling gully erosion. The necessity of erosion control measures, however, decreases with the adoption of good farm practices and appropriate farming systems. Erosion control measures are necessary if the land use system is inappropriate for the soil and ecological environment. Mechanical measures (such as terracing, diversion channels, drop structures) are secondary and complementary to more effective biologic control measures based on appropriate land use and soil surface management.

Clearly soil erosion is less of a problem with good farming. Before resorting to elaborate and expensive erosion control measures, therefore, it is logical to consider using simple, good farm practices that are also erosion prevention techniques. In soil erosion management there is no substitute for good farming. Although suitable erosion control measures vary according to soils, slopes, rainfall regimes, and farming systems, the principles of good farming practices have much wider applicability. Any cropping system that ensures high and sustained yields over a long period also causes less soil erosion. The data in Fig. 11.1 from Hudson and Jackson (1959) are a relevant example showing that cultural practice that produces high maize yield also causes less erosion.

An important underlying theme is to ensure an early ground cover. The schematic diagram in Fig. 11.2 indicates the relative effectiveness of different systems in erosion control. Farming system A attains 50 percent canopy cover much earlier than systems B and C. Other factors remaining the same, soil erosion risks are in the order of $A < B < C$. Cultural practices and farming systems that facilitate establishing an early canopy/ground cover include the following subsystems:

1. Appropriate land use
2. Choice of appropriate crops
3. Early planting

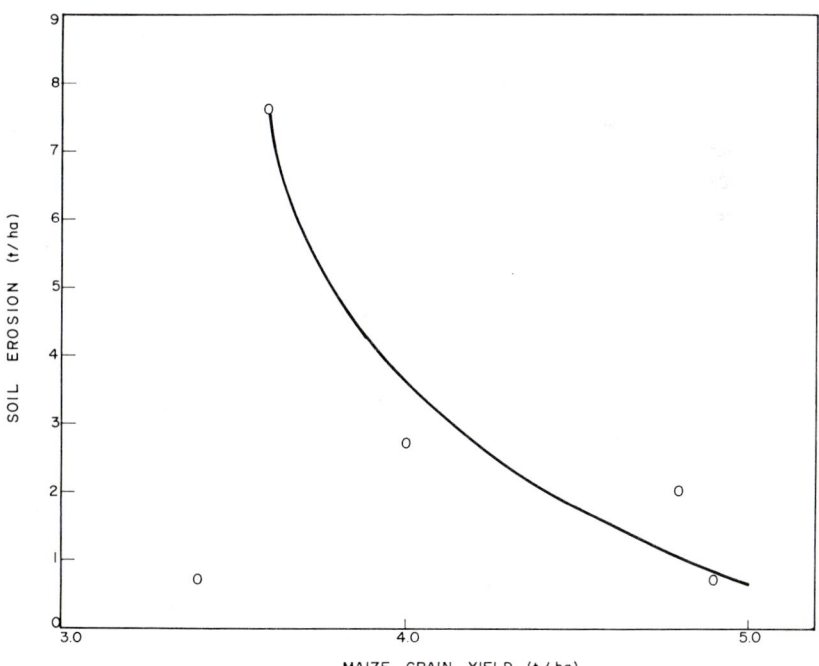

Figure 11.1 Relation between soil erosion and crop yield in Zimbabwe. (Adapted from Hudson and Jackson, 1959.)

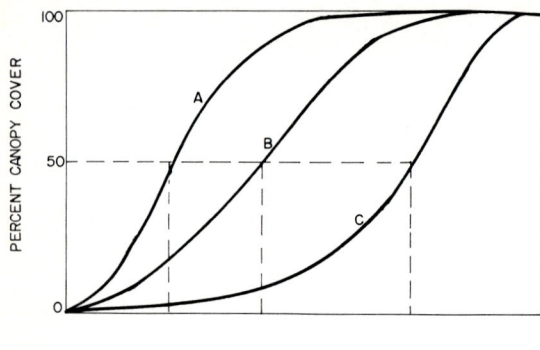

Figure 11.2 Relative erosion hazard in different farming systems A, B, and C.

4. Suitable variety

5. Good crop stand and an optimum plant population

6. Balanced fertilizer application

7. Adequate weed control

8. Control of insects and diseases

9. Crop-harvesting methods

10. Crop rotations and cropping systems

11. Crop residue and management

The above list of cultural practices indicates that the roots of soil erosion management lie in the farming system. Erosion management, in fact, is a holistic concept that involves adopting an integrated approach. Erosion control measures are often unnecessary if all the points listed above are duly considered and scientifically implemented.

Canopy Cover and Soil Erosion

Upland erosion is caused by impacting raindrops on exposed soil. Crop canopy intercepts rainfall and wind pulses and decreases the kinetic energy which reaches the soil surface. The *relative energy*—the ratio of kinetic energy which reaches the soil to the maximum kinetic energy at the top of the canopy—can be calculated by Eq. (11.1) which was developed and used by many researchers (De Tar et al., 1980; Quinn and Laflen, 1983; Gregory, 1984):

$$E_c = 1 - \frac{E_t - E_h}{E_t} F_c \tag{11.1}$$

Here E_c is relative energy under a canopy, E_t is kinetic energy per area of rainfall at terminal velocity, E_h is kinetic energy per area of rainfall dropping from a canopy of height h, and F_c is fraction of canopy cover.

TABLE 11.1 Effect of Low-Growing Canopy Cover on Soil Erosion

Treatment	Soil erosion, t/3 yr	Relative erosion*
Permanent grass cover (protection from raindrop impact and reduced runoff)	7.4	1.1
Two layers of mosquito gauze suspended 15 cm above bare soil (minimizing drop impact, no reduction of runoff)	6.7	1
Bare soil	780	114.9

*Relative to the erosion under mosquito gauze.
SOURCE: Williams and Joseph, 1970.

Vegetative cover close to the ground surface breaks the raindrop impact and dissipates its energy more effectively than tall canopies. In Zimbabwe, Hudson and Jackson (1959) observed average soil erosion 85.8 ton/acre from the bare plot vs. 0.7 ton/acre from the gauzed plot—a 122.6-fold difference. The data in Table 11.1 by Williams and Joseph (1970) also strengthen the argument in favor of preventing splash for controlling upland erosion. The effect of a low-growing grass in preventing splash was similar to that of two layers of mosquito gauze suspended 15 cm above the ground—reducing erosion by 115 times in comparison with the bare unprotected soil. Effects similar to those of mosquito gauze are also produced by a crop cover. For example, the effects of maize canopy in reducing runoff and soil erosion are demonstrated by the data of Wilkinson (1975b) in southern Nigeria (Table 11.2). The rainfall of 24.9 mm received on September 5, 1972, occurred when maize was 7 days old and 99 percent of the soil surface was exposed to rain. The soil erosion caused by this rain was 165 kg/ha. A rainfall of similar amount and intensity recorded on September 25, 1972, when 62 percent of the soil was exposed caused erosion of only 69 kg/ha.

TABLE 11.2 Effect of Canopy Cover of Maize and of Proportion of Soil Exposed on Soil Erosion and Runoff in Southern Nigeria

Storm date in 1972	Rainfall amount, mm	Time after maize emergence, days	Height maize, cm	Exposed soil, %	Runoff, %	Soil loss, kg/ha	
						Measured	Predicted
Sept. 5	24.9	7	13	99	4.4	165	158
Sept. 8	12.2	10	21	97	4.9	29	40
Sept. 21	22.9	23	67	71	3.5	127	274
Sept. 25	24.4	27	86	62	6.2	69	263
Sept. 27	12.4	29	98	59	4.0	27	67
Sept. 30	17.5	32	129	54	0.6	8	170
Oct. 6	28.2	38	190	48	0.7	5	735
Total	142.5					430	1712

SOURCE: Modified from Wilkinson, 1975b.

TABLE 11.3 Effect of Percentage of Ground Cover by *Aristida* Grass on Runoff and Erosion in Lake Alaostra, Madagascar

Slope, %	Soil covered, %	Runoff, %	Erosion, t/ha
20	20	29	12
24	20	24	12
35	40–60	45	4
36	100	6.9	0.025

SOURCE: Fournier, 1967.

A 38-mm storm on October 6, 1972, when only 48 percent of the soil was exposed, caused erosion of only 5 kg/ha. Furthermore, Wilkinson observed that measured and predicted soil losses were similar during the initial stages of seedling establishment when the soil was relatively bare. At later stages, however, measured soil loss was drastically less than that predicted by the USLE. Erosion prediction using the USLE was not good enough when crop cover was established. This discrepancy between the measured and predicted erosion loss is attributable to some other soil-conserving effects of crop growth than just the protective vegetative cover. Root growth and root development hold the soil together and probably increase the infiltration rate. Therefore, a cropping system that provides an early canopy cover and a dense root system is likely to cause less erosion than a slow-growing crop with sparse root systems.

Another relevant example of the effect of canopy cover on soil erosion is shown by the data of Fournier (1967) from Madagascar. Table 11.3 shows that increasing ground cover by *Aristida* pasture decreased soil erosion despite the increase in slope steepness. A 100 percent ground cover provided perfect erosion control even on a steep slope of 36 percent.

In Rhodesia, Elwell and Stocking (1976) reported that the percentage of vegetal cover is the major factor determining the erosion hazard from crops and grasslands. They considered the possibility of adopting the percentage of area

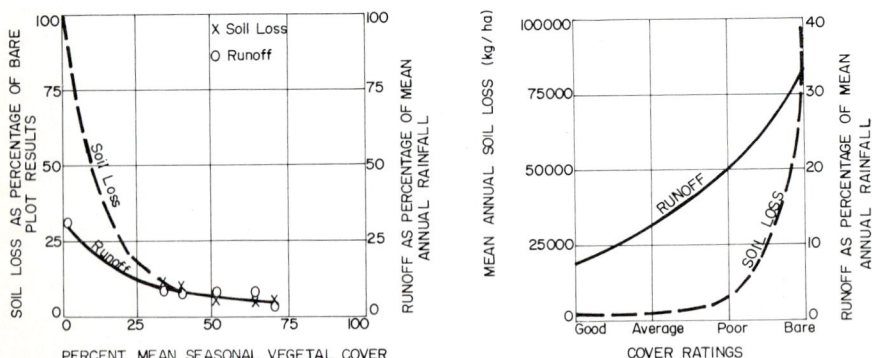

Figure 11.3 Effects of vegetal cover on runoff and erosion in Zimbabwe, southern Africa (Elwell and Stocking, 1976).

TABLE 11.4 Regression Equations Relating Vegetal Cover and Soil Erosion

Cropping	Correlation coefficient r*	Regression equation
Soybean-soybean	0.63	$Y = 5.38e^{-0.04x}$
Pigeon pea–pigeon pea	0.94	$Y = 3.27e^{-0.01x}$
Maize-cassava (mixed cropping)	0.84	$Y = 2.20e^{-0.01x}$
Cassava (monoculture)	0.90	$Y = 2.71e^{-0.00x}$

x = percentage of vegetal cover and Y = rain [t/(ha · cm²)].
*Significant at 1% level of probability.
SOURCE: Aina et al., 1979.

of soil protected by vegetation as a soil loss estimation parameter (Fig. 11.3). On the basis of more quantitative data, Elwell and Stocking (1976) observed that runoff and soil loss are significantly reduced when the vegetal cover exceeds 30 percent (Fig. 11.3).

In southwestern Nigeria, Aina et al. (1979) observed significant correlation between percentage of vegetal cover and soil erosion. Soil erosion decreased exponentially with an increase in vegetal cover (Table 11.4). Among the four cropping systems studied, the rate of vegetal cover development was in the order of soybean > pigeon pea > corn + cassava mixed cropping > monoculture cassava (Fig. 11.4). The soil erosion followed the same trend (Fig. 11.5). In another experiment involving a range of crops and cropping systems, Huke and Lal (unpublished data) observed significant differences in soil erosion among crops (Table 11.5). Soil erosion was also related to the total leaf area index (Table 11.6), indicating the importance of ground cover in erosion control.

The closer the canopy to the soil surface, the more effective it is in breaking the raindrop impact and in dissipating its energy. The erosivity of leaf drip from tall tree canopy, as is discussed in Chap. 13, may be even greater than that of natural rain.

Land Use

A continuous ground cover close to the soil surface can be achieved through an appropriate land use. Erosion is caused by land abuse. Land use capability

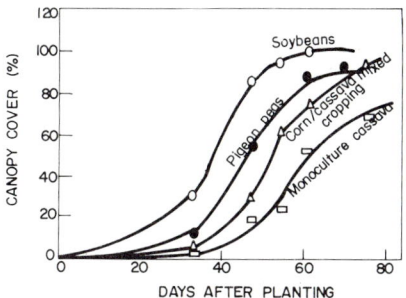

Figure 11.4 Canopy cover development for different crops and cropping combinations in southern Nigeria (Aina et al., 1979).

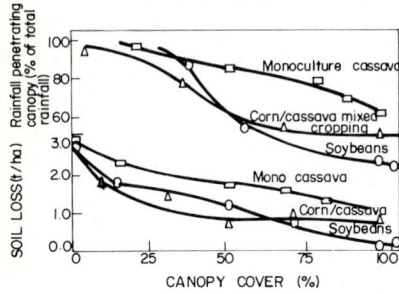

Figure 11.5 Relation between canopy cover and soil erosion of an alfisol in southern Nigeria (Aina et al., 1979).

should be assessed on the basis of its soil, slope, vegetation, and water resources. Some soils, by virtue of their slope, topsoil depth, or general fertility status, should be maintained under natural cover. Others are suitable for pastures with controlled grazing or for created forests. Only those lands with low risk of soil erosion should be used for cultivation of food crop annuals.

If all land use systems are properly implemented, erosion risks are in the order of natural forest ≤ created forest ≤ ungrazed pastures < controlled grazing < mixed cropping and crop rotation < monoculture crop rotation < excessive and uncontrolled grazing. If the land use system is not properly implemented, however, even forest and pastures can cause more severe erosion than arable land use.

Crop Husbandry

Crop management includes a range of cultural practices that affect the economic returns. In general, it is the package of agronomic practices that opti-

TABLE 11.5 Effect of Crop Cover and Methods of Seedbed Preparation on Soil and Sand Splash under Different Cropping Systems from 14 June to 12 October 1982

Cropping system	Soil splash		Sand splash	
	Absolute, t/ha	Relative	Absolute, t/ha	Relative
Cassava (ridges)	186.8	42.4	61.3	65.8
Maize	11.0	25.2	53.5	57.5
Yam (mounds)	203.6	46.3	81.2	87.2
Sweet potato	237.2	53.9	86.1	92.5
Cassava + sweet potato (ridges)	60.2	13.7	62.3	66.9
Sweet potato + maize	140.0	31.8	76.5	82.2
Yam + maize (mounds)	175.0	39.8	74.2	79.7
Cassava + maize (ridges)	65.4	14.9	68.0	73.0
Ridges (bare)	440.1	100.0	93.1	100.0

SOURCE: Unpublished data of S. Huke and R. Lal.

TABLE 11.6 Regression Equations Relating Soil Splash with Leaf Area Index, Rainfall Amount, and 30-min Maximum Intensity

Cropping system	Regression equation	r
Cassava	$E = 0.56 + 0.70I_{30} + 0.43A - 0.46\text{LAI}$	0.93
Yam	$E = 0.08 + 0.88I_{30} + 0.51A - 0.18\text{LAI}$	0.76
Maize	$E = 0.15 + 0.06I_{30} + 0.53A - 0.05\text{LAI}$	0.77
Sweet potato	$E = 0.91 + 0.30I_{30} + 0.38A - 0.91\text{LAI}$	0.53
Cassava + sweet potato	$E = 0.45 + 0.18I_{30} + 0.08A - 0.28\text{LAI}$	0.63
Maize + sweet potato	$E = 0.31 + 0.03I_{30} + 0.54A - 0.07\text{LAI}$	0.58
Yam + maize	$E = 0.14 + 0.32I_{30} + 0.65A - 0.07\text{LAI}$	0.80
Cassava + maize	$E = 0.02 + 0.11I_{30} + 0.23A - 0.01\text{LAI}$	0.87

E = splash (kg/m^2), I_{30} = maximum 30-min intensity (in/h), LAI = leaf area index, A = rainfall amount per storm (in).

SOURCE: Unpublished data of S. Huke and R. Lal.

mizes the economic returns that also causes minimal soil erosion. This package comprises a range of simple but important techniques.

Crop species and cultivars

Choosing an appropriate crop species is important. An appropriate crop cultivar is the one suited to the soil and environmental conditions that can establish a quick ground cover. Agronomists and soil scientists should work with plant breeders to develop crop cultivars that have the desirable plant attributes for providing a quick cover, product sufficient biomass, and can be grown as mixed and relay crops with other species. An example of intervarietal differences in canopy cover and susceptibility to erosion is shown by the data in Table 11.7. The cassava cultivar 30001 produces an early and a dense canopy and therefore causes less erosion than cultivar 30572. Although it is a creeping and low-growing crop, the erosion risks can be high in slow-growing sweet potato cultivar (Table 11.7).

TABLE 11.7 Effects of Crop Rotations and of Cultivars on Runoff and Erosion from a Tropical Alfisol at Ibadan, Nigeria

Treatments	5% Slope		10% Slope		15% Slope	
	Runoff, mm	Erosion, t/ha	Runoff, mm	Erosion, t/ha	Runoff, mm	Erosion, t/ha
Cassava (var. 30001)	3.9	0.6	5.3	1.9	7.5	3.0
Cassava (var. 30572)	6.9	1.0	6.6	4.4	8.0	9.6
Sweet potato	5.6	1.7	6.3	4.6	10.2	6.5
Maize-cowpea	3.2	0.8	3.3	2.3	6.5	4.5
LSD$_{.05}$	3.4					

LSD = least significant difference.
SOURCE: Unpublished data of Maduakor and Lal, 1979.

Unless chemical amendments are added, some crops and crop varieties do not grow well in soils with a pH of less than 5. Crops susceptible to low soil pH include cotton (*Gossypium* spp.), tomato (*Lycopersicon* spp.), alfalfa (*Medicago sativa* L. subsp. *sativa*), celery [*Apium graveolens* L. var dulce (Mill.) Pers.], corn (*Zea mays* L.), grain sorghum [*Sorghum bicolo* L. (Moench)], and sugar beet (*Beta vulgaris* L.). Aluminum-tolerant plants include peanuts (*Arachis hypogaea* L.), pearl millet [Pennisetum americanum (L.) Leeke], and Bermuda grass [Cynodon plectostchyus K. Schum) (Donahue and Miller, 1977). Conway (1985) cites an example of growing coconut rather than rice on acid sulfate soils in Kalimantan, Indonesia. In another example from Java, introduction of potato caused severe erosion on uplands, whereas mixed home gardens and multistory crops increased production and reduced erosion. Wheat varieties tolerant to high aluminum concentrations have been successfully introduced in Brazilian Cerrado. The tolerance or susceptibility to aluminum toxicity of some species, e.g., wheat (*Triticum aestivum* L.) (Foy et al., 1965; Lafever et al., 1977), barley (*Hardeum vulgare* L.) (Foy et al., 1965; MacLean and Chiasson, 1966), and soybeans (*Glycine max* L.) (Armiger et al., 1968; Hanson and Kamprath, 1979), depends on the variety. Soil testing is an important practice for determining the adaptability of a particular species or cultivar to a soil where aluminum toxicity is a potential problem. Introduction of improved farming systems on previously underutilized vertisols has decreased erosion losses from these soils highly susceptible to erosion. Improved farming systems involve cultivation immediately after the past-rainy-season crop, improved drainage with broad-bed and furrow system, dry seeding, improved seeds, and appropriate placement of seeds and fertilizers (Dillon and Virmani, 1985). Introduction of short-duration rice varieties and direct sowing have increased cropping intensity and decreased runoff and erosion in many rainfall areas of the Philippines (Dillon and Virmani, 1985).

A survey of the literature shows many appropriate examples of drastic reduction in soil erosion simply by choosing an appropriate crop (ICAR, 1973, 1984; Bertoni and Lombardi, 1985). The data in Table 11.8 from Malaysia show differences in soil susceptibility to erosion among three leguminous crops. Erosion and runoff were less from cowpea than from land planted to mung bean or groundnut (Sulaiman et al., 1981). In Indonesia, Abdurachman et al. (1984) reported a wide range in the C factor for different crops (Table 11.9). Simply

TABLE 11.8 Soil Erosion and Water Runoff under Different Crops Grown in Malaysia

Crop treatment	Erosion, t/(ha · yr)	Runoff, % of rainfall
Bare fallow	44.9	32.7
Mung bean	17.0	26.1
Cowpea	8.2	13.0
Groundnut	9.5	13.6
$LSD_{.05}$	11.2	9.6

Rainfall = 2107 mm; LSD = least significant difference.
SOURCE: Modified from Sulaiman et al., 1981.

TABLE 11.9 The *C* Factor for Different Crops Grown in Java and Lumpung Regions of Indonesia

Crop	C factor
Brachiaria decumbens, fully established	0.002
Red beans	0.161
Sorghum	0.242
B. decumbens, first year	0.287
Cassava	0.363
Soybean	0.399
Lemon grass	0.434
Groundnut	0.452
Upland rice	0.561
Maize	0.637
Bare fallow	1.000

SOURCE: Abdurachman et al., 1984.

choosing different crops changed the *C* factor from 0.002 for fully established *Brachiaria* to 0.637 for the open-row maize.

Crop rotations, sequences, and mixtures

After an appropriate soil-conserving crop is chosen, erosion can be further curtailed by following suitable cropping systems. Cropping systems are designed to ensure maximum ground cover. Compatible mixtures for successful multiple and mixed cropping systems include growing an early-maturing crop with the late-ripening one, a tall-growing crop with a short-statured and shade-tolerant crop, and a deep-rooted crop with surface-feeder crops. In addition to compatible species, varieties and cultivars should be specifically selected for growing with mixtures. Selection from a monoculture breeding nursery may not always provide the best cultivar for a mixed cropping system. Appropriate and compatible mixtures optimize resource utilization, provide a continuous ground cover throughout the rainy season, improve soil physical conditions, and yield stable production.

In addition to the ground cover, crop rotations and mixed cropping systems influence soil erosion through their effects on soil properties. A good crop rotation involves growing soil-degrading crops in rotation with soil-conserving crops, or growing a shallow-rooted cereal in sequential cropping with deep-rooted legumes. These rotations improve the soil's structural stability, macroporosity, and water infiltration rate. Fournier (1967) reported from Sefa, Senegal, that the structural instability index increased from 0.40 to 0.57 for soil under forest to 0.74 to 0.88 one year after cropping to groundnut, to 1.21 two years after cropping to groundnut, and to 1.45 to 1.77 six years after continuous cropping to groundnut. There also occurred a corresponding decline in water transmission properties. The saturated hydraulic conductivity declined from

TABLE 11.10 Effects of Crop Cover on Surface Runoff at Samaru, Nigeria

Season	Crop cover	Total rainfall, mm	Rainfall contributing to runoff, mm	Runoff, mm	Percentage of total rainfall	Percentage of rainfall contributing to runoff
1964	Cotton	1019	828	318	31.2	38.4
1965	Cotton	879	671	216	24.6	32.2
1966	Sorghum	1304	927	239	18.1	25.8
1967	Groundnut	1085	795	218	20.0	27.3
1968	Cotton	1024	607	140	13.7	23.1

SOURCE: Kowal, 1970.

3.1 to 4.7 cm/h under forest to 2.09 to 3.07 cm/h with 1 year of cropping to groundnut, to 2.5 cm/h with 2 years of cropping to groundnut, and to 1.42 to 1.77 cm/h with 6 years of cropping to groundnut. It is the continuous monocropping with soil-depleting crops that causes severe structural degradation and accelerates the erosion problem. Also in Sefa, Senegal, Fauck et al. (1967) reported severe sheet erosion [9 t/(ha · yr)] even on a gentle slope of 2 percent following 15 years of continuous cropping. Soil erosion on these gentle slopes ranged from 7.0 to 14.2 t/(ha · yr) under sorghum, from 6.4 to 9.5 t/(ha · yr) under rice, and from 3.1 to 4.3 t/(ha · yr) under groundnuts (Fournier, 1967).

Results of experiments to measure the effects of cropping systems on erosion conducted in Nigeria lead to conclusions similar to those in Senegal. In northern Nigeria, Kowal (1970) observed that mean runoff losses under cotton were greater than those under sorghum and groundnut (Table 11.10). In southwestern Nigeria, Lal (1976) observed more erosion and runoff under maize than under cowpea. The mean soil erosion under maize was about double that under cowpea (Table 11.11). The high rate of erosion under maize compared with cowpea was attributed to the open canopy of maize. In contrast, however, the data

TABLE 11.11 Effect of Cropping Systems on Erosion under Maize in Southwestern Nigeria during the First Season 1973

Slope, %	Soil erosion, t/(ha · season)		
	Bare fallow	Maize followed by maize	Cowpea followed by maize
1	7.5	1.2	0.6
5	80.4	8.2	5.6
10	152.9	4.4	3.3
15	155.3	23.6	7.6
Mean	90.0	9.4	4.3

SOURCE: Lal, 1976.

TABLE 11.12 Effect of Cropping Systems on Erosion under Maize in Southwestern Nigeria during the Second Season 1973

	Soil erosion, t/(ha · season)		
Slope, %	Bare fallow	Maize after maize	Maize after cowpea
1	3.7	0.4	0.3
5	75.8	2.8	4.0
10	79.7	2.8	3.0
15	73.9	17.1	35.4
Mean	58.3	5.8	10.7

SOURCE: Lal, 1976.

in Table 11.12 show more erosion from maize following cowpea than maize following maize. The erosion from maize following cowpea was about double that from maize following maize. The low amount of crop residue left by cowpea and the high rate of its decomposition (because of low C/N ratio) are the factors responsible for high erosion rates in maize grown after cowpea.

The agronomic benefits of mixed and relay cropping are well established, especially in low-input traditional agriculture (Willey, 1985). By providing continuous canopy cover, mixed cropping decreases soil erosion risks. In western Nigeria, Aina et al. (1979) observed that maize plus cassava mixed cropping caused less soil erosion than monoculture cassava (Table 11.13). In comparison with monocropped cassava, maize plus cassava mixed cropping decreased erosion by 40 percent in the first season and by 87 percent in the second. Cassava, a slow-growing crop, leaves considerable open ground in between the widely spaced cassava seedlings during the initial stages. Growing maize in between the open spaces therefore reduces risks of soil erosion. Once the cassava canopy is fully grown, it offers an effective protection against erosion (Plate 11.1). In this experiment, Aina et al. observed, as one would expect, an exponential decrease in erosion with an increasing canopy cover of different cropping systems (Table 11.14). Similar conclusions are supported by a follow-up study at IITA, as shown by the data in Table 11.15; mixed cropping of cassava with maize or melons reduces losses of water runoff. In Madagascar, east Africa, Roche and

TABLE 11.13 Effects of Cropping Systems on Crop Management Factor C for Soil Erosion on Alfisol in Western Nigeria

Cropping system	First season	Second season
Bare fallow	1.00	1.00
Cassava (monocropping)	0.72	0.39
Maize + cassava (mixed cropping)	0.43	0.05
Soybean	0.19	0.02

SOURCE: Aina et al., 1979.

Plate 11.1 A canopy structure of different crops: (*a*) soybean, (*b*) cowpea, (*c*) corn, and (*d*) cassava. (Adapted from Aina et al., 1977.)

Dubois (1959) observed that pasture oats caused less erosion than a plowed-under green-manure crop (Table 11.16).

Many experiments have been conducted in India that also substantiate the benefits of crop rotation in reducing soil erosion. Verma et al. (1968) recommended cropping patterns containing cereals and legumes to reduce risks of erosion. In Uttar Pradesh, Goel and Khanna (1969) reported 2 to 3 times more erosion under uncropped fallow during monsoon season than from the cropped land. In the hill regions of Dehra Dun in northern India, Bhatt et al. (1971) observed that erosion losses under maize-wheat rotation were 76.1 t/(ha · yr) in comparison with one-half to one-third this rate from san hemp-wheat and sorghum-wheat rotations. In semiarid Haryana, Nijhawan and Garg (1974) reported improvements in structural stability and proportion of water-stable ag-

TABLE 11.14 Relationship between Canopy Cover *x* and Soil Erosion *y* for Different Cropping Systems

Cropping system	Correlation coefficient*	Regression equation
Soybean-soybean	0.63	$y = 5.38e^{-0.04x}$
Pigeon pea–pigeon pea	0.94	$y = 3.27e^{-0.01x}$
Maize + cassava	0.84	$y = 2.20e^{-0.01x}$
Cassava	0.90	$y = 2.71e^{0.01x}$

y = erosion, t/(ha · cm)
x = canopy cover, %
*Significant at 1% level of probability.
SOURCE: Aina et al., 1979.

TABLE 11.15 Effects of Canopy Cover by Different Crops and Crop Combinations on Runoff (mm) from a Tropical Alfisol at Ibadan, Nigeria

Month	Yam	Cassava	Maize	Melon	Sweet potato	Yam + melon	Cassava + maize	Cassava + melon	Bare	$LSD_{.05}$
July	7.3	14.3	14.5	9.8	12.4	11.5	9.8	7.6	18.5	10.5
Aug.	1.3	11.0	7.5	5.7	2.8	8.7	3.6	3.9	11.4	5.0
Sept.	3.8	8.2	6.1	2.8	3.4	3.0	9.2	9.7	13.8	5.9
Oct.	2.8	2.2	2.6	3.0	2.2	2.7	2.6	2.3	2.8	0.9

LSD = least significant difference.
SOURCE: Unpublished data of Maduaker and Lal, 1980.

gregates larger than 0.2 mm by irrigated cropping system vs. rain-fed cropping. Improvements in soil structure were probably due to more biomass produced by supplementary irrigation and a resultant increase in soil organic matter content. In Dehra Dun, India, Rama Rao (1974) observed significant differences in soil erosion among different cropping system treatment. Soil erosion was reduced from 28.5 t/ha for maize-wheat rotation to 19.3 t/ha for maize + cowpea–wheat rotation. Grass cover provided the best protection against erosion (Table 11.17).

A large number of research examples from different regions in southeast Asia indicate drastic reductions in erosion by choosing an appropriate cropping system. The data in Table 11.18 from Indonesia regarding the C factor for different cropping systems show significantly less erosion from mixed cropping than sequential crop combinations. Similar to the findings of Aina et al. (1979) from Nigeria, Abdurachman et al. (1984) observed that growing cassava in combination with soybean or groundnut reduced erosion risks. Soil erosion was further decreased by using crop residue mulch. Similar results are reported from experiments conducted in Malaysia (Tables 11.19 and 11.20). Leguminous crops (mung bean, groundnut) caused less erosion than cereals (maize, sugarcane). Erosion was further decreased by mulching with crop residue.

In Hawaii, El-Swaify and Cooley (1980) observed that soil erosion from watersheds growing sugarcane and pineapple ranged from 2.2 to 6.2 t/(ha · yr)—the rate which is within the acceptable limits of soil erosion. Serious erosion,

TABLE 11.16 Effect of Crop Rotations on Soil Erosion in Madagascar

Crop	Runoff, mm	Runoff, %	Erosion, t/(ha · yr)
Potato	246.7	15.5	8.6
Oats	269.0	16.9	6.8
Pasture oats (first year)	323.2	20.3	17.6
Pasture oats (second year)	227.6	14.3	1.0
Green manure (plowed under)	364.6	22.9	21.4

Rainfall from Jan. 9, 1958, to Jan. 5, 1959 = 1592 mm.
SOURCE: Roche and Dubois, 1959.

TABLE 11.17 Effects of Crops and Cropping Systems on Runoff and Soil Erosion Measured on Dhulkot Silty Clay Loam Soil on 8% Slope at Dehra Dun, Northern India

Treatments	Runoff, mm	Erosion, t/(ha · yr)
1. Maize–wheat (cultivated upslope and downslope)	670	28.5
2. Maize + cowpea–wheat (contour cultivation)	511	19.3
3. Cowpea–wheat (contour cultivation)	405	28.3
4. Giant star grass	31	1.3
5. Bare fallow	445	44.0

SOURCE: Rama Rao, 1974.

however, is caused by field roads (Plate 11.2). Annual erosion rates approached tolerable limits when 20 percent of the field was in the road. In cane fields in Trinidad, Georges (1977) also observed that planted cane fields have significantly less erosion than bare soil. Differential effects of crops and crop rotations on soil erosion have been reported from tropical America. In Brazil, Marques et al. (1961) observed the least erosion under sweet potato and the highest under

TABLE 11.18 Comparative Effects of Cropping Systems and Mulching on C Factor for Some Soils in Indonesia

Cropping and system	C factor
Cassava + soybean	0.181
Cassava + groundnut	0.195
Upland rice fb sorghum	0.345
Upland rice fb soybean	0.417
Groundnut + pigeon pea	0.495
Groundnut + cowpea	0.571
Groundnut with straw mulch @ 4 t/ha	0.049
Upland rice with straw mulch @ 4 t/ha	0.096
Groundnut + maize with straw mulch @ 4 t/ha	0.128
Intercropping rice + maize + cassava with rice straw mulch @ 6 t/ha	0.079
Sequential cropping of rice fb maize with crop residue mulch	0.347
Sequential cropping of rice fb maize	0.496
Intercropping rice + maize + cassava with residue mulch	0.357
Intercrop rice + maize + cassava	0.588
Bare fallow	1.000

fb = followed by
SOURCE: Modified from Abdurachman et al., 1984.

TABLE 11.19 *C* Factor for Different Crops and Cropping Systems Observed in Different Regions of Malaysia

Cropping system and ground cover	*C* factor
Maize	0.39
Maize with imperata mulch	0.02
Fallow with maize residue	0.09
Fallow with maize residue and mulch	0.01
Mung bean	0.04
Cowpea	0.27
Groundnut	0.28
Fallow with mung bean residue	0.25
Fallow with cowpea residue	0.03
Fallow with groundnut residue	0.01
Chili	0.33
Sugarcane	0.39

SOURCE: Sulaiman et al., 1983.

castor bean (Table 11.21). Crops that established a quick and a dense cover close to the soil surface caused less erosion.

While these principles of crop cover in relation to erosion hold true in most cases, there is always the danger of overgeneralization because soil erosion under natural conditions is influenced by many other interacting factors such as rainfall, slope (gradient, length, aspect, shape), cover, soil type, and management. There are, therefore, some exceptions to these general rules. Since cropped land generally undergoes less erosion than bare fallow land, there may exist an interaction between soil type and cropping system. In Puerto Rico, Barnett et al. (1972) reported an interaction between soil texture and cropping system. Whereas all cropping treatments studied reduced runoff and soil erosion compared with fallow land, this was not the case on a clayey soil. Contrary to what one would expect, the fallowed treatment on a clayey soil caused the least erosion. Interaction between soil type and cropping system in relation to runoff and erosion is also evidenced by the data from Brazil by Lombardi et al. (1975)

TABLE 11.20 Effects of Crop Cover and Management on Runoff and Erosion from a Typic Paleudult at Serdang, Malaysia

Treatment	Runoff, % of rainfall	Erosion, t/(ha · yr)
Bare fallow	28	19.5
Maize	29	7.5
Maize with mulch	4	0.5
Groundnut	21	10.1

SOURCE: Adapted from Mokhtaruddin and Maene, 1979.

Plate 11.2 Soil erosion from a pineapple plantation in Puerto Rico caused by poorly planned access roads and waterways.

(Table 11.22). Continuous maize caused about 25 percent more erosion on a podzolic soil than on red latosol (Table 11.23). Maize grown in rotation with mucuna and with manure had similar levels of erosion regardless of soil type.

The effect of the cropping system on runoff and erosion is also influenced by the interaction between the cropping system, tillage method, and rainfall

TABLE 11.21 Effects of Different Crops on Runoff and Soil Erosion in Brazil

Crop	Runoff, %	Erosion, t/ha
Castor	12.0	10.3
Beans	11.1	9.7
Cassava	11.2	8.5
Peanut	9.2	6.8
Rice	11.1	6.2
Cotton	9.9	6.2
Soybean	7.6	5.6
Potato	6.6	4.7
Sugarcane	4.2	3.1
Sorghum	5.0	3.0
Sorghum–bean	4.5	3.0
Sweet potato	4.0	1.6

SOURCE: Marques et al., 1961.

TABLE 11.22 Effects of Crop Rotation and Management on Runoff and Soil Erosion on a Podzolic Soil at Pidorama Experiment Station, Brazil

Management	Erosion, t/ha	Runoff, % of rainfall
Maize with residue burning	25.2	8.1
Maize in rotation with soybean and cotton	24.1	8.2
Maize continuously	21.8	7.0
Maize + beans	18.4	6.3
Maize + mucuna (mulch)	10.6	3.5
Maize + mucuna (incorporated)	8.5	3.3
Maize + barnyard manure	4.5	2.1

SOURCE: Adapted from Lombardi et al., 1975.

amount. In Trinidad, Gumbs and Lindsay (1982) reported that at a lower level of rainfall a significant increase in soil loss (but not runoff) occurred in bare plots with an increase in slope from 11 to 22 to 52 percent, but a nonsignificant increase occurred in both runoff and soil loss from cropped plots. At the higher level of rainfall, however, there was a significant increase in erosion with an increase in slope gradient for both bare and cropped plots. The effects of tillage methods on runoff and erosion also differed according to cropping systems, slope gradient, and rainfall regime.

Another example of the interaction between the rainfall amount and cropping system is evident from the data from Thailand shown in Table 11.24. In 1982 with a low rainfall amount of 550 mm, the runoff and erosion under groundnut–mung bean rotation were only 33 and 46 percent, respectively, of the amounts that occurred under rice. In 1983 with high rainfall at 1093 mm, however, erosion under groundnut–mung bean rotation was 41 percent greater than that under rice. The differences in runoff and erosion between two contrasting rainfall regimes are probably due to differences in crop growth and

TABLE 11.23 Effects of Crop Rotation and Management on Runoff and Soil Erosion on a Red Latosol at the Experiment Station of Ribeirao Preto

Management	Erosion, t/ha	Runoff, % of rainfall
Maize in rotation with cotton and groundnut	17.8	8.7
Maize continuously	16.5	12.3
Maize mucuna (in rotation with cotton)	15.1	9.3
Maize in rotation with cotton and mucuna	13.7	7.2
Maize + continuous mucuna	11.7	7.7
Maize with liming	8.9	5.7
Maize with manure	4.9	4.0

SOURCE: Lombardi et al., 1975.

TABLE 11.24 Effects of Crop Rotations on Runoff and Soil Erosion from a Soil at Hang Chat, Thailand

Year	Rainfall, mm	Crop rotation	Runoff, % of rainfall	Erosion, t/(ha · yr)
1982	550	Rice	43	5.2
		Groundnut–mung bean	14	2.4
1983	1093	Rice	25	3.7
		Groundnut–mung bean	24	5.2

SOURCE: Marston et al., 1984.

vegetal cover. Crop stand and pest incidence are greatly influenced by the rainfall amount. The cropping system–tillage intersection is discussed in Chap. 12.

Early planting

Spring in the tropics is hot and dry whereas that in the temperate-zone climate is cold and wet. The initial rains in the beginning of the rainy season in the tropics come when the soil is devoid of vegetation cover. A system of planting early, immediately after the beginning of the rainy season, will thus establish a quick ground cover and reduce the risks of soil erosion. The benefits of early planting have been widely documented in the tropics and subtropics for a wide range of reasons, e.g., to overcome the constraints of erratic rainfall distribution, an early-maturing variety, unfavorable soil moisture and soil temperature regimes, loss of mineralized nitrogen, and accelerated soil erosion (Lal, 1986). In addition to high erosion risks, yields of late-planted crops are always less and more variable than those of early-sown crops. In Zimbabwe, Hudson (1971) measured soil losses of 5.2 and 9.6 t/ha from early- and late-planted crops, respectively. In Trinidad, Georges (1977) recommended an early planting of sugarcane to reduce soil erosion. Early sowing in temperate-zone climate may also reduce the risks of erosion provided that soil management systems are specifically chosen to alleviate the constraints of suboptimal soil temperatures and poor soil drainage.

Adequate stand and optimum plant population

Adequate crop stand is essential for producing satisfactory yield and reducing erosion. The canopy cover depends on the crop stand. Uneven germination and a poor stand create large areas of exposed soil surface prone to accelerated erosion. Low crop stand is caused by many factors including poor seed quality, faulty seeding equipment, unfavorable (too high or too low) soil temperature and soil moisture regimes, and damage to young seedlings by birds, rodents, insects, and disease. Use of good-quality seed treated with appropriate chemicals, appropriate methods of seedbed preparation, and suitable seeding equipment are all soil-conserving practices.

In Zimbabwe, Hudson and Jackson (1959) reported that 40 percent of the soil was left exposed when the maize population was 25,000 plants per hectare. In comparison, only 10 percent of the soil was exposed when the plant population was increased to 37,000 plants per hectare. Increasing plant population alone was responsible for about a 5 percent decrease in soil erosion and 175 percent decrease in water runoff. On steep slopes with a gradient of 8 percent and above, soil erosion ranged from 12.3 to 48.0 ton/acre for 100 percent stand to 14.7 to 54.8 ton/acre for 60 percent stand. Soil erosion losses for a gentle slope of less than 5 percent were 5.5 to 21.9 ton/acre for 100 percent stand and 9.1 to 26.2 ton/acre for 60 percent stand. The recommended plant population of maize for an optimum grain yield in southwestern Nigeria is 50,000 plants per hectare. In another experiment in Zimbabwe, Stocking and Elwell (1973) studied the effects of plant population on soil erosion. There were significant differences in soil erosion due to population-caused differences in vegetal cover. In Trinidad, Mohammed and Gumbs (1982) studied the effects on erosion of maize population at 41,500 and 69,500 plants per hectare. The high population significantly reduced soil loss in comparison with low population.

Weeding

Weed growth has a mixed effect on soil erosion. Both clean-weeding and excessive weed growth can increase the erosion risk. Whereas clean-weeding increases soil exposure, excessive weeding adversely affects crop growth. Poor growth means low vegetal cover, low biomass production, low residue return, low soil organic matter content, and more erosion. A certain minimum weed growth may be useful in reducing soil erosion. In Kenya, Pereira and Jones (1954) observed more erosion from clean-weeded coffee than from coffee that had been mulched or had some weed growth. Similar observations were made on grain crops in Tanzania by Temple (1972). In Malawi, Weil (1982) observed that soil losses by erosion under maize during one growing season were reduced from 12.1 t/ha on weeded plots to 4.5 t/ha on unweeded plots. The corresponding water losses by runoff were reduced from 20 percent of rainfall on weeded plots to 15 percent from the unweeded plots.

Chemical weeding can also accelerate soil erosion in comparison with slashing or manual weeding. Slashing the weeds and leaving the biomass on the surface may be the best erosion prevention measure. If weed growth is luxurious, however, chemically killed weeds provide mulch and protect the soil against splash. Mechanical weeding, however, disturbs the soil, increases exposure, and decreases the relative amount of weed residue left on the surface. The data from CENICAFE (1975) showed that soil erosion was the least from plots with slashed weeds and extremely high from plots where weed growth was suppressed by herbicides (Table 11.25).

Disease and insect control

Similar to weed eradication, an adequate disease and insect control is essential for obtaining a good crop stand, high biomass, and satisfactory yield. Indiscriminate use of pesticides, however, can increase the risks of soil erosion by elimi-

TABLE 11.25 Effects of Weed Control Methods on Soil Erosion for Coffee at Chinchina, Colombia

Treatment	Erosion, t/(ha · yr)	Relative erosion*
Machete slashing	0.05	1
Hoeing	2.77	52.0
Herbicides	6.5	121.2

*Relative erosion compared with machete slashing as control.
SOURCE: CENICAFE, 1975.

nating or reducing the population and activity of soil fauna. Soil-applied pesticides (e.g., furadan) have drastic adverse effects on soil macrofauna such as earthworms and termites. A soil devoid of macrofauna has low macroporosity, low infiltration rate, and high incidence of runoff and erosion (Lal, 1984, 1985).

Harvesting and vehicular traffic

Harvesting and carrying away the farm produce affect erosion through compaction caused by the vehicular traffic and through removal of biomass. Sometimes harvesting of the first-season crop has to be done in a short time so that the second crop can be sown quickly. The soil is often wet during this time, and vehicular traffic involved in mechanized harvesting causes severe compaction and accelerates erosion (Lal, 1984, 1985).

There are special harvesting problems in the case of sugarcane and some root crops. While burning the sugarcane trash has some merits, the practice can severely increase soil erosion. This topic is discussed at length in Chap. 12 which describes the effects of mulches and conservation tillage on soil erosion. Harvesting root crops (such as cassava, yam, sweet potato, sugar beet) creates considerable soil disturbance and soil heterogeneity. Planting is purposely done on a raised seedbed, such as ridges and mounds, to facilitate harvesting. Cassava is harvested when the soil is wet at the onset of the next rainy season. Trampling and vehicular traffic on a wet soil cause soil compaction and increase erosion risks.

Soil fertility maintenance and balanced fertilizer application

Maintaining a high level of soil fertility is essential for satisfactory crop yields, production of large biomass, and creation of a favorable soil biologic environment. A soil of low fertility causes poor and stunted crop growth, low vegetal cover, high proportion of exposed soil, and high risks of accelerated soil erosion.

Whitaker et al. (1961) reported that the more rapid development of crops receiving proper fertilization reduced the time of direct exposure of the soil to raindrop impact, improved root systems, increased plant residues, and promoted resistance to erosion on steeplands. Fertility maintenance with organic fertilizers has even more beneficial effects in improving soil structure and re-

(a)

(b)

Plate 11.3 (*a*) Mixed cropping of corn and cowpea. (*b*) Mixed cropping of corn with live mulch.

Plate 11.4 Grain crops grown in alternate strips with legume cover.

ducing erosion than that with chemical fertilizers. The beneficial effects of adding compost and organic manures in improving structure and reducing erosion risks are well established (Lee, 1978; Lal and Kang, 1982; Harrison, 1981; Chandra and De, 1982; Obi, 1982).

Buffer strips and cover crops

The role of grasses and leguminous covers in regenerating soil structure has long been recognized for both temperate and tropical regions (Martin, 1944). Cover crops, densely grown leguminous or grass pastures, are grown in association with open-row seasonal crops for the purpose of reducing soil erosion risks. Grain crops are grown in association with cover crops in many ways, e.g., mixed crops as live mulch (Plate 11.3), sequential cropping or growing cover crops in rotation with grain crops, and in alternate strips (Plate 11.4). All three methods are effective in reducing soil erosion and have different levels of economic return. The first two methods (e.g., live mulch and sequential cropping) are discussed in detail in Chap. 12. These methods have been helpful in reducing erosion in diverse soils and ecological regions (Babalola and Chheda, 1975; Wilkinson 1970, 1975a; Herrera, 1973; Landencia, 1972; Kannegieter, 1969; Eltz et al., 1977; Moore et al., 1979; Xiaoliang Experimental and Extension Station of Soil Conservation, 1977; Keane, 1977; Bajpai et al., 1975; Ghildyal and Sinha, 1969).

Cover crops are also grown in alternate strips as runoff absorbers and to encourage sedimentation of entrained sediments from the cropped strip upslope. These buffer strips are laid out on the contour (Fig. 11.6) and are effectively used in controlling erosion on gently sloping to rolling lands. Buffer strips are even more important for relatively steep slopes (Plate 11.5). In east Africa, the

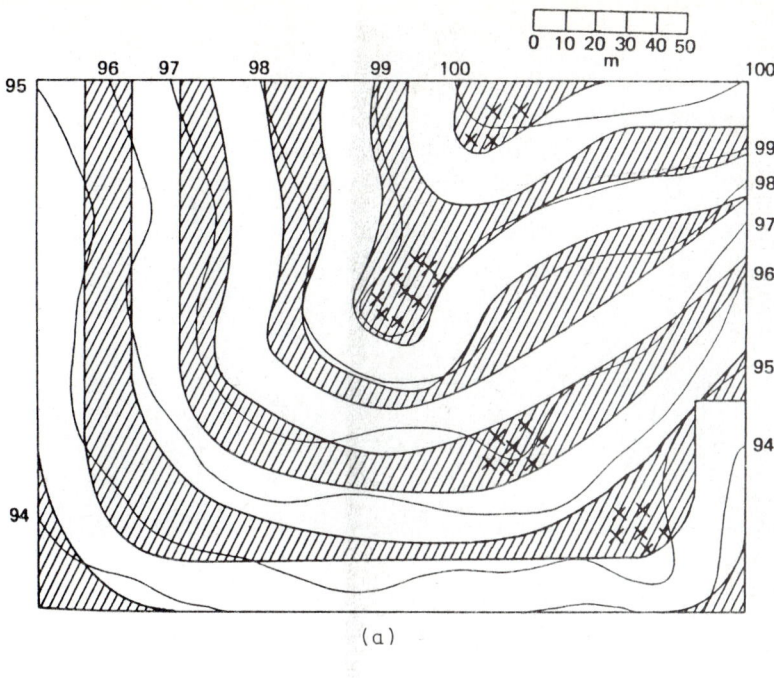

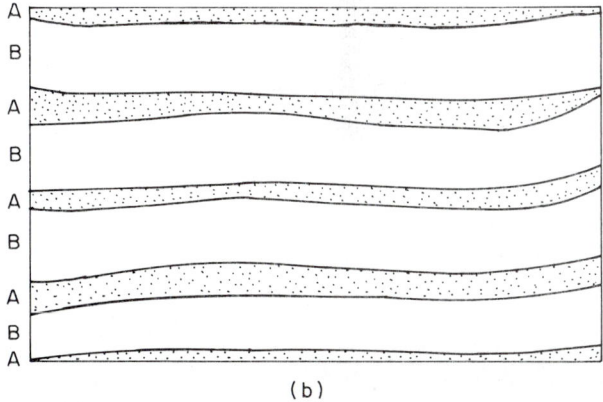

Figure 11.6 Schematic layout of strip-cropping technique. (*a*) Buffer strip cropping design on contour map. Numbers represent elevations in meters above an arbitrary base level. Xs represent rocky areas that are uncroppable (after Troeh et al., 1980). (*b*) Alternate strips of (A) soil-conserving and (B) open-row crops grown on the contour.

use of elephant grass (*Pennisetum* sp.) for resting strip was recommended in Uganda in the early 1940s (Kerr, 1942). In addition to controlling erosion, grass strips are used for producing mulch and for possible grazing or stall-feeding of cattle with consequential manuring. More recently, grass strips are recommended for erosion control on steeplands (Moore et al., 1980). In west Africa,

344 Erosion Control

(a)

(b)

Plate 11.5 Buffer strips of grass and legumes are used on contour to reduce erosion on steep slopes in (*a*) Kenya and (*b*) Dominican Republic.

TABLE 11.26 Effects of Contour Grass Strips on Runoff and Soil Erosion from an Inceptisol on 15 to 22% Slope in Darmaga, West Java

Treatment	Years after establishing the grass strips			
	1 (1976–1977)	2	3	4
	Soil Erosion, t/ha			
Control	452.5	340.7	209.7	193.5
Brachiaria decumbens (0.5 m)	23.2	10.6	41.9	0
Paspalum notatum (1 m)	—	78.8	0	0
	Runoff, m³/ha			
Control	4305	2979	2882	—
B. decumbens (0.5 m)	4359	2148	2001	—
P. notatum (1 m)	—	3598	1978	—

SOURCE: Abujamin et al., 1984.

Roose and Bertrand (1971) evaluated the effectiveness of buffer strip cropping for three ecological regions with annual rainfalls of 2100, 1250, and 520 mm. Permanent grass strips 2 to 4 m wide proved effective in reducing erosion and encouraging development of natural terraces and vegetation regrowth on them within about 4 years. Similar observations have been made in Kenya.

Buffer strips have also been effectively used for erosion control on sloping lands, in southeast Asia. In Darmaga, west Java, Abujamin et al. (1984) observed that 0.5-m-wide strips of *Brachiaria decumbens* and 1-m-wide strips of *Paspalum notatum* significantly reduced soil erosion in the second and third years after grass establishment (Table 11.26). While a large proportion of sediments are deposited within the grass strip, only 30 to 40 percent of water runoff is absorbed. In northern Thailand, Ryan et al. (1984) observed that strip cropping was more effective in reducing runoff and erosion than the ridge-furrow method of seedbed preparation or terracing (Table 11.27). This topic is also discussed at length in Chap. 12.

TABLE 11.27 Effects of Strip Cropping, Ridge-Furrow Seedbed, and Terrace Spacing on Runoff and Soil Erosion in Northern Thailand

Treatment	Runoff, m³/ha	Erosion, t/ha
Double-space bank	1429	4.2
Single-space bank	1229	3.3
Strip cropping	971	2.1
Single-space grass strip	1491	4.5
10-m bed and furrow	1163	4.7

SOURCE: Ryan et al., 1984.

Plate 11.6 Stone lines or diggets used to reduce water runoff in Burkina Faso.

Quick-growing hedges recommended for arid regions comprise *Euphorbia balsamifera,* grasses, and some multipurpose trees such as *Acacia albida.*

Establishing grass strip and hedges on degraded lands and in harsh environments (semiarid or arid climates) is facilitated by installing stone lines on the contour. Stones, often collected from the field, are put on contour in one or two rows (Plate 11.6). These stone lines retard the runoff velocity, encourage sedimentation, and facilitate grass establishment. Such stone lines, also called *diggets,* are widely recommended for conserving soil and water in the west African Sahel, e.g., Niger and Burkina Faso. Roose (1985) reported their usefulness in establishing natural terraces. Stone lines are now widely used by farmers in the Mossi Plateau region of Burkina Faso and in the southern parts of Mali (Roose, 1984; Hallam and Van Campen, 1985). The data in Table 11.28 from Allokoto, Niger, show significant reductions in runoff and erosion by installation of stone lines and earth dykes with stones.

TABLE 11.28 Comparative Effects of Stone Lines and Buffer Strip Cropping on Runoff and Erosion from a Soil at Allokoto, Niger

Treatment	Runoff, % of rainfall	Erosion, t/(ha · yr)
Control (traditional hoe)	17.6	9.5
Buffer strips (contour plowing, ridging, weeding)	5.2	1.1
Stone lines (contour plowing, ridging, weeding)	3.8	0.5
Earth dykes with stones (contour plowing, ridging, weeding)	0.9	0.2

SOURCE: Delwaulle, 1973.

In recent years, there has been a considerable emphasis on the usefulness of vetiver (or khus) grass (*Vetiveria zizanioides*) in controlling soil erosion in a range of climactic environments (World Bank, 1988). Vetiver is a densely tufted, awnless, wiry, and glabrous perennial grass. It does not produce rhizomes or stolons, and it is propagated by root divisions or slips. It has a deep, strong, and fibrous root system. It grows in large clumps from the root stock and is propagated vegetatively. The grass can be planted on the contour to establish protective contour hedges. These hedges retard runoff velocity, encourage sedimentation, increase infiltration, and facilitate formation of natural terraces. The data from experiments conducted at Akola, Maharashtra, India, have indicated significant reductions in runoff and soil erosion (World Bank, 1990). The average total soil loss over the cropping season was 26, 17, and 8 t/ha for across-the-slope cultivation, contour cultivation with a contour hedgerow of *Leucaena*, and contour cultivation with a contour hedgerow of vetiver, respectively. In addition, the grass can be established on earth banks or bunds and on terraces to reinforce and stabilize these structures. Its thick root system prevents slope failure due to rilling, gullying, or tunneling. Establishment of continuous hedges of vetiver with no gaps provides maximum protection. While controlling soil erosion, vegetative contour hedges of vetiver also conserve soil water by enhancing infiltration and decreasing losses due to runoff. Over and above its benefits as a soil-conserving plant, vetiver grows under a wide range of environments and is therefore relatively tolerant of moderate levels of drought or inundation, and damage by fire. It reportedly grows in the rainfall regime ranging from 200 to 6000 mm per year, and from sea level to about 2500 m in altitude. It also has some economic value because an aromatic oil is extracted from its roots.

Residue management

Crop residue management is a powerful tool for soil erosion control, soil fertility maintenance, and regulation of pest incidence. Crop residue is managed differently to achieve different objectives, e.g., left on the surface, carried away, incorporated into the soil, and burned. Although each of these practices has merit, soil erosion is best achieved by leaving the crop residue on the soil surface. This can be done by adopting a range of tillage techniques. The suitability of different tillage methods in relation to soil erosion is described in Chap. 12.

Conclusions

Scientific crop management is an important and effective tool for soil conservation. All crop husbandry and management techniques that increase crop yields also decrease the risks of soil erosion. A well-managed farm runs lower risks of erosion than a poorly managed farm.

Simple agronomic techniques of soil and crop management are usually just as effective (or more so) in erosion control as some expensive and complex engineering devices installed for water management. It is important not to neglect these simple practices such as the use of improved crop varieties, good and viable seed, early planting, crop rotations involving legumes and cereals, balanced fertilizer, pest control, residue management, and buffer strips.

The soil erosion problem is exacerbated when the principles of good crop husbandry are neglected.

References

Abdurachman, A., Abujamin, S., and Kurnia, U. 1984. Soil and crop management practices for erosion control. *Pembr. Pen. Tanah Dan Pupuk* 3:7–12.

Abujamin, S., Abdurachman, A., and Suwardjo, M. 1984. Contour grass strip as a low cost conservation practice. In: *Proc. Soil Erosion and Its Countermeasures,* 11–19 November 1984, Chiangmai, Thailand.

Abujamin, S., Abujamin, A., and Kurnia, U. 1983. Permanent grass strip as one of soil conservation methods. *Pembr. Pen. Tanah Dan Pupuk* 1:16–20.

Aina, P. O., Lal, R., and Taylor, G. S. 1977. Soil and crop management in relation to soil erosion in the rainforest region of western Nigeria. In: *Soil Erosion: Prediction and Control,* SCSA Publ. 21, pp. 75–84, SCSA, Ankeny, Iowa.

Aina, P. O., Lal, R., and Taylor, G. S. 1979. Effects of vegetal cover on soil erosion on an Alfisol. In: *Soil Physical Properties and Crop Production in the Tropics* (R. Lal and D. J. Greenland, eds.), pp. 501–508, Wiley, Chichester, U.K.

Armiger, W. H., Foy, C. D., Fleming, A. L., and Caldwell, B. E. 1968. Differential tolerance of soybean varieties to an acid soil high in exchangeable aluminum. *Agron. J.* 60: 67–70.

Babalola, O., and Chheda, H. R. 1975. Influence of crops and cultural practice on soil and water loss from a western Nigerian soil. *Ghana J. Sci.* 15(1):93–99.

Bajpai, M. R., Gujar, R. G., and Singh, G. 1975. A note on efficacy of vegetative covers and slopes on runoff and fertility losses in a semi-arid tract of Rajasthan. *Ann. Arid Zone* 14(4):376–378.

Barnett, A. P., Carreker, J. R., and Abruna, F. 1972. Soil and nutrient losses in runoff with selected cropping treatments in tropical soils. *Agron. J.* 64(3):391–395.

Bertoni, J., and Lombardi, F. Neto. 1985. *Conservacao do solo,* Livruceres Ltda, Brasil.

Bhatt, P. N., Gupta, O. P., and Tejwani, K. G. 1971. Influence of cropping pattern and land use on plant nutrient losses in Doon Valley. *Proc. International Symposium on Soil Fertility Evaluation,* vol. 1, pp. 541–547.

CENICAFE, 1975. *Manual de Conservation de Sudes de Ladera,* Chinchina, Caldas, Colombia.

Chandra, S., and De, S. K. 1982. Effect of cattle manure on soil erosion by water. *Soc. Sci.* 133(4):228–231.

Conway, G. R. 1985. Agricultural ecology and farming systems research. In: *Agricultural Systems Research for Developing Countries,* ACIAR Proc. 11, pp. 43–59, Canberra, Australia.

Delwaulle, J. C. 1973. Resultats de six années d'observations sur l'erosion au Niger. *Bois For. Trop.* 150:15–37.

De Tar, W. R., Ross, J. J., and Cunningham, R. L. 1980. Estimating the C factor in the Universal Soil Loss Equation for landscaped slopes. *J. Soil Water Conserv.* 35:40–41.

Dillon, J. L., and Virmani, S. M. 1985. The farming systems approach. In: *Agro-Research for the Semi-Arid Tropics: North-West Australia* (R. C. Muchow, ed.), pp. 507–532, University Queensland Press, St. Lucia, Australia.

Donahue, R. L., and Miller, R. W. 1977. *Soil: An Introduction to Soils and Plant Growth,* Prentice-Hall, Englewood Cliffs, N.J.

El-Swaify, S. A., and Cooley, K. R. 1980. Soil loss from sugarcane and pineapple lands in Hawaii. In: *Assessment of Erosion* (M. de Boodt and D. Gabriels, eds.), pp. 327–340, Wiley, Chichester, U.K.

Eltz, F. L. F., Cogo, N. P., and Mielniczuk, J. 1977. Erosion losses under different management systems and crop cover in a reddish brown dystrophic lateritic soil (Sao Jeronimo). I. First year results. *Rev. Bras. Cienc. Solo* 1:123–127.

Elwell, H. A., and Stocking, M. A. 1976. Vegetal cover to estimate soil erosion hazard in Rhodesia. *Geoderma* 15:61–70.
Fauck, R., Moureaux, C., and Thomanu, C. 1967. Evolution of SEFA (Senegal) soils during fifteen years of continuous cropping. *C. R. Acad. Sci. Agric. Fr.* 53:698–703.
Fournier, F. 1967. Research on soil erosion and soil conservation in Africa. *Afr. Soils* 12:53–96.
Foy, C. D., Armiger, W. H., Briggle, L. W., and Reid, D. A. 1965. Differential aluminum tolerance of wheat and barley varieties to acid soils. *Agron. J.* 57:413–417.
Georges, J. E. W. 1977. Soil erosion in cane fields in hill lands in Trinidad. *Int. Sugar J.* 81:147.
Ghildyal, B. P., and Sinha, A. K. 1969. Effect of vegetal cover on runoff, soil loss and soil fertility. *J. Indian Soc. Soil Sci.* 17:449–455.
Goel, K. N., and Khanna, M. L. 1969. Effect of crop rotations in reducing the nutrient loss in alluvial tracts of Uttar Pradesh. *J. Soil Water Conserv. India* 17:42–46.
Gregory, J. M. 1984. Prediction of soil erosion by water and wind for various fractions of cover. *Trans. ASAE* 97:1345–1350, 1354.
Gumbs, F. A., and Lindsay, J. I. 1982. Runoff and soil loss in Trinidad under different crops and soil management. *Soil Sci. Soc. Am. J.* 46(6):1264–1266.
Hallam, G., and Van Campen, W. 1985. Reacting to farmers' complaints of soil erosion on intensive farms in southern Mali. *Fourth ISCO Conference,* Maracay, Venezuela, 3–9 November 1985.
Hanson, W. D., and Kamprath, E. J. 1979. Selection for aluminum tolerance in soybeans based on seedling root growth. *Agron. J.* 71:581–586.
Harrison, P. 1981. Food, fuel and zero tillage. *Dev. Coop.* 6:13–14.
Herrera, B. J. A. 1973. The mechanization of agriculture on the slopes as a contribution to rural development in Colombia. *Rev. Inst. Colombiamo Agropecuario (Colombia)* 8(2):197–227.
Hudson, N. W. 1971. *Soil Erosion,* Batsford, London.
Hudson, N. W. 1981. A research project on hydrology and soil erosion in watersheds in Sri Lanka. In: *Tropical Agricultural Hydrology* (R. Lal and E. W. Russell, eds.), Wiley, Chichester, U.K.
Hudson, N. W., and Jackson, D. C. 1959. Results achieved in the measurement of erosion and runoff in southern Rhodesia. *Third Inter-African Soils Conference, Dalaba,* vol. 2, pp. 575–595.
ICAR. 1973. *Soil and Water Conservation Research 1956–1971,* ICAR, New Delhi, India.
ICAR. 1984. *Soil Conservation,* Annual report, Central Soil and Water Conservation Research and Training Institute, Dehra Dun, India.
Kannegieter, A. 1969. The combination of a short term *Pueraria* fallow, zero cultivation and fertilizer application: Its effect on a following maize crop. *Trop. Agric. Ceylon* 15:77–94.
Keane, P. A. 1977. Native species for soil conservation in the Alps, New South Wales. *Soil Conserv. Serv.* (Looma, New South Wales, Australia) 33(3):200–217.
Kerr, A. J. 1942. A new system of grass-fallow strip cropping for the maintenance of soil fertility. *Emp. J. Exp. Agric.* 10:125–132.
Kowal, J. 1970. The hydrology of a small catchment basin at Samaru, Nigeria. III. Assessment of surface runoff under varied land management and vegetation cover. *Niger. Agric. J.* 7:120–133.
Lafever, H. N., Campbell, L. G., and Foy, C. D. 1977. Differential response of wheat cultivars to Al. *Agron. J.* 1969:563–568.
Lal, R. 1975. Soil-conserving versus soil-degrading crops and soil management for erosion control. In: *Soil Conservation and Management in the Humid Tropics* (D. J. Greenland and R. Lal, eds.), Wiley, Chichester, U.K.
Lal, R. 1976, *Soil Erosion Problems on an Alfisol in Western Nigeria, and Their Control,* IITA Monogr. 1, IITA, Ibadan, Nigeria.
Lal, R. 1984. Mechanized tillage systems: Effects on soil erosion from an Alfisol in watersheds cropped to maize. *Soil Tillage Res.* 4:349–360.
Lal, R. 1985. Mechanized tillage systems: Effects on properties of a tropical Alfisol in watershed cropped to maize. *Soil Tillage Res.* 6:149–161.

Lal, R. 1986. Soil surface management in the tropics for intensive landuse and high and sustained production. *Adv. Soil Sci.* 5:1–109.
Lal, R., and Kang, B. T. 1982. Management of organic matter in soils of the tropics and sub-tropics. In: *Transactions of the 12th International Congress of Soil Science,* 8–16 February 1982, New Delhi, India.
Landencia, P. N. 1972. Soil and water conservation through the improvement of soil cover. *Philippine Geogr. J.* 16:42–52.
Lee, J. H. N. 1978. Safe disposal of pigger effluent assists erosion control. *J. Soil Conserv. Serv. NSW* 34(3):160–164.
Lombardi, F. N., Bertoni, J., and Benatti, Jr., R. 1975. Practicas conservacionistas en cafezal e perdes por exosao en Latossolo Roxo. *Anais Do XV Congresso Brasileira De Ciencia Do Solo,* pp. 581–583.
MacLean, A. A., and Chiasson, T. C. 1966. Differential performance of two barley varieties to varying aluminum concentrations. *Can. J. Soil Sci.* 46:147–153.
Marques, J., Quintiliano, A., Bertoni, J., and Barreto, G. 1961. Perdas por e rosao no estado de Sao Paulo. *Bragantia* 20:1144–1181.
Marston, D., Aneckamphant, C., and Chirasathaworn, R. 1984. The use of mulching and no-tillage for soil conservation in tropical upland crops. In: *Fifth ASEAN Soil Conference,* 10–23 June 1984, Bangkok, Thailand, pp. E1.1–E1.12.
Martin, W. S. 1944. Grass covers in their relation to soil structure. *Emp. J. Exp. Agric.* 12:21–33 (Uganda).
Mohammed, A., and Gumbs, F. A. 1982. The effect of plant spacing on water runoff, soil erosion and yield of maize (*Zea mays* L.) on a steep slope of an Ultisol in Trinidad. *J. Agric. Eng. Res.* 27(6):481–488.
Mokhtaruddin, A. M., and Maene, L. M. 1979. Soil erosion under different crops and management practices. In: *Proc. International Conference on Agricultural Engineering in National Development,* September 1979, Serdang, Malaysia, University of Pertania, Malaysia.
Moore, T. R., Thomas, D. B., and Barber, R. G. 1979. The influence of grass cover on runoff and soil erosion from soils in the Machakos area, Kenya. *Trop. Agric.* 56:339–344.
Moore, T. R., Thomas, D. B., and Barber, R. G. 1980. The influence of grass cover on runoff and soil erosion from soils in the Machakos area, Kenya. Department of Agricultural Engineering, University of Nairobi, Kenya.
Nijhawan, S. D., and Garg, K. C. 1974. A comparative study of structure of soils of Rohtak under different cultural practices. *Indian J. Agric. Res.* 8(4):223–237.
Obi, M. E. 1982. Runoff and soil loss from an Oxisol in southeastern Nigeria under various management practices. *Agric. Water Manage.* 5:192–203.
Pereira, H. C., and Jones, P. A. 1954. A tillage study in Kenya coffee. I. The effects of tillage practices on coffee yields. *Emp. J. Exp. Agric.* 22:231–240.
Quinn, N. W., and Laflen, J. M. 1983. Characteristics of raindrop throughfall under corn canopy. *Trans. ASAE* 26:1445–1450.
Rama Rao, M. S. V. 1974. *Soil Conservation in India,* ICAR, New Delhi.
Rapp, A., Murray-Rust, D. H., Christiansson, C., and Berry, C. 1972. Soil erosion and sedimentation in four catchments near Dodoma, Tanzania. In: *Studies of Soil Erosion and Sedimentation in Tanzania* (A. Rapp, L. Berry, and P. Temple, eds.), *Geografiska Annaler* 54A:255–318.
Roche, P., and Dubois, B. 1959. Mesures de ruissellement et d'erosion realisées a Madagascar. *Third Inter-African Soils Conference,* Dalaba, Africa, vol. 2, pp. 601–614.
Roose, E. J. 1967. Dix années de mesure de l'erosion et du ruissellement au Senegal. *Agron. Trop.* 22:123–152.
Roose, E. J. 1984. Causes et facteurs de l'erosion hydrique sous climat tropical, consequences sur les methodes antierosives. *Mach. Agric. Trop.* 87:4–18.
Roose, E. J. 1985. Diversion channel terraces or permeable approaches on the little farms of soudano-sahelian area of western Africa. In: *IVth International Conference on Soil Conservation,* 3–9 November 1985, Maracay, Venezuela.
Roose, E. J., and Bertrand, R. 1971. Contribution to the study of the buffer strip cropping method to control water erosion in west Africa. Experimental results and observations in the field. *Agron. Trop. (France)* 26:1270–1283.

Roose, E. T., and Piot, J. 1984. *Runoff, Erosion and Soil Fertility Restoration on the Mossi Plateau, Burkina Faso.* Wallingford, U.K., IAHS Publ. 144, pp. 485–498.

Ryan, K., Anecksamphant, and Marston, D. 1984. Soil and water conservation in uplands of northern Thailand. In: *Proc. Soil Erosion and Its Countermeasures,* 11–19 November 1984, Chiangmai, Thailand.

Saraiva, O. F., Cogo, N. P., and Mieluiczuk, J. 1981. Erosion by rainfall and erosion losses under different soil management systems and plant cover in a reddish brown distrophic lateritic soil. I. Second year results. *Presquisa Agropecuaria Bras.* 16(1):121–128.

Stocking, M. A., and Elwell, H. Q. 1973. Prediction of subtropical storm soil losses from field plot studies. *Agric. Meteorol.* 12:193–201.

Sulaiman, W., Maene, L. M., and Mokhtaruddin, A. M. 1981. Runoff, soil and nutrient losses from an Ultisol under different legumes. In: *Proc. South-East Asian Regional Symposium on Problems of Soil Erosion and Sedimentation,* 27–29 January 1981, Bangkok, Thailand, Asian Institute of Technology.

Sulaiman, W. H. W., Mok, C. K., Maesschalck, and Jamal, T. 1983. Advances in soil and water conservation research in Malaysia. *Pertanika* 6(Rev. suppl.):115–132.

Temple, P. H. 1972. Measurement of runoff and soil erosion at an erosion plot scale with particular reference to Tanzania. In: *Studies of Soil Erosion and Sedimentation in Tanzania* (A. Rapp, L. Berry, and P. Temple, eds.), *Geografiska Annaler* 54A:203–220.

Thomas, D. B. 1975. Land use and soil erosion in part of Kalama location, Machakos District. *Kijani* (Kenya) 1(1):16–17.

Troeh, F. R., Hobbs, J. R., and Donahue, R. L. 1980. *Soil and Water Conservation.* Prentice-Hall, Englewood Cliffs, N.J.

Valet, S. 1974. Notes on the observations and measurements of climatic, hydrologic, and pedologic factors and their effect on agricultural production at the Dschang station (Cameroon). *Agron. Trop. (France)* 29:1266–1287.

Vasudevaiah, R. D., Teotia, S. P., and Guha, D. P. 1965. Runoff–soil loss determination studies at Deochanda Experiment Station. II. Effect of annually cultivated grain crops and perennial grasses on 5 p.c. slope. *J. Soil Water Conserv. India* 13(3–4):36–46.

Weil, R. R. 1982. Maize-weed competition and soil erosion in unweeded maize. *Trop. Agric.* 59(3):207–213.

Whitaker, F. D., Jamison, V. C., and Thornton, J. F. 1961. Runoff and erosion losses from Mexico silt loam in relation to fertilization and other management practices. *Soil Sci. Soc. Am. Proc.* 25:401–403.

Wilkinson, G. E. 1970. The infiltration of water into Samaru soils. *Samaru Agric. News.* 12(5):81–83.

Wilkinson, G. E. 1975a. Effect of grass fallow rotations on the infiltration of water into a savanna soil of northern Nigeria. *Trop. Agric.* 52:97–103.

Wilkinson, G. E. 1975b. Canopy characteristics of maize and the effect on soil erosion in western Nigeria. *Trop. Agric.* 52:289–297.

Williams, C. N., and Joseph, K. T. 1970. *Climate, Soil and Crop Production in the Humid Tropics,* Oxford University Press, Kuala Lumpur.

Willey, R. W. 1985. The prospects of intercropping. In: *Agro-Research for the Semi-Arid Tropics: Northwest Australia* (R. C. Muchow, ed.), pp. 435–449, Queensland University Press, St. Lucia, Australia.

World Bank. 1988. *Vetiver Grass: A Method of Vegetative Soil and Moisture Conservation,* 2d ed. New Delhi, India.

World Bank. 1990. *Vetiver Newsletter,* no. 3, March. Washington, D.C.

Xiaoliang Experimental and Extension Station of Soil Conservation. 1977. Effects of artificial vegetation on soil conservation of littoral hilly slopes. *Acta Bot. Sinica* 19:182–189.

Chapter 12

Conservation Tillage

Introduction

Mechanical soil disturbance is a major factor that influences susceptibility of a soil to erosion. Soil turnover by plowing removes the vegetation cover and exposes the soil to climatic elements, rainfall and wind, and fluvial processes. Achieving a weed-free seedbed is one of the reasons for mechanical soil disturbance and its turnover. Another reason is the need to incorporate fertilizer and soil amendments. The effects of tillage-induced soil disturbance on the magnitude of erosion depend on a range of interacting factors (Fig. 12.1). Frequency, intensity, type of tillage implements, and the source of power used alter soil properties to different degrees and influence the susceptibility of a soil to erosion. The degree of soil erosion is also influenced by properties of the surface soil and the profile characteristics. In addition to soil characteristics, cropping systems and crop management practices strongly interact with the prevailing climate in determining the magnitude of soil erosion.

The effects of tillage methods on soil properties and on the magnitude and trends in erosion are hard to generalize. The effects vary depending on the antecedent soil properties. There are two important scenarios to be considered. First, the soil has favorable structure comprising a high proportion of water-stable aggregates, a large percentage of transmission or macropores, no impermeable or slowly permeable layer within the soil profile, and a crust-free surface layer. When these conditions prevail, mechanical soil disturbance involving primary and secondary tillage is likely to increase the risks of soil erosion. Second, given contrasting soil conditions, the antecedent soil properties render a soil highly vulnerable to erosion. These conditions include smooth soil

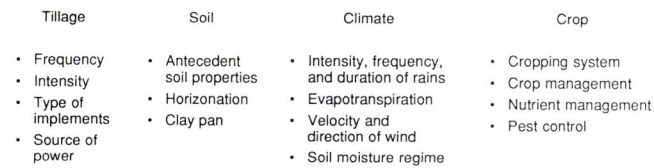

Figure 12.1 Factors influencing the magnitude of soil erosion by tillage-induced soil disturbance.

surface devoid of vegetation, crusted surface and compacted subsoil horizons, massive structure with low total porosity and low percentage of macropores, soil structure unstable to quick wetting, and a low percentage of water-stable aggregates. When these conditions prevail, a right type of mechanical tillage, with appropriate frequency and intensity, is likely to decrease the risks of soil erosion, at least temporarily. Because the effects of tillage methods on soil erosion depend on antecedent soil properties, the term *conservation tillage* means different sets of tillage practices under different conditions. This vague use of one term for different situations has created confusion and misunderstanding. The term *conservation tillage* therefore encompasses a broad spectrum of tillage practices ranging from no-tillage to intensive tillage, depending on the soil-related constraints that increase the risks of erosion.

Conservation Tillage: Basic Concepts and Definitions

Because no single tillage system seems to control soil erosion for diverse soils and climatic situations, the definition of conservation tillage must include all possible tillage options that may have even a remote application in controlling or reducing soil erosion. *Conservation tillage* has been defined as "any tillage sequence that reduces loss of soil or water relative to conventional tillage; often a form of non-inversion tillage that retains a productive amount of mulch on the surface" (*Resource Conservation Glossary,* 1982). This definition is rather restricted because it considers only noninversion and residue mulch. The key concepts to be included in conservation tillage for soil and water conservation are (1) residue mulch and (2) an increase in random roughness. While criterion 1 is satisfied by no-tillage, minimum tillage, or mulch farming systems, an increase in random roughness can be achieved by chisel plowing, strip tillage, ridge-furrow system, and tillage methods that cause soil inversion. If done at the right soil moisture content and with right equipment, inversion tillage can, in fact, produce the desired rough seedbed and improve water infiltration. Although using mulch is a useful concept, it is argued that a desired amount of mulch may not be available in all ecological regions and for all farming systems. In these situations, conservation tillage may include other tillage techniques such as contour ridges, tied ridges, camber bed system, and broad-bed furrow system. Therefore, a broad definition of conservation tillage, such as that proposed by Wittmus et al. (1973), may be appropriate. According to this definition, "conservation tillage includes tillage systems that create as good an environment as possible for the growing crop and that optimize the conservation of soil and water resources, consistent with sound economic practices."

Because conservation tillage covers a broad range of tillage practices, we suggest that this term be used as an umbrella term to cover all tillage practices that conserve soil and water under diverse conditions (Mannering and Fenster,

(a)

(b)

Plate 12.1 A no-till system for growing crops at IITA, Ibadan, Nigeria. (*a*) Cowpea, (*b*) soybean, (*c*) maize.

1983). Some of the tillage practices included under the conservation tillage umbrella are discussed now.

Reduced tillage systems

A broad range of reduced tillage systems are used to minimize soil and water losses. Commonly used reduced tillage systems include no-tillage, minimum tillage, mulch tillage, and ridge tillage.

No-tillage. A tillage system whereby a crop is seeded directly into a seedbed that has not been tilled since the previous seedbed is called a *no-till,* or *no-tillage, system* (Plate 12.1). A no-tillage system has the following characteristics: All preseeding mechanical tillage is zero, the maximum amount of crop residue is retained on the surface, and weeds are controlled by chemicals, by using an aggressive cover crop, or by a combination of these methods. Some researchers argue that if a crop requires soil disturbance for harvesting (e.g., a root crop including yam, cassava, sweet potato, potato, or sugar beet), the system is no longer a no-till system. According to the definition given here, however, soil disturbance at harvest is not a necessary characteristic of the no-till system. Root crops can thus technically be grown with a no-till system.

Minimum tillage. The term *minimum tillage* has caused the greatest confusion because the minimum tillage required to successfully grow a crop varies from zero to a complete range of primary and secondary tillage operations depending on soil properties and crops to be grown. It is commonly defined as "the

(c)

Plate 12.1 (*Continued*)

minimum soil manipulation necessary for crop production or meeting tillage requirements under the existing soil and climatic conditions" (*Resource Conservation Glossary*, 1982). It often means any system that has fewer tillage operations than the conventional one. However, "conventional" tillage also varies from place-to-place. The conventional tillage system in the tropics, based on a manual source of power or animal traction, is certainly different from the motorized equipment used in western Europe and North America. The most widely used conventional tillage in the tropics is either the ridge-furrow system (Plate 12.2) or the hillocks (Plate 12.3). In some cases, traditional tillage simply means dropping seeds in a hole made with a digging stick and sowing immediately following the first rain. In North America, western Europe, and other industrialized societies, conventional tillage means the combined primary and secondary tillage operations normally performed in preparing a seedbed for a given crop in a given geographical area. Because of these regional and ecological differences, therefore, there is a need to standardize the terminology to avoid confusion and misunderstanding.

Minimum tillage may also refer to a "stale-bed" system whereby the soil inversion by moldboard plowing is done at the end of the previous crop cycle. The next crop is seeded with a minimum of seedbed preparations such as disk harrow performed at the onset of the next rains. This is commonly recommended for soils of the semiarid tropics in west Africa (Charreau and Nicou, 1971). Another variant of this system commonly observed in the temperate-zone climate is that the primary tillage operations are performed in the fall or early spring followed by secondary tillage operations at the time of seeding.

Plate 12.2 The ridge-furrow system commonly used as a traditional method of seedbed preparation in the tropics.

Conservation Tillage 357

Mulch tillage. A tillage system that ensures a maximum retention of crop residue on the soil surface is called *mulch tillage* or *stubble mulch farming*. It is defined as preparation of the soil in such a way that plant residue or other mulching materials are specifically left on or near the surface. Mulch tillage is also a broad term and includes practices such as no-tillage, disk plant system,

(a)

(b)

Plate 12.3 Hillocks or mounds are constructed by scraping topsoil with a handhoe. (*a*) The top of mound is mulched to regulate soil temperature. (*b*) Large mounds are constructed in poorly drained soil.

chisel plant system, and strip tillage system. When a grain crop is seeded through the mulch of a chemically killed cover crop, it is called *sod seeding*. If the cover is only temporarily suppressed or not killed, the system is called *live mulch*.

A cover crop, usually a legume, is specifically grown to break the cropping cycle, to produce mulch material, to improve soil structure, and to enhance soil organic matter content. This system is called *planted fallow*. Another variant of planted fallowing, practiced in North America, is referred to as *summer fallow* or *ecofallow*. The latter is a system of fallowing in which weed growth is restricted by shallow cultivation or by using herbicides to conserve soil moisture. Crops are grown every other year or once in 3 years. This type of "cropless" fallow is mostly used in arid climates to conserve soil moisture.

No-tillage is also practiced in association with some cultural practices involving an agroforestry system—growing food crop annuals in association with woody shrubs and perennials. A commonly used practice is alley cropping whereby food crop annuals are grown within the widely spaced hedges of perennial shrubs (Plate 12.4). The hedges are planted on the contour and are regularly pruned to provide mulch (see Chap. 13).

Ridge tillage. The practice of planting or seeding crops on ridges is a widely adapted system in both temperate and tropical climates. The crop row may be planted on the ridgetop, along both ridge sides, or in the furrow. Ridge tillage facilitates the mixed cropping systems—a practice of growing more than one crop simultaneously in the same plot of land, a system commonly practiced throughout the tropics and subtropics (Bradfield, 1970). The drawing in Fig. 12.2 shows different possibilities of seeding crops in a ridge-furrow system. The system can also be used with motorized equipment (Fig. 12.3a and b).

The ridges may be made every season. Alternatively, in a semipermanent

Plate 12.4 Growing no-till maize by using an alley cropping system.

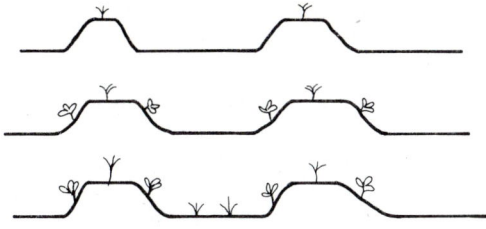

Narrow beds and furrows are adapted to 75cm rows only (e.g., maize)

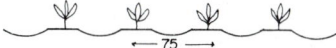

Broad beds and furrows are adapted to many row spacings

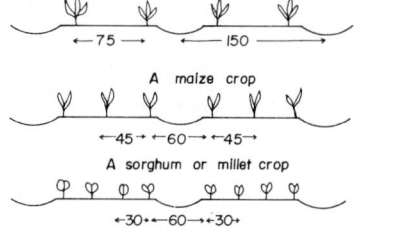

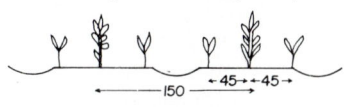

Figure 12.2 Ridge tillage is performed to facilitate mixed cropping (Kampen et al., 1981).

ridge-furrow system, only the necessary repair is done at the onset of a new cropping cycle. The ridges may be on the contour with graded furrows draining into a grassed waterway, or the ridges may have short cross-ties to create a series of basins to store water. The latter system with cross-ties is called the *tied-ridge system* (Plate 12.5).

The outline in Fig. 12.4 is an attempt to classify different types of conserva-

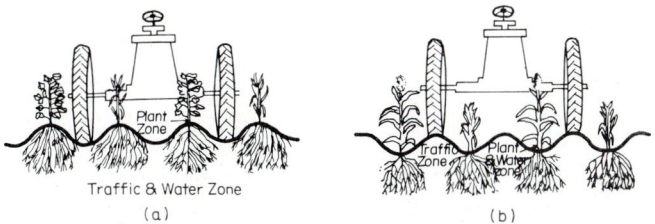

Figure 12.3 A permanent ridge-furrow system with planting (*a*) on ridge top with tractorized equipment and (*b*) in furrows with tractorized equipment. (After Johnston, 1973.)

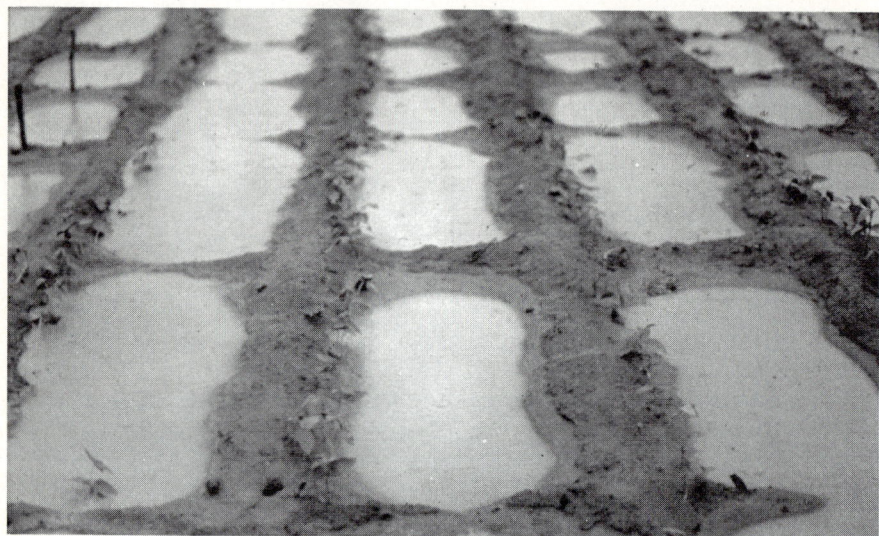

Plate 12.5 The tied-ridge or basin tillage system used for water conservation in arid and semiarid regions.

tion tillage used. There are three broad categories: mulch tillage, reduced tillage, and ridge-furrow system. Most commonly used variants of conservation tillage can be classified within these three broad categories. In addition, there are tillage methods specifically performed for some problem soils, e.g., salt-affected soils, poorly drained soils, or soils with clay pan in the subsurface horizons. Tillage systems labeled 1, 2, 3, and 4 in Fig. 12.4 are generally referred to as no-tillage systems because in all these systems the plant residue is left on the soil surface and crops are seeded with complete elimination of all preplant mechanical tillage operations.

Erosion Control with Conservation Tillage

Different forms of conservation tillage have been used, with varying degree of success, to conserve soil and water in different agroecological regions. Similar

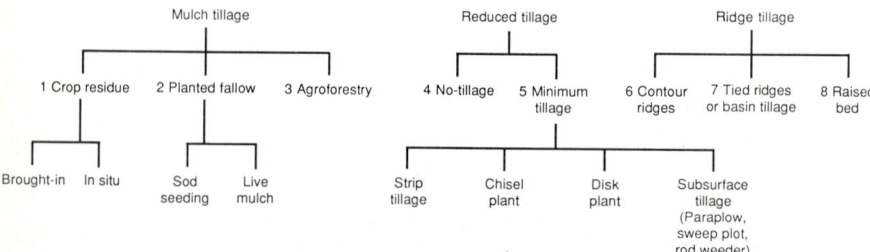

Figure 12.4 Types of conservation tillage.

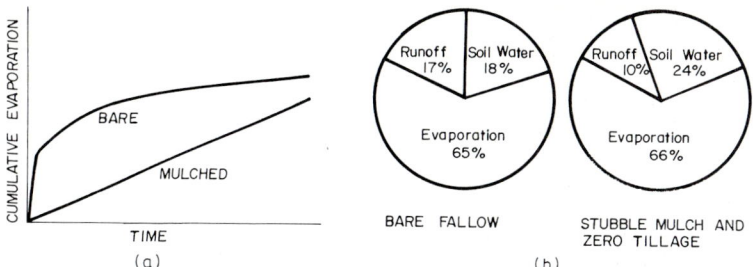

Figure 12.5 (*a*) Schematic diagram showing cumulative evaporation from a bare and a mulched soil as influenced by time (Unger, 1973). (*b*) Effects of stubble mulch on summer fallow water balance (black earth) for a vertisol in Queensland, Australia (Queensland Dept. of Primary Industry, 1984–1985).

to water erosion, conservation tillage is used for controlling wind erosion (Bodolay, 1974). The following examples are specifically chosen to highlight the contrasting requirements for conservation tillage to alleviate soil- and climate-related constraints of different agroecological environments.

Conservation tillage based on crop residue mulch

As defined in Fig. 12.4, the term *no-tillage* denotes a wide range of practices based on using crop residue mulch, mulch produced in situ by specifically growing cover crops, and by procuring mulch by pruning perennial shrubs used in an agroforestry system, e.g., alley cropping.

Mulch farming. Crop residue mulch is an effective system to control rill-interrill erosion. Mulches reduce water erosion by reducing raindrop impact, increasing the soil infiltration rate, decreasing crusting and surface sealing, increasing surface storage of water runoff, decreasing the runoff velocity, im-

TABLE 12.1 Effects of Mulch Rate on Soil Water Conservation during Fallow and on Subsequent Sorghum Grain Yield at Bushland, Texas

Mulch rate, t/ha	Precipitation storage, cm	Sorghum yield, kg/ha
0	7.2 c	1780 c
1	9.9 b	2410 b
2	10.0 b	2600 b
4	11.6 b	2980 b
8	13.9 a	3680 a
12	14.7 a	3990 a

Figures followed by the same letter are statistically identical.
SOURCE: Unger, 1978.

TABLE 12.2 Increase in Available Soil Moisture (mm per 1-m-Deep Soil) due to Rice Straw Mulch

Soil management	Year of observation		
	1976	1977	1978
Intertillage	+6	−17+22	+33
Subsoiling	+7	−15	+10
Rice straw mulch	+61	+30	+46
Rainfall, mm	782	475	1000

SOURCE: Kubota, 1982.

proving soil structure and porosity, and improving the biological activity related to soil cover and its influence on porosity.

The effects of residue mulch in improving the infiltration rate and soil structure are validated by experimental data from many soils and agroecological regions. Use of residue mulch on the soil surface increases plant-available water reserves by decreasing cumulative evaporation on a short-term basis (Fig. 12.5). The beneficial effects of mulch in reducing cumulative evaporation are not substantial on a long-term basis (Bond and Willis, 1971). Consequently, mulched soil has more soil water reserves than unmulched soil for the time soil wetness is at the first and the second stages of evaporation (Fig. 12.5a). The usefulness of surface mulch in water conservation and in improving sorghum yield is demonstrated by the data in Table 12.1 from a semiarid part of the United States. Increasing the mulch rate increased the amount of plant-available water in the root zone and the sorghum grain yield (Unger, 1978). The data from Thailand in Table 12.2 show that the maximum available water capacity was measured in plots receiving rice straw mulch (Kubota et al., 1982). The data in Table 12.3 from Taiwan show that in comparison with clean cultivation the use of rice straw mulch increased the soil moisture content by 22 percent. In northern India, Prihar et al. (1979) reported significant and positively beneficial effects of mulch in soil water conservation. In Nigeria, Lal (1975, 1978) observed more soil water in the root zone of mulched than unmulched plots.

For the soils of the semiarid region in northern Nigeria, Lawes (1962) observed significantly high infiltration rates of mulch in comparison with plots

TABLE 12.3 Effects of Mulching Practices on Moisture Conservation under Pineapple Plantation in Taiwan

Treatment	Soil moisture content, % by weight
Complete rice straw mulch	17.7
Interrow rice straw mulch	15.5
Polyethylene sheet	15.0
Clean cultivation	14.5

SOURCE: Taiwan Agricultural Research Institute, 1983.

TABLE 12.4 Effects of Mulch and Mechanical Treatments on Infiltration Rate and Preventing Runoff in Northern Nigeria

Soil surface treatment	Percentage of total rainfall penetration	Maximum infiltration rate, cm/h
Bare soil		
Undisturbed	31	1.3
Hoed at fortnightly intervals	49	2.3
Hoed after every storm	57	3.1
Mulches		
Dead grass	90	>12.5
Groundnut shells	89	>12.5
Sorghum stalks	98	>12.5

SOURCE: Lawes, 1962.

where the crust was broken by mechanical means. The data in Table 12.4 show that the maximum infiltration rate and the percentage of total rainfall penetrating the soil surface were the highest for mulched treatments. On alfisols developed on basement complex rocks in southern Nigeria, Lal (1976c,d) observed a high infiltration rate of mulched compared with unmulched plots. Furthermore, the infiltration rate normally increases with the increasing mulch rate. Both sorptivity and transmissivity increase with an increasing mulch rate, as evidenced by the data in Fig. 12.6 and Table 12.5.

In principle, therefore, the infiltration rate is governed more by the proportion of surface covered by mulch than by the mulch rate. It is presumed, however, that the greater the quantity of mulch material, the greater the proportion of soil surface covered (Fig. 12.7). Similar to the data from North America

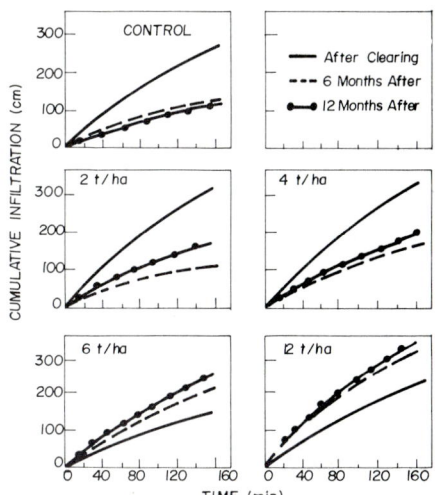

Figure 12.6 Cumulative water infiltration into a tropical alfisol as affected by mulch rate (Lal et al., 1980).

TABLE 12.5 Effects of Mulch Rate on Sorptivity and Transmissivity at Different Times after Deforestation

Mulch rate, t/ha	Time after clearing, months								
	0			6			12		
	A	S	r	A	S	r	A	S	r
0	0.88	13.2	0.99*	0.33	7.27	0.99*	0.32	5.56	0.98*
2	0.91	18.0	0.99*	0.16	9.27	0.99*	0.57	7.81	0.95*
4	1.30	13.2	0.90*	0.57	7.03	0.99*	0.67	7.50	0.94*
6	0.71	3.7	0.99*	0.60	12.25	0.98*	0.84	10.21	0.97*
12	0.91	8.2	0.99*	0.77	19.24	0.98*	1.05	15.36	0.97*

A = transmissivity, S = sorptivity, and r = correlation coefficient.
*Significant at the 0.01 level.
SOURCE: Lal et al., 1980.

shown in Fig. 12.7, Barber and Thomas (1981) reported from Kenya that the percentage of ground cover increases exponentially with an increasing rate of surface-applied mulch (Fig. 12.8).

The protective effect of mulch in reducing runoff and soil erosion is generally related to the quantity of mulch or the mulch rate. For permeable or well-drained soils, both runoff and erosion generally decrease exponentially with an increasing mulch rate up to a maximum ground cover. The data in Table 12.6 from a tropical alfisol in southern Nigeria show that the mean annual runoff losses were 393, 81, 30, and 13 mm for mulch rates of 0, 2, 4, and 6 t/ha, respectively. The corresponding effect on soil loss followed a similar trend for different mulch rates (Fig. 12.9a and b). For a soil of low permeability, however, the

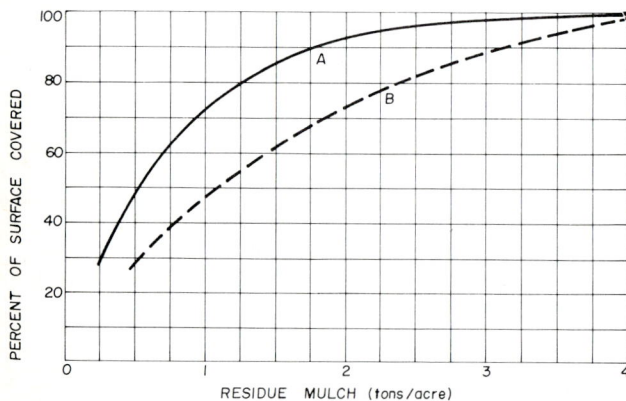

Figure 12.7 Relation of percentage of surface covered to mulch rates. The percentage of surface covered depends on whether the residue is standing or is chopped. Curve A is for small-grain straw, and curve B is for chopped corn stalk (Wischmeier, 1973).

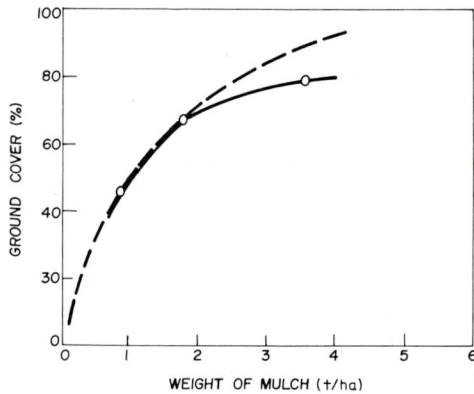

Figure 12.8 Relationship between percentage of ground cover and dry weight of uniformly distributed grass mulch. Dashed and solid lines show the projected and actual values, respectively, of ground cover. (Barber and Thomas, 1981.)

mulch rate may have no significant effect on water runoff. Soil erosion may nevertheless decrease with an increasing mulch rate even for a slowly permeable soil. Experiments conducted in Tanzania, east Africa, indicated that residue mulch applied at 6 t/ha effectively controlled runoff and erosion even on steep slopes of up to 22 percent (Tables 12.7 and 12.8). Other examples of the usefulness of crop residue and organic materials as mulch for erosion control are seen in the results of experiments conducted in the Ivory Coast, west Africa, by Roose (1977, 1980) and in Malawi, east Africa, by Shaxson (1975). The beneficial effects of mulch in controlling wind and water erosion also have been validated in India by Misra and Bhattacharya (1963), Rama Mohan Rao (1973), and Manipura (1972) and in Chile by Pena (1981). In Thailand, application of crop residue mulch decreased water runoff from 19.0 to 2.6 mm per 30 min, decreased the runoff ratio from 63.3 percent of total rainfall to 8.6 percent, and decreased soil erosion from 852.3 to 5.8 g/m^2 (Anonymous, 1986).*

---

*The data from Queensland, Australia, for vertisol showed that stubble mulch left on the surface reduced runoff by only 39 percent but soil movement by as much as 92 percent (Fig. 12.10). The data showing 6-year average effects on runoff and soil loss in relation to residue management showed similar trends (Fig. 12.11).

TABLE 12.6 Effect of Mulch Rate and Slope Steepness on Runoff (mm)

Slope, %	Mulch rate, t/ha			
	0	2	4	6
1	411.7	36.2	6.7	0.0
5	483.0	126.1	28.3	10.7
10	302.9	73.8	34.7	21.1
15	374.7	86.8	50.6	19.9
Mean	393.1	80.7	30.1	12.9

SOURCE: Adapted from Lal, 1976c.

366 Erosion Control

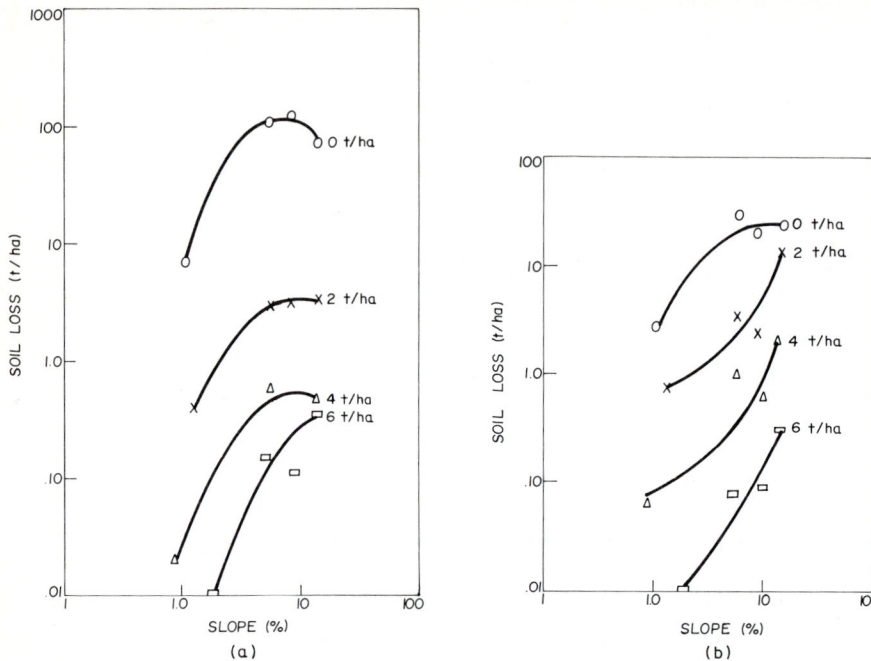

Figure 12.9 Effects of mulch rate and slope steepness on soil erosion of a tropical alfisol at IITA, Ibadan, Nigeria. (*a*) Data of first season 1974, (*b*) data of first season 1975. (Lal, 1976.)

Protective effects of mulch in reducing runoff and erosion have been reported for different soils in North America (Mannering and Meyer, 1961; Swanson et al., 1965; Barnett et al., 1967; SCSA, 1979; Lindstrom et al., 1979; Laflen and Colvin, 1981). In North Dakota, Gilley (1980) observed that application of straw mulch on mined sites reduced erosion by 66 percent over bare soil conditions. At Coshocton, Ohio, Harrold (1972) measured erosion from a rainulator experiment used on runoff plots. He observed soil erosion of 62 t/ha with no mulch, 20 t/ha for 0.56 t/ha of mulch, 19.5 t/ha for 1.1 t/ha of mulch, 11.4 t/ha for 2.24 t/ha of mulch, 2.5 t/ha for 4.5 t/ha of mulch, and 1.6 t/ha of soil loss for 9 t/ha of mulch.

The effects of mulch rate on runoff and soil erosion are usually expressed in the form of the *mulch factor*—the ratio of soil and/or water loss with mulch to a corresponding loss with no mulch (Wischmeier, 1973). For permeable and well-

TABLE 12.7 Effects of Mulching on Runoff and Soil Erosion in Zanzibar, Tanzania

Observations	Unmulched	Mulched
Runoff, mm	108.6	6.9
Runoff, % of rainfall	10.2	0.01
Soil erosion, t/ha	5.5	0.2
Nutrient loss in runoff, kg/ha	28.0	2.0

SOURCE: Khatibu et al., 1984.

TABLE 12.8 Mulching Effects on Runoff and Soil Erosion at Tanga, Tanzania

Season	Slope, %	Runoff, mm		Soil erosion, t/ha	
		Bare fallow	Mulched	Bare fallow	Mulched
Vuli	10	9.6	0.7	10.1	0.04
	19	10.8	1.2	26.1	0.06
	22	15.1	0.8	21.2	0.08
Masika	10	14.0	2.1	27.8	0.08
	19	48.9	1.7	66.7	0.11
	22	40.2	1.1	62.9	0.10
Annual total	10	23.6	2.8	37.9	0.12
	19	59.7	2.9	92.8	0.17
	22	55.3	1.9	88.1	0.18

SOURCE: Ngatunga et al., 1984.

drained soils, the mulch factor generally decreases exponentially with an increasing mulch rate. The data in Fig. 12.12 from field runoff plots under natural rainfall conditions in southern Nigeria show a logarithmic decline in the mulch factor with an increasing mulch rate. Similar results were obtained under laboratory conditions by using a rainulator for some soils in Nigeria (Fig. 12.13) and Kenya (Barber and Thomas, 1981).

Cover crops. Appropriate cover crops, usually low-growing legumes, are grown in rotation with food crop annuals specifically to procure mulch in situ. Once established, these covers effectively control runoff and soil erosion. Cover crops are used widely in both tropical and temperate climates. Some cover crops used for soil and water conservation and fertility maintenance are listed in Table 12.9. A wide range of cover crops are available for different regions, and lists of recommended covers have been prepared for different regions, such as that compiled for Hawaii by Evans et al. (1983). What constitutes a suitable cover crop differs according to soil, rainfall amount and distribution, ecological regions, farming systems, and other factors. An appropriate cover crop is the

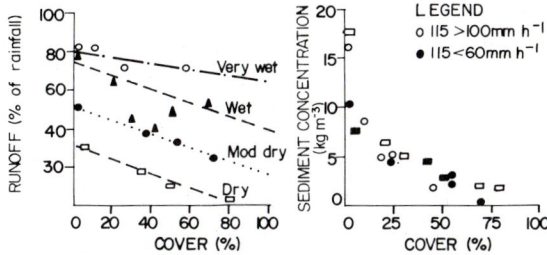

Figure 12.10 Effect of surface cover on runoff and sediment concentration (Queensland Dept. of Primary Industry, 1984–1985).

368 Erosion Control

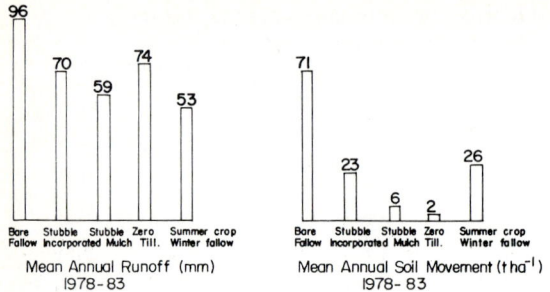

Figure 12.11 Annual runoff and soil loss at Greenmount for the 6 years 1978–1983 (Freebairn and Wockner, 1983).

one that is easily established, provides a quick ground cover, and effectively eliminates other weeds and undesirable vegetation. Once the cover is established, soil erosion is usually minimal.

The effectiveness of a cover in controlling runoff and erosion varies for different crops depending on the canopy cover, biomass produced, etc. The data in Table 12.10, for example, show that *Calopogonium* was the most effective cover in reducing erosion in Malaysia. *Calopogonium* may not, however, be suitable for other soil and climatic conditions. In southwestern Nigeria, *Mucuna* has proved the most effective cover. The effects of growing mucuna cover on runoff and soil erosion for an alfisol at IITA, Ibadan, Nigeria, are shown in Fig. 12.14 and Table 12.11. Mucuna is an aggressive cover crop (Plate 12.6), and once established, it effectively reduces runoff and soil erosion. Ripley et al. (1961) observed from a Canadian soil that water runoff and soil erosion from plots growing alfalfa and timothy cover were significantly less than from maize or oat crops (Table 12.12).

Although soil erosion is controlled by the prevention of raindrop impact through the protective effect of the cover canopy, runoff losses from the land

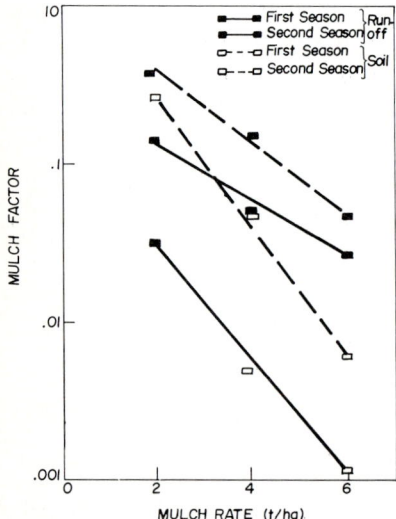

Figure 12.12 Mulch factor for runoff and soil erosion using field runoff plots and natural rainfall conditions in southern Nigeria (Lal, 1976).

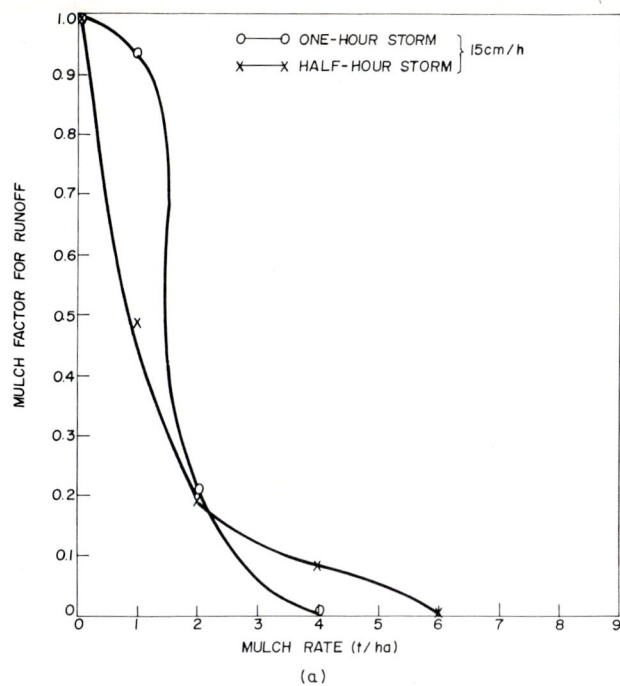

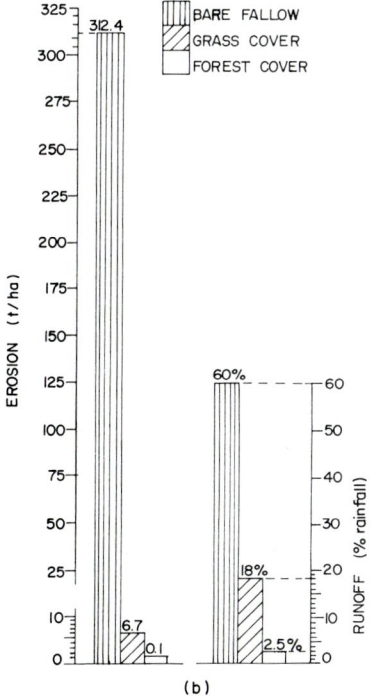

Figure 12.13 (a) Effect of mulch rate on mulch factor for runoff with simulated rainstorm under laboratory conditions (Okigbo and Lal, 1977). (b) Results of experiment by Barber and Thomas (1981) in Kenya.

TABLE 12.9 Commonly Used Cover Crops for Soil and Water Conservation, for Improving Soil Fertility, and for Green Manuring

Common name	Scientific name
Tropics	
Grasses	
Carpet grass	*Axonopus compressus*
Buffel	*Cenchrus ciliaris*
Molasses	*Melinis minutiflora*
Sabi	*Urochloa masambicensis*
Elephant	*Pennisetum purpureum*
Pangola	*Digitaria decumbens*
Guinea	*Panicum maximum*
Rhodes	*Chloria gayana*
Gamba grass	*Andropogon gayanus*
Bahia	*Paspalpum notatum*
Congo	*Brachiaria ruziziensis*
Bermuda	*Cynodon dactylon*
Giant cynodon	*C. nlemfuensis*
Legumes	
Pigeon pea	*Cajanus cajan*
Huban clover	*Arachis prostrata*
Phasey bean	*Phaseolus lathyroides*
Spanish clover	*Desmodium ucinatum*
Saladin lucerne	*Medicago sativa*
San hemp	*Crotalaria juncea*
Kudzu	*Pueraria phaseoloides*
Centro	*Centrosema pubescens*
Calopo	*Calopogonium mucunoides*
Stylo	*Stylosanthes gracilis*
Velvet bean	*Stizolobium deeringianum*
Townsville stylo	*Stylosanthes humilis*
Mucuna	*Mucuna utilis*
Glycine (perennial soybean)	*Glycine wightii*
Temperate	
Grasses	
Timothy	*Phleum pratense*
Orchard	*Dactylis glomerata*
Legumes	
Alfalfa	*Medicago sativa*
Arrowleaf clover	*Trifolium vesiculosum*
Big flower vetch	*Vicia grandflora*
Common vetch	*Vicia sativa*
Hairy vetch	*Vicia villosa*
Crimson clover	*Trifolium incarnatum*
Ladino clover	*Trifolium repens*
Lupines	*Lupinus alba*
Red clover	*Trifolium pratense*
White clover	*Trifolium repens*
Tall fescue	*Festuca arundinacea*

TABLE 12.10 Comparative Effectiveness of Different Cover Crops in Controlling Runoff and Erosion in Malaysia

Cover crop	Soil deposited on terraces over 20-month period, cm
Calopogonium	5.64
Pueraria	11.07
Crotalaria	15.69
Tephrosia	14.02
Grasses	12.67
Bare	19.04

SOURCE: Rubber Research Institute of Malaysia, 1977.

growing a cover crop are low because of the high infiltration rate. Growing grass and leguminous covers is known to improve the soil structure and water infiltration rate (Pereira et al., 1954, 1958; Wilkinson, 1975). In northern Australia, Bridge et al. (1983) reported that *Stylosanthes* improved infiltration and macroporosity.* Lal et al. (1978, 1979) in Nigeria observed significant improvements in the infiltration rate of an eroded and compacted alfisol by growing grass and leguminous covers (Fig. 12.15). The data in Fig. 12.16 are used to

---

*In Chinchina, Colombia, De Castro (quoted by Primavesi, 1982) reported that grass cover was almost as effective in reducing runoff and soil erosion as the undisturbed forest cover. Because of its dense cover close to the ground surface, grass cover can be more effective in erosion control than forest with a high canopy.

Plate 12.6 Mucuna cover, once established, provides an effective protection against runoff and erosion.

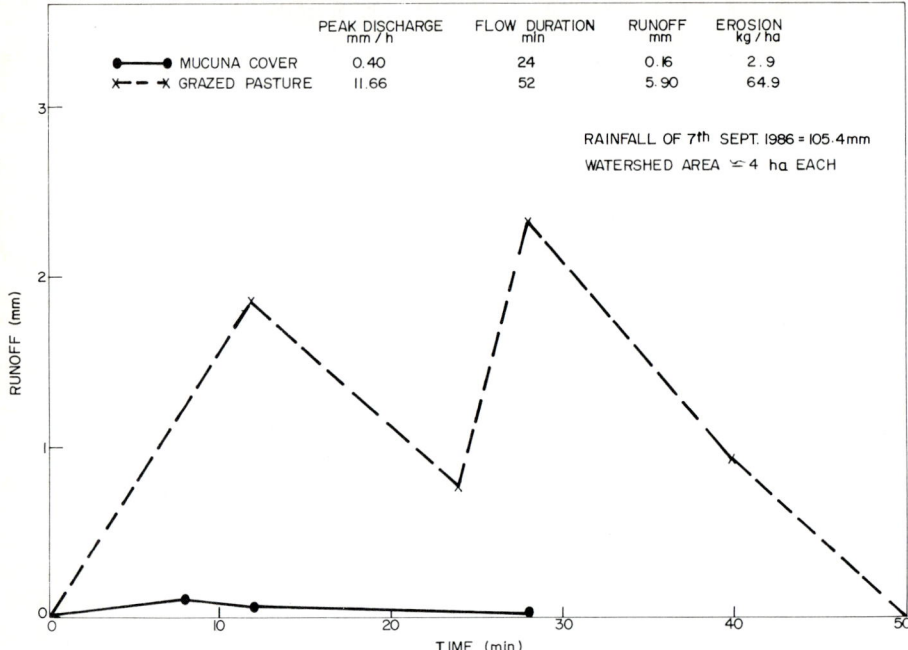

Figure 12.14 Runoff and soil erosion from watersheds planted to *Mucuna utilis* compared with the same measurements of grazed pastures.

compare the water infiltration rate of two watersheds cleared in 1979. Both watersheds were used for growing maize-cowpea in an annual rotation cycle. In 1986 one of the watersheds was sown to mucuna while the other was maintained under maize-cowpea rotation. The cumulative infiltration under mucuna cover was about double the amount under continuous maize-cowpea rotation. Similar to the beneficial effects of fallowing with grass or leguminous crops' soil structure and infiltration rate observed in east and west Africa, Touchton et al. (1982) observed significant improvements in the infiltration rate of a Typic Paleudult near Alabama by growing clover and vetch (Fig. 12.17). Because of their widely recognized beneficial effects on the soil structure

TABLE 12.11 Effect of Cover by *Mucuna utilis* on Runoff and Soil Erosion on 3- to 4-ha Watersheds at IITA, Ibadan, Nigeria (Rainfall of 29 August 1982)

Parameter	With mucuna cover	Without mucuna cover
Runoff, mm	4.04	10.01
Erosion, kg/ha	76	560
Duration of surface flow, h	1.30	2.12
Sediment density, g/L	1.88	5.6

TABLE 12.12 **Effect of Timothy Grass and Alfalfa Legume on Runoff and Soil Erosion from a Canadian Soil**

Crop/cover	Runoff, mm/yr	Soil erosion, t/(ha · yr)
Bare summer fallow	194.4	228.1
Maize (planted up and down the hill)	267.9	264.2
Maize (planted on contour)	55.6	55.5
Oats (18 cm apart)	25.5	15.0
Alfalfa	6.2	0.2
Timothy	13.6	0.7

SOURCE: Modified from Ripley et al., 1961.

and infiltration rate, planted fallows are increasingly being used as an integral component of improved rotations for seasonal food crop production.

A fallow management system of growing grain or food crops through the unkilled mulch is termed *live mulch*. Although crop yields can be suppressed by aggressive and climbing legumes (Plate 12.7; Wilson et al., 1982), maintenance

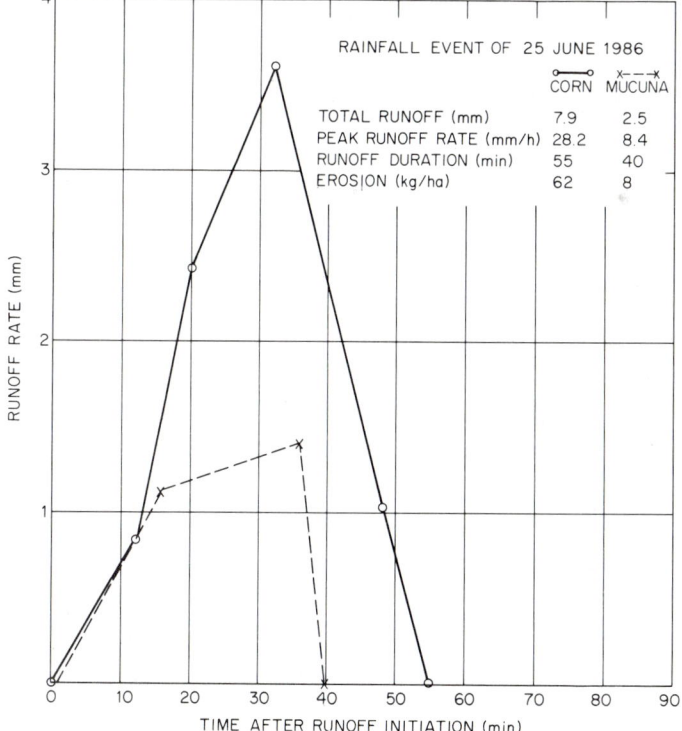

Figure 12.15 Effects of *Mucuna* cover on runoff and soil erosion compared with maize-cowpea rotation grown with a no-till system.

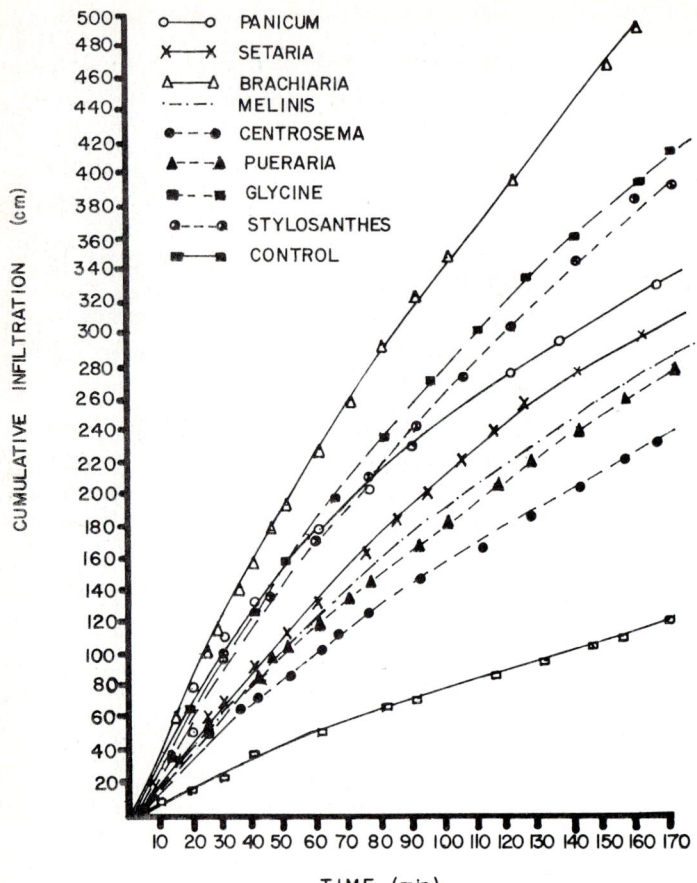

Figure 12.16 Effects of different cover crops on infiltration rate of an eroded tropical alfisol (Lal et al., 1979).

of a continuous ground cover has definite beneficial effects in controlling water runoff and soil erosion. The live mulch system may, however, be suited more for humid than arid or semiarid environments, ecological regions where soil moisture is adequate. Under humid conditions, live mulch may not compete aggressively with the grain crop grown through it. The data from Table 12.13 show no differences in moisture reserves among cover crops and residue mulch for the humid climate in Taiwan. For subhumid and semiarid regions, however, aggressive cover crops can deplete soil moisture reserves up to about 1-m depth (Fig. 12.18). The live mulch system is also more suited for perennial than seasonal crops, e.g., growing a legume cover underneath oil palm (Plate 12.8), rubber (Plate 12.9), or citrus plantation. The live mulch system is gaining momentum for growing some perennial crops in the United States. The disadvantage, if any, of the live mulch system lies in higher incidence of diseases and insects. A lea farming system based on the no-tillage technique has been developed for northern Australia to enable growing of food crops through lightly grazed cover crops (McCown et al., 1985).

Agroforestry systems. Growing food crop annuals in association with perennials is a useful system to minimize the risk of soil erosion (see Plate 12.4). If improperly managed, however, any system is vulnerable to accelerated erosion, and forestry and perennial crops are no exception. In fact, severe erosion and soil degradation are observed in poorly managed plantations and from logging roads in natural and planned forests (see Chap. 13). Given the same level of management and inputs, however, the risks of soil erosion are greater from land cultivated to seasonal crops than to perennials, plantations, or agroforestry-based systems.

Agroforestry is, in a way, a special case of mixed cropping. It involves growing of deep-rooted perennial shrubs and trees with annuals in an orderly fashion so as to maximize the resource use and minimize risks of soil erosion. Perennial crops that bear fruit annually, such as the banana, are more suitable for intercropping with annuals than plantation crops. Shade-tolerant annuals can be

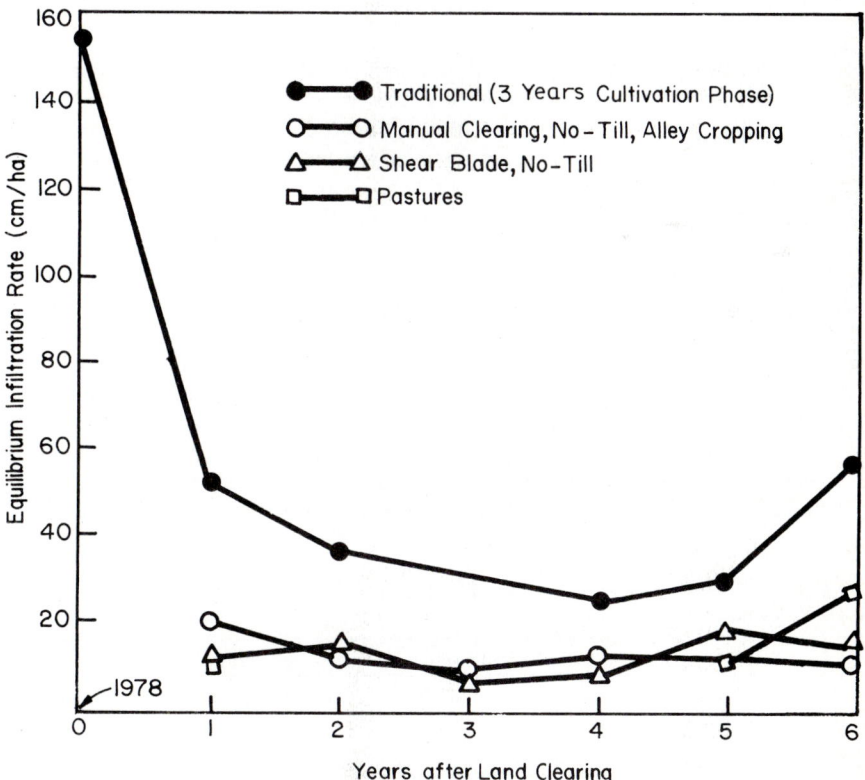

Figure 12.17 Effect of pastures and cultivation on infiltration rate of an alfisol.

Plate 12.7 A climbing legume cover, if not properly suppressed, can drastically reduce the yield of fallowing grain crops such as maize.

grown under the tall-canopy perennials. If perennial shrubs happen to be leguminous, they may also contribute to the nitrogen in the soil.

Alley cropping is one of many agroforestry systems. It involves growing food crop annuals between rows of specially planted woody shrubs and trees. The woody shrubs are regularly pruned during the cropping season to prevent shading. This practice is discussed at length in Chap. 13.

No-tillage farming. No-tillage farming involves seeding through the previous crop residue mulch in an untilled soil in which weeds are controlled by herbicides, cover crops, and mulch. With an adequate amount of residue mulch, the system has proved an effective erosion control practice for a wide range of soils and agroecological regions. Satisfactory crop yields are obtained if proper seeding equipment is available to obtain a good crop stand, weeds are adequately controlled, and soil is well drained. The system has been successfully used for row crops in the temperate zone, grain crops and root tubers in the tropics, and sugarcane in Australia and the Caribbean (Sprague and Triplett, 1986; Lal, 1989).

TABLE 12.13 Soil Moisture Conservation by Mulching and Growing Cover Crops in Taiwan

Soil depth, cm	Soil moisture content, % by weight					
	Centrosema	Indigofera	Bahia grass	Guinea grass	Rice straw mulch	Clean cultivation
15–20	9.3	9.8	8.6	8.4	10.2	8.3
25–30	10.2	11.2	9.6	10.0	11.3	9.9

SOURCE: Wang, 1984.

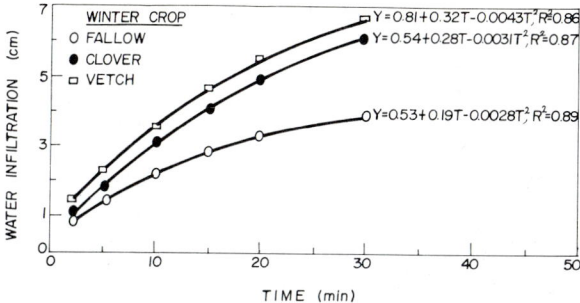

Figure 12.18 Water infiltration in no-tillage cotton after 2 years of double cropping with various winter cover crops (Touchton et al., 1982).

No-till farming decreases runoff and soil erosion by minimizing the raindrop impact and increasing the water infiltration rate. A no-till system also reduces the amount and rate of runoff, decreases the runoff velocity, improves soil structure, and increases macroporosity by improving the organic matter content of the surface layer and the activity of soil fauna. Rosewall and Marston (1978) reported from Australia that plowing destroyed soil structure built by 100 years of grassland leas. In Indonesia, Suwardjo and Sutono (1984) observed that tillage decreased aggregate stability and caused degradation of soil physical properties.

The utility of the no-till system in controlling runoff and reducing soil erosion by both water and wind depends on the quantity of mulch available. Crop residue is the most effective in erosion control when it is kept on the soil surface

Plate 12.8 A live mulch of *Pueraria phaseoloides* provides a good cover within the open space of young oil palm seedlings.

Plate 12.9 Soils in rubber plantations of Asia and west Africa are protected from erosion by growing of leguminous cover crop between the trees.

rather than burned or incorporated. The quantity of water stored in the root zone also depends on the quantity of residue on the soil surface (see Chap. 9). Different tillage implements used with no-till techniques permit a variable quantity of crop residue on the surface. Moldenhauer et al. (1983) reported that in the central and northern Great Plains the water loss from the soil surface is often in the order of the following tillage tools: one-way disk > chisel > sweep plow > sod weeder with semichisels. The greater the residue, the greater the available water.

In the Ivory Coast, west Africa, Roose and Asseline (1978) observed that burning or plowing under the pineapple residue drastically increased erosion in comparison with maintaining the residue as mulch. In São Paulo, Brazil, Bertoni and Lombardi (1985) observed the maximum runoff and erosion from plots in which crop residue was burned (Table 12.14). In Queensland, Australia, Prove and Truog (1984) observed that erosion losses in cane fields were the most severe when the cane trash was burned. Many experiments conducted in

TABLE 12.14 Effect of Sorghum Residue Management on Runoff and Soil Erosion

Residue management	Soil erosion, t/ha	Water runoff, %
Burning	20.2	8.0
Incorporation	13.8	5.8
Mulch	6.5	2.5

SOURCE: Bertoni and Lombardi, 1985.

TABLE 12.15 **Effects of Cropping Practice on Relative Erosion at Gunnedah Research Centre, New South Wales, Australia**

Treatment	Relative soil hold
Wheat, long fallow stubble burned	100
Annual wheat stubble burned	40
Annual wheat stubble incorporated	14
Permanent pasture	1

SOURCE: Rosewall and Marston, 1978.

Queensland and New South Wales have conclusively demonstrated beneficial effects of residue retention in erosion control (Marston and Doyle, 1978; Freebairn and Wockner, 1984–1985; Queensland Department of Primary Industries, 1984–1985; Sallaway et al., 1984). In New South Wales, Australia, Rosewall and Marston (1978) observed that stubble burning increased erosion by 3 to 7 times in comparison with stubble incorporation (Table 12.15). Burning of residue was also shown to cause more severe erosion than incorporation or mulching in Darling Downs, Queensland, Australia (Table 12.16). The effectiveness of a no-till system, therefore, lies in maintaining the residue mulch on the soil surface.

Although less practical, the use of residue mulch after plowing also controls erosion. Mulching after plowing, however, is usually less effective in erosion control than mulching combined with a no-till system (Table 12.17). The effectiveness of mulch in erosion control also varies among soils. Mulching controls erosion more on well-drained or coarse-textured soils than slow-draining clayey soils (Table 12.18). For well-drained soils that have not already been compacted by previous tillage traffic and cultivation, the infiltration rate is often greater in no-till than in plowed plots. The plowed soil is more susceptible to surface sealing and crusting than the no-till soil. The impacting raindrops from an intense rain break up unprotected structural aggregates. Consequently, the surface soil develops a semipermeable crust that impedes water and air movement. The data in Fig. 12.19 show that cumulative infiltration into an alfisol at

TABLE 12.16 **Effects of Residue Management on Runoff and Soil Erosion at Darling Downs, Queensland, Australia**

Treatment	Runoff, mm	Erosion, t/ha	Wheat grain yield, t/ha
Stubble burned	96	71	2.82
Stubble incorporated	70	23	2.82
Stubble mulched	59	6	2.95
No-tillage	74	2	2.76
Summer crop	53	26	

SOURCE: Adapted from Freebairn and Wockner, 1984.

TABLE 12.17 Effects of Tillage Methods and Residue Management on Runoff and Soil Erosion from Vertisols at Two Sites in Queensland, Australia

Treatment	Runoff, % of rainfall	Erosion, t/(ha · yr)
Bare fallow	33	105.3
Conventional tillage	22	35.1
Stubble mulch	3	1.2
No-tillage	3	0.9

Values represent mean of 3 years of measurements from 1981 to 1983.
SOURCE: Queensland Department of Primary Industries, 1984–1985.

Ibadan was greater for no-till than for plowed seedbed. Similar results have been reported for some soils from Brazil. The data in Table 12.19 show that the infiltration rate in no-till soil was as good as or better than that in pasture. The effects of no-tillage on the infiltration rate, however, depend on the quantity and distribution of mulch. The infiltration rate in a no-till system can decline rapidly if the mulch material is not available (Fig. 12.20). Similar observations regarding the effects of no-tillage on infiltration and erosion have been reported from Brazil by Derpsch et al. (1986). Their data, for an oxisol from Londrina in Parana State, showed that the infiltration rate was in the order of no-tillage > chisel plow > conventional tillage. These conclusions are also supported by the examples discussed below from different soils and agroecological regions.

For alfisols in Nigeria, Lal (1976a, b, c, d, 1984) reported significant reductions in soil erosion by the no-tillage system compared with the plow-based system. The data in Table 12.20 show that runoff and soil erosion from maize plots were drastically reduced by no-till compared with plowed treatment. In fact, the no-till system was as effective in controlling erosion as the plot that received 6 t/ha of crop residue mulch. Similar results were obtained from twin watersheds, about 5 ha each, managed with no-till and plowed systems of seedbed preparation (Fig. 12.21). In Ghana, Baffoe-Bonnie and Quansah (1975) observed that both runoff and soil erosion increased with an increasing intensity

TABLE 12.18 Effects of Residue Management on Erosion from Different Soil Types in Southeast Queensland, Australia

	Erosion, t/(ha · yr)		
Soil type	Bare fallow	Stubble incorporation	Stubble mulch contour cultivation
Alluvial fertile, self-mulching, 1–2% slope	60	24	12
Colluvial clay, red-brown, 1 m deep, 1–3% slope	76	31	12
Colluvial clay, red-brown, 1 m deep, 5–8% slope	270	110	10

SOURCE: Cummings et al., 1973; quoted by Marston and Doyle, 1978.

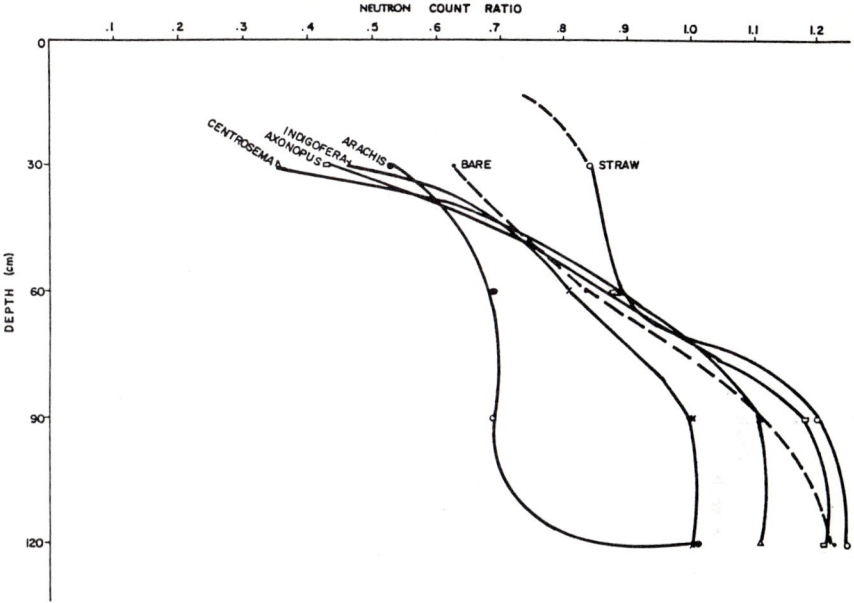

Figure 12.19 Soil moisture reserves under different cover crops, bare soil, and crop residue mulch (Lal, 1986).

of tillage operations. Also in Ghana, Bonsu and Obeng (1979) observed that the no-till system was as effective in controlling erosion and runoff as contour ridges made across the slope (Table 12.21). Because no-till treatments had less mulch, erosion was better controlled with mulched plots than with no-till treatments.

TABLE 12.19 Effects of Tillage Methods on Water Infiltration into Soil

Time, h	Forestral, mm/h	Pasture, mm/h	No-tillage, mm/h	Conventional tillage, mm/h
1	136.8	96.1	113.1	48.0
2	92.9	66.3	78.9	33.0
3	82.6	63.0	74.5	31.5
4	82.0	52.7	62.7	25.5
5	77.0	51.8	61.0	24.0
6	75.0	46.7	54.8	23.0
7	73.0	44.2	51.5	22.0
8	73.0	42.5	50.4	21.0
9	72.3	41.6	49.5	20.5
Mean	84.9	56.1	66.3	27.6

Each value is mean of six replicates.
SOURCE: Machado, 1976.

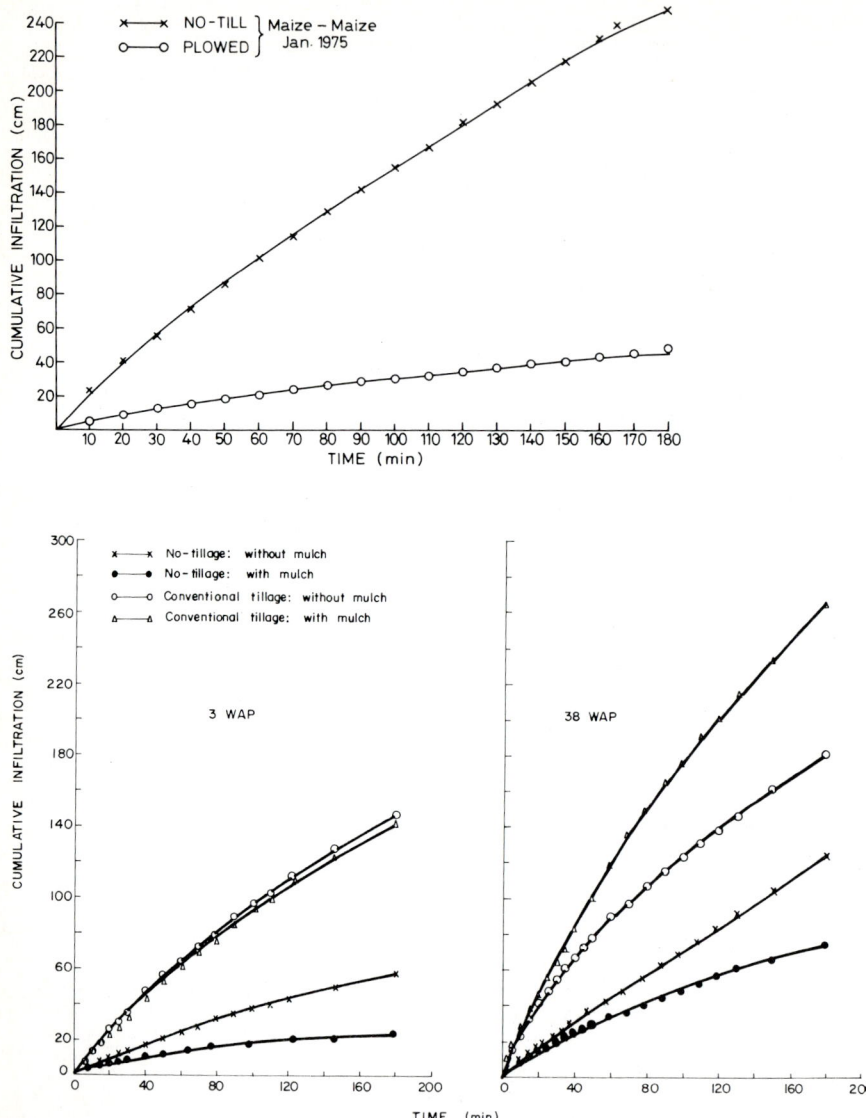

Figure 12.20 Effects of tillage systems on accumulative water infiltration into a tropical alfisol. WAP = weeks after planting. (Lal, 1976.)

The no-till system has also proved effective for erosion control in tropical America. Many studies in Brazil using a rainfall simulator have shown definite advantages of the no-till system for erosion control in a wheat-soybean rotation (Second National Soil Erosion Conference, 1978; Primavesi, 1982). In Parana State, Mondardo et al. (1977) and Sidiras et al. (1982) reported that, in comparison with direct drilling and no-tillage systems, plowing increased erosion and

TABLE 12.20 Soil Erosion and Runoff from Maize Grown on a Tropical Alfisol with No-Tillage and Plowed Method of Seedbed Preparation, First Season 1973*

Slope, %	Runoff, mm				Soil erosion, t/ha			
	Bare fallow	Mulch	No-till	Plowed	Bare fallow	Mulch	No-till	Plowed
1	315.7	0.0	11.4	55.1	7.5	0.0	0.0	1.2
5	347.3	6.9	11.8	158.7	80.4	0.0	0.2	8.2
10	311.0	20.3	20.3	52.4	152.9	0.1	0.1	4.4
15	316.5	16.8	21.0	89.9	155.3	0.0	0.1	23.6
Mean	322.6	11.0	16.1	89.0	99.0	0.025	0.1	9.4

*Data based on field runoff plots of 25 × 4 m.
SOURCE: Lal, 1976a, b.

decreased aggregate stability and water-holding capacity. In Ceara, Brazil, Oliveira and Silva (1982) observed degradation of soil structure and formation of surface sealing crust on plowed red-yellow podzolic soil. In Rio Grande Do Sul, southern Brazil, Wunsche and Denardin (1980) observed that soil losses are decreased by incorporation of straw rather than burning and by using the

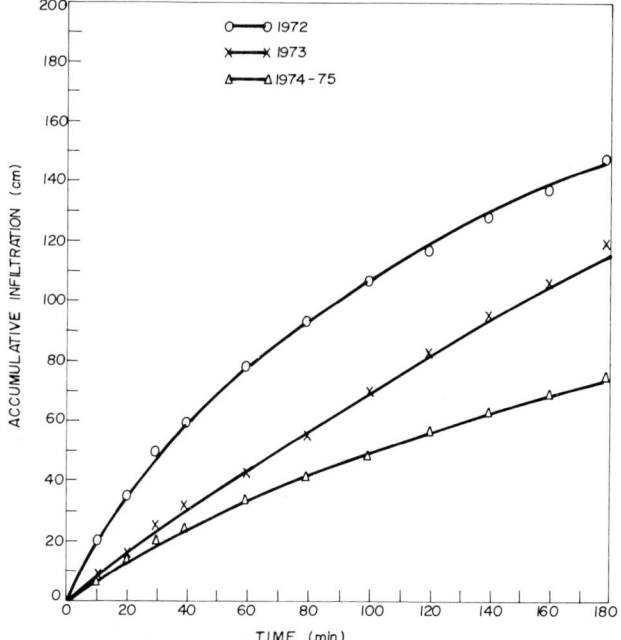

Figure 12.21 Effect of lack of crop residue mulch on decline in cumulative infiltration with cultivation in a no-till system using a tropical alfisol (Okigbo and Lal, 1977).

TABLE 12.21 Effects of Tillage Methods on Soil Erosion and Water Runoff from Two Sites in Ghana

	Mean of 1974 to 1976			
	Runoff, % of rainfall		Erosion, t/(ha · yr)	
Treatment	Kwadaso*	Ejura	Kwadaso	Ejura
Bare fallow	43.9	36.7	200.1	18.8
No-tillage	6.7	3.7	3.8	6.4
Mulch	2.1	0.8	0.4	1.3
Ridges (across slope)	7.4	3.1	3.6	3.4
Minimum tillage	8.8	4.8	3.2	2.8
Mixed cropping	13.5	4.5	18.8	2.1

*Kwadaso: 7.5% slope, forest ochrosol; Ejura: 3% slope, forest ochrosol.
SOURCE: Modified from Bonsu and Obeng, 1979.

straw as mulch through direct drilling into the stubbles rather than plowing it under. Barker and Wunsche (1977) observed significantly low soil erosion under no-tillage and minimum tillage than under conventional tillage systems (Table 12.22). Also in Rio Granda Do Sul, Vieira et al. (1978), using simulated rainfall, compared the effects of four tillage systems for soil losses under soybean. The no-tillage system reduced erosion by 75 percent of that from plowed treatments. On a latosol on a slope of 6.3 percent in São Paulo, Bernatti et al. (1977) observed that the no-tillage system decreased soil losses by 20 percent. On a podzolized soil with a 10.8 percent slope, no-tillage reduced soil losses by 63 percent. The no-till system has been widely adapted in wheat-soybean or maize-soybean rotations in central and southern Brazil.

Soil erosion is a serious problem in northern and central Queensland, Australia, where sugarcane is grown on undulating to rolling lands and even on steeply sloping lands (Dawson et al., 1983). Mullins et al. (1984) used the universal soil loss equation (USLE) to predict the erosion hazard in these lands. The crop management factor C was estimated to be 0.242 to 0.342 for conven-

TABLE 12.22 Soil Erosion from a Brazilian Oxisol under Different Tillage Systems, as Measured with a Rainfall Simulator

	Soil erosion,* t/ha			
Rainfall intensity	Conventional tillage 1	Conventional tillage 2	Minimum tillage	No-tillage
(60 mm/h)/30 min	26.8	13.3	3.6	1.4
(120 mm/h)/18 min	46.5	19.7	6.7	4.1

*Conventional tillage 1: plowed once, disked twice, kept bare; conventional tillage 2: plowed once, disked twice, crop residue incorporated and sown to wheat; minimum tillage: cultivated once with duck-foot-tined cultivator, disked twice, previous residue incorporated, sown to wheat; no-tillage: no plowing or disking, previous crop residue left on surface.
SOURCE: Barker and Wunsche, 1977.

TABLE 12.23 Soil Erosion from Different Sites in Cane Fields in North, Central, and South Queensland, Australia

Treatment	Soil erosion in 1982–1983 for different sites, t/(ha · yr)						
	Babinda	Cowley	Liverpool Creek	Nerada	Palmerston	Pinnacle	Victoria Plains
CP	135	82	72	74	150	165	30
GCTB	2	4	−50	−17	NA	NA	9
GCTI	NA	NA	NA	NA	NA	NA	43
BCTB	22	−17	−10	5	8	24	NA
BCTI	NA	NA	NA	NA	NA	72	NA
No-till	16	−11	5	6	NA	NA	NA

NA = Not measured.
CP = Conventional practice.
GCTB = Green cane harvest, trash blanket, chemical weed control.
GCTI = Green cane harvest, trash incorporated, chemical weed control.
BCTB = Burned cane harvest, trash blanket, chemical weed control.
BCTI = Burned cane harvest, trash incorporated, chemical weed control.
No-till = Burned cane harvest, burned trash, no-till, chemical weed control.
Minus sign means soil deposition.
SOURCE: Bureau of Sugar Experiment Stations and Department of Primary Industries, 1984.

tional tillage and trash burning compared with 0.025 when cane residue was kept on the surface as mulch. The mean annual soil erosion in these two systems was estimated to be 58 and 6 t/ha, respectively. The Bureau of Sugar Experiment Stations conducted field experiments on slopes ranging from 3 to 16 percent on soils classified as red earth, krasnozems, red podzolic, and yellow

Plate 12.10 High runoff and accelerated erosion are severe problems during monsoon season for vertisols.

podzolic. The data in Table 12.23 show that soil erosion in conventional tillage and residue burning treatments ranged from 72 to 150 t/ha. In comparison, the soil erosion from no-tillage with residue mulching was reduced to 20 t/ha. In Guyana, where sugarcane is grown on clayey soils, mechanical soil manipulation causes breakdown of structure and soil compaction (Paul, 1982). These experiments lead one to conclude that soil erosion from sloping cane lands can be controlled effectively by management practices such as no-tillage and trash retention as mulch.

Vertisols—soils containing a high percentage of high-activity clays—occur widely in semiarid regions of India, Australia, and tropical Africa. Because of their poor trafficability when wet, these soils are often left unused and fallow during the rainy season when accelerated erosion becomes a severe problem (Plate 12.10). The no-tillage system is ideally suited for these hard-to-manage soils because it eliminates the need for vehicular traffic. The presence of crop residue mulch decreases splash by preventing raindrop impact and by reducing the heat of wetting (see Chap. 3). These conclusions are supported by field experiments conducted in Queensland, Australia. Freebairn and Wockner (1982) reported that the tillage systems that leave the greatest amount of residue intact on the surface were the most effective in reducing soil erosion. In another study, Freebairn et al. (1986) reported the results of experiments involving the use of different types of reduced-tillage systems and crop residue management on soil erosion. The data in Table 12.24 show that soil erosion was drastically reduced by no-tillage and stubble mulch systems in comparison with the bare fallow treatments. These results indicate that soil erosion values from a 70-mm rainfall recorded on May 2, 1983, at Greenmount, Queensland, were 57, 27, 12, and 5 t/ha for bare fallow, stubble incorporation, stubble mulch, and no-tillage

TABLE 12.24 Effects of Tillage Methods and Residue Management on Runoff and Soil Erosion from Vertisols at Two Sites in Queensland, Australia

Soil and rainfall conditions	Tillage and residue management systems			
	Bare fallow	Stubble incorporation	Stubble mulch	No-tillage
Greenmount: May 2, 1983, Rainfall = 70 mm				
Cover, %	2	20	25	57
Runoff, mm	56	57	50	52
Peak runoff rate, mm/h	36	37	24	28
Soil erosion, t/ha	57	27	12	5
Greenwood: Feb. 5, 1980, Rainfall = 63 mm				
Cover, %	5	25	55	56
Runoff, mm	29	22	7	<7
Peak runoff rate, mm/h	45	27	11	Not determined
Soil movement, t/ha	30	5	0.2	Negligible

SOURCE: Freebairn et al., 1986.

treatments, respectively. No-tillage and stubble mulch treatments were even more effective in erosion control at Greenwood. In addition to controlling erosion, the no-tillage treatment reduced the peak runoff rate (Figs. 12.22, 12.23, and 12.24). The no-till system, however, may not be as effective as the stubble mulch system, especially if the amount of mulch is inadequate and the soil is compacted by vehicular traffic. The following water balance for vertisols in this region

	Bare fallow	No-tillage
Runoff, %	17.0	10.0
Evaporation, %	65.0	66.0
Soil water, %	18.0	24.0

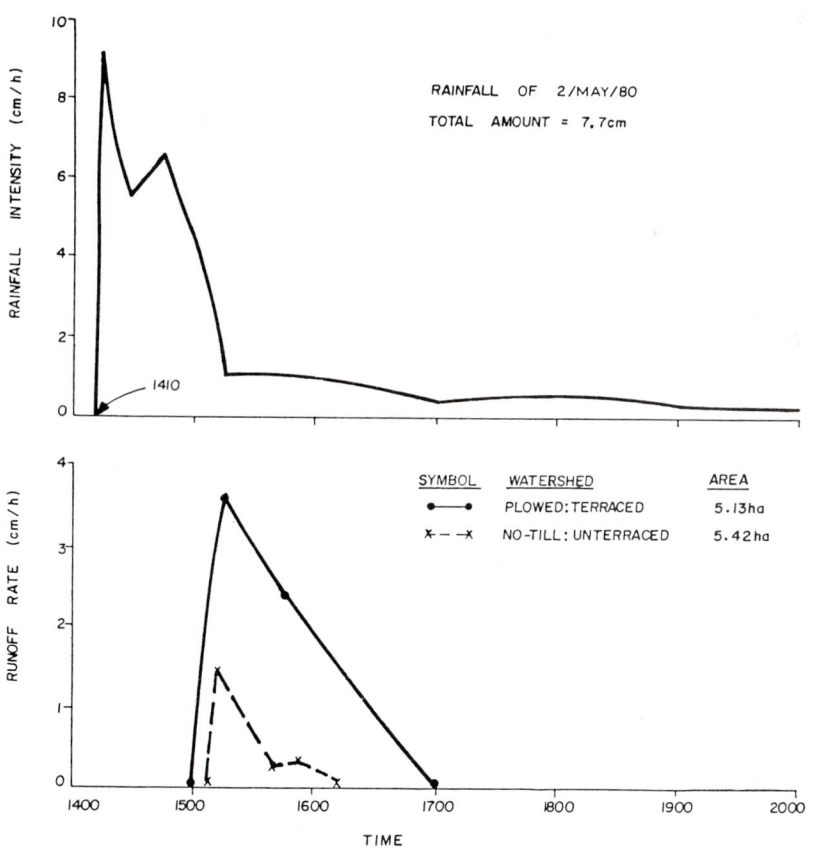

Figure 12.22 Runoff and soil erosion from twin watersheds managed with no-till and plow-based systems of seedbed preparation (Lal, 1984).

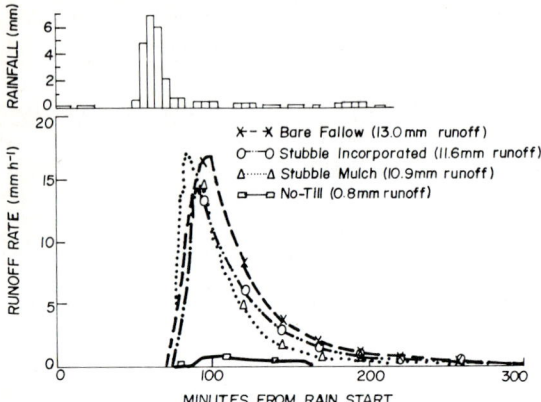

Figure 12.23 Hyetograph and hydrograph of the rainfall event on 3 December 1982 (total rainfall 24.5 mm) at Greenwood, showing responses to surface cover and tillage for 1-ha catchment on 5 to 7% slope (Freebairn et al., 1986).

indicates more water stored in the root zone for crop use with no-tillage and stubble mulch farming than with the bare fallow system of soil management. Sallaway et al. (1983) observed that runoff varied from zero for the no-tillage treatment to 46.4 mm for disk plowing treatment. Soil erosion also varied from zero for the no-tillage treatment to 13.3 t/ha for bare fallow planted to sorghum. Marston (1978) observed in New South Wales, Australia, that the erosion potential of conventional tillage systems generally increases with the removal of crop residue by burning or grazing and from an excessive fallow cultivation. For a reddish brown lateritic soil in Thailand, Kubota et al. (1982) also reported that stubble mulch left on the soil surface was more effective in reducing runoff and controlling erosion than subsoiling or intertillage.

The no-tillage system has been widely adapted in North America, especially in the corn belt region of the midwestern United States where soil erosion has been a severe problem for grain crop production on sloping lands. Watershed management experiments conducted at Coshocton, Ohio, have shown that the no-tillage treatment reduces soil erosion by several orders of magnitude. The data in Table 12.25 show that soil erosion from no-till watershed with contour-sown corn rows was less than 1 percent of that from plowed watershed, also with contour-sown corn rows, and only 0.1 percent of that from plowed watershed with sloping corn rows. Soil erosion from the no-tillage watershed was low in spite of the fact that the slope for the no-till watershed was 20.7 percent in comparison with slopes of 5.8 and 6.6 percent for plowed watersheds. Also in Coshocton, Ohio, Harrold (1972) compared soil erosion from 2-acre twin watersheds growing corn with conventional and no-tillage practices from 1964 to 1971. His data showed that whenever heavy rains caused a large amount of soil erosion from conventionally tilled watershed, there was no or much less erosion from the no-tillage corn. In Guelph, Ontario, Canada, Ketcheson (1977) observed that the effectiveness of the no-tillage system in reducing water runoff and soil erosion by summer rains depended on the presence of residue mulch.

TABLE 12.25 Runoff and Sediment Yield from Plowed and No-Tillage Watershed Growing Corn at Coshocton, Ohio

Treatment	Slope, %	Rainfall, cm	Runoff, cm	Sediment yield, t/ha
1. Plowed, clean-tilled, sloping rows	6.6	13.97	11.18	50.7
2. Plowed, clean-tilled, contour rows	5.8	13.97	5.84	7.2
3. No-tillage, contour rows	20.7	12.88	6.35	0.7

Rainfall event of July 5, 1969.
SOURCE: Harrold and Edwards, 1972.

Erosion was equally severe in no-till and plowed soils if the stover was removed (Table 12.26). For the winter rains and snow melt, however, no-till plots suffered less erosion than plowed treatments regardless of the mulch. The effects of various conservation tillage systems on runoff and soil erosion were investigated for a Typic Fragiudalfs or loessial origin, at Holy Spring, Miss., by McDowell and McGregor (1984). Their data (shown in Table 12.27) indicate that although water runoff was only slightly affected, soil erosion was drastically reduced by no-till and reduced tillage systems compared with conventional tillage systems.

Similar to the tillage × rainfall and tillage × soil interactions discussed in Chap. 9, there exist significant interactions in tillage × cropping systems in determining the erosion risks. Some relevant examples of such interactions can be seen in the data from the United States. The influence of the crop and cropping system depends on the amount, quality, and persistence of crop residue. Leguminous crops produce a low quantity of residue and have a low $C:N$ ratio. A soil after legume does not possess as much residue cover as that after cereals.

TABLE 12.26 Effects of Plowing and Maize Stover on Runoff and Soil Loss from Guelph Loam Soil (8% Slope) at Guelph, Ontario

	Mean annual loss			
	May to October, 9-yr period		November to April, 6-yr period	
Treatment	Runoff, cm	Soil, t/ha	Runoff, cm	Soil, t/ha
Stover left on field				
Not plowed	1.1	3.1	3.1	0.2
Fall-plowed	2.5	24.6	8.2	8.3
Stover removed				
Not plowed	3.7	38.1	10.6	7.6
Fall-plowed	2.5	40.3	8.3	21.8

SOURCE: Ketcheson, 1977.

TABLE 12.27 Mean Annual Runoff and Soil Erosion in Corn Grown on a Typic Fragiudalfs at Holy Spring, Mississippi*

Treatment	Runoff, mm/yr	Soil erosion, t/(ha · yr)
Conventional tillage, silage	408	24.1
Conventional tillage, grains	373	17.5
Reduced tillage, grains	206	1.4
No-tillage, silage	259	0.8
No-tillage, grains	328	0.8

*Data values are mean of 3 years, 1975 to 1977.
SOURCE: Modified from McDowell and McGregor, 1984.

Consequently, erosion after leguminous crops, such as soybeans, can be greater than that after corn. These principles are substantiated by the data in Tables 12.28 and 12.29. Erosion was greater after single-cropped soybeans than when soybeans were grown in rotation.

Similar to the experiences in North America, many experiments conducted in Europe have indicated the usefulness of the no-tillage system with residue mulch in reducing runoff and soil erosion. In Italy, Chisci and Zanchi (1980) observed lower runoff and soil loss from wheat grown with the reduced tillage system than with conventional tillage system. In Kiev, Ukraine, U.S.S.R., Shikula (1981) measured soil erosion and runoff of snow melt water on a podzolic loam soil from 1967 to 1979. The data in Table 12.30 show that the mean annual erosion caused by the snow melt was 13.3, 2.6, and 0.3 t/ha for moldboard plowing across slope, no-tillage with sowing along slope, and no-tillage with sowing across slope, respectively.

While reducing water losses by runoff and decreasing evaporation, the no-tillage system is likely to conserve more plant-available water reserves in the root zone than the plow-based system. High soil moisture reserves in the no-till system have been observed in west Africa by Lal (1976a, 1976b, 1982) and in Kentucky by Phillips et al. (1980). The data in Fig. 12.25 from Thailand also show more water available in no-till and mulch tillage systems than in a control treatment. The no-tillage systems improve the organic matter content of the topsoil layers (Lal, 1982) and thus influence soil moisture characteristics. The

TABLE 12.28 Crop × Tillage Interaction in Relation to Soil Erosion for a Morley Clay Loam on a 4% Slope

Tillage method	Percentage of cover		Soil erosion, t/ha, after:	
	Soybean	Corn	Soybean	Corn
No-till	26	69	13.4	2.4
Chisel (up and down plow)	12	25	30.3	15.0
Plow	1	7	40.0	21.8

SOURCE: Mannering and Fenster, 1977.

TABLE 12.29 Effects of Tillage and Cropping Systems on Soil Loss from a 63-mm Rainfall at Milan, Tenn., on June 11, 1981

Treatment	Soil loss, t/ha
Conventional tillage, single-cropped soybeans, no winter cover	62
Disk tillage, single-cropped soybeans, no winter cover	0.3
No-tillage, single-cropped soybeans	84
No-tillage, double-cropped wheat-soybeans	0.3

SOURCE: Moldenhauer et al., 1983.

data in Fig. 12.26 from Thailand are a relevant example of the beneficial effects of the no-till system on an increase in soil moisture retention, especially at low suctions. As a consequence of better soil moisture reserves, crops grown with a no-till system can withstand drought better than those grown with a plow-based system.

Examples from different ecological regions presented above show that the practice of no-tillage farming with crop residue mulch and seeding across the slope effectively reduces soil erosion compared with a plow-based system. The effects of the no-tillage method on reducing water runoff, however, are not as conclusive as those on soil erosion. Furthermore, the no-tillage system may not be applicable for all soils and crops. Specific modifications are required to conserve water and to reduce risks of soil erosion under those soil and climatic conditions where the no-tillage system is not applicable.

TABLE 12.30 Effects of Tillage Methods on Soil Erosion of a Podzolic Loam Soil at Kiev, Ukraine, U.S.S.R.

	Soil erosion, m³/ha		
	Moldboard plowing across slope	No-tillage, sowing along slope	No-tillage, sowing across slope
1967	2.5	3.2	1.8
1968	5.2	0	0
1969	0.9	0	0
1970	3.2	0	0
1971	32.8	11.9	0
1972	46.0	6.9	1.3
1973	21.7	5.5	0
1977	14.6	0	0
1978	5.5	0	0
1979	11.3	0	0.2
Mean	13.3	2.6	0.3

SOURCE: Shikula, 1981.

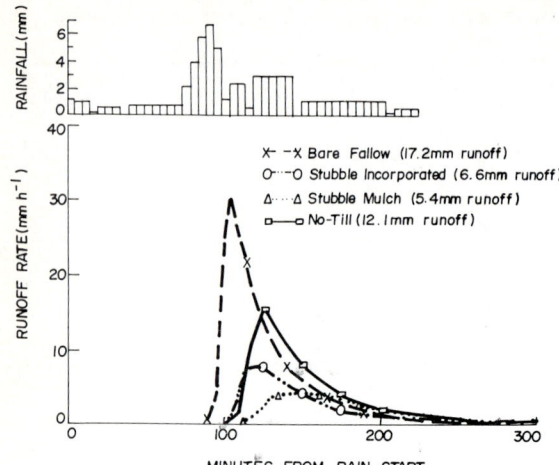

Figure 12.24 Hyetograph and hydrograph of the rainfall event on 24 December 1981 (total rainfall 38.1 mm) at Greenwood, showing higher runoff from no-till catchment compared to stubble mulch (Freebairn et al., 1986).

Minimum tillage. Reduced or minimum tillage is often used for soils with a low infiltration rate caused by surface crust or seal or by compacted surface horizon. The aim is to improve water infiltration by surface tillage systems that will break the crust and yet permit a maximum amount of residue retention on the soil surface. A range of tillage implements are used to achieve this, e.g., disk plow, chisel plow, strip tillage, and paraplow (Plate 12.11). Shallow plowing

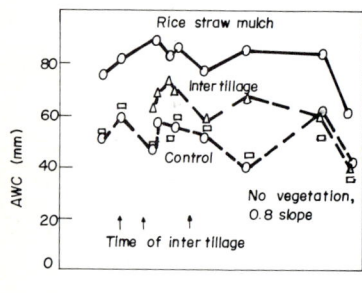

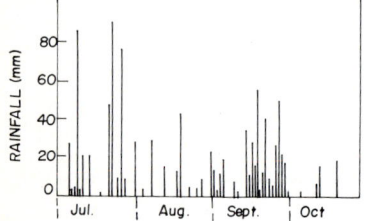

Figure 12.25 Effects of mulching and tillage methods on soil water conservation in Thailand. AWC = available water capacity. (Kubota, 1982.)

Plate 12.11 Paraplow is a noninversion tillage implement that can loosen the soil up to about 50-cm depth.

without inversion is also practiced to break the surface crust. The effectiveness of the minimum tillage system in reducing erosion depends on the type of minimum tillage, quantity and distribution of crop residue mulch, land use history, and structural stability of the soil.

Near Lafayette, Ind., Mannering and Johnson (1966), using a rainulator, observed the effects of minimum and conventional tillage methods on infiltration, runoff, and soil erosion. The relative erosion-reducing effectiveness of minimum tillage as compared with conventional tillage depended on the different forms of minimum tillage (plow-plant vs. wheel-track-plant), the proportion of

TABLE 12.31 Soil Erosion from Watersheds Growing Corn, with Minimum and Conventional Tillage Methods, at Coshocton, Ohio, for Storms of High Rainfall Energy

Date	Tillage	Slope, %	Soil erosion, t/ha
June 12, 1957	Conventional	6	10.8
	Minimum	11	4.5
June 28, 1957	Conventional	6	4.4
	Minimum	11	0.3
July 5, 1969	Conventional	6	7.2
	No-tillage	9	0.01
	No-tillage	21	0.07

SOURCE: Harrold, 1972.

TABLE 12.32 Influence of Quantity of Crop Residue on Soil Erosion with Chisel Plow and Strip Tillage Systems for Tama Silt Loam in Central Illinois, 6% Slope under Corn

Residue, kg/ha	Chisel plow system		Strip tillage system	
	t/(ha · yr)	%	t/(ha · yr)	%
	31	100	31	100
1120–2240	22	71	29	93
2240–3360	16	50	18	57
3360–4480	9	29	13	43
4480–6725	4	14	9	29
>6725	2	7	7	21

SOURCE: Oschwald, 1974.

soil surface covered by mulch, and the presence or absence of crust. The effectiveness also decreased with the increasing number of cropping cycles of corn cultivation following several years of alfalfa and orchard grass meadow. At Coshocton, Ohio, Harrold (1972) observed that a minimum tillage method involving plow-plant plant system caused a sizable reduction in erosion compared to that of conventional tillage (Table 12.31). In both cases, tillage was performed on the contour. Another example of how the effectiveness of a minimum tillage system depends on the quantity of crop residue mulch is given in the data from central Illinois (Table 12.32). In both chisel plow and strip tillage systems, soil erosion decreased with an increasing amount of crop residue mulch.

In Kiev, Ukraine, U.S.S.R., Shikula observed that soil erosion by snow melt water was decreased and snow water storage increased by the presence of mulch in subsurface tillage compared with plowed treatments. He also observed that the subsurface treatment performed across the slope effectively reduced soil erosion in comparison with plowing along or across the slope (Table 12.33).

For some soils and crops, use of a minimum tillage system implies a reduction in frequency, intensity, or proportion of land surface tilled. In São Paulo,

TABLE 12.33 Influence of Plowing and Subsurface Treatment on Erosion from a Podzolic Soil in Kiev, Ukraine, U.S.S.R.

Treatment	Soil erosion, m^3/ha	
	Erosion on slope	Soil accumulation at slope base
Plowing along slope	86.9	75.1
Plowing across slope	46.4	52.0
Subsurface treatment downslope	14.7	15.9
Subsurface treatment across slope	1.7	1.1

SOURCE: Shikula, 1981.

TABLE 12.34 Effects of Plowing Intensity on Runoff and Soil Erosion

Tillage treatment	Erosion, t/ha	Runoff, %
Two plowings	14.6 (122)*	5.7 (104)
One plowing	12.0 (100)	5.5 (100)
Subsoiling	8.6 (72)	5.0 (94)

*Figure in parentheses refers to values relative to one plowing.
SOURCE: Bertoni and Lombardi, 1985.

Brazil, Bertoni and Lombardi (1985) observed that the reduction of tillage intensity decreased soil erosion. In comparison with one plowing, soil erosion was increased 22 percent by two plowings and was decreased 28 percent by subsoiling (Table 12.34). In India, Khybri et al. (1984) observed that a reduction in the proportion of land surface tilled to one-third decreased erosion by 20 percent compared with plowing the whole field (Table 12.35).

A minimum tillage system involving a reduction in intensity and frequency of tillage operations can reduce both runoff and soil erosion. The appropriate minimum tillage package, for both effective erosion control and satisfactory crop growth, however, varies among soils, crops, climatic regimes, past land use, and the management systems. It is difficult to make any generalizations regarding the minimum tillage system most desirable for a soil type or a region.

Contour ridges. Ridge farming is a traditional method of seedbed preparation in many regions, especially in the tropics and subtropics. On steeplands, ridges are deliberately made up and down the slope, probably to facilitate surface drainage so as to minimize the risks of landslides and mass movement (Plate 12.12). Ridges made upslope and downslope may, however, cause more runoff and soil erosion than sowing on the flat (Kowal, 1970). The data in Table 12.36 from northern Nigeria show the highest runoff and erosion with ridges made upslope and downslope. Under these conditions even tie ridges and broad beds were not more effective than planting on the flat. Experiments conducted at

TABLE 12.35 Effects of Mulch and Minimum Tillage on Maize and on Runoff and Erosion from a Soil on 8% Slope at Dehra Dun, Northern India*

Treatment	Runoff, % of rainfall	Erosion, t/(ha · yr)
Conventional tillage (plowing whole field)	49.4	36.5
Conventional tillage + mulch @ 4 t/ha	22.3	6.2
Strip tillage (one-third of field plowed)	50.3	29.2
Grass cover (*Cynodon dactylon*)	2.0	0.2
Bare fallow	66.5	104.9

*Each data value is a mean of 3 consecutive years from 1978 to 1980. Mean annual rainfall = 1325.6 mm. Plot size = 22 × 1.8 m.
SOURCE: Khybri et al., 1984.

Plate 12.12 Ridges are made up and down the slope to provide quick drainage on steeplands susceptible to landslides and mass movement.

IITA, Ibadan, also showed the maximum runoff and erosion from plots with ridges made up and down the slope (Table 12.37). On rolling to undulating slopes, however, contour ridges increase the time for water to infiltrate into the soil (Plate 12.13). The surface detention capacity for runoff storage is drastically increased, and both the amount and the rate of runoff are decreased by ridges constructed on the contour. The ridge-furrow system, however, is usually stable for gentle slopes of up to about 7 percent and for soils that have a relatively stable structure. Severe erosion can be caused by the failure of a ridge-

TABLE 12.36 Effects of Ridges on Runoff in Northern Nigeria

Treatment	Estimated amount, cm	Runoff, % of rainfall
Bare fallow (nonridged land)	4.45	47
Maize planted on flat (0.6 × 4.2 m)	4.29	45
Maize planted on 3-ft ridges (0.3 × 0.9 m)	5.46	57
Same as treatment directly above, but alternately tied ridges	2.92	30
Maize planted on flat with plant residue incorporated into soil (0.6 × 0.375 m)	3.05	31
Maize planted on flat with plant residues left on surface (0.6 × 0.375 m)	2.79	29
Catchment basin	4.04	56

SOURCE: Kowal, 1972.

TABLE 12.37 **Effects of Ridges up and down the Slope on Erosion and Runoff under Cassava, Maize, and Uncropped Plot**

Treatment	Runoff, mm	Mean sediment density, g/L	Soil erosion, t/ha
Uncropped (plowed, bare)	250.9	1.74	4.4
Maize (flat)	197.2	3.0	4.4
Cassava (ridged)	461.8	2.36	8.0
Yam (mound)	185.6	1.46	2.0

SOURCE: Unpublished data of R. Lal, July–November 1980.

furrow system, when such a system is installed on steep slopes and on soils of unstable structure (Plate 12.14).

The International Crop Research Institute for the Semiarid Tropics (ICRISAT) based at Hyderabad, central India, has developed a broad-bed and furrow (BBF) system for controlling runoff and erosion on vertisols (Plate 12.15). Different row arrangements can be used on these beds to maximize the land use and to facilitate mixed or relay cropping. Compared with uncropped fallow, this system of conservation tillage is believed to reduce erosion, provide surface drainage, reduce soil compaction in the plant row zone, lend itself to a semipermanent system for use with minimum tillage, and allow an intensive land use. Watershed management experiments conducted at ICRISAT showed that water runoff was 53.0 mm in fallow land vs. 1.5 mm in the BBF system during 1979 and 410 mm in fallow land vs. 273 mm in the BBF system

Plate 12.13 Contour ridges allow more time for water infiltration into the soil.

Plate 12.14 Failure of ridge-furrow system on soils of unstable structure can cause severe erosion.

during the 1978 season (Kampen et al., 1981). The data in Table 12.38 show that both the runoff amount and the peak runoff rate are decreased by the BBF system. The mean annual soil erosion (average of 6 years from 1973 to 1978) was 5.6 t/(ha · yr) for the fallow watershed in comparison with 1.6 t/(ha · yr) for the watershed where the BBF system was installed. The corresponding mean

Plate 12.15 The broad-bed and furrow (BBF) system developed for vertisols at ICRISAT, Hyderabad, India.

TABLE 12.38 Rainfall, Runoff, and Soil Loss, t/ha, Measured at Watershed Outlets of Cropped Deep Vertisol Watershed (BW1) and Rainy Season Fallow Deep Vertisol Watershed (BW4C) from 1973 to 1978

Year	BW1				BW4C			
	Rainfall, mm	Runoff, mm	Peak rate, $m^3/(s \cdot ha)$	Soil loss, t/ha	Rainfall, mm	Runoff, mm	Peak rate, $m^3/(s \cdot ha)$	Soil loss, t/ha
1973	697.0	51.2	0.03	3.0	734.6	58.7	0.06	3.9
1974	810.4	116.1	0.20	1.3	806.9	223.4	0.22	6.8
1975	1041.6	162.2	0.06	0.7	1055.0	253.2	0.15	2.1
1976	687.3	73.1	0.09	0.8	710.1	238.1	0.16	9.2
1977	585.6	1.5	0.01	0.1	585.9	53.0	0.06	1.7
1978	1125.2	272.5	0.11	3.4	1116.7	410.1	0.15	9.7

SOURCE: Kampen et al., 1981.

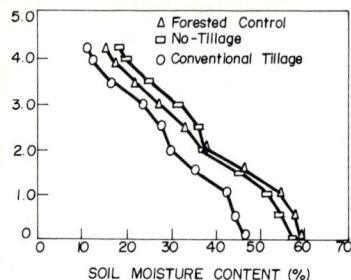

Figure 12.26 Effects of mechanical tillage methods on soil moisture characteristics of an upland soil in Thailand (Tulaphitak et al., 1983).

annual losses of water runoff were 206.1 and 112.8 mm for the fallow and BBF systems, respectively. Both runoff and soil erosion can be decreased even more by combining the BBF system with crop residue mulch, as shown by the results of experiments on watershed management for vertisols conducted in Queensland, Australia (Table 12.17). In the semiarid tropics with a high cattle population, as in central India, the mulch material is, however, difficult to procure.

The broad-bed system has also been used in semiarid regions of northern Negev, Israel, to conserve water for wheat cultivation on loess soils (Morin et al., 1984). The data in Table 12.39 show that in comparison with fallow and sowing on the flat, both 0.6- and 1.6-m-wide beds reduced runoff and soil erosion.

The broad-bed system is generally used on poorly drained soils to provide surface drainage, such as with the cambered-bed technique used in Central America and the Caribbeans (Simpson and Gumbs, 1982). The BBF system can also be used for supplementary irrigation, by using the furrows for supplying the irrigation water. The system has wide applicability and can be adapted to a broad range of soil-related management objectives.

The system of summer fallow is widely used to increase production in arid regions to conserve water in the root zone for cropping in alternate years. The fallow system is applicable for some soils and rainfall regimes but not for others. In addition, the summer fallow system has a major drawback of severe erosion during the summer rains, as shown by the data from India (Tables 12.40 and 12.41) and Israel (Table 12.42). The system can therefore be made to work to improve production per unit input wherever soil erosion is not severe, as in

TABLE 12.39 Soil and Water Conservation with Broad-Bed and Furrow System on a Loess Soil in Northern Negev, Israel, Winter 1980–1981

Treatment	Runoff, mm	Erosion, kg/ha
Control, sown on flat	17.8	293
Fallow plots	21.6	397
1.6-m-wide beds	7.9	156
0.6-m-wide bed	7.4	72

SOURCE: Modified from Morin et al., 1984.

TABLE 12.40 Rainfall and Runoff on a Cropped Deep Vertisol with Broad-Bed and Furrow System (BW1) and on a Rainy Season Fallowed Watershed (BW4C)

Date	Rainfall,* mm	Runoff, mm	
		Deep vertisol cropped (BW1)	Deep vertisol fallow (BW4C)
June 17	36.8	5.0	2.0
18	24.7	2.1	1.9
21	24.7	3.3	4.4
July 12	44.2	0.7	16.5
16	33.6	1.9	10.5
18	32.2	10.9	21.5
Aug. 7	27.2	0.1	10.1
8	14.6	0.3	5.9
13	52.7	6.8	28.6
14, 15	219.5	170.0	190.6
19	16.7	1.5	10.7
21	17.3	—	5.8
22	65.5	49.5	58.9
23	13.5	9.3	10.8
28	25.9	8.2	16.5
Small storms	57.0	2.9	15.4
Total	706.1	272.5	410.1

*Runoff-producing storms only; total seasonal rainfall (June–October) was 1120 mm.
SOURCE: ICRISAT, 1978; Kampen et al., 1981.

the western United States, west of about 100° meridian (USDA-ARS, 1974). The advantages of summer fallow, in terms of water storage and soil conservation, can be substantially increased by combining summer fallow with crop residue mulch (Greb et al., 1967). Furthermore, soil water storage by summer fallow can be improved by using herbicides rather than plowing to control weeds. Smika and Wicks (1968) reported that the storage efficiency was 35.4 percent for conventional tillage compared with 42.4 percent for herbicide treatment. The importance of soil conservation by growing a cover crop during the fallow phase has also been recognized in east Africa (Pereira et al., 1958) and west Africa (Okigbo and Lal, 1977).

Tied ridges or basin tillage. Ridging with the addition of cross-ties in the furrows is an improvement over the traditional ridge-furrow system because it holds surplus water in individual basins (Plate 12.16) and allows more time for water to infiltrate the soil. The practice of tie ridging, widely used in east Africa since early 1940s (Prentice, 1946), is similar to the "basin-listing" technique used in the semiarid regions of the southeastern United States (Stewart et al., 1981). The tied-ridge system is also known by other names including *furrow blocking, furrow damming,* and *furrow diking.* This system was introduced in the southeastern United States in the early 1930s (Unger, 1984).

A survey of the literature provides contradictory reports on the effectiveness of the tie-ridge system on soil and water conservation and on improving the crop yield. On some soils tied ridges do not provide satisfactory erosion control (Pereira et al., 1967), although the risks of gully erosion are minimized. In fact, there may be cases of greater erosion with tied ridges during periods of heavy rain (Clark and Hudspetch, 1976). The tied-ridge system may also cause yield reductions during periods of heavy rain (Plate 12.17). Formation of a thin clay film in the furrow decreases the water infiltration rate (Fig. 12.27) and causes water to stagnate for days after heavy rains. Crops such as maize and cowpea are adversely affected by the waterlogging and poor aeration thus created. Weed control can also be a serious problem with a tied-ridge system.

In spite of the drawbacks outlined above, the tied-ridge system has been used widely to conserve soil and water in arid and semiarid regions (Unger, 1984; Lal, 1986). An example of erosion control by the tied-ridge system in Tanzania is seen in Table 12.43. The plots with tied ridges made on the contour had lower runoff and soil loss than the fallow or the flat-sown treatments. The advantages of the tied-ridge system in conserving water are greater in seasons with below-normal rainfall than when rainfall is adequate. In Israel, Rawitz et al. (1983) observed the least runoff from the basin tillage or tied-ridge treatment during

TABLE 12.41 Rainfall and Runoff on a Vertisol Watershed with a Broad-Bed and Furrow System at 0.6% Slope (BW1) and a Rainy Season Fallowed Watershed (BW4C)

Date	Rainfall,* mm	Runoff, mm Cropped	Fallow
June 15	34	0	0.2
July 4	37	0.7	1.8
17	54	0.5	3.2
Aug. 10	75	0.3	25.4
22	30	0	3.2
23	15	0	4.4
24	5	0	2.9
28	21	0	3.6
Sept. 1	14	0	5.7
Small storms (5)	52	0	2.6
Total	349	1.5	53.0

*The total rainfall from the 42 rainy days of the season (June–October) was only 519 mm. The runoff was unusually low in 1977, due to below-normal total precipitation and lack of high-intensity storms.

SOURCE: ICRISAT, 1979; Kampen et al., 1981.

TABLE 12.42 Runoff and Eroded Material from the 1980–1981 Winter Fallow Preceding Cotton

Treatment*	Storm of 10/12/80		Storm of 11/01/81		Total for growing season	
	Runoff, mm	Erosion, kg/ha	Runoff, mm	Erosion, kg/ha	Runoff, mm	Erosion, kg/ha
Deep plowed along slope	0	0	0	0	0	0
Deep plowed on contour	0	0	0	0	0	0
Ridges after plowing	12†	130 ± 33	0.1 ± 0.90	0	12.1	131
Basin tillage after plowing	10.6 ± 1.1	68 ± 42	0	0	10.6	68
Basin tillage after subsoiling	0	0	2.9 ± 0.2	0	2.9	Traces
Minimum tillage ridging	9.0 ± 2.7	238 ± 24	3.8 ± 1.5	29 ± 5	12.8	267
Minimum tillage— basin tillage	0.8 ± 0.8	4 ± 2	0	0	0.8	4

*An eighth treatment (direct ridging) is not represented in the table because tillage was carried out after the storm of 10/12/80 and no runoff was measured following this operation.
†Amount measured before the barrel floated. The true amount was certainly much higher.
SOURCE: Rawitz et al., 1983.

the 1979–1980 rainy season (Fig. 12.28). In accord with the runoff losses, the cumulative erosion was also the least from the basin-tilled treatments (Fig. 12.29). One severe drawback of a ridge-based system in the arid and semiarid tropics is the high soil temperature and low soil moisture conditions on the

TABLE 12.43 Effects of Tillage Methods and Tied Ridges on Runoff and Erosion from a Sandy Red Loam Soil on 7% Slope at Mpwapwa Research Station, Tanzania

Treatment	Runoff, %	Erosion, t/ha
1. Bare, no cultivation	48	118.7
2. No cultivation, allowed vegetation regrowth	27	24.6
3. Bare, flat cultivation	27	121.0
4. Flat cultivation, allowed vegetation regrowth	16	47.0
5. Bare, tied ridges on contour	0	0
6. Bare, ridges downhill	17	60.5
7. Cropped with ridges downhill	30	125.4
8. Cropped, cultivated on flat	30	96.3
9. Perennial tufted grass	1.5	0
10. Deciduous thicket	Negligible	0

SOURCE: Adapted from Rounce, 1946.

Plate 12.16 Tied-ridge system used in semiarid west Africa.

ridgetop. Supraoptimal soil temperatures occur on ridgetops, causing severe yield reductions in crops such as maize and soybeans (Lal, 1986).

Conservation Tillage for Erosion Control on Problem Soils

A range of tillage practices are used to alleviate some soil-related constraints on problem soils and to improve crop productivity. Specific problems encountered include poorly drained soils, soils with claypan or fragipan, soils with compacted surface or subsurface horizons, and salt-affected soils. Tillage methods used to alleviate these constraints are discussed below with specific examples.

Poorly drained soils

In the tropics and subtropics, poorly drained and low-lying soils are profitably used for cultivation of rice. Soil is *puddled*—a mechanical tillage operation is

TABLE 12.44 Effects of Tillage System, Crop Management, and Soil Drainage on Runoff and Erosion of a Soil near Pisa, Italy

	Undrained		Tile-drained	
Tillage system and crop	Runoff, %	Soil loss, t/ha	Runoff, %	Soil loss, t/ha
Conventional tillage, wheat	3.6	4.0	2.5	3.7
Minimum tillage, wheat	5.8	1.6	3.8	1.5
Lawn pasture, forage	3.0	0.18	2.2	0.15

SOURCE: Chisci and Zanchi, 1980.

Plate 12.17 Poor drainage conditions created by tied ridges can reduce yield of crops sensitive to waterlogging.

performed when the soil moisture content is near the saturation point, to destroy structural porosity and reduced percolation. Puddling is also done to reduce macroporosity and to control aquatic weeds. Once the "paddy profile" has been developed, however, puddling is not necessary (Lal, 1986).

Special tillage practices and soil surface management techniques are needed to grow upland crops on poorly drained soils. Because poorly drained soils generally exist on nearly level land, soil erosion is often not a severe problem. Risks of erosion increase, however, even with a gentle gradient and on soils of high erodibility.

In Italy, Chisci and Zanchi (1980) studied the effects of different tillage methods on runoff and soil erosion from poorly drained land with and without artificial drainage. The experiment was conducted on a silty clay loam soil with 12 percent slope near Pisa, Italy, and the test crops were wheat and pasture forage. The conventional tillage involved moldboard plowing in comparison with disk plowing plus chemical weed control for minimum tillage. The data in Table 12.44 show that although the runoff was high from the minimum tillage treatment, the soil loss from conventionally tilled plots was more than double that of the minimum tillage treatment. The least runoff and erosion were recorded in case of the pasture forage treatment. The data support the conclusion that minimum tillage or no-tillage systems can be profitably used once the water table is regulated by installing an appropriate drainage system. These conclusions are also supported by the data of Fausey (1984) from experiments conducted on poorly drained soils in northern Ohio (Table 12.45).

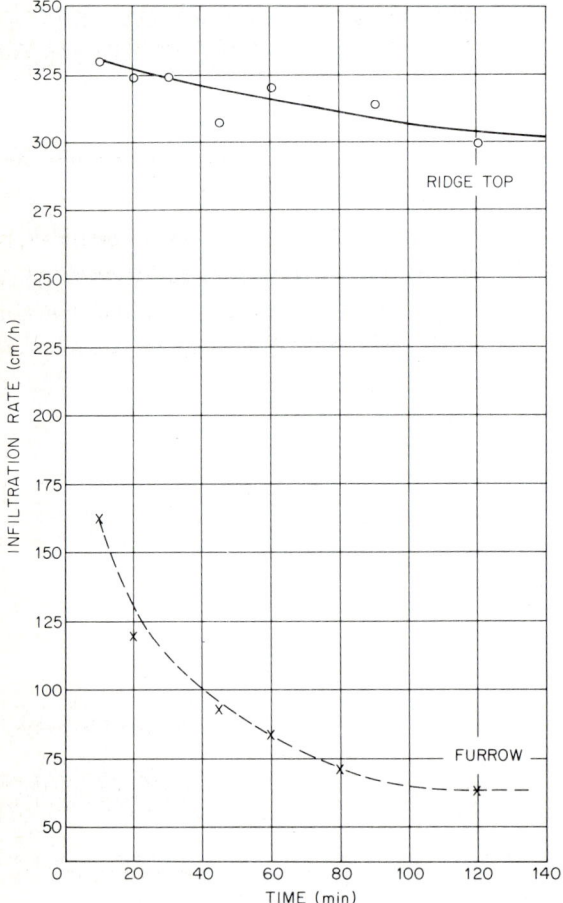

Figure 12.27 Effects of ridge-furrow system on water infiltration rate in the furrow. (Unpublished data of R. Lal.)

Claypan soils

Claypan in the subsurface layers often occurs due to natural soil-forming factors that lead to clay eluviation. The adverse effects of such a natural process are greatly accentuated by the smearing action of tillage tools, leading to the formation of a "plow pan." Although the no-tillage or minimum tillage system can be used on such soils, their effects on soil and water conservation are not as satisfactory as are those normally observed for soils with free drainage and those without an impeding layer in the profile. Based on their experiments in central Missouri, Smith et al. (1974) observed more yearly runoff from no-tillage corn grown on a claypan soil than from plots that had been plowed—17.0-cm runoff vs. 12.7 cm of yearly runoff. The runoff from no-till soil is likely to increase if the infiltrating water cannot be conducted through the soil profile. Subsequent experiments conducted by Steichen (1984) showed that tillage

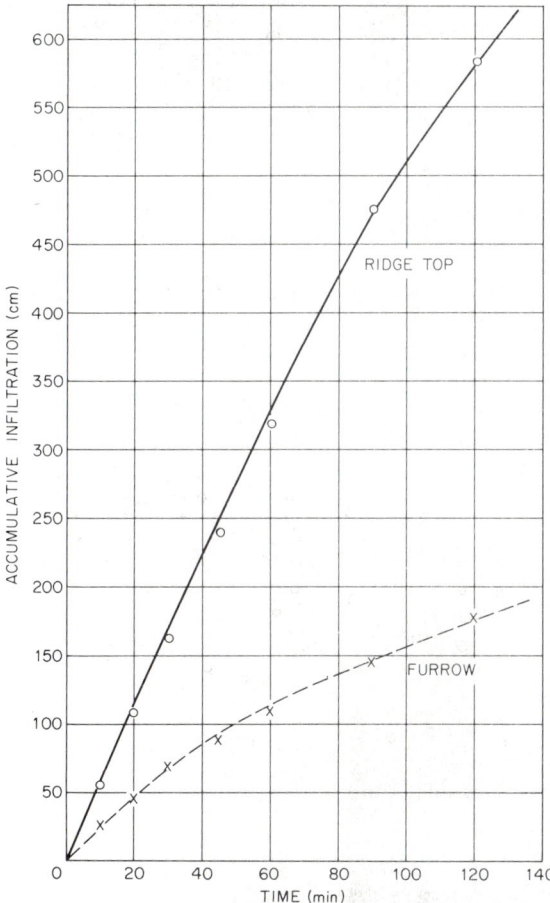

Figure 12.27 (*Continued*)

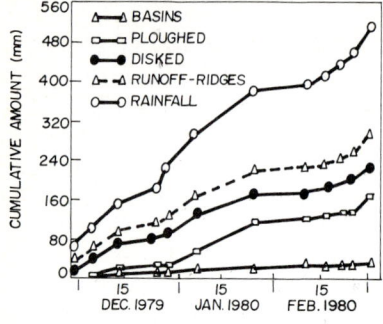

Figure 12.28 Effects of tied ridges on cumulative runoff during the 1979–1980 rainy season (Rawitz et al., 1983).

treatments that caused the highest random roughness also had the highest water infiltration and the lowest runoff. In this regard, the bare no-tillage plot had the lowest infiltration. The high random roughness was created by the chiseled plow treatment.

Once the surface soil conditions have been improved and the clay pan has been loosened, the no-tillage system can be profitably used for one or many consecutive cropping seasons. Depending on the frequency of claypan formation, periodic chiseling or the paraplow operation may be required to restore the soil infiltration rate.

Soils with compacted surface and subsoil horizons

Soil compaction is a complex problem. Some soils are naturally hard-setting, such as soils low in organic matter content and soils containing predominantly low-activity clays. Other soils are easily compacted when used for continuous and intensive cultivation with tractor-driven equipment. Above all, there are soils that slake on wetting and form surface seal or crust. Different tillage systems are used to alleviate various types of compaction.

Use of motorized equipment causes a rapid compaction of the surface horizon of structurally inert soils of the tropics and subtropics. In Bahia, Brazil, Silva (1981) observed severe compaction with mechanized cultivation of an oxisol. The mechanized cultivation decreased the total porosity of 0- to 30-cm soil depth from a mean value of 55 to 27 percent. In Nigeria, Lal (1985) observed

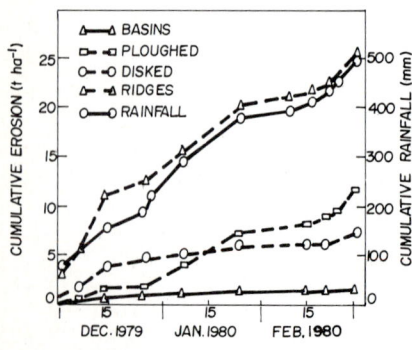

Figure 12.29 Effects of tied ridges on cumulative soil erosion during the 1979–1980 rainy season (Rawitz et al., 1983).

TABLE 12.45 Average Corn Yield (Mg/ha) by Year, Tillage Treatment, and Drainage Level

Tillage	Year	Distance from drain, m					Average
		0–3	3–6	6–9	9–12	12–15	
Beds	1978	9.88	9.52	9.00	—	9.25	9.41
	1979	9.52	9.33	9.52	9.04	8.80	9.24
	1980	10.57	9.98	9.66	9.83	9.73	9.95
	Average	9.99	9.61	9.39	9.44	9.26	
Chisel	1978	9.29	8.64	8.43	—	7.66	8.50
	1979	9.90	9.81	9.23	9.40	9.10	9.49
	1980	11.33	10.96	10.36	10.18	9.58	10.48
	Average	10.17	9.80	9.34	9.79	8.78	
No-till	1978	9.30	8.50	8.03	—	8.38	8.55
	1979	9.95	9.42	9.25	9.64	9.23	8.50
	1980	9.45	6.62	6.17	6.55	7.47	7.25
	Average	9.57	8.18	7.82	8.10	8.36	

SOURCE: Fausey, 1984.

significant reductions in the water infiltration rate at the head points. The data in Fig. 12.30 show that the cumulative infiltration was 28 percent greater within the plot than at the turning point. Significant increases in bulk density and drastic reductions in infiltration rate were also observed under the wheel tracks of a track-type tractor used for deforestation and land clearing (Lal and Cummings, 1979). Indiscriminate use of machinery and heavy vehicular traffic when the soil is wet cause severe compaction of the surface horizon. In Nigeria, Lal (1985) observed that continuous use of mechanized cultivation for 4 consecutive years caused severe compaction, resulting in a collapse of the soil structure. Consequently, the infiltration rate severely declined with time in both the no-tillage and conventional tillage treatments (see Fig. 3.10).

Soil compaction is also caused by the use of heavy machinery during deforestation or land clearing. The data in Fig. 12.31 indicate that deforestation per se decreases infiltration. However, the magnitude of decline is more severe with mechanized than with manual clearing. Furthermore, adverse effects of defor-

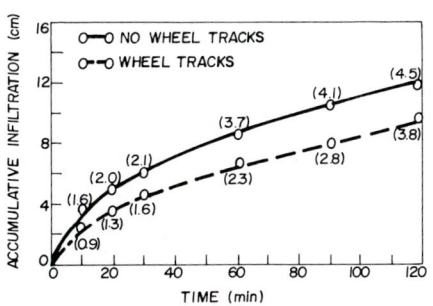

Figure 12.30 Decrease in water infiltration rate of a tropical alfisol at the turning points and wheel tracks of the farm machinery (Lal, 1985).

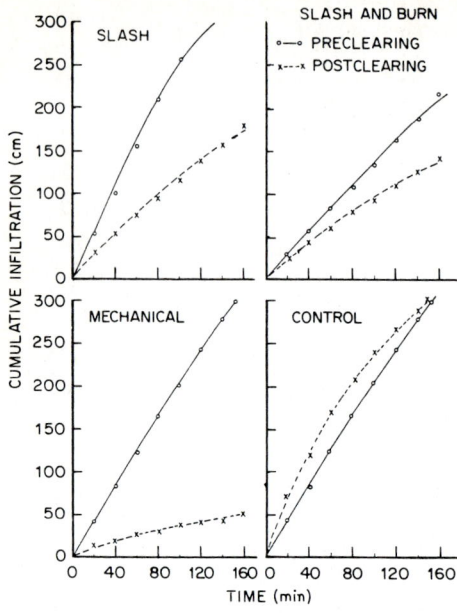

Figure 12.31 Effects of land clearing methods on accumulative infiltration for a tropical alfisol (Lal and Cummings, 1979).

estation are accentuated by the vehicular traffic during tillage and harvesting operations.

The problem of traffic-induced soil compaction is equally severe in soils of the temperate regions. In East Shropshire, England, Fullen (1985) observed severe erosion because of soil structural changes caused by compaction. The infiltration rate saw a drastic reduction (2638 times less) in wheel tracks (Table 12.46) compared to the uncompacted site. A severe reduction in the infiltration rate was also observed in the soil with a plow pan in Morris, Minn. Lindstrom and Voorhees (1980) and Lindstrom et al. (1981) used simulated rainfall to evaluate runoff and erosion from wheel-tracked and non-wheel-tracked interrows in corn planted by different tillage methods. The wheel-tracked interrow required less kinetic energy to start runoff from each tillage system. The kinetic energy required to initiate runoff was less for the no-tillage and conservation tillage systems than for conventional tillage. The ranking of the treatments for kinetic en-

TABLE 12.46 Effects of Compaction Caused by Vehicular Traffic on Infiltration Rate of a Soil in East Shropshire, England

Condition	Infiltration rate, mm/h				No. observations
	Maximum	Minimum	Mean	Standard deviation	
Pasture	807	147	343	192	23
Crusted	78.6	5.15	30.16	17.85	88
Plow pan	5.74	1.57	3.30	1.30	32
Tractor wheeling	0.37	0.016	0.13	0.07	126

SOURCE: Fullen, 1985.

ergy required to initiate runoff was conservation tillage > conventional tillage > no-tillage for the non-wheel-tracked interrow and conservation tillage > conventional tillage > no-tillage for the wheel-tracked interrow. The cumulative infiltration was the lowest in the no-till wheel-tracked zone (Fig. 12.32).

Soils prone to surface seal

Soils prone to crust formation and development of surface seal are widespread in low and middle latitudes. These soils are particularly widespread in the semiarid and arid tropics. Soils low in organic matter content and containing predominantly low-activity clays are highly vulnerable to crust formation. Soils in the west African Sahel develop an algal crust. This crust makes these soils virtually impermeable to water because the algal crust has hydrophobic properties. The crusted surface may be discontinuous, but where the crust exists, as much as 90 percent of the rainfall received is lost as surface runoff. Stroosnijder and Hoogmoed (1984) conducted water-balance studies of natural vegetation in the Sahel. Even under the protective natural vegetation cover, runoff measured 40 percent of the total growing season rain of 415 mm. The available water left in the root zone was about 50 mm. In comparison, the runoff from the pearl millet field was 50 percent of the 300-mm rainfall received, leaving a maximum of 40 mm of plant-available water in the root zone. Crust formation was found to be a major factor in the high percentage of water runoff. The data in Fig. 12.33 by Hoogmoed and Stroosnijder (1984) show that crust removal drastically increases the infiltration rate. The beneficial effect of crust removal, however, hardly lasts more than 2 weeks. Crust formation is also a serious limitation in cultivated soils of the sub-Sahelian region, e.g., Senegal (Charreau, 1969, 1972; Nicou, 1974; Kalms, 1977). High runoff losses and poor crop establishment on plowed alfisols in the subhumid of west Africa are also attributed to crusting (Lal, 1986). The soils in pasturelands of northern Australia are also prone to developing surface seal. The seal causes low infiltration and high runoff and erosion (Mott et al., 1979).

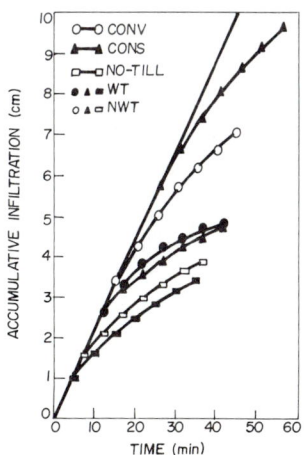

Figure 12.32 Effects of wheel tracks on cumulative infiltration for a soil in Minnesota. CONV = conventional tillage, CONS = conservation tillage, NO-TILL = no-tillage, WT = wheel-tracked interrow, NWT = non-wheel-tracked interrow. (Lindstrom et al., 1981.)

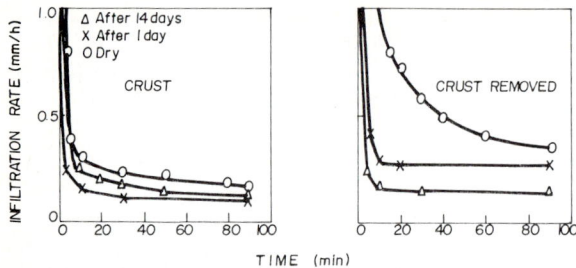

Figure 12.33 Effects of crust on water infiltration into an arid region soil in the west African Sahel (Hoogmoed and Stroosnijder, 1984).

Crust formation is minimized by biologic, mechanical, or chemical treatments. The biologic treatments involve use of biomass (residue, twigs, branches, etc.) so as to encourage the activity of termites and other soil fauna. Termite activity plays a major role in creating channels for water conductance in arid and semiarid regions (Plate 12.18). In the subhumid and humid regions, use of crop residue mulch minimizes the risk of crust formation by preventing raindrop impact and increasing biotic activity of earthworms (Plate 12.19). The beneficial effects of planted fallows and no-tillage farming are also attributed to the activity of earthworms and other soil fauna (Low, 1972; Wilkinson, 1975; Lal, 1976a; Ehlers, 1975). Unless a regular supply of the biomass and mulch material is maintained, the beneficial effects of fallowing and mulching are

Plate 12.18 Residue mulch increases termite activity, breaks crust, and improves infiltration rate in arid regions.

Plate 12.19 Earthworm activity is greatly enhanced by mulch farming and no-till system in humid regions.

transient and easily eliminated during the first cropping cycles using plow-based tillage operations (Wilkinson, 1975; Pereira et al., 1954). If feasible, therefore, it is logical to use mulch-based biologic techniques continuously and to develop farming systems to grow those crops that produce adequate biomass to enable the adoption of mulch farming techniques.

Mechanical soil manipulations are widely practiced to temporarily break the surface seal, improve infiltration, and decrease runoff and soil erosion. In the Sahel region of Mali, Stroosnijder and Hoogmoed (1984) reported that mechanical tillage destroys the crust and increases the surface storage of rainwater. The crust-breaking effect of tillage, however, lasts for only a few rain showers. The effects of tillage practices on water runoff for two surface storage values are shown in Table 12.47 for a soil at Niono, Mali. The mean annual runoff loss for

TABLE 12.47 Effects of Tillage Methods on Water Runoff at Niono, Mali

| | | Runoff, mm | | | |
| | Rainfall, | Surface storage = 1.0 mm | | Surface storage = 10.0 mm | |
Year	mm	No-tillage	Tilled	No-tillage	Tilled
1977	368	155	76	48	11
1978	271	104	49	33	19
1979	361	141	80	70	38
Mean	333.3	133.3	68.3	50.3	22.7

SOURCE: Adapted from Stroosnijder and Hoogmoed, 1984.

1.0-mm surface storage capacity was 39.3 percent for the no-tillage treatment and 19.3 percent for the tilled treatment. For the 10-mm surface storage capacity, the mean annual runoff decreased in both tillage treatments and was 14.7 percent for no-tillage soil and 7.0 percent for tilled soil. These results also imply that if the surface storage capacity were increased even more, e.g., by use of residue mulch, the runoff losses would decrease regardless of the tillage treatments. There are, however, severe practical limitations in procuring the mulch material for a region with unreliable and low annual rainfall.

Salt-affected soils

Sodic soils develop a massive structure due to the dispersion of clay and other colloids by the high concentrations of Na^+ on the exchange complex. Consequently these soils have low permeability. The restorative processes include replacement of Na^+ by Ca^{2+} on the exchange complex by liming. The lime must, however, be incorporated in the soil profile for which both deep tillage and soil inversion become necessary.

Conservation Tillage for Controlling Wind Erosion

Wind erosion and its control measures are discussed at length in Chap. 15. The principles of controlling wind erosion are similar to those of controlling water erosion—increasing the surface roughness, providing mulch cover, and reducing the velocity of wind near the soil surface. The conservation tillage for wind

TABLE 12.48 Effects of Tillage Systems and Residue Management on Soil Erosion by Wind in Northwestern Ohio

Tillage	Surface residues, t/ha	Soil loss, t/ha
Experiment I*		
Fall (autumn) plow	0.28	26.1
Spring plow	0.12	8.5
No-tillage	5.60	1.2
Experiment II†		
Plow, normal residue	0.14	3.5
Disk, normal residue	0.54	5.1
Disk, double residue	1.76	0.8
No-tillage, no residue	0	3.0
No-tillage, normal residue	1.82	0.6
No-tillage, double residue	2.85	0.5

*Comparison of autumn and spring plowing and no-tillage.
†Comparison of plowing with normal residues and disk and no-tillage with no, normal, or double residues.
SOURCE: Woodruff, 1972.

TABLE 12.49 Effect of Land Preparation on Soil Erosion by Wind on Newly Planted Maize Fields in Northwestern Ohio

Land preparation	Soil type	Soil loss, t/ha
Plowed and planted	Ottokee loamy find sand	403.0
Power-disked and planted	Oakville loamy fine sand	7.6
No-tillage and planted	Spinks loamy fine sand	1.3
Untilled maize stalk field	Oakville loamy fine sand	0.8

SOURCE: Woodruff, 1972.

erosion is thus similar to that for water erosion. In general, a soil with a fine seedbed is more prone to wind erosion than one with a rough seedbed or mulch tillage.

There is little experimental information available regarding the effects of tillage methods on wind erosion. Relevant examples are, however, available from experiments by Woodruff conducted to control wind erosion in Ohio and Wisconsin. The data in Tables 12.48, 12.49, and 12.50 show the relative effectiveness of different types of conservation tillage used in controlling wind erosion. In Ohio, wind erosion was greater in plowed and disk-plowed fields than in no-tillage fields (Table 12.48). Leaving the maize stalks standing in an untilled field provided the best defense against wind erosion in northwestern Ohio (Table 12.49). Similar results were obtained in Wisconsin where plowed and disk-plowed fields suffered more wind erosion than no-till and mulch fields (Table 12.50). Wind erosion was the least from land where residues were left standing.

Conclusions

Conservation tillage methods—tillage techniques required to effectively control wind and water erosion and conserve water and nutrients in the root zone—differ among soils, crops, rainfall regimes, and agroecological environments. For a

TABLE 12.50 Effect of Land Preparation on Soil Erosion by Wind on Newly Prepared Planted Maize Fields in Central Wisconsin

Land preparation	Soil type	Soil loss, t/ha
Plowed and planted	Plain field loamy sand	188.0
Disked and planted	Boone-Hixton loamy sand	62.7
Plowed and planted—crust broken	Plainfield loamy sand	44.8
No-tillage and planted	Richfield loamy sand	33.6
Untilled maize-killed field	Plainfield loamy sand	6.7
Disked winter-killed oats	Plainfield loamy sand	1.8
Standing chemically killed rye	Plainfield loamy sand	0.09

SOURCE: Woodruff, 1972.

wide range of soil and climates in both temperate and tropical regions, however, the no-tillage and mulch farming systems are synonymous with conservation tillage. For arid and semiarid regions, some form of mechanical tillage is necessary to loosen the compacted and crusted soils, to increase the surface roughness and surface detention capacity, and to increase the water infiltration rate. The frequency and intensity of mechanical tillage required to conserve soil and water, however, vary among soils and climates. In some soils, contour plowing may be all that is necessary. In others, minimum tillage necessary to break the crust is enough to control erosion. In still others, periodic loosening by deep ripping and chiseling, but without soil inversion, is necessary. Such treatment may, however, be required once every three to five cropping cycles. Deep tillage with soil inversion is required only for some problem soils, e.g., sodic soil where incorporating lime in subsoil horizons is a necessary restorative measure.

The basin tillage or tied-ridge system is a widely adapted conservation tillage system for many soils in semiarid and arid regions of the tropics and subtropics. For clayey vertisols, broad-bed and furrow systems and mulch tillage are useful conservation tillage measures.

References

Allmaras, R. R., and Holt, R. F. 1968. Water conservation in the subhumid, north latitude part of the corn belt of USA. *Ninth International Congress of Soil Science,* Adelaide, Australia. vol. 1, pp. 291–302.

Anonymous, 1986. Effects of mulching on crop growth in fields with reddish brown lateritic soils in central Thailand. In: *Proc. International Symposium on Red Soils,* pp. 671–675, Elsevier, Holland.

Baffoe-Bonnie, E., and Quansah, C. 1975. The effect of tillage on soil and water loss. *Ghana J. Agric. Sci.* 8:191–195.

Barber, R. G., and Thomas, D. B. 1981. *Infiltration, Surface Runoff and Soil Loss from High Intensity Simulated Rainfall in Kenya,* Faculty of Agriculture, University of Nairobi, Kenya.

Barker, M. R., and Wunsche, W. A. 1977. Plantio direto in rio Grande do sul, Brazil. *Outlook Agric.* 9:114–120.

Barnett, A. P., Diseker, E. G., and Richardson, E. G. 1967. Evaluation of mulching methods for erosion control on newly prepared and seeded highway back slope. *Agron. J.* 59:83–85.

Benatti, R. Jr., Bertoni, J., and Moreira, C. A. 1977. Erosion losses in a maize crop under conventional and no-tillage systems of cultivation in two soils of Sao Paulo. *Rev. Bras. Cienc. Solo* 1(2–3):121–123.

Bertoni, J., and Lombardi, Neto F. 1985. *Conservacao do solo,* Livruceres Ltda, Brasil.

Birot, P. 1968. *The Cycle of Erosion in Different Climates,* Batsford, London.

Bond, J. J., and Willis, W. O. 1971. Soil-water evaporation: Long-term drying as influenced by surface cover. *Soil Sci. Soc. Am. Proc.* 35:984–987.

Bonsu, M., and Obeng, H. B. 1979. Effects of cultural practices on soil erosion and maize production in the semi-deciduous rainforest and forest-savanna transitional zones of Ghana. In: *Soil Physical Properties and Crop Production in the Tropics* (R. Lal and D. J. Greenland, eds.), pp. 504–520, Wiley, Chichester, U.K.

Bradfield, R. 1970. Multiple cropping systems for the tropics. In: *Traditional Agricultural Systems and Their Improvements,* 16–20 November 1970, IITA, Ibadan, Nigeria.

Bridge, B. J., Mott, J. J., Winter, W. H., and Hartigen, R. J. 1983. Improvement in soil structure resulting from sown pastures on degraded areas in the dry savanna woodlands of Northern Australia. *Aust. J. Soil Res.* 21:83–90.

Bureau of Sugar Experiment Stations and the Department of Primary Industries. 1984.

A review of results of trials with trash management for soil conservation. *Proceedings Australian Society Sugarcane Technology 1984*, Brisbane, pp. 101–106.
Charreau, C. 1969. Effect of cultural techniques on runoff and erosion in casmance (Senegal). *Agron. Trop. (Paris)* 24:826–842.
Charreau, C. 1972. Problèmes posés par l'utilisation agricole des sols tropicaux par des sols tropicaux par des cultures annueles. *Agron. Trop.* 27:905–929.
Charreau, C., and Nicou, R. 1971. L'amelioration du profil cultural dans les sols sableux et sablo-angilleux de la zone tropicale seche Ouest Africaine et ses incidences agronomiques. *Agron. Trop. (Paris)* 26:1183–1247.
Chisci, G., and Zanchi, C. 1980. The influence of different tillage systems and different crops on soil losses on hilly silty-clayey soil. In: *Soil Conservation: Problems and Prospects* (R. P. C. Morgan, ed.), pp. 212–217, Wiley, Chichester, U.K.
Clark, R. N., and Hudspetch, E. B. 1976. Runoff control for summer crop production in the Southern Plains. *Trans. Am. Soc. Agric. Eng.*, Paper 76-2008, St. Joseph, Mich.
Constantinesco, I. 1976. *Soil Conservation for Developing Countries*. Rome, Italy, FAO Soils Bull. 30.
Das, D. C. 1976. Soil conservation practices and erosion control in India. In: *Soil Conservation and Management in Developing Countries*. Rome, Italy, FAO Soils Bull. 33, pp. 11–50.
Dawson, N., Berndt, R., and Venz, B. 1983. Landuse planning—Queensland canelands. *Proc. Australian Society Sugarcane Technology Conference,* pp. 43–52.
Derpsch, R., Sidiras, N., and Roth, C. H. 1986. Results of studies made from 1977 to 1984 to control erosion by cover crops and no-tillage techniques in Parana, Brazil. *Soil Tillage Res.* 8:253–263.
Dunne, T. 1976. Studying patterns of soil erosion in Kenya. In: *Soil Conservation and Management in Developing Countries*. Rome, Italy, FAO Soils Bulletin, 33, pp. 109–122.
Ehlers, W. 1975. Observation on earthworm channels and infiltration on tilled and untilled loess soil. *Soil Sci.* 199:242–249.
Evans, D. O., Yost, R. S., and Lundgren, G. W. 1983. *A Selected and Annotated Bibliography of Tropical Green Manures and Legume Covers,* University of Hawaii, Research Extension Series 028.
Fausey, N. R. 1984. Drainage-tillage interaction on Clermont soil. *Trans. ASAE* 27:403–406.
Freebairn, D. M., Ward, L. D., Clarke, A. L., and Smith, G. D. 1986. Research and development of reduced tillage systems for Vertisols in Queensland., Australia. *Soil Tillage Res.* 8:211–230.
Freebairn, D. M., and Wockner, G. H. 1982. The influence of tillage implements on soil erosion. In: *National Conference Publication 82/8*. Conference on Agricultural Engineering, Armidale, New South Wales, Australia, pp. 186–188.
Freebairn, D. M., and Wockner, G. H. 1983. Soil erosion control research provides management answers. *Qld. Agric. J.* 109:227–234.
Freebairn, D. M., and Wockner, G. H. 1984. Soil erosion and water balance studies, Darling Downs, Queensland, *Proc. National Soils Conference,* 13–18 May 1984, Brisbane, Australia, CSIR.
Freebairn, D. M., and Wockner, G. H. 1984–1985. Soil erosion and water balance studies, Darling Downs. *Soil Conservation Research Branch, Biennial Report,* Department of Primary Industrial, pp. 20–21.
Fullen, M. A. 1985. Compaction, hydrological processes and soil erosion on loamy sands in East Shropshire, England. *Soil Tillage Res.* 6:17–30.
Gilley, J. E. 1980. Runoff and erosion characteristics of a vegetated surface mined site in western North Dakota. *N. Dakota Farm Res.* 37(6):17–20.
Greb, B. W., Smika, D. E., and Black, A. L. 1967. Effect of straw mulch rates on soil water storage during summer fallow in the Great Plains. *Soil Sci. Soc. Am. Proc.* 31:556–559.
Harrold, L. L. 1972. Soil erosion by water as affected by reduced tillage systems. *Proc. No-Tillage Symposium 21–22 February 1972,* Ohio State University, pp. 21–28.
Harrold, L. L., and Edwards, W. M. 1972. A severe rainstorm test for no-tillage corn. *J. Soil Water Cons.* 27:30–36.

Heinonen, R. 1977. *Soil Management and Crop Water Supply,* Department of Soil Science, University of Uppsala, Sweden.
Hoogmoed, W. B. 1979. Tillage, rainwater infiltration and runoff, sowing and emergence of millets on a sandy soil near Niono, Mali. I. Report on the research activities summer. Report No. 1, Soil Tillage Laboratory, Agricultural University, Wageningen, Netherlands.
Hoogmoed, W. B., and Derpsch, R. 1985. Chisel ploughing as an alternative tillage system in Parana, Brazil. *Soil Tillage Res.* 6:53–68.
Hoogmoed, W. B., and Stroosnijder, L. 1984. Crust formation on sandy soils in the Sahel. I. Rainfall and infiltration. *Soil Tillage Res.* 4:1–4.
Johnston, J. R. 1973. Conservation and production systems for a better environment. In: *Conservation Tillage,* pp. 125–131, SCSA, Ankeny, Iowa.
Kalms, J. M. 1975. Studies on cultivation techniques at Bouake, Ivory Coast. In: *Soil Conservation and Management in the Humid Tropics* (D. J. Greenland, and R. Lal, eds.), pp. 195–200, Wiley, Chichester, U.K.
Kampen, J., Hari-Krishna, J., and Pathak, P. 1981. Rainy season cropping on deep Vertisols in the semiarid tropics—Effects on hydrology and soil erosion. In: *Tropical Agricultural Hydrology* (R. Lal and E. W. Russell, eds.), pp. 257–272, Wiley, Chichester, U.K.
Ketcheson, J. 1977. Conservation tillage in eastern Canada. *J. Soil Water Conserv.* 32:57–60.
Khatibu, A. I., Lal, R., and Jana, R. K. 1984. Effects of tillage methods and mulching on erosion and physical properties of a sandy clay loam in an equatorial warm humid region. *Field Crops Res.* 8:239–254.
Khybri, M. L., Bhardwaji, S. D., Prasad, S. N., and Ram, S. 1984. Effect of minimum tillage and mulch on soil and water loss in maize on 8% slope. *Conf. Proc. Soil Erosion and Counter Measures,* November 1984, Chiangmai, Thailand, Soil Conservation Society of Thailand.
Kowal, J. 1970a. The hydrology of a small catchment basin at Samaru, Nigeria. III. Assessment of surface runoff under varied land management and vegetation cover. *Niger. Agric. J.* 7:120–133.
Kowal, J. 1970b. The hydrology of a small catchment basin at Samaru, Nigeria. IV. Assessment of soil erosion under varied land management and vegetation cover. *Niger. Agric. J.* 7:134–147.
Kubota, T. 1982. Improvement of the moisture regime of upland soils in Thailand by soil management. In: *International Symposium on Distribution, Characteristics and Utilization of Problem Soils,* Japanese Society of Soil Science and Plant Nutrition, Tokyo, Japan, pp. 351–372.
Kubota, T., Varakpatananirund, P., Piyapongse, P., and Phetchawee, S. 1982. Improvement of the moisture regime of upland soils by soil management. In: *Proc. International Symposium on Distribution, Characteristics and Utilization of Problem Soils,* pp. 351–372, Japanese Society of Soil Science and Plant Nutrition, Tropical Agricultural Research Series 15, Tokyo.
Laflen, J. M., and Colvin, T. S. 1981. Effect of crop residue on soil loss from continuous row cropping. *Trans. Am. Soc. Agric. Eng.* 24:605–609.
Lal, R. 1975. Soil management systems and erosion control. In: *Soil Conservation and Management in the Humid Tropics* (D. J. Greenland, and R. Lal, eds.), pp. 93–97, Wiley, Chichester, U.K.
Lal, R. 1976a. *Role of Mulching Techniques in Tropical Soil and Water Management,* IITA Tech. Bull. 1, IITA, Ibadan, Nigeria.
Lal, R. 1976b. Soil erosion on Alfisols in Western Nigeria. I. Effects of slope, crop rotation and residue management. *Geoderma* 16(5):363–375.
Lal, R. 1976c. Soil erosion on Alfisols in Western Nigeria. II. Effects of mulch rates. *Geoderma* 16(5):377–387.
Lal, R. 1976d. *Soil Erosion Problems on an Alfisol in Western Nigeria and Their Control* IITA Monogr. 1, p. 160, Ibadan, Nigeria.
Lal, R. 1978. Influence of within and between-row mulching on soil temperature, soil

moisture, root development and yield of maize in tropical soil. *Field Crops Res.* 1:127–139.
Lal, R. 1982. Tillage research in the tropics. *Soil Tillage Res.* 2:305–309.
Lal, R. 1984. Mechanized tillage system effects on soil erosion from an Alfisol in a watershed cropped to maize. *Soil Tillage Res.* 4:349–360.
Lal, R. 1985. Mechanized tillage system effects on properties of a tropical Alfisol cropped to maize. *Soil Tillage Res.* 6:149–162.
Lal, R. 1986a. Effects of 6 years of continuous no-till or puddling systems on soil properties and rice yield of a loamy soil. *Soil Tillage Res.* 8:181–200.
Lal, R. 1986b. Conversion of tropical rainforest: Agronomic potential and ecological consequences. *Adv. Agron.* 39:173–269.
Lal, R. 1989. Conservation tillage for sustainable agriculture: Tropics vs. temperate environments. *Adv. Agron.* 42:85–197.
Lal, R., and Cummings, D. J. 1979. Clearing a tropical forest. I. Effects on soil and microclimate. *Field Crops Res.* 1:91–107.
Lal, R., De Vleeschauwer, D., and Nganje, R. Malafe. 1980. Changes in properties of newly cleared tropical Alfisols as affected by mulching. *Soil Sci. Soc. Am. J.* 44:827–833.
Lal, R., Wilson, G. F., and Okigbo, B. N. 1978. No-tillage farming after various grasses and leguminous cover crops in tropical Alfisols. I. Crop performance. *Field Crops Res.* 1:71–84.
Lal, R., Wilson, G. F., and Okigbo, B. N. 1979. Changes in properties of an Alfisol produced by various cover crops. *Soil Sci.* 127:377–382.
Lawes, D. A. 1962. The influence of rainfall conservation on the fertility of the loess plain soils of northern Nigeria. *J. Geogr. Assoc. Niger.* 5:33–38.
Lindstrom, M. J., Gupta, S. C., Onstad, C. A., Larson, W. E., and Holt, R. F. 1979. *Tillage and Crop Residue Effects on Soil Erosion in the Corn Belt.* SCSA Spec. Publ. 25, pp. 10–12, SCSA, Ankeny, Iowa.
Lindstrom, M. J., and Voorhees, W. B. 1980. Planting wheel traffic effects on interrow runoff and infiltration. *Soil Sci. Soc. Am. J.* 44:84–88.
Lindstrom, M. J., Voorhees, W. B., and Randall, G. W. 1981. Long-term tillage effects on interrow runoff and infiltration. *Soil Sci. Soc. Am. J.* 45:945–948.
Loch, R. J., and Donnollan, T. E. 1982. Field rainfall simulator studies on two clay soils of the Darling Downs Old. I. The effects of plot length and tillage orientation on erosion processes and runoff and erosion rate. *Aust. J. Soil Res.* 21:33–46.
Lombardi, Neto, F., Silva, I. R., Dechen, S. C. F., and Castro, O. M. 1980. Manejo de solo em replação à erosão e à produçao na cultura de Milho. In: *Congresso Brasileiro De Conservaçao do Solo,* 3, Brasilio.
Low, A. J. 1972. The effect of cultivation on the structure and other physical characteristics of grasslands and arable soils. *J. Soil Sci.* 23:363–380.
Machado, J. A. 1976. Efeito des sistemes de cultivco reduzido e convencional ne alteracao de algumas propriededed fisiscas e quimicas do solo. Tese de Livre doc. UFSMIR, Res. report. EMBRAPP, Brazilia, Brazil.
Manipura, W. B. 1972. Influence of mulch and cover crops on surface runoff and soil erosion on tea lands during the early growth of replanted tea. *Tea Q.* 43(3):95–102.
Mannering, J. V., and Fenster, C. R. 1977. Vegetative water erosion control for agricultural areas. In: *Soil Erosion and Sedimentation,* American Society of Agricultural Engineers, St. Joseph, Mich.
Mannering, J. V., and Fenster, C. R. 1983. What is conservation tillage? *J. Soil Water Conserv.* 141–143.
Mannering, J. V., and Johnson, C. B. 1966. Infiltration and erosion as affected by minimum tillage for corn. *Soil Sci. Soc. Am. Proc.* 30:101–105.
Mannering, J. V., and Meyer, L. D. 1961. The effects of different methods of cornstalk residue management on runoff and erosion as evaluated by simulated rainfall. *Soil Sci. Soc. Am. Proc.* 25:506–510.
Marston, D. 1978. Conventional tillage systems as they affect soil erosion—in northern New South Wales. *J. Soil Conserv. Serv. NSW* 34(4):194–198.
Marston, D., and Doyle, A. D. 1978. Stubble retention systems as soil conservation management practices. *J. Soil Cons. Serv. NSW* 34:210–217.

Marston, D., Anecksamphant, C., and Chirasalthaworu, R. 1984. The use of mulching and no-tillage for soil conservation in tropical upland crops. Fifth ASEAN Soil Conference, 10–23 June 1984, Bangkok, Thailand, pp. E1.1–E1.12, Asian Institute of Technology.

McCown, R. L., Jones, R. K., and Peake, D. C. I. 1985. Evaluation of a no-till, tropical legume ley-farming strategy. In: *Agro-Research for the Semi-arid Tropics: Northwest Australia* (R. C. Muchow, ed.), pp. 450–469, University Old Press, St. Lucia, Australia.

McDowell, L. L., and McGregor, K. C. 1984. Plant nutrient losses in runoff from conservation tillage corn. *Soil Tillage Res.* 4:79–92.

Misra, D. K., and Bhattacharya, B. B. 1963. A review of stubble mulching. *Indian J. Agron.* 7:256–268.

Moldenhauer, W. C., Langdale, G. W., Frye, W., McCool, D. K., Papendick, R. I., Smika, D. E., and Fryrear, W. R. 1983. Conservation tillage for erosion control. *J. Soil Water Conserv.* 38:144–151.

Mondardo, A., Henklain, J. C., Fairas, G. S., De Rufino, R. L., Juncksch, I., and Vieira, M. J. 1977. *Control of Erosion in Parana State,* Circular IAPAR No. 3, Londrina, Paraná, Brazil.

Morin, J., Benyamini, Y., Rawitz, E., and Hoogmoed, W. B. 1984. Tillage practices for soil and water conservation in the semi-arid zone. III. Runoff modeling as a tool for conservation tillage design., *Soil Tillage Res.* 4:215–224.

Mott, J., Bridge, B. J., and Arndt, W. 1979. Soil seals in tropical tall grass pastures of Northern Australia. *Aust. J. Soil Res.* 30:483–494.

Mullings, J. A., Truog, P. N., and Ptove, B. G. 1984. Options for controlling soil loss in canelands—some interim values. *Proc. Australian Society Sugar Cane Technology 1984 Conference,* Brisbane, pp. 95–100.

Ngatunga, E. L. N., Lal, R., and Uriyo, A. P. 1984. Effects of surface management on runoff and soil erosion from some plots at Mlingano, Tanzania. *Geoderma* 33:1–12.

Nicou, R. 1974. Contribution on the study and improvement of the porosity of sand and sandy-clay soil in the dry tropical zone. Agricultural consequences. *Agron. Trop.* 29:110–127.

Okigbo, B. N., and Lal, R. 1976. Role of cover crops in soil and water conservation. In: *Soil Conservation and Management in Developing Countries,* FAO Soils Bull. 33, pp. 97–108.

Okigbo, B. N., and Lal, R. 1977. Role of cover crops in soil and water conservation. In: *Soil Conservation and Management in Developing Countries.* Rome, Italy, FAO Soil Bull. 33, FAO, Rome.

Oliveira, J. B. De, and Silva, J. R. C. 1982. Effects of soil management on erosion of entropic red-yellow podzolic and solodized planosol soils of the homogeneous micro-region 68 of Ceara, Brazil. *Rev. Bras. Cienc. Solo* 6(3):231–235.

Oschwald, W. R. 1974. *Conservation Tillage, A Handbook for Farmers,* Soil Conservation Society of America, Ankeny, Iowa.

Paul, C. L. 1982. Influence of physical properties of clay soils on their management for sugar production of Guyana. *Trop. Agric.* 59(2):162–166.

Pena, M. L. 1981. Contour cropping and minimal stubble residue for erosion control on smooth trmao slopes. *Agric. Tech. (Chile)* 41(4):243–247.

Pereira, H. C., Chenery, E. M., and Mills, W. R. 1954. The transient effects of grasses on the structure of tropical soils. *Emp. J. Exp. Agr.* 22:148–160.

Pereira, H. C., Hosegood, P. H., and Dagg, M. 1967. Effects of tied ridges, terraces and grass leys on a lateritic soil in Kenya. *Exp. Agric.* 3:89–98.

Phillips, R. E., Thomas, G. W., and Blevins, R. L. 1980. *No-Tillage Research: Research Reports and Reviews,* University of Kentucky, Lexington.

Prentice, A. N. 1946. Tie-ridging, with special reference to semi-arid areas. *E. Afr. Agric. J.* 12:101–108.

Prihar, S. S., Singh, N. T., and Sandhu, B. S. 1979. Response of crops to soil temperature changes induced by mulching and irrigation. In: *Soil Physical Conditions and Crop Production in the Tropics* (R. Lal and D. J. Greenland, eds.), pp. 305–315, Wiley, Chichester, U.K.

Primavesi, A. 1982. *O Manejo Ecologico Do Solo,* Livraria Nobel, S.A., Brasil.

Prove, B. G., and Truog, P. N. 1984. Crop residue management for soil conservation in Queensland Canelands. *Proc. National Soils Conference 13–18 May 1984,* Brisbane, Australia, CSIRO.

Queensland Department of Primary Industries. 1984–1985. Soil Conservation Research Branch, Biennial Report, Brisbane.

Rama Mohan Rao, M. S. 1973. Effect of vertical mulch on moisture conservation and crop performance under dryland conditions in black cotton soils of Bellary. *J. Indian Soc. Soil Sci.* 21(2):237–239.

Rapp, A. 1976. Soil erosion and reservoir sedimentation-case studies in Tanzania. In: *Soil Conservation and Management in Developing Countries.* Rome Italy, FAO Soils Bull. 33, pp. 123–132.

Rawitz, E., Morin, J., Hoogmoed, W. B., Margolin, M., and Etkin, H. 1983. Tillage practices for soil and water conservation in the semi-arid zone. I. Management of fallow during the rainy season preceding cotton. *Soil Tillage Res.* 3:211–231.

Resource Conservation Glossary. 1982. Soil Conservation Society of America, Ankeny, Iowa.

Ripley, P. O., Kalbeleisch, W. M., Bourget, S. J., and Cooper, D. Y. 1961. *Soil Erosion by Water,* Canadian Department Agricultural Publ. 1084, Ottawa, p. 34.

Roose, E. J. 1977. Adaptation of soil conservation techniques to the ecological and socioeconomic conditions of West Africa. *Agron. Trop.* 32(2):132–140.

Roose, E. J. 1980. Dynamique actuelle de sols ferrallitiques et ferrugineux tropicaux d'Afrique Occidentale. Etude experimentale des transferts hydrologiques et biologiques de matières sous vegetations naturelles ou cultivées. Thèse Docteur en Sciences, University Orleans, France.

Roose, E. J., and Asseline, J. 1978. Measurement of erosion phenomenon under simulated rainfall at Adiopodoume; soluble and solid loads in runoff water from pineapple plantations. *Cah. ORSTOM Pedol.* 16:43–72.

Roose, E. J., and Fauck, R. F. 1980. Climatic conditions limiting the cultivation of ferrallitic soils in the humid tropical regions of the Ivory Coast. *Cah. ORSTOM Pedol.* 18(2):153–157.

Rosewall, C. J., and Marston, D. 1978. The erosion process as it occurs within cropping systems. *J. Soil Conserv. Serv. NSW* 34:186–192.

Rounce, N. V. 1946. *The Agriculture of the Cultivation Steppe,* Longman, Green & Co., Cape Town, Union of South Africa.

Rubber Research Institute of Malaysia (RRIM). 1977. *Mucuna cochinchinensis*—A potential short term legume cover plant. *Planters Bull.* 150:78–82.

Sallaway, M. M., Yule, D. F., and Ladeing, J. H. 1983. Effect of agricultural management on catchment runoff and soil loss in a semi-arid tropical environment. In: *Hydrology and Water Resources Symposium 1983,* pp. 252–256, Canberra, Australia.

Sallaway, M. M., Yule, D. F., Lawson, D., and Nickson, D. J. 1984. Runoff and soil loss study. *Central Old Proceedings National Soil Conference,* May 13–18, Brisbane, Australia, CSIRO.

SCSA. 1979. *Effects of Tillage and Crop Residue Removal on Erosion, Runoff, and Plant Nutrients,* SCSA Special Publ. 25, pp. 4–28, SCSA, Ankeny, Iowa.

Second National Soil Erosion Conference. 1978. EMBRAPA, Pass Fundo, Brazil.

Shaxson, T. F. 1975. Soil erosion, water conservation and organic matter. *World Crops UK* 27(1):6–10.

Shikula, N. K. 1981. Non-plowing land tillage and soil protection against erosion, Agricultural Academy, Kiev, Ukraine.

Sidiras, N., Henklain, J. C., and Derpsch, R. 1982. Comparison of three different tillage systems with respect to aggregate stability, the soil and water conservation and the yields of soybean and wheat on an Oxisol. In: *Proc. of the 9th Conference of the Soil Tillage Research Organization,* Osijek, Yugoslavia, 1982, pp. 537–544.

Silva, L. F. Da. 1981. Edaphic changes in "tabuleiro" soils (Haplorthoxs) as affected by clearing, burning and management systems. *Rev. Theobroma* 11:5–9.

Simpson, L. A., and Gumbs, F. A. 1982. Soil physical properties and sugar-cane root and shoot growth on conisiana banks after conversion from cambered beds field layout in Trinidad, West Indies. *Trop. Agric.* 59:38–42.

Smika, D. E., and Wicks, G. A. 1968. Soil water storage during fallow in the Central Great Plains as influenced by tillage and herbicide treatments. *Soil Sci. Soc. Am. Proc.* 32:591–595.

Smith, G. E., Whitaker, F. D., and Heinemann, H. G. 1974. *Losses of Fertilizers and Pesticides from a Claypan Soil.* Environmental Protection Technology Service, EPA-660/2-74-068, Washington, D.C.

Sprague, M. A., and Triplett, G. B. (eds.). 1986. *No-Tillage and Surface Tillage Agriculture,* Wiley, New York.

Steichen, J. M. 1984. Infiltration and random roughness of a tilled and untilled claypan soil. *Soil Tillage Res.* 4:251–262.

Stewart, B. A., Dusek, D. A., and Musick, J. T. 1981. A management system for the conjunctive use of rainfall and limited irrigation of graded furrows. *Soil Sci. Soc. Am. J.* 45:413–419.

Stewart, B. A., Woolhizer, D. A., Wischmeier, W. H., Caro, J. H., and Frere, M. H. 1975. *Control of Water Pollution from Cropland,* vol. 1, *A Manual for Guideline Development.* USDA Rept. ARS-H-5-1, Washington.

Stroosnijder, L., and Hoogmoed, W. B. 1984. Crust formation on sandy soils in the Sahel. II. Tillage and its effect on the water balance. *Soil Tillage Res.* 4:321–327.

Suwardjo, A., and Sutono, A. 1984. Effect of mulch and tillage on soil productivity of a Lampung Red Yellow Podsolic, Indonesia. *Pembr. Pen. Tanah Dan Pupuk* 3:12–16.

Swanson, A. P., Derick, A. P., Weekly, H. E., and Haise, H. R. 1965. Comparing mulches—scientists check effects of four mulching materials on 6% slope. *Agric. Res.* 13(8):195–208.

Taiwan Agricultural Research Institute. 1983. *Experiments on Soil Conservation Practices for Slopeland of Pineapple Plantation,* Annual report 1983, pp. 98–100.

Touchton, J. T., Gardner, W. A., Hargrove, W. L., and Duncan, R. R. 1982. Reseeding rimson clover as a N source for no-tillage grain sorghum production. *Agron. J.* 74:283–287.

Tulaphitak, T., Pairintra, T., and Kyuma, K. 1983. Soil fertility and tilth. In: *Shifting Cultivation* (K. Kyuma and T. Pairintra, eds.), pp. 63–83, Ministry of Science, Technology, and Energy, Thailand.

Unger, P. W. 1978. Straw-mulch effect on soil water storage and sorghum yield. *Soil Sci. Soc. Am. J.* 42:486–491.

Unger, P. W. 1984. *Tillage Systems for Soil and Water Conservation.* Rome, Italy, FAO Soils Bull. 54.

Unger, P. W., and Phillips, R. E. 1973. Soil water evaporation and storage. In: *Conservation Tillage.* Soil Conservation Society of America, Ankeny, Iowa, pp. 42–54.

USDA-ARS. 1974. *Summer Fallow in the Western United States,* USDA Conservation Rept. 17, Washington.

Vieira, M. J., Cogo, N. P., and Cassol, E. A. 1978. Erosion losses under different tillage systems for soyabeans [*Colycine max* (L) Merr.] using simulated rainfall. *Rev. Bras. Cienc. Solo* 2(3):209–214.

Wang, S. T. 1984. Management of problem soils in Taiwan, ROC. In: *Ecology and Management of Problem Soils in Asia,* FFTC Book Series 27, FFTC, Taipei, Taiwan, pp. 74–88.

Wilkinson, G. E. 1975. Effect of grass fallow rotations in Northern Nigeria. *Trop. Agric.* 57:97–103.

Wilson, G. F., Lal, R., and Okigbo, B. N. 1982. Effects of cover crops on soil structure and on yield of subsequent arable crops grown under strip tillage on an eroded alfisol. *Soil Tillage Res.* 2:233–250.

Wischmeier, W. H. 1973. Environmental considerations. In: *Conservation Tillage,* SCSA, Ankeny, Iowa, pp. 133–142.

Wittmus, H. D., Triplett, G. B. Jr., and Greb, B. W. 1973. Concepts of conservation tillage systems using surface mulches. In: *Conservation Tillage,* SCSA, Ankeny, Iowa, pp. 5–12.

Woodruff, N. P. 1972. Wind erosion as affected by reduced tillage systems. In: *Proc. No-Tillage Systems Symposium, Columbus, Ohio, February 1972,* pp. 5–20, Ohio State University, Columbus.

Wunsche, W. A., and Denardin, J. E. 1980. Soil conservation and management. I. The uplands of Rio Grande do Sul. General consideration. *Circ. Tec.* no. 2, EMPRAPA, Brazil.

Chapter 13

Trees and Soil Erosion

The natural forest cover, defined as a mixture of naturally growing diverse tree species and shrubs with its characteristic undergrowth and a layer of leaf litter on the soil surface, supposedly protects the soil against the erosive action of various agents. The antierosive effects of a natural forest are often greater than those of a planned forest. The latter is defined as a monoculture stand of a single specie of trees planted and managed for commercial use. An example of a successful planned forest is the dipterocarp forest of *Shorea javanica* in Sumatra (Torquebian, 1984). The monoculture tree system lacks the diversity of the natural forest. The planned forest does not have the multistory canopy and the undergrowth. The soil under monoculture planned forests may not have as favorable a level of physical properties as is normally the case for a soil found under a natural forest.

The forest cover, both planned and natural, provides more protection to soil against erosion than seasonal or annual crops do. The soil surface is more or less continuously covered by tree canopy and the leaf litter. Furthermore, the soil surface is neither disturbed nor exposed as is usually the case for cultivation of seasonal or annual crops that require seedbed preparation for a satisfactory stand establishment. In addition to the lack of soil disturbance, the deep root system with a dense root mat immediately beneath the leaf litter binds the soil and provides additional strength against the shearing force of runoff or the splashing effect of through-fall. Some undisturbed forests are closed ecosystems with efficient mechanisms involved in water and nutrient recycling (Turvey, 1975; Nortcliff et al., 1979; Cole, 1981; Lal, 1987).

Soil erosion is usually less under forest cover than from agricultural land regardless of the climatic conditions. Risks of soil erosion may, however, be greater under tropical than temperate forests. Tropical rain forest is confined to the equatorial region and is a more diverse and complex ecosystem than that of temperate-zone forests (Plate 13.1). Because of the prevailing high temperatures, however, the decomposition rate of leaf litter and biomass on the forest floor is higher in the tropics than in temperate climate. Therefore, the leaf litter does not accumulate as readily under tropical conditions as in a temperate-zone forest. There are also differences in rainfall and soil conditions in the temperate and tropical regions, as explained in Chaps. 2 and 3.

Soil Erosion under Forest

The predominant erosion processes under undisturbed natural forest cover are splash, creep, and sheet flow. Rill erosion and gully erosion occur only under special conditions, particularly when the natural ecosystem is disturbed by logging road construction and other anthropogenic perturbations.

Splash

Forest canopy, especially multistory canopy and dense undergrowth with a thick layer of leaf litter and debris on the soil surface, protects the soil against raindrop impact. These ideal conditions, however, are not always present, as is often the case in a primary forest. Shrub undergrowth is sparse in primary forests because there is insufficient light for a satisfactory undergrowth. Under tropical rain forest, though, the amount of leaf litter is low because of the high decomposition rate (Thaiutsa and Granger, 1980). Furthermore, the erosivity of

Plate 13.1   A tropical rain forest with a multistory, diverse, and evergreen canopy.

canopy drip, comprising large coalesced drops falling from a height of 10 m or more, is extremely high (see Chap. 10). Splash erosion therefore occurs in a mature or a primary forest due to high energy and low litter cover on the soil surface.

Soil creep

Another process of soil erosion on the forest floor is soil creep. *Soil creep* is defined as "slow mass movement of soil and soil material down relatively steep slopes, primarily under the influence of gravity but facilitated by saturation of water and by alternate freezing and thawing" (SCSA, 1982). The rate of soil creep depends on the soil, age of the forest, amount of leaf litter, slope, rainfall characteristics, etc. The measured rate of soil creep ranges from 0.5 cm/yr (Eyles and Ho, 1970) to about 1.0 cm/yr (Williams, 1973; Lewis, 1974).

Rill erosion

Rill erosion is caused by concentrated overland flow. The flow concentration may be caused by natural depressions, animal tracks, and the stem flow. The concentrated stem flow from large trees during heavy rains can initiate rill erosion near the tree base. Birot (1968) observed that, in primary forest with high rainfall, rainwater streaming down tree trunks of medium girth has sufficient energy to cause rill erosion.

Water balance

The quantity of runoff depends on soil properties including the infiltration rate, activity of microfauna, and presence of cracks. Generally, the infiltration rate is greater in forest than in savanna soil, as is also true for the activity of microfauna (Daubenmire, 1971). There are more soil cracks, however, in a savanna than in a forest soil.

The water balance of tropical rain forests has been studied by many (Hopkins, 1960; Clegg, 1963; Sheng and Koh, 1965; Nieuwolt, 1965; Kerfoot and McCulloch, 1967; Flenley, 1971; Brinkman, 1972; Low, 1972; Jackson, 1975; Chanphaka et al., 1976; Garcia and Bernollia, 1979; Manokarau, 1980; Parfait and Lallmahomed, 1980; Lockwood, 1980; Lawson et al., 1981). The throughfall varies from 70 to 90 percent, stem flow from 10 to 30 percent, and canopy interception from 10 to 20 percent. High variation in different components is attributed to tree density, canopy cover, and rainfall characteristics.

The proportion of rainfall lost as runoff from forest depends on the soil, canopy cover, and terrain. The stem flow usually combines with surface runoff to cause a noticeable rill erosion at the forest floor. The proportion of surface runoff under natural forest is low for regions with rainfall of less than 2000 mm and where interflow or subsurface flow is not very pronounced. In the forested catchments in Kumaun, Himalaya, India, Pandey et al. (1983) measured the average overland flow in the forested catchment to be only 0.50 percent of the total incident rainfall. In regions of high annual precipitation, however, 10 to 20 percent of the rainfall received may be lost as runoff. For example, in Puerto

Rico, Odum (1970) observed that 44.6 percent of the total rainfall of 3759 mm in a forested catchment was lost as surface and subsurface runoff. In humid tropical regions of northeastern Australia, storm runoff has been measured to be as much as 63 percent of the mean annual rainfall of 4175 mm (Gilmour and Bonell, 1977; Bonell and Gilmour, 1978; and Bonell et al., 1979). Excessive runoff in the wet tropical forests is caused by saturated flow in the top few centimeters of the forest floor (Gilmour et al., 1980).

Erosion

High runoff losses from a forested catchment can also cause high sediment movement depending on the soil properties, slope gradient, and amount of leaf litter. More erosion is likely to occur under semideciduous rain forest conditions with a pronounced dry season of 3 to 5 months than in a continuously wet climate.

Erosion by landslides and gully can occur on steep slopes and where the ecological balance has been disturbed. Under lowland mature rain forest, however, mass movement and landslides are rare. Erosion under natural rain forests, measured for a range of ecological environments (Dyrness, 1967; Douglas, 1967; Lal, 1986), substantiates this claim.

In Malaysia, the annual suspended and dissolved sediment load from a natural forest was measured to range from 0.20 to 0.65 t/ha (Sim and Hock, 1986). Hatch (1981) measured erosion rates of 0.05 to 0.12 t/(ha · yr) from primary and secondary forests in Malaysia. In Indonesia, it was observed that erosion was 0.0 t/(ha · yr) under undisturbed natural forest, 0.4 t/(ha · yr) when a few trees were removed, and 48.0 t/(ha · yr) when leaf litter and undergrowth were also removed. In northern Thailand with a mean annual rainfall of about 1500 mm, Takahashi et al. (1983) measured erosion rates under natural forest ranging from 2.9 to 4.7 m^3/(ha · yr). In the Colombian rain forest zone, McGregor (1980) observed wide fluctuations in runoff under the forest cover, but the sediment load was extremely low. In Hawaii, Ahuja and El-Swaify (1979) and Doty et al. (1980) evaluated water flow and sediment transport in forested catchments. The total annual sediment transport from forested catchments in the Koolan Mountains ranged from 0.084 to 6.17 t/ha. Doty et al. (1981) observed that about 90 percent of the total suspended sediment was produced in less than 2 percent of the time.

Logging and removal of leaf litter are among major reasons for increasing erosion from forested catchments. Gilmour (1971, 1977) observed from a north Queensland rain forest catchment that sediment source areas are linked with logging activities. Similar observations were reported for the Tawan Hills Forest Reserve of Malaysia by Lieu (1974) and for the Philippines by Gananpin (1978).

Conversion of Natural Forest to Planned Forest

Risks of soil erosion increase when natural vegetation is replaced even by quick-growing tree crops, e.g., timber, rubber, oil palm, coconut, etc. The magnitude of the increase depends on the degree of soil disturbance, method of de-

TABLE 13.1 Through-fall and Erosive Power under Tree Canopies in Java

	Tree height, m	Canopy cover, %	Through-fall, % incident rainfall	Erosive power,* % incident rainfall
Tree plantations				
Acacia auriculiformis	13	80	81	119
Albizzia falcataria	20	40	80	102
Anthocephalus chinensis	10	80	80	147
Bamboo sp.			80	180
Agroforestry systems				
Hamegarden			79	202
Mixed garden			81	196

*Erosive power measured by splash cups.
SOURCE: Lembaga Ekologi, 1980.

forestation, time required for the planned forest to reestablish complete ground cover, and management system adopted (e.g., growing cover crops). One of the principal sources of sediments in planned forests is the access roads and footpaths. Improperly constructed and poorly maintained access roads cause severe gully erosion. Another reason is the high erosivity of canopy drip because of the big size and high terminal velocity of the coalesced drops. In Indonesia, Lembaga Ekologi (1980) reported that the erosive power of canopy drip may be

Plate 13.2 A cover of kudzu (*Pueraria phaseoloides*) established in a coconut plantation in Indonesia.

as much as 200 percent of the natural rain. He observed that the erosive power of through-fall and canopy drip under planned forest differed among tree species (Table 13.1).

In Malaysia, Sim and Hock (1986) observed that the annual erosion rate of 0.20 to 0.39 t/ha under natural forest increased to 1.6 to 4.1 t/ha immediately after deforestation and declined again to 1.1 t/ha when oil palm was established. The annual sediment load increased by about 14-fold during the first year after deforestation. The corresponding concentration of suspended sediments was 28 to 112 mg/L under forest. The maximum sediment concentration recorded after clear felling was 8126 mg/L. Low erosion rates and low sediment concentration under oil palm were related to the establishment of a cover crop (Plate 13.2).

Rubber and teak are also important crops that have replaced tropical rain forest in Asia, Africa, and tropical America. Erosion under young rubber plan-

TABLE 13.2 Effects of Conversion of Rain Forest to Plantation Crops on Erosion in Java, Indonesia

	Average runoff of small watersheds per year, %	Erosion, t/(ha · yr)
1. Natural forest	1.7	0
Idem with cut trees	7.5	0.4
Idem with removed litter and undergrowth	46.4	48.0
2. *Albizzia falcataria* plantation	4.0	0.8
Idem without litter and undergrowth	71.0	79.8
Imperata grassland	6.0	2.0
Agricultural land	23.0	39.1
3. Mahogany and teak plantation	5.2–7.1	0.8–1.1
Planted teak:		
1st year with taungya	13.75	5.2
2d year	8.15	0.6
3d–4th years	6.0	1.0
Secondary vegetation with *Imperata*	6.6–7.8	1.3–1.9
4. *Imperata* vegetation	5.0	3.5
Lantana vegetation	2.1	5.1
Agriculture	11.9	345.0
5. Taungya system:		
Cassava	20.3	56.8
Cassava and corn	15.5	37.5
Cassava and peanut	15.7	31.5
Cassava and rice	17.8	31.5
Cassava and soybean	12.9	49.8

tation can be more severe than under oil palm because of the practice of clean weeding between trees. In Malaysia, Leigh (1982) reported that sediment transport under 6- to 8-year-old rubber was about 16 times more than that in the natural forest. High erosion rates of 80 to 100 t/(ha · yr) are observed under clean-weeded mature rubber (RRI, 1974). Mature rubber has a low canopy cover and little, if any, litter falls.

In Java, Indonesia, erosion rates were measured under the plantations of *Albizzia falcataria* and mahogany and teak. In comparison with no erosion under natural forest, erosion under mature plantations ranged from 0.8 to 1.1 t/(ha · yr). Erosion rates, however, increased drastically when trees were cut, the leaf litter was removed, or food crops were grown in association with trees (Table 13.2). Similar to the forest cover, the data in Table 13.2 also show that even the *Imperata* vegetation provides an effective protection against erosion. Soil erosion increased 100 times when *Imperata* vegetation was converted to agriculture. Erosion rates were also very high under *Taungya* system (Table 13.2).

Experiments conducted in east Africa have shown that the magnitude of the increase in stream flow by deforestation of a catchment depends on the proportional reduction in forest cover (Pereira, 1965). Furthermore, the water balance of a fully developed tea plantation was found to be similar to that of the forested catchment. There were significant differences, however, during the initial phases of tea canopy establishment.

In the Philippines, it has been observed that soil erosion on steep slopes under pure stand of *Leucaena leucocephala* was greater than that under natural forest. Densely planted *Leucaena* suppresses undergrowth by shading and by possible allelopathic effects. High water runoff originating on steep slopes

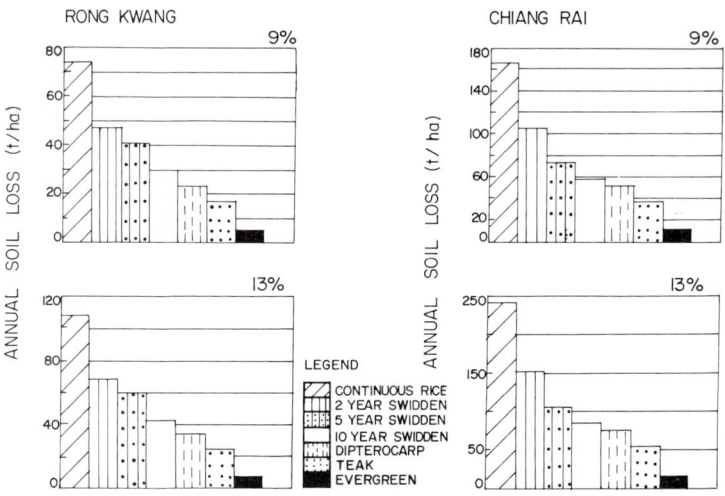

Figure 13.1 Effects of tree crops, natural forest cover, and swidden cultivation on soil erosion (Attaviroj, 1986).

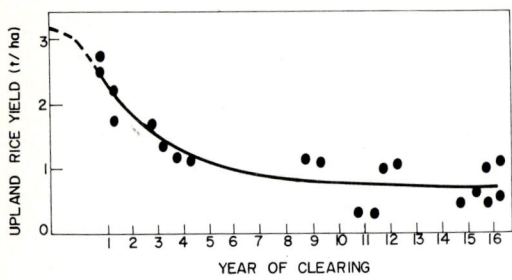

Figure 13.2 Decline in yield of rice with continuous cultivation (Attaviroj, 1986).

under *Leucaena* can cause severe erosion (IDRC, 1982). Also in the Philippines, Garcia (1981) observed less erosion in recently cut *Leucaena* stand because of biomass on the surface and vegetation regrowth than in an old stand of *Leucaena*. With good management, however, *Leucaena* and other trees can be used as an antierosive measure on croplands (Jones, 1983). In Thailand, Attaviroj (1987) observed that soil erosion at Rong Kwang and Chiang Rai was the least under evergreen forest and that erosion increased with increasing intensity of cultivation (Fig. 13.1). Soil erosion was also less under teak and dipterocarp plantations than under swidden cultivation for different cycles of fallowing. With swidden cultivation, the yield of rice and maize declines sharply with years after first clearing (Figs. 13.2 and 13.3). The rate of decline in yield depends on the application of fertilizers and management (Figs. 13.4 and 13.5).

In Nigeria a watershed management experiment was done on a gently sloping land of 3 to 5 percent slope whereby a high rain forest was converted to oil palm, trees, banana, and coconut plus pasture. Even during the first year after deforestation, erosion was generally high under plantation crops compared with the forested control (Table 13.3).

The following conclusions can be drawn from the literature surveyed above:

1. The surface runoff from catchments with natural forest is generally low but increases with increasing annual rainfall amount.

2. Soil erosion and sediment transport from the catchments with natural forests are also low and of the order of about 1 t/(ha · yr).

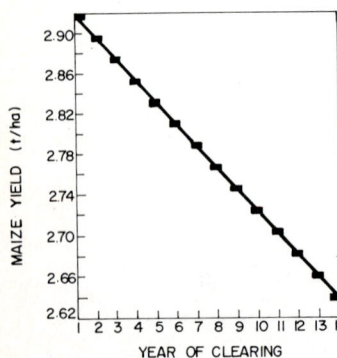

Figure 13.3 Decline in yield of maize with continuous cultivation (Attaviroj, 1986).

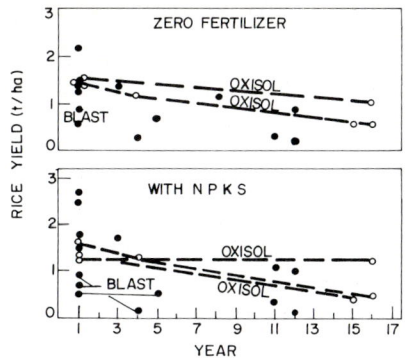

Figure 13.4 Effect of continuous cultivation on the yield of rice with and without fertilizers (Attaviroj, 1986).

3. Soil erosion increases when natural forest is changed to planned forest. Soil erosion is greater during the initial stages of tree establishment than when tree canopy is fully developed. The risks of soil erosion under plantation crops and planned forests differ among crop species depending upon their canopy characteristics and the soil management systems used.

Similar conclusions are drawn from the data available from temperate-zone forests. Because of the low risks of soil erosion under forest cover, establishing trees and afforestation are recommended measures for erosion control on arable lands.

Erosion Control in Tree Crops and Planned Forests

Because of slow initial growth, a considerable ground surface is exposed to raindrop impact during the initial 2 to 4 years until the tree canopy is fully developed. There are three general considerations. First, removal of existing vegetation should be done so as to not disturb the soil or remove the leaf litter. Second, a good general management practice includes the establishment of a suitable

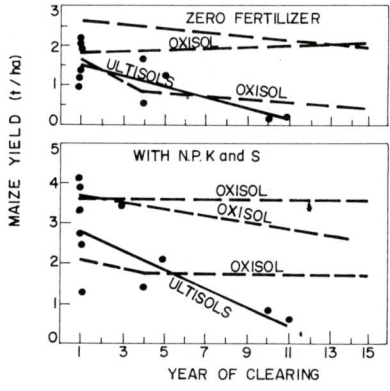

Figure 13.5 Yield of maize (corn) with and without fertilizer application as influenced by years of cultivation (Attaviroj, 1986).

TABLE 13.3 Effects of Forest Conversion on Runoff and Erosion in Nigeria

Treatment	Runoff, mm/4 months	Erosion, kg/(ha · 4 mo)
1. Forest	0.25	0.32
2. Oil palm + maize	66.4	170.1
3. Plantain	33.3	157.3
4. Coconut + pasture	39.1	183.1
5. Tree crops	39.8	172.4

Rainfall = 1285 mm from July to October 1985.
SOURCE: IITA-UNU, 1985.

creeping legume soon after deforestation, to establish a quick ground cover. Third, the management of forest plantation is as important for erosion control as is soil and crop management on arable lands. Special water management practices may be required for steep terrains.

Land clearing

If feasible, land should be cleared manually so as to minimize adverse effects on soil physical and chemical characteristics. If land must be cleared mechanically, it is important not to remove litter, roots, and stumps. The time of land clearing should be carefully chosen so that the soil is not too wet, and use of heavy machinery should be avoided. Mechanical equipment used should not

Plate 13.3 Clearing a rain forest in Sumatra by chain saw.

Trees and Soil Erosion 433

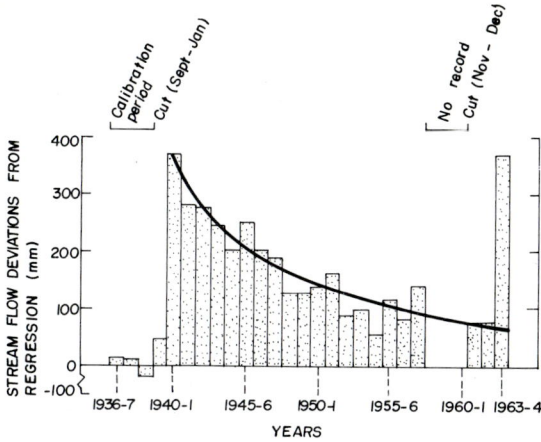

Figure 13.6 Effects of clear felling on stream flow measured at the Coweeta Hydrological Laboratory, Pennsylvania State University (Hibbert, 1967).

scrape off the topsoil. Windrows should be established 40 to 50 m apart so that trees and stumps are not dragged over long distances. A chain saw is appropriate equipment for manual land clearing. This equipment has been widely used in Sumatra, Indonesia, for establishing large areas of tree crops (Plate 13.3).

Large-scale deforestation by clear felling increases total water yield from the watershed. The total water yield comprises both surface runoff and the seepage flow. Hibbert (1967) compared water runoff from twin watersheds, one cleared and another maintained under forest, at the Coweeta Hydrological Laboratory, Pennsylvania State. The data in Fig. 13.6 show a drastic increase in water runoff from the cleared watershed immediately after clear felling in 1940. Water runoff decreased with time, however, as the forest regrew from 1940 to 1962. The runoff increased once again in 1962 to a level similar to that of 1940, when the forest was cut once again. Effects of large-scale intensive fire on the water yield from a watershed are similar to that of clear felling. The data in Fig. 13.7 show the effects of fire on discharge recorded from a stream draining the reser-

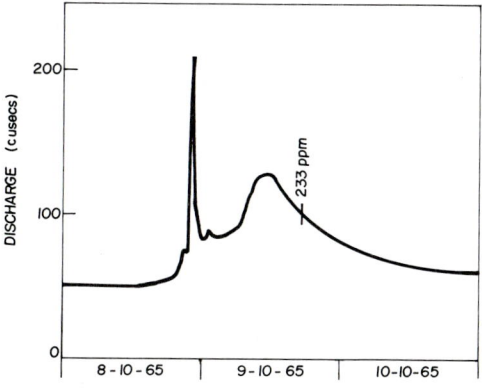

Figure 13.7 Effects of forest fire on stream flow measured in the Snowy Mountains Hydro-Electric Authority Project in the Australian Alps. (Adapted from Pereira, 1973.)

voir of the Snowy Mountains Hydro-Electric Authority Project in the Australian Alps. There was a sharp increase in the peak flow from the burned watershed—from 60 to 80 m³/s before fire to 370 m³/s after fire. There was a 100-fold increase in the suspended sediment load. The sediment concentration measured a few months after the fire was 14.4 percent by weight.

The results of experiments conducted on twin watersheds at IITA, Ibadan, Nigeria, showed that the total water yield increased drastically after the watershed was cleared for seasonal crop production. Lawson et al. (1981) observed that an intermittent stream was converted to a perennial stream following de-

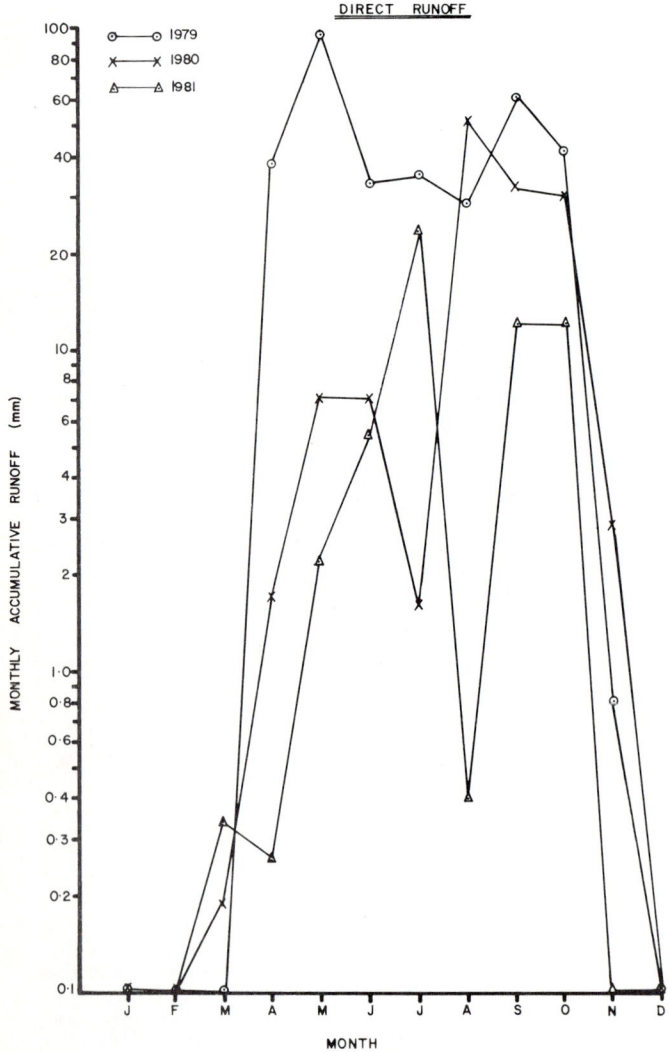

Figure 13.8 Total water yield in 1986 from a tropical watershed cleared in 1979.

forestation with continuous flow throughout the year. A measurable flow was continued for cleared watershed throughout the dry season. In addition to changing the base flow patterns, the experiments at IITA showed that deforestation completely changed the surface hydrology. The data of total water yield from a 44-ha cleared watershed are shown for 1986 in Fig. 13.8. Compared with no or slight water yield from the forested watershed, the water yield increased to about 30 to 40 percent of the total rainfall received. Similar results were reported from the Ivory Coast. Roose (1979) observed that runoff and erosion were 50 and 1000 times greater on cleared than uncleared forested land.

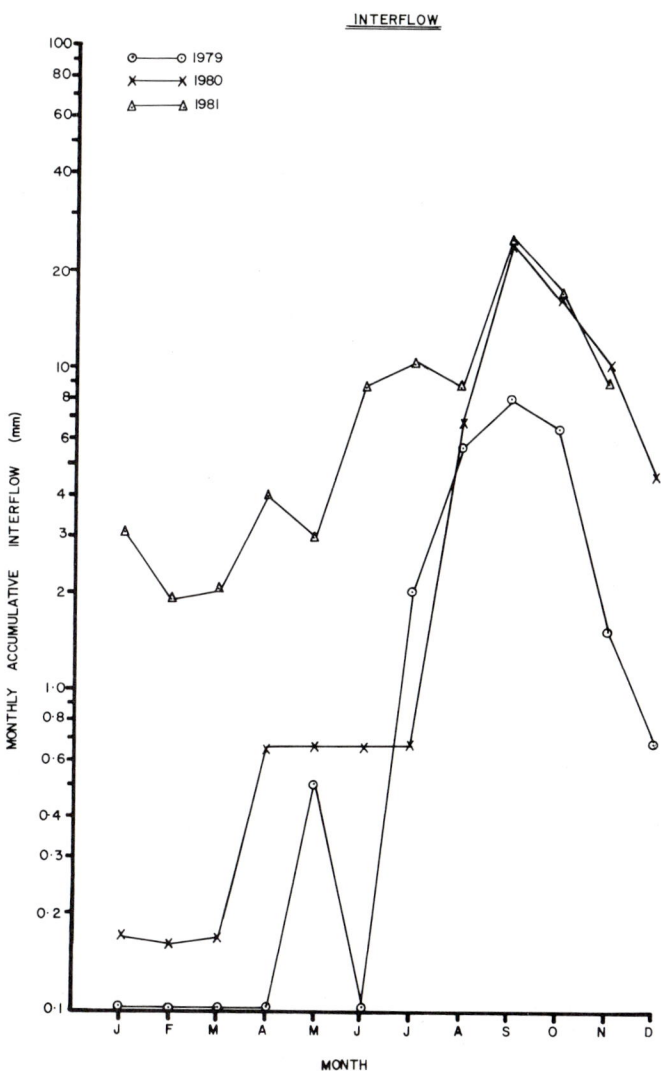

Figure 13.8 (*Continued*)

TABLE 13.4 Effects of Methods of Deforestation on Runoff and Soil Erosion

Method of land clearing	Runoff, mm/yr	Soil erosion, t/(ha · yr)
Traditional*	3	0.01
Manual†	35	2.5
Shear blade*	86	3.8
Tree pusher/root rake†	202	17.5

*No-tillage only (each mean is an average of two independent measurements).
†Mean of no-tillage and plowed treatments (each figure is a mean of four independent measurements).
SOURCE: Modified from Lal, 1981.

Field experiments conducted in Nigeria and elsewhere have shown that, among different types of mechanical equipment used, use of shear blade (Plate 13.4) causes less runoff and erosion than use of tree pusher/root rake (Plate 13.5). The data in Table 13.4 show that runoff and soil erosion were considerably less from land cleared with shear blade than with tree pusher/root rake. Even though they are labor-intensive, slow, manual methods of land clearing are ecologically more sound than mechanical methods.

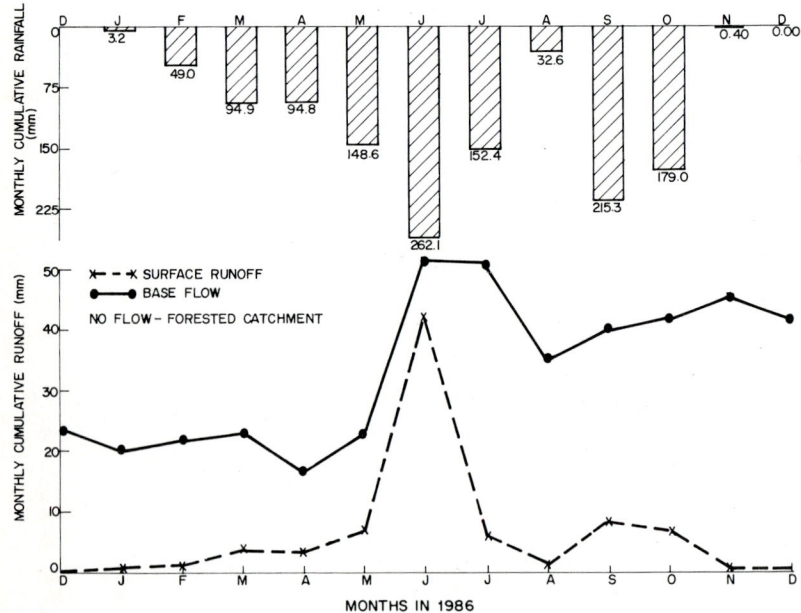

Figure 13.8 (*Continued*)

Plate 13.4 A front-mounted shear blade.

Establishing a cover crop

The land cleared of its vegetation cover is prone to soil compaction, a reduction in the infiltration rate, and an increase in runoff and erosion. Some degree of soil compaction is inevitable regardless of the land clearing method. Soil compaction, however, is more severe in tropical than in temperate climates, in

Plate 13.5 The front-mounted tree pusher/root rake attachment.

structurally inert than in structurally active soils, in mechanically than in manually cleared land, and in seasonals/annuals than in perennials or tree crops.

There are many examples of field measurements indicating soil compaction following land clearing (Cunningham, 1963; Nye and Greenland, 1964; Van der Weert, 1974; Seubert et al., 1977; Lal and Cummings, 1979; Hulugalle et al., 1984; Lal et al., 1979, 1986; Barus et al., 1986). The data in Table 13.5, also from an alfisol at IITA, Ibadan, Nigeria, indicate a drastic increase in soil bulk density and penetrometer resistance after mechanical land clearing. As a conse-

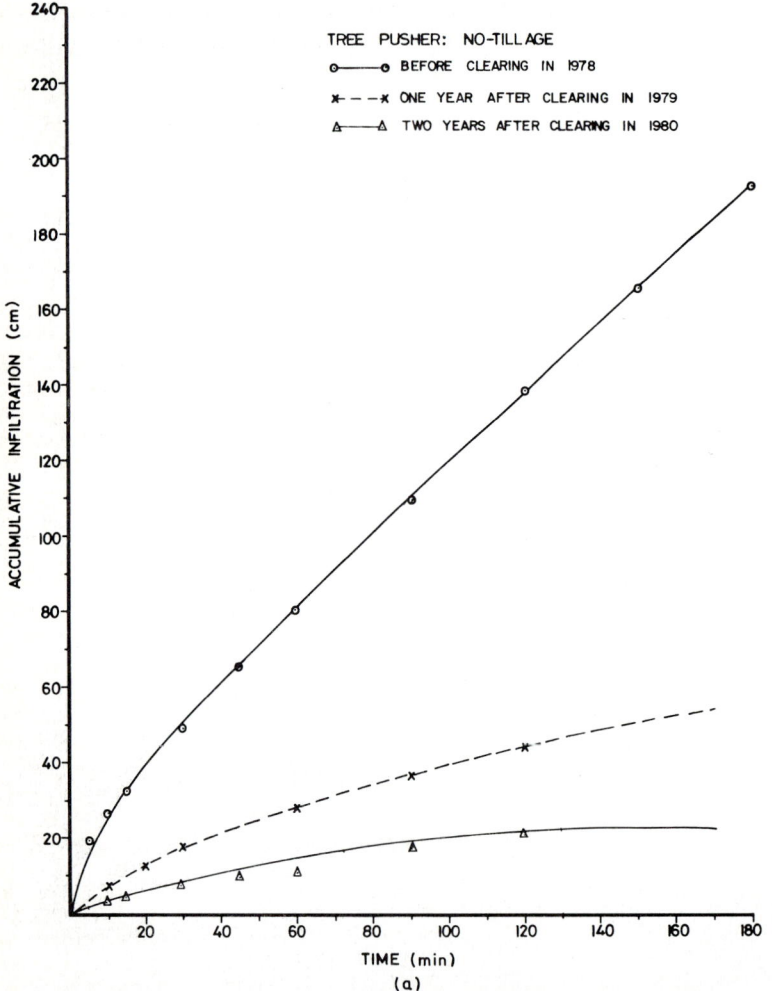

Figure 13.9 Effects of deforestation on reduction in infiltration rate with (a) tree pusher/root rake clearing, (b) shear blade clearing, (c) manual clearing, (d) clearing for shifting cultivation, (e) traditional farming.

quence of compaction and the decrease in the proportion of macropores, there is a drastic reduction in both the cumulative infiltration and the infiltration rate of the mechanically compared with manually cleared land (Fig. 13.9).

An important step following the deforestation is the alleviation of soil compaction and restoring of the soil structure. These are achieved preferably by biologic rather than mechanical means. Experiments conducted around the world have shown that deep-rooted legumes improve soil physical properties and increase macroporosity. Growing 1 to 2 years of a suitable cover crop increases the soil organic matter content and macroporosity and decreases runoff and erosion. The data in Table 13.6 from IITA, Ibadan, Nigeria, show that the porosity in plots cleared manually, by shear blade, and by tree pusher was increased by

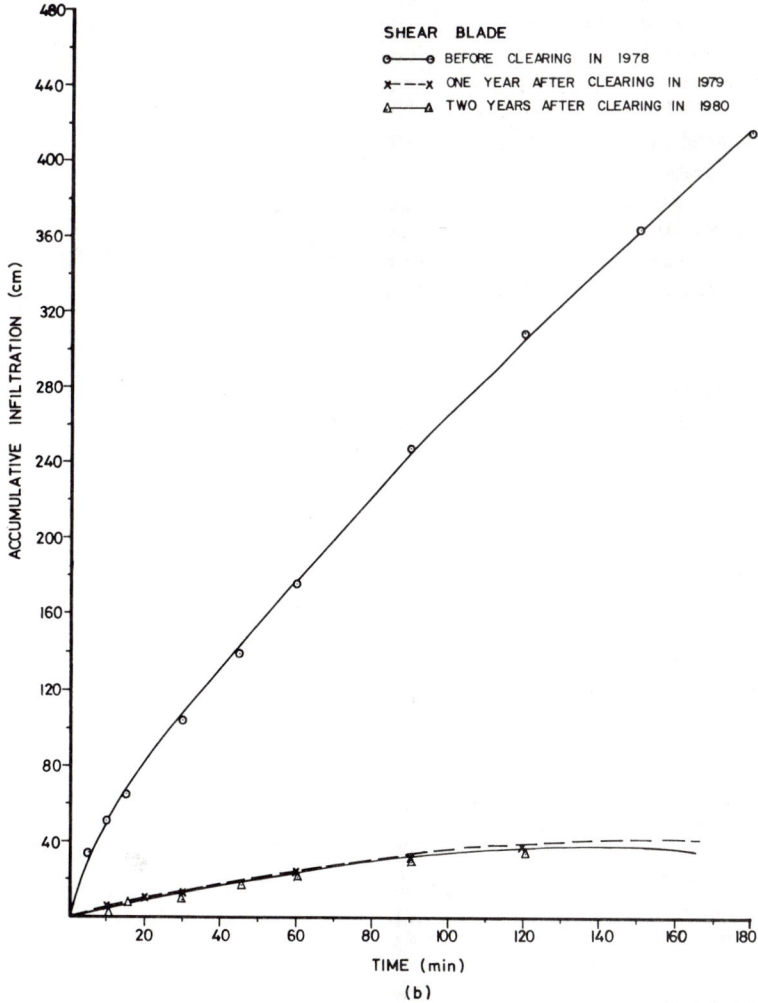

Figure 13.9 (*Continued*)

planting the mucuna cover. The increase in porosity was caused by the increase in the relative proportion of the transmission pores greater than 5 micrometers (μm). As a result, planting the mucuna cover also brought about an increase in infiltration rate. Similar results have been reported by Lal et al. (1979).

In Guinea, Fournier (1967) observed a satisfactory erosion control in citrus by growing cover crops. Soil erosion was decreased to merely 14.3 and 28.6 percent of the control by growing citrus in association with *Pueraria* and *Dolicus lablab,* respectively (Table 13.7). Growing a cover crop is considered to be an integral part of good soil management for growing rubber and oil palm in Malaysia. In Colombia, use of shade tree and cover crop or mulch reduced erosion from 8.2 to 0.1 to 1.3 t/(ha · yr) (Rodriguez, 1958). Similar results were reported

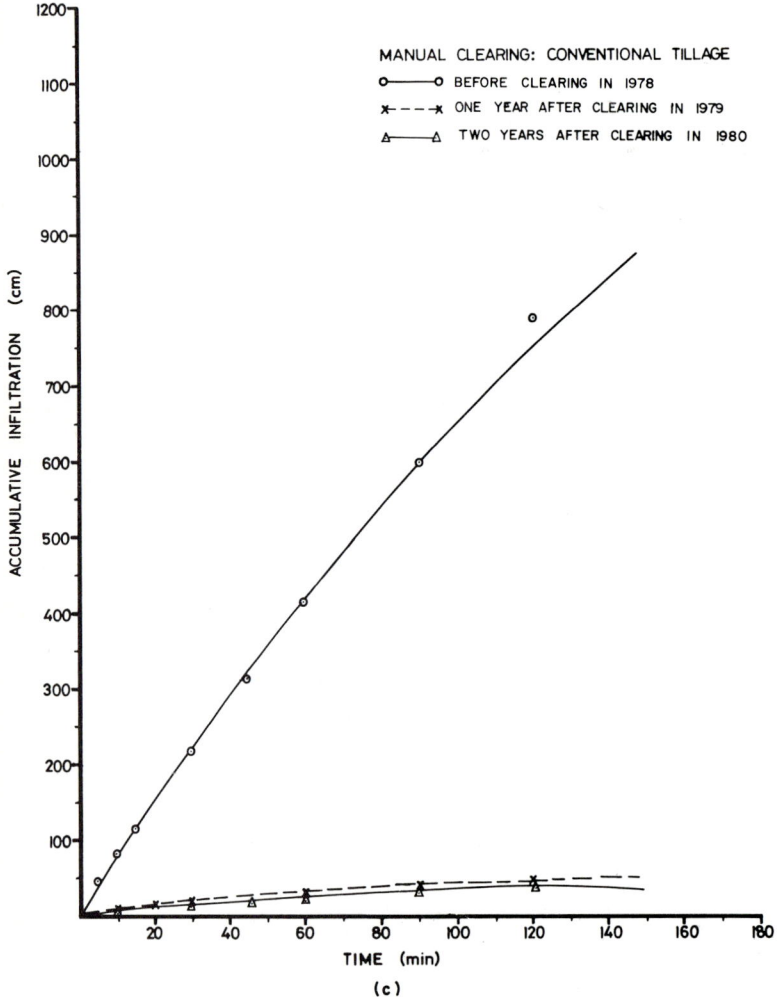

Figure 13.9 (*Continued*)

from Costa Rica by Bermudez Mendez (1980). In Tanzania, use of residue mulch and cover crops decreased erosion under plantation crops from 34 to 54 t/(ha · yr) to 0.75 to 5.6 t/(ha · yr) (Mitchell, 1965).

In addition to establishing cover crops, crop residue mulch and microwatersheds are also widely used to control erosion in some high-valued plantation crops, e.g., fruit orchards, coffee and tea plantations, rubber, oil palm, and citrus (Plates 13.6 and 13.7). In Kenya, mulch from specially grown grasses has been profitably used to control erosion in fields of young tea (Othieno, 1974, 1975, 1982). A combination of cover crops, mulching, and terracing is recommended for erosion control in tea gardens of Darjeeling, India (Sarkar, 1974). Mulching and growing Bahia grass have proved useful in controlling erosion in

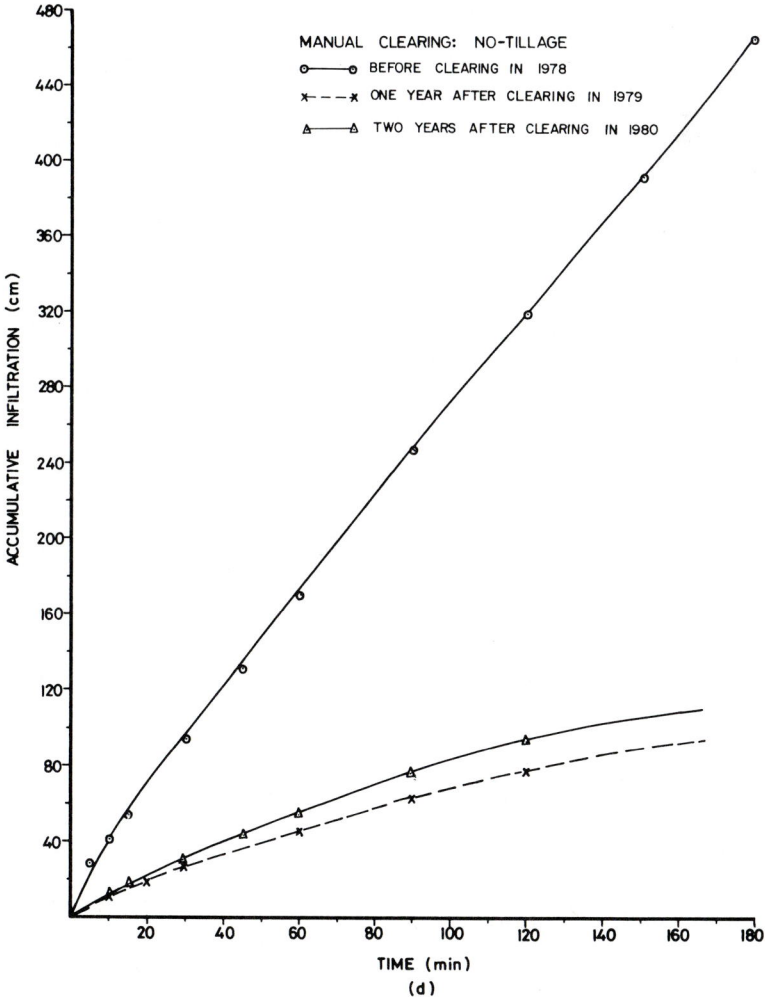

Figure 13.9 (*Continued*)

442 Erosion Control

orange plantations in Taiwan (Liao and Chang, 1975). Use of grass strips and mulch have been recommended for pineapple cultivation in Trinidad (Alleyne and Percy, 1966) and Martinique (Ducreux et al., 1980).

A relevant example of erosion control under tree crops is seen in the experiments conducted in Indonesia. The data in Table 13.8 show that soil erosion levels under plantations, undisturbed forest, and tree crops grown in association with a cover crop or mulch were similar and low. Soil erosion, however, increased drastically when tree crops were clean-weeded, forest plantations were burned, or leaf litter was removed. Although erosion was severe during the cropping phase of shifting cultivation, it was relatively less under the Taungya system—an agroforestry system of growing food crops in association with annu-

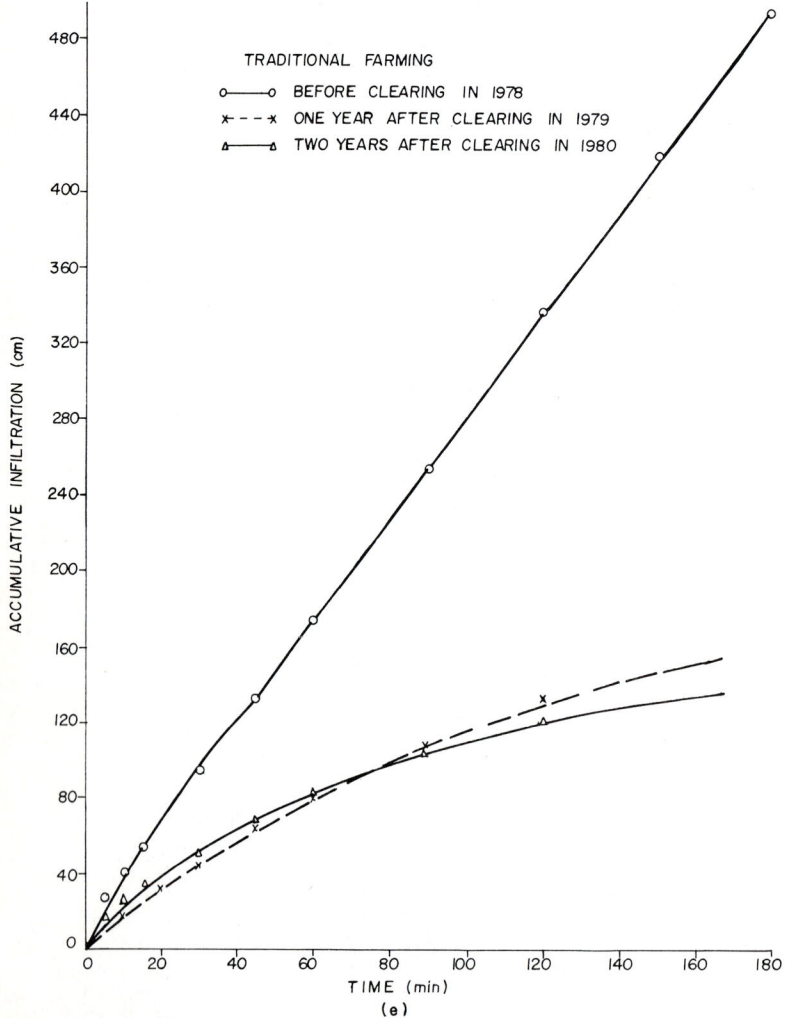

Figure 13.9 (*Continued*)

TABLE 13.5 Effects of Methods of Deforestation on Soil Bulk Density and Penetrometer Resistance of 0- to 5-cm Layer

Treatment	Bulk density, g/cm^3				Penetrometer resistance, kg/cm^2			
	Preclearing				Preclearing			
	1978	1979	1980	1981	1978	1979	1980	1981
Traditional farming	0.64	1.06	1.07	1.27	0.21	0.96	0.52	1.32
Manual clearing	0.68	1.17	1.17	1.39	0.20	1.4	0.75	1.19
Shear blade	0.70	1.19	1.37	1.38	0.26	1.0	1.84	2.19
Tree pusher/root rake	0.60	1.24	1.32	1.42	0.20	1.3	0.73	1.23

Each figure is a mean of 25 separate analyses.
SOURCE: Unpublished data of R. Lal.

als. This practice is discussed at length in another section of this chapter. Another example of establishing ground cover in controlling erosion under rubber plantation is shown by the data in Table 13.9 from Malaysia. In addition to legumes, ferns and grasses are used as cover crops to control erosion. The fern *Nephrolepis* was even more effective than the grass cover in controlling erosion under rubber in Malaysia.

TABLE 13.6 Effect of Clearing Methods and Cropping Treatments on Porosity (0- to 100-mm depth), Penetrometer Resistance (50- to 70-mm depth), and Infiltration Characteristics*

Treatments	Porosity, m^3/m^3	Penetrometer resistance,† kPa	Infiltration rate, mm/s		Cumulative infiltration in 3 h,† mm	
			1 min	180 min		
Manual/cropped	0.55b	426.4b (8.2a)	0.51b	0.02cd	607.5bc	(1.1c)
Manual/mucuna cover	0.59a	408.7d (8.6a)	0.72a	0.08a	1422.4a	(7.2a)
Shear blade/cropped	0.54c	425.4b (8.6a)	0.34c	0.02cd	492cd	(1.8c)
Shear blade/mucuna cover	0.57b	417.6c (8.8a)	0.45bc	0.04b	764.1b	(3.5b)
Tree pusher/cropped	0.57b	436.2a (8.0a)	0.17d	0.02cd	287.0de	(1.8c)
Tree pusher/mucuna cover	0.59a	398.0f (8.8a)	0.35c	0.004d	328.9de	(7.1a)
Tree pusher/root rake/cropped	0.53c	411.7d (8.6a)	0.20d	0.002d	193.0e	(1.4a)
Tree pusher/root rake/mucuna cover	0.52c	402.9e (8.3a)	0.25cd	0.03bd	556.8bcd	(4.5b)

*Values within the same column followed by the same letter do not differ significantly at the 5% level of probability (Duncan's multiple-range test).
†Values in parentheses show soil water content [m^3/(m^3 · %)] at time of measurement.
SOURCE: Hulugalle et al., 1984.

TABLE 13.7 Effects of Cover Crops on Erosion Control in Citrus Plantation Established on a 7% Slope in Kindra, Guinea

Year	Observation	Citrus on bare soil	Citrus plus *Pueraria*	Citrus plus *Dolichos lablab*	Pineapple
1956	Runoff, %	9.4	2.1	2.2	1.6
	Erosion, t/ha	17.9	3.1	5.6	2.05
1957	Runoff, %	9	2.8	4.2	4.4
	Erosion, t/ha	30.7	4.25	9.3	9.9
1958	Runoff, %		2.4	2.4	4.2
	Erosion, t/ha	8.1	0.75	1.2	1.15
Mean	Runoff, %	9.2	2.4	2.9	3.4
	Erosion, t/ha	18.9	2.7	5.4	1.4

SOURCE: Fournier, 1967.

Establishing perennial crops on small farms, such as those in the tropics, can be facilitated by the following series of steps, suggested by Von Uexkull (1984).

1. Underbrush the land during the rainy season, pile up the debris and biomass in separate heaps, and leave big trees standing.
2. Seed an appropriate legume cover (e.g., *Mucuna* or *Pueraria*) without any soil disturbance.

Plate 13.6 Mulching a young coffee bush in Cameroon, west Africa.

TABLE 13.8 Erosion under Different Tropical Moist Forest and Tree Crops Systems

Systems	Soil erosion, t/(ha · yr)		
	Minimal	Median	Maximal
Multistory tree gardens (4/4)*	0.01	0.06	0.14
Natural forests (18/27)	0.03	0.30	6.16
Shifting cultivation, fallow period (6/14)	0.05	0.15	7.40
Forest plantations, undisturbed (14/20)	0.02	0.58	6.20
Tree crops with cover crop or mulch (9/17)	0.10	0.75	5.60
Shifting cultivation, cropping period (7/22)	0.40	2.78	70.05
Taungya cultivation (2/6)	0.63	5.23	17.37
Tree crops, clean-weeded (10/17)	1.20	47.60	182.9
Forest plantations, burned/litter removed (7/7)	5.92	53.40	104.8

*Expression in parentheses is fraction x/y, where x = number of locations and y = number of treatments/observations.
SOURCE: After Wiersum, 1984.

3. Ring-bark the large trees so that they will die slowly and shed their leaves.

4. Fell standing trees manually about 3 months after ring barking. Cover crops are usually fully established within 3 to 4 months and provide a good ground cover.

5. Transplant tree seedlings after localized clearing of the cover crop, either manually or by using herbicides.

The key components of this approach are planning ahead of time, manual clearing, no burning, and establishing an appropriate cover crop.

Plantation management for erosion control

Adoption of judicious cultural practices for plantation management is important to minimize erosion risks. Some forest plantations are deliberately started at high tree population. The initial high tree population is thinned to an optimum tree stand at a later stage. Once the plantation is established, however,

TABLE 13.9 Effects of Ground Cover Management on Soil Erosion under Rubber

Soil series	Slope, deg	Rainfall, mm	Soil erosion, t/(ha · yr)		
			Bare	Grass cover	*Nephrolepis**
Rengam	4–5	2920	82	35	Negligible
Serdang	3–4	3250	106	94	47

*A fern.
SOURCE: Rubber Research Institute, 1974.

Plate 13.7 Microcatchments for tree establishments in eroded lands in Niger.

pruning and thinning are important not only to improve the timber quality but also to control erosion. The overcrowded plantation with dense stand may be suitable for some species (such as *Pinus patula*) but not for others. The unthinned stand of trees such as teak, gmelina, and *Cupressus lusitanica* usually results in bare soil surface that can cause severe sheet and rill erosion (Plate 13.7).

TABLE 13.10 Engineering Devices Recommended for Erosion Control on Plantations

Land class	Maximum slope,* deg	Minimum soil depth, mm	Conservation treatment	Maximum intensity of land use†
1	7		0 to 2° contour cultivation	Any
			2 to 7° channel terraces	Any
2	15	1000	Bench terraces	Any
3	20	500	Step terraces	Close cover crops and semiperennials
4	25	500	Step terraces or hillside ditches	Tree crops with ground cover
5	33	250	Orchard terraces or platforms	Tree crops with ground cover (no cultivation)
6	>33	—	None	Forest only

*Equivalent slopes are 12, 27, 36, 42, and 65 percent.
†Minimum soil depths are required when terraces are to be cut into the hillside.
SOURCE: Hudson, 1977.

Road construction in plantations is another important factor affecting soil erosion. Runoff from roads constructed up and down the hill causes severe gully erosion. Logging roads often lead to severe gullying.

Grazing and fire are other causes of soil erosion in plantations. Both uncontrolled grazing and voluntary fires cause severe damage to plantations and accelerate soil erosion. Forest fire denudes vegetation, alters soil structure, and accelerates soil erosion. In Australia, Clinnick (1984) and Atkinson (1984) observed that erosion from forested catchments increased from 2.5 to 8 t/(ha · yr) before fire to as much as 32.5 t/ha in a single rainstorm recorded about 10 weeks after a wild fire destroyed a plantation.

Water management

Installation of some engineering structures may be necessary for safe disposal of excess water runoff, especially on steep slopes. The necessity of installing such devices, however, is minimized if the land is properly cleared and a cover crop is established soon after deforestation. Table 13.10 lists the guidelines for selecting an appropriate land use and engineering structures on different slopes and soil types. Although growing food crop annuals is recommended for cultivation on slopes of less than 7°, these slope limits can be exceeded if food crop annuals are grown with appropriate soil and crop management systems, discussed in Chaps. 11 and 12. Many comprehensive manuals have been prepared such as that by Liao and Wu (1987), among others, to install engineering structures for orchards and tree crops grown on steeplands. Check dams and

TABLE 13.11 Effects of Cultural Practices on Runoff and Soil Erosion from Coffee Grown on a Podzolic Soil in Brazil

Management	1 to 5 yr		6 to 10 yr	
	Erosion, t/ha	Runoff, %	Erosion, t/ha	Runoff, %
Mechanical cultivation	49.4	7.9	0.09	0.16
Herbicides	26.2	6.0	0.09	0.14
Disk harrow	19.1	4.0	0.40	0.36
Without contour layout	13.0	4.7	0.11	0.31
Planting in squares (small)	11.2	3.9	0.07	0.26
Hedge planting (small)	10.7	2.9	0.03	0.09
Planting on contour (small)	8.0	2.6	0.05	0.11
Planting on contour (large)	7.8	3.5	0.04	0.24
Hoe weeding	7.1	2.9	0.09	0.15
Hedge planting (large)	6.7	1.9	9.04	0.16
Contouring	4.5	2.0	1.13	0.49
Forage crop	0.6	0.3	0.01	0.09
Green manuring the forage crop	0.4	0.2	—	0.03

SOURCE: Lombardi et al., 1975a.

TABLE 13.12 Effects of Type of Conservation Practices on Soil and Water Loss from a Coffee Plantation Established on a Podzolic Soil in Brazil

Type of practices	Soil erosion, t/ha		Water runoff, % of rainfall	
	1–5 yr	6–10 yr	1–5 yr	6–10 yr
Superficial soil cultivation	26.9	0.17	5.7	0.2
Mechanical engineering devices	6.9	0.04	2.6	0.2
Vegetative biologic techniques	6.3	0.04	2.0	0.1
Soil management techniques	2.5	0.57	1.1	0.3

SOURCE: Lombardi et al., 1975b.

spillways have been successfully used for controlling erosion on mountainous watersheds in northwest Pakistan (Hanif, 1979).

Similar to the erosion control in arable lands, biologic and soil management techniques of erosion control are also more effective in plantation tree crops than engineering and mechanical crop techniques. The effectiveness of the engineering devices is increased when they are used in association with biologic measures. Quick and effective ground cover, suitable crop management, manuring and balanced fertilizer application, pest control, and cover crops are all as necessary for tree crops as for seasonal crops. In Brazil, Lombardi et al. (1975a, 1975b) compared the effects of different management systems on soil and water loss from coffee. The data in Table 13.11 show the following:

1. Soil erosion is more severe in the first 5 years of establishment than in the next 5 years.
2. Mechanical soil disturbance by disk harrowing, tillage, etc., causes the most severe soil erosion and water runoff.
3. Biologic measures (e.g., cover crops, green manuring) caused the least erosion. These conclusions are supported by the data in Tables 13.12 and 13.13 for podzol and red latosolic soils, respectively. The least soil erosion in coffee

TABLE 13.13 Effects of Type of Conservation Practices on Soil and Water Loss from a Coffee Plantation Established on a Red Latosol Soil in Brazil

Type of practices	Soil erosion, t/ha		Water runoff, % of rainfall	
	1–5 yr	6–10 yr	1–5 yr	6–10 yr
Superficial soil cultivation	1.13	0.19	1.7	1.2
Mechanical engineering devices	1.10	0.06	1.7	1.0
Vegetative biologic techniques	1.15	0.06	1.7	1.1
Soil management techniques	0.60	0.01	1.3	1.2

SOURCE: Lombardi et al., 1975b.

plantation occurred when appropriate soil management and biologic erosion control practices were used in conjunction with the engineering devices.

Afforestation for Erosion Control

Establishing a tree cover is a widely used, effective, and usually inexpensive measure for controlling erosion, stabilizing steep slopes, and rehabilitating eroded and degraded lands. Whereas tree canopy provides protection against wind and water erosion, tree roots stabilize the soil and increase its resistance by forming a network of root mat (Plate 13.8). Afforestation has long been considered a possible solution to environmental problems and energy difficulties in many regions, e.g., Africa. Weinstabel and Zech (1982) recommended that afforestation with fast-growing trees may, to some extent, alleviate the severe problems of environmental degradation in semiarid Africa. Experiments conducted

Plate 13.8 There exists a thick mat of roots below the leaf litter under the forest canopy. These roots anchor the soil even on steep terrain.

TABLE 13.14 Reduction in Stream Flow Caused by Reforestation in Southern United States

River and gaging station	Ocmulgee at Macon	Oconee at Milledgeville	Flint at Culloden	Oconee at Greensboro	Chattahoochee at West Point	Savannah at Augusta	Tallapoosa at Wodley	Chattahoochee at Norcross	Saluda at Columbia	Saluda at Silverstreet
Basin area, km²	5800	7640	4790	2820	9195	19,450	4300	3030	6500	4200
Dates of collection	1900–1910, 1912, 1929–1940, 1955–1975	1904–1932, 1955–1975	1912–1922, 1929–1930, 1938–1940, 1955–1975	1904–1932, 1937–1940, 1955–1975	1900–1940, 1955, 1959–1975	1900–1906, 1926–1940, 1955–1960, 1962–1970, 1972–1975	1923–1940, 1955–1975	1902–1940, 1955–1975	1929–1939, 1955–1975	1927–1939, 1955–1966
Percentage of area reforested	27.5	27.5	25.7	21.3	20.0	15.4	11.8	11.0	10.5	9.7
Average runoff, cm	43.1	40.4	46.7	47.0	57.0	47.5	53.3	70.0	41.0	49.6
Regression analysis										
Decrease of water yield, cm	3.6	5.8	6.35	9.9	3.8	8.9	2.5	2.8	6.6	4.8
Decrease, %	8.4	14.4	13.6	21.0	6.7	18.7	4.7	4.0	16.1	19.6
Probability*	0.09	0.01	0.01	0.01	0.07	0.05	0.06	0.20	0.05	0.28
Double-mass analysis										
Decrease of water yield, cm	3.8	6.1	4.8	9.4	4.3	5.6	2.8	6.1	6.9	6.1
Decrease, %	8.8	15.1	10.3	20.0	7.5	11.8	5.3	8.7	16.8	12.3
Probability†	0.01	0.01	0.01	0.01	0.01	0.01	0.05	0.06	0.05	0.01

*Analysis of covariance, dummy variable.
†Analysis of covariance.
SOURCE: Trimble and Weirich, 1987.

in the southeastern United States have shown that afforestation reduces stream flow. Trimble and Weirich (1987) observed that ten large river basins totaling 54,000 km^2 had 10 to 88 percent of their respective areas reforested during the period from 1919 to 1967, decreasing the water yield from 3 to 10 cm. These reductions constituted a 4 to 21 percent decrease in annual stream discharge (Table 13.14). Manuals for afforestation are available for different regions in the United States (Weber, 1977) and in Commonwealth countries (Commonwealth Science Council, 1986).

A wide range of tree species are available for afforestation. For example, the principal tree species recommended for afforestation in Ethiopia are listed in Table 13.15. The choice of tree species is an important consideration. Among other factors (ease of establishment, high growth rate, canopy cover, multipurpose uses), rooting patterns of trees play an important role in soil erosion control and in stabilizing steep slopes. A dense and extensive root system is very useful for erosion control and slope stabilization.

Although there are many criteria for choosing an appropriate tree species for erosion control, little is known about their root systems and their effects on soil properties and on erosion control. In Taiwan, Yen (1984) studied the root system of about 100 tree species commonly used for afforestation. From the point of view of erosion control, he classified the root system into five categories (Fig. 13.10):

1. *Horizontal or H type.* The maximum depth of root development is shallow to medium, with about 80 percent of the roots confined to the top 60-cm layer. Most roots extend laterally, and the horizontal spread is greater than the vertical distribution.

2. *Right or R type.* The maximum depth of root development is deep, and at least 20 percent of the roots extend beyond the 60-cm depth. In addition to deep vertical distribution, the R-type root system has an oblique lateral distribution.

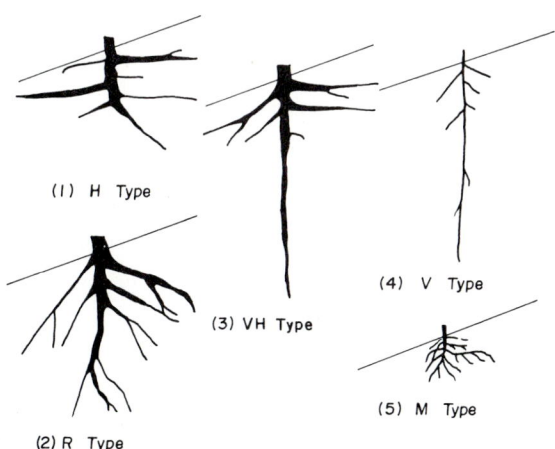

Figure 13.10 Major types of tree root systems (Yen, 1984).

Erosion Control

TABLE 13.15 Some Forest Species Recommended for Afforestation in Semiarid and Arid Ethiopia

Species (indigenous and introduced)	Elevation, m	Rainfall, mm
Acacia abyssinica	2200–3000	800–1500
Acacia albida	1500–2000	600–1000
Acacia decurrens	1500–2500	900–1600
Acacia melanoxylon	1500–1500	900–2700
Acacia mollissima	1500–250	900–1600
Acacia saligna	0–2500	300–700
Acacia senegal	0–1500	200–500
Acacia tortolis	800–1800	300–900
Adhalhota species	1700–2500	600–1300
Albizzia lebbeck	0–1400	500–1000
Azadirachta indica	0–500	450–1000
Aningeria adolfifreiderici	1600–2600	800–220
Casuarina equisetifolia	0–1400	750–1100
Cordia africana	1000–2000	600–1500
Cupressus lusitanica	1300–330	1000–1500
Cupressus macrocarpa	1200–3500	700–1600
Cupressus toruloso	1500–2800	650–1000
Ekebergia capensis	1600–2600	900–1400
Eucalyptus camaldulensis	500–2000	400–1000
Eucalyptus citriodora	0–1800	650–1600
Eucalyptus delegatensis	2000–3000	1000–2000
Eucalyptus darympleana	2000–3500	750–1500
Eucalyptus fastigata	1600–2500	750–1100
Eucalyptus grandis	0–2100	1000–4000
Eucalyptus globulus	1500–3000	900–1800
Eucalyptus maidenii	1000–2100	760–2000
Eucalyptus maculata	1000–2000	620–1250
Eucalyptus nitens	2000–35000	750–1500
Eucalyptus saligna	500–2100	1000–4000

TABLE 13.15 Some Forest Species Recommended for Afforestation in Semiarid and Arid Ethiopia (*Continued*)

Species (indigenous and introduced)	Elevation, m	Rainfall, mm
Eucalyptus viminalis	2000–3000	750–2500
Hygenea abyssinica	2500–3200	1000–1500
Grevilea robusta	800–2100	700–1200
Juniperus procera	1800–3000	600–2000
Leucaena leucocephala	0–1600	600–1000
Parkinsonia aculeata	0–1400	250–400
Moringa stenopetala	800–1600	600–1000
Parkinsonia aculeata	0–1400	250–400
Pinus patula	1400–3200	750–2000
Pinus pseudastrobus	1300–2800	1000–1500
Pinus michoacana	1000–2300	1000–1700
Pinus montezymae	1400–3000	900–1600
Podocarpus gracillar	1700–2200	800–2000
Prosopis juliflora	0–2000	200–600
Sesbania aculeata	800–1500	700–1000
Schinus molle	1000–3500	300–620
Tamaris aphylla	0–1400	200–500
Ziziphus spinachristi	0–800	>100

Fruit Trees in Soil and Water Conservation*

1. *Annona squamosa*   Custard apple
2. *Caria papaya*   Papaya
3. *Citrus sinensis*   Orange
4. *Mangifera indica*   Mango
5. *Musa paradisiaca*   Banana
6. *Passidium guajava*   Guava
7. *Passiflora edulis*   Passion fruit
8. *Persea americana*   Avocado
9. *Prunus persica*   Peach

*From Ato Berhanu Eka, Forest Research Institute, Addis Ababa, Ethiopia.
SOURCE: Chadhokar, 1986.

3. *Vertical-Horizontal or VH type.* The maximum depth of root development is medium to deep, and about 80 percent of the roots are confined to the top 60-cm layer. This root system is characterized by a deep taproot or pendent root. Although lateral roots also grow profusely, they are mostly confined to the shallow or medium depth.

4. *Vertical or V type.* This root system is characterized by a taproot system, with the maximum depth of root penetration either medium or deep. There are a few lateral roots.

5. *Massive or M type.* This root system is shallow, and about 80 percent of the roots are confined to the top 30-cm layer. The lateral spread is also narrow, creating a dense root system within a small soil volume.

Both H- and VH-type root systems are good for slope stabilization and offer resistance against being uprooted by strong winds. Steeplands, susceptible to mass movement, and terrace walls can be stabilized with trees characterized by H- and VH-type root systems. Soil binding characteristics are exhibited by H- and M-type root systems, whereas shelter belts for wind erosion control are favored by the V-type root system. Sheet erosion and rill erosion are better controlled by the M-type root system.

On a volcanic soil in Kenya, Pereira and Hosegood (1962) observed the root system of radiata pine, Monterey cypress, and bamboo thicket. Roots of all three species were confined to the top 3-m layer of this deep soil. There were, however, notable differences in the root systems of different species (Fig. 13.11). Whereas the root system of cypress and bamboo are of M types, that of pine is H or R type. All three forests effectively extracted water from up to 3.2-m depth. That is why, similar to the observations by Trimble and Weirich in the United States, the total water yield from a watershed following afforestation with these species was also drastically decreased.

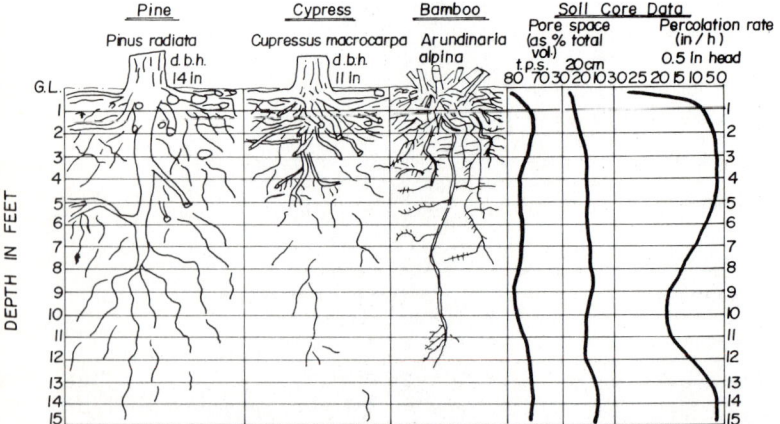

Figure 13.11 Root systems of radiata pine, monterey cypress, and bamboo thicket grown in Kenya, east Africa. G.L. = ground level, d.b.h. = diameter at breast height, t.p.s. = total pore space. (Pereira and Hosegood, 1962.)

Trees and Soil Erosion

While establishing tree cover can improve soil physical properties and reduce risks of soil erosion, it is difficult to generalize about whether afforestation can also bring about improvements in the soil nutrient capital. Trees require plant nutrients for satisfactory growth, just as do seasonal or annual crops. Trees can recycle some nutrients from deeper to surface horizons, if there exist some nutrients in deeper layers to be recycled. Some highly weathered and deeply leached acid tropical soils are devoid of essential nutrients even in the deep subsoil layers. Whereas leguminous trees may be able to contribute some nitrogen or at least grow without substantial need for additional nitrogen, trees on eroded and degraded lands cannot be successfully established without adding at least the basal dose of essential nutrients.

It is widely recognized that deforestation increases the total water yield and sediment discharge from a watershed (Fig. 13.12). In addition, there are many examples indicating a drastic decline in water runoff and erosion following reforestation. One of the most successful is the Tennessee Valley Authority (TVA, 1962). An experiment on afforestation was conducted on a 40-ha small watershed that had been severely eroded by overgrazing, burning, and general misuse. The stream flow and sediment yield on eroded watershed were measured from 1941 to 1945, and afforestation for land restoration began in 1946. The watershed was planted to pines, and eroded gullys were planted with black locust. The total tree population was established at 100,000 trees/40 ha. Once the trees were established, floods and soil erosion were completely eliminated, the peak runoff rate was reduced by 90 percent, and the sediment load was reduced by 96 percent. The total water yield decreased by about 50 percent.

Very successful afforestation programs have been conducted in many European countries. Some successful projects in the U.S.S.R. have been summarized by Molchanov (1960). He recommended that trees be planted on the contour as a shelter belt covering about 6 percent of the watershed area under contoured forest strips. These contoured forest strips can reduce overland flow by about 50 percent. Molchanov (1960) estimated that water runoff and erosion from wa-

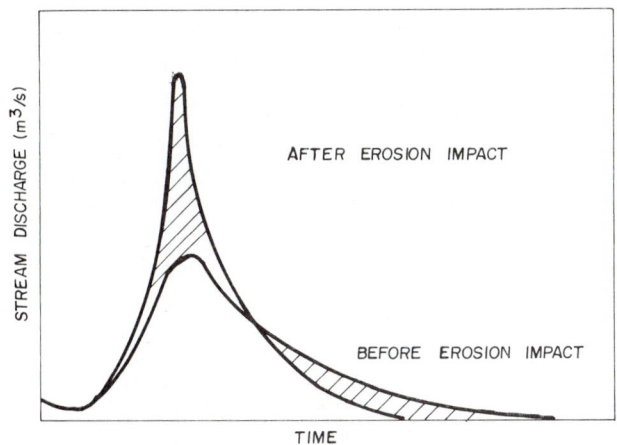

Figure 13.12 Effects of afforestation on stream flow and sediment discharge in a TVA project (TVA, 1962).

tersheds can be eliminated by growing 30 to 40 percent of the watershed areas to trees.

Afforestation therefore is a proved technique for controlling flash floods, decreasing runoff amount, minimizing sediment transport and soil erosion, and restoring eroded and degraded lands. This restorative measure, however, is slow and becomes effective over many years, depending on the rate of tree establishment, antecedent soil properties, and management inputs. Special terraces and microcatchments may need to be constructed for establishing trees on degraded lands (Plates 13.7 and 13.9). These microcatchments are especially useful in arid and semiarid regions. The water collected can greatly increase the chances of tree establishment. There are vast areas of denuded lands around the world, devoid of any vegetation cover, that are being severely eroded. Runoff and erosion from these lands are sources of pollution of natural waters. The world community should therefore lead a coordinated campaign toward afforestation of these wasted lands. Merely planting a few ceremonious trees is not enough. It requires dedication and the commitment of the nations concerned.

Improperly managed soil under planned forests also suffers from severe erosion. A relevant example is a teak (*Tectona indica*) forest in Narmada Valley, India. Teak does not support undergrowth. The soil beneath is, therefore, devoid of vegetative cover. The problem is further aggravated by overgrazing. Runoff and erosion are serious in such planned forests (Greenfield, 1986). Erosion under teak forests in India is being controlled by planting contour strips of Khus grass (*Vetiveria zizanioides*) and sisal (*Agave* spp.).

Plate 13.9 A terrace system facilitates water conservation for tree establishment in arid regions.

Agroforestry as an Antierosive Technique

Discussion in the previous section supports the conclusion that trees cause less erosion than cultivation of seasonal crops. It is logical, therefore, that growing trees in association with seasonal and annual crops will make the land less vulnerable to erosion. In general, this principle holds. No system, however, is a guarantee against land misuse. Trees and tree crop-based systems are no exception. The simple fact that a system involves forestry or agroforestry does not make it an antierosive or sustainable system. Both erosion control and sustainability depend on how the system and land are managed. It is the management that makes all the difference. All other conditions remaining the same, however, a tree-based system should be less susceptible to erosion than one based on annuals.

Agroforestry involves those land use systems and technologies in which woody perennials (trees, shrubs, palms, bamboos) are deliberately grown on the same land management units as agricultural crops and/or animals, either in the same spatial arrangement or in temporal sequence. In agroforestry systems there are both ecological and economical interactions between the different components (Lundgren and Raintree, 1983). Nair (1987) outlined the following criteria for a land use system to be classified as an agroforestry system:

1. The land use system involves two or more species of plants (or plant and animals), at least one of which is a woody perennial.
2. An agroforestry system has two or more outputs.
3. The cycle of an agroforestry system is always more than 1 year.

Plate 13.10 Food crops grown underneath randomly distributed trees in a traditional shifting cultivation system.

458   Erosion Control

4. Even the most simple agroforestry system is more complex ecologically (structurally and functionally) and economically than a monocropping system.

Both productivity and the risk of soil erosion depend on the type of agroforestry system. There are five possible methods by which food crop annu-

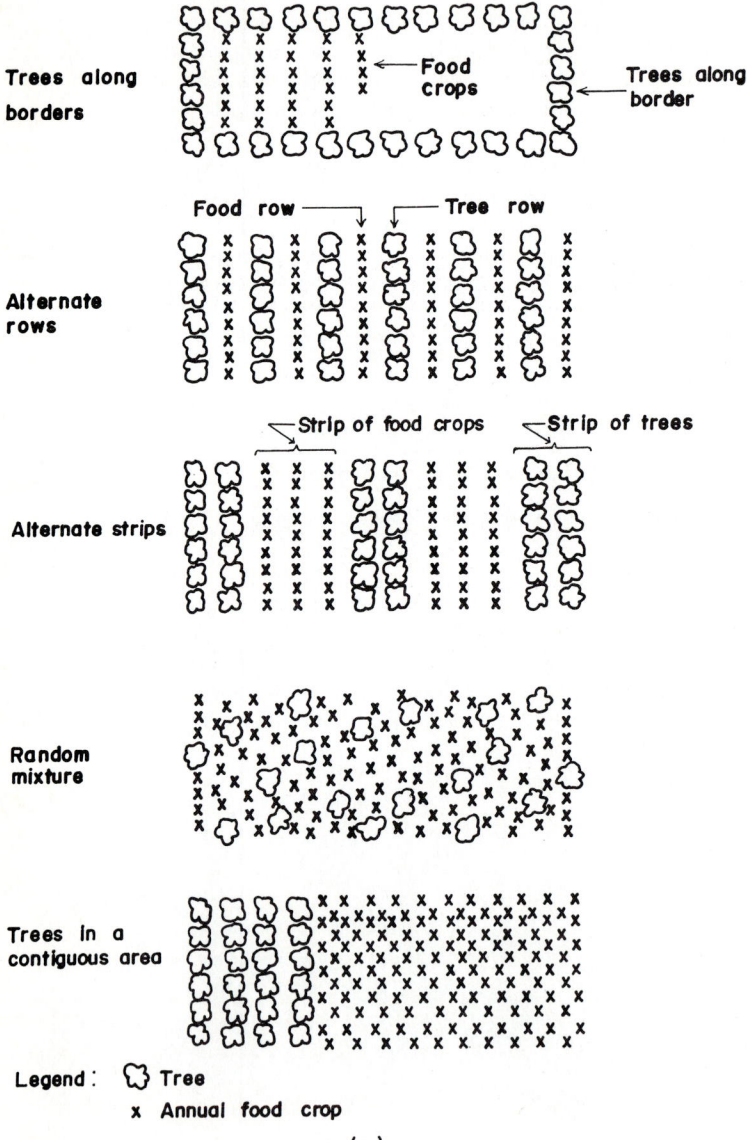

Figure 13.13 (*a*) Five possible methods of growing food crop annuals in association with trees and woody perennials (Vergara, 1982, 1987).

Trees and Soil Erosion 459

Plate 13.11 Perennial shrubs and trees are commonly grown on field boundaries. Eucalyptus is grown widely on plot boundaries in India.

als can be grown in association with woody shrubs and perennials. The diagram in Fig. 13.13a outlines these five methods. Traditionally, trees are grown at random, scattered within the field, and food crops are grown underneath, as is done by the shifting cultivator (Plate 13.10). Another common practice is to grow perennials on field boundaries at rather wide spacings (Plate 13.11). A row of trees can also be grown on contour lines or on terrace walls to stabilize them by a system that may be called *contour management*. Trees may be grown in strips, with alternate strips of trees and food crops. The latter is called a *corridor system*, and it was used by INEAC in Zaire (Jurion and Henry, 1969). Trees may also be grown in a single-row hedge planted on the contour, as is done with alley cropping (Plates 12.4 and 13.12). Different types of tree-based systems are

Pastures	Food crop annuals	Perennials
• Trees grown at random • Live fences • Alley cropping • Contiguous area for stall feeding	• Trees grown at random • Contiguous area for trees • Trees grown on field boundaries • Trees grown in alternate strips • Contour hedges (single row, multiple rows)	• Taungya/shahba • Shade trees

(b)

Figure 13.13 (*Continued*) (*b*) Farming systems based on different types of spatial/temporal arrangements of trees, with pastures and food crop annuals.

shown in Fig. 13.13b. On the basis of these spatial arrangements, however, Wiersum (1984) distinguished four basic groups of agroforestry systems:

1. Shifting cultivation and related bush fallow system widely used in the tropics and subtropics. These extensive land use systems were also used in Europe and Asia in the recent past. In these systems, annual crops are rotated with natural forest regrowth and tree fallows.
2. Taungya system, used in south and southeast Asia, refers to temporary intercropping of annual crops in young tree plantations before tree canopy is fully established.
3. Multistory tree gardening, around homestead and household compounds, in which various perennials and sometimes a few annual crops are cultivated simultaneously with trees.
4. Alley cropping in which food crop annuals are grown within the alleys formed by contour hedges of regularly pruned perennial shrubs and trees (Plates 13.12 and 13.13).

Soil erosion and shifting cultivation

If the ratio of the fallow period to the cultivation period is large enough, soil erosion and water runoff are often less from land under shifting than intensive cultivation. If the regrowth is easily established, soil erosion can be especially low during the fallow phase. Soil erosion during the cultivation phase may be several times greater than during the fallow phase. In Nigeria, Lal (1981) reported negligible runoff and soil loss from treatment with shifting cultivation even

Plate 13.12 Development of natural terraces by *Leucaena* hedges on steeplands is an indication of severe interterrace erosion. (Courtesy of P. K. Nair.)

Plate 13.13 An erosion-monitoring experiment established at IITA, Ibadan, Nigeria, in 1982 to study the effects of *Leucaena* and *Gliricidia* hedges on runoff and erosion in a corn-soybean rotation.

during the cultivation phase (Table 13.4). These plots were, however, cleared after about 20 years of fallowing and were cultivated for only 3 consecutive years. Consequently, there was no runoff or erosion during the fallow phase. Experiments conducted by Takahashi et al. (1983) in northeast Thailand also indicated that runoff and erosion under shifting cultivation were lower than under continuous cropping (Table 13.16). At Chiang Mai, Thailand, Naprakob et al. (1975) observed that soil erosion during the cultivation phase was 1.7 to 3.1 t/(ha · yr) in comparison with 0.05 to 0.07 t/(ha · yr) during the fallow phase.

Runoff and soil erosion, however, increase with an increase in the duration of cultivation. The data in Table 13.17 from the Philippines show that both water runoff and sediment loss increased drastically with an increase in duration of

TABLE 13.16 Soil Erosion (m^3/ha) under Shifting Cultivation and Continuous Cropping in Khon Kaen, Thailand

Treatment	Period of measurement	
	1980	1981
Shifting cultivation	87.6	6.9
Continuous cropping	107.0	6.4
Bare fallow	154.3	25.9
Forested control	4.7	2.9
Total rainfall, mm	1542	1009

SOURCE: Takahashi et al., 1983.

TABLE 13.17 Erosion and Runoff from Soil under Shifting Cultivation in the Philippines

	Runoff, %		Sediment loss, g/day	
	Cropping period	Postharvest	Cropping period	Postharvest
Logged-over forest	—	—	—	—
New maize swidden	1.52	0.86	3.03	0.65
New rice swidden	1.08	0.42	1.45	0.3
2-year-old maize swidden	1.78	0.69 (4.08)*	12.05	9.81
12-year-old rice swidden	11.64	6.73 (14.15)	119.31	6.32

*Numbers in parentheses refer to postcropping period.
SOURCE: Kellman, 1969.

cultivation. Results similar to those from the Philippines were reported from the Shillong Hills of northeast India. Toky and Ramakrishnan (1981) observed that runoff and percolation losses were related to the length of the fallow cycle. Erosion and water runoff losses were greater under a short than a long fallow cycle (Table 13.18). Similar conclusions were arrived at by Mishra and Ramakrishnan (1983) who observed more erosion with a shorter fallow and longer cultivation cycle than with longer fallow and shorter cultivation phases (Table 13.19). In this study, soil erosion during the cultivation phase was, however, decreased by installing terraces (Fig. 13.14). This means that plots that were cleared after 5 years of fallowing had greater erosion and runoff losses than plots cleared after 10 or 30 years of fallowing. Similarly, lower runoff and less erosion were observed from plots that had been fallowed for 10 years than for 5 years.

The risks of soil erosion under shifting cultivation thus depend on the length of the cultivation period, length of fallow period, soil type, and management. Good management and inputs, including use of mulches and short cultivation cycle, cause less erosion under shifting cultivation.

Additional and backup soil conservation measures are needed in traditional

TABLE 13.18 Effects of Length of Cultivation and Fallow Periods on Runoff, Percolation, and Soil Erosion in Shillong, India

Cultivation/fallow cycle	Runoff water, cm	Percolation water, cm	Erosion, t/(ha · yr)
Clearing after 5 years of fallowing	36.6	22.9	30.1
Clearing after 10 years of fallowing	33.9	19.0	23.1
Clearing after 30 years of fallowing			
5-year fallow	29.4	14.4	22.5
10-year fallow	18.5	14.2	0.8

Rainfall = 142.1 cm.
SOURCE: Toky and Ramakrishnan, 1981.

TABLE 13.19 Soil and Water Loss and Infiltration Characteristics in Relation to the Cycle of Shifting Cultivation (Jhum) in Northeastern India

Category of loss	Jhum agroecosystem		Jhum fallows of different age, yr			Terrace agro-ecosystem
	10-yr cycle	5-yr cycle	1	5	10	
Sediment	49.7 ± 1.0	54.9 ± 1.1 (56.3 ± 1.3)*	7.4 ± 0.2	3.5 ± 0.1	1.9 ± 0.1	33.8 ± 0.9 (42.3 ± 1.1)
Runoff water	54.2 ± 1.4	59.3 ± 1.7 (84.5 ± 1.5)	44.9 ± 1.2	38.2 ± 1.3	23.5 ± 1.1	49.4 ± 2.0 (63.1 ± 1.8)
Percolation water	12.1 ± 0.8	13.3 ± 0.5 (10.6 ± 0.9)	17.8 ± 0.8	30.8 ± 1.0	17.7 ± 0.5	16.1 ± 0.4 (12.7 ± 0.5)

*Values within parentheses indicate losses during the second year of observation.
SOURCE: Mishra and Ramakrishna, 1983.

farming and shifting cultivation whenever the fallow land is needed for cultivation due to the problem posed by a high demographic pressure. In Malawi, soil erosion of 16.4 t/ha was measured during the 1981–1982 rainy season for traditionally farmed watershed in comparison with 0.15 t/ha for the one where a recommended land husbandry plan was implemented. The data in Table 13.20 show drastic reductions in runoff, erosion, and nutrient loss by installing a complete physical conservation and land use plan. Installation of physical structures (including bench terracing, grass strips, contour bunding, stick and stone bunding, broad-based banks) for erosion control on traditional farms was also recommended in Sierra Leone by Millington (1981). The author argued that the land area of 2 to 5 percent taken by these structures and the installation cost may be partly recovered by gains in productivity within the first 2 or 3 years.

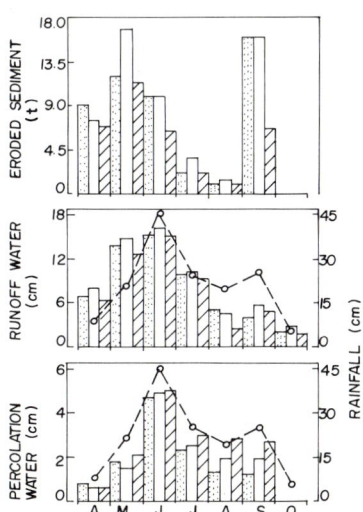

Figure 13.14 Effects of terraces on erosion control under shifting cultivation in Shillong Hills of northeastern India (Toky and Ramakrishnan, 1983).

TABLE 13.20 Runoff, Soil Erosion, and Nutrient Loss from Watersheds in Malawi under Different Management Systems

Parameter	Rainy season	Traditional farming	Complete land use plan	Physical conservation only	Eucalyptus plantation
Rainfall, cm	1981–1982	890	957	—	—
	1982–1983	910	822	975	951
Runoff, cm	1981–1982	81.0	29.7	—	—
	1982–1983	156.7	19.0	54.8	7.4
Soil erosion, t/ha	1981–1982	16.4	0.15	—	—
	1982–1983	14.6	0.20	2.53	0.13
N loss, kg/ha	1981–1982	0.75	0.06	—	—
	1982–1983	0.86	0.21	0.45	0.051
P loss, kg/ha	1981–1982	0.07	0.04	—	—
	1982–1983	0.28	0.05	0.12	0.018
K loss, kg/ha	1981–1982	4.48	1.38	—	—
	1982–1983	0.13	0.02	0.04	0.006
Area, ha		5.3	7.8	6.7	18.0
Gradient, %		8.8	7.2	8.1	—

SOURCE: Amphlett and Tucker, 1984.

Soil erosion under taungya (Shambe) system and multistory tree garden

The practice of growing food crops during the initial stages (first 1 to 3 years) of tree crop establishment is called the *taungya* system in Asia and the *Shambe* system in Kenya. Soil erosion under these agroforestry systems is usually less than if the same land were cultivated to food crop annuals (Wiersum, 1984; Young, 1986). In spite of these apparent advantages, however, it is difficult and even dangerous to generalize the erosion control potential of these systems, because the erosion depends on the soil management and the landscape characteristics. If the management practices adopted favor erosion, erosion risks can be extremely high. Cultivation and tillage within the tree rows are bound to increase the risks of erosion.

Taungya cultivation may decrease soil organic matter content and therefore increase soil erodibility. The decline in soil structure brought about by cultivation decreases the infiltration rate and increases water runoff. An example of a decline in soil organic matter content with the taungya system is reported from Nigeria in a system of growing food crops through young seedlings of *Gmelina arborea* (Ojeniyi and Agbede, 1980). In Kerala, India, Alexander et al. (1980) observed a decline in soil physical and nutritional properties by even 2 years of taungya cultivation. They observed that the surface horizon had been partially eroded and the subsurface horizon was being exposed. These changes were attributed to disturbance of the soil during cultivation. In the Philippines, the infiltration rate measured under pure stand *Leucaena leucocephala* was greater than that under intercropped *Leucaena* plots (Baconguis and Rondilla, 1979; Baconguis and Punzalan, 1983). High erosion rates of 0.6 to 17.4 t/(ha · yr) under the taungya system with *Gmelina arborea* were reported by Watnaprateep (1980) from north Thailand. In Java, erosion rates of 5.2 t/(ha · yr) were reported in a taungya system based on teak, *Tectona grandis* (Coster, 1983), and of 1.0 to 2.0 cm/yr from a system based on *Agathis dammara* grown on a volcanic soil (Wiersum, 1984).

In Java, Noer (1981) observed more splash erosion under home garden and mixed gardens than under an adjacent patch of bamboo forest. Human traffic under garden causes compaction and exposes the soil to erosion. In Tanzania, Lundgren (1980) observed a decline in the soil organic matter content, aggregate stability, and infiltration rate when a natural mountain forest was changed to a multistory tea garden. With good management, however, the sediment loss from home gardens can be as low as 0.05 t/(ha · yr), as reported from Kandy, Sri Lanka, by Krishnarajah (1984).

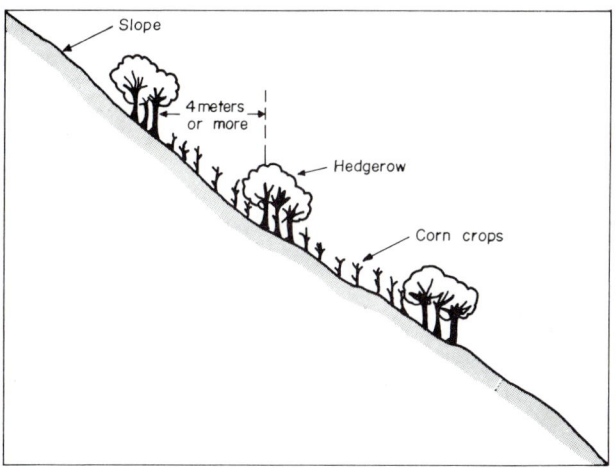

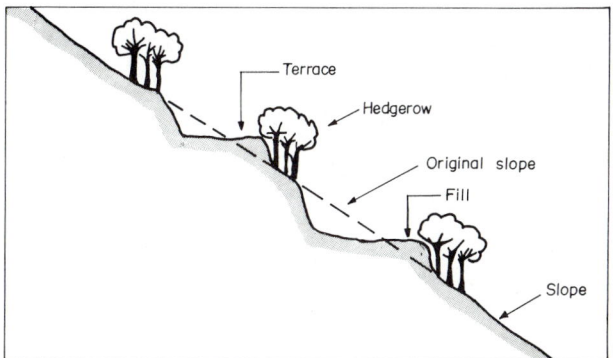

Figure 13.15 A schematic of the formation of natural terraces by *Leucaena* hedgerows (Pacardo, 1985).

Alley cropping

Having concluded that established tree crops reduce the risks of soil erosion, we know that it is logical to believe that establishing tree crops with or without grass and legume pastures is likely to reduce erosion risks on arable lands. This logic is the basis for recommending tree-based systems as erosion prevention measures. In central India, for example, Chinnamani and Singh (1971) recommended the use of trees and perennial grasses to control erosion on alfisols. In Indonesia, Sukmana et al. (1985) advocated the use of *Flamingia congesta* for reclamation and conservation of volcanic skeletal soils. The challenge lies in integrating trees with food crop annuals so as to conserve soils and yet grow food crops.

The practice of growing food crop annuals within the hedges formed by regularly pruned perennial shrubs originated in the Philippines and in Indonesia. For countries with hilly regions and steeplands, there is a need to develop a system that may enable a relatively intensive use of marginal steeplands for food crop production. The need for such a system has been felt, especially for densely populated regions such as the Philippines (Gwyer, 1978; De La Rosa, 1984), Sri Lanka (Wijewardene, 1982), and Rwanda (CTFT, 1979; Dressler and Neumann, 1982; Lipman, 1986). The package of agronomic practices for this system has been outlined for the Philippines (Celestino, 1984, 1985) and northeastern India. The practice of hedgerow intercropping is being widely adapted for farming on steeplands (Benge, 1979; Fonzen and Oberholzer, 1984; Ssekabembe, 1985). Its agronomic applications to agroecological environments of Africa have been evaluated at IITA, Ibadan, Nigeria (Kang et al., 1984, 1985) and at ICRAF, Nairobi, Kenya (Nair, 1987; Young, 1986).

Similar to the taungya system, alley cropping may increase the risk of soil

Plate 13.14 Growing coffee under the shade of *Leucaena* trees in Cameroon.

erosion compared with pure stand shrubs, especially if the alley space is to be mechanically tilled for food crop establishment. The risk of accelerated erosion is especially high for steeplands. Loch (1985) reported from the Philippines that sheet wash within alley cropping is often severe on steeplands, as is evidenced by the development of terracing in *Leucaena* hedges (Fig. 13.15, Plates 13.12 and 13.14). Terraces are apparently developed by the deposition of eroded material on upslope sides of the hedgerows. The mere fact that sediments are being trapped by hedgerows is, of course, a significant advantage of such a system. It would be better, however, if the sediments did not originate and entrain in the first place. For steeplands, Loch (1985) observed that a single row of well-spaced stems planted from cuttings is less effective in controlling erosion than a strip of perennial shrubs. A strip of densely planted shrubs comprising two or three rows reduces runoff velocity and amount and causes sediment deposition to form natural terraces.

Field experiments have been conducted to compare runoff and soil erosion under seasonal crops grown alone with that grown with the alley cropping system, but for a few sites only. Field data on water balance and erosion are not available from comprehensive, well-planned, long-term experiments. Loch (1985) presented a conceptual basis of the effects of hedgerows on runoff and soil erosion. The analysis presented in Fig. 13.16 depicts two cases:

1. Hedges have little beneficial effect in reducing erosion from within alleys. This situation prevails if the sediment-carrying capacity of water runoff is rapidly filled.
2. Erosion is significantly curtailed by the hedgerows. This desirable situation occurs if the overland flow does not load to capacity.

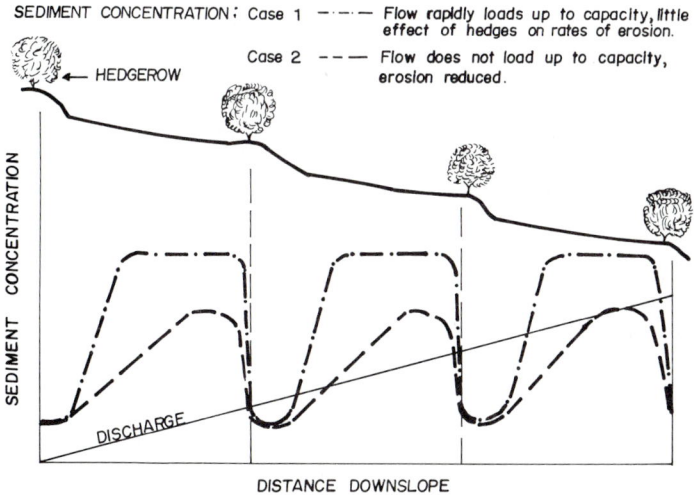

Figure 13.16 A conceptual basis of runoff and erosion control with the alley cropping system (Loch, 1985).

TABLE 13.21 Runoff and Soil Losses during 7 Months at Carcar and Barili in Cebu Province in the Philippines

	Carcar		Barili	
Treatment	Runoff, mm	Soil loss, g/plot	Runoff, mm	Soil loss, g/plot
Bare	87	3156	56	5667
Corn alone (stubble removed)	30	680	33	2214
Corn/ipil-ipil* (stubble retained)	2	14	13	712
Corn/ipil-ipil (stubble removed)	4	8	16	820

*Ipil-ipil = *Leucaena leucocephala*.
SOURCE: Pacardo and Montecillo, 1983.

This analysis, similar to one on the effects of terracing, supports the argument in favor of appropriate soil management within the alleys to reduce soil erosion risks. Only the residue mulch and no-till systems prevent the runoff from being loaded by sediments to capacity and will, therefore, increase the effectiveness of hedgerows in controlling erosion. This implies that soil management within alleys plays a more important role in erosion control than hedges do. And then there is the question of the stability of natural terraces formed by hedgerows. A breach in the terraces made by burrowing animals may cause severe gullying. If implemented properly, however, runoff and soil losses from cropped land can be drastically reduced by growing annuals in association with perennials. The data in Table 13.21 from the Philippines are a relevant exam-

TABLE 13.22 Surface Runoff from Maize in First Growing Seasons for the Period 1982 through 1987

Treatment			Runoff, % of rainfall					
Species		Spacing, m	1982	1983	1984	1985	1986	1987
A	Plow-till	—	1.1	19.1	29.9	18.1	23.4	10.5
	No-till	—	0.1	0.2	0.8	1.8	3.3	1.6
B	*Leucaena*	4	3.0	13.3	1.2	1.4	5.9	4.7
	Leucaena	2	3.9	6.0	1.3	1.1	2.7	4.8
C	*Gliricidia*	4	0.1	5.3	4.9	3.6	7.1	4.6
	Gliricidia	2	0.4	2.8	2.2	1.2	5.8	2.0
Rainfall, mm			516	589	727	914	486	711
				LSD*				
			(0.05)			(0.10)		
(i) Systems (S)			2.97			2.45		
(ii) Treatments (T)			3.64			3.00		
(iii) Years (Y)			5.15			4.24		
(iv) $T \times Y$			8.93			7.35		
(v) $T \times S$			5.15			4.24		

*LSD = least significant difference.
SOURCE: Lal, 1989.

ple of the beneficial effects of alley cropping in reducing losses due to runoff and soil erosion. With stubble retention, growing corn with *Leucaena* reduced runoff and soil erosion to 6.7 and 2.1 percent of that under corn alone at Carcar and to 39.4 and 32.2 percent, respectively, at the Barili site (Table 13.21).

Factors affecting soil erosion by alley cropping

The literature surveyed in the previous section indicates that the alley cropping system causes more erosion than pure stand planned forest but less erosion than pure stand seasonal crops. As in other cropping systems, the magnitude of erosion with alley cropping depends on many factors, some of which are described below with appropriate examples from a long-term experiment conducted at IITA. The experiment was established in April 1982 when *Leucaena* and *Gliricidia* hedges were established at 2- and 4-m spacings on a plot of about 8 percent slope (Plate 13.13). Prior to establishing this experiment, the land was under fallow for about 5 years. The soil structure immediately after clearing in 1982 was, therefore, good. There was thus low runoff and erosion in 1982 in all plots.

Age of tree species grown in hedges. In humid and subhumid regions, most species become effective in trapping sediments within 2 or 3 years after establishment. In the subhumid region with the mean rainfall between 1200 and

TABLE 13.23 Surface Runoff from Cowpea in First Growing Seasons for the Period 1982 through 1987

	Treatment		Runoff, % of rainfall					
Species		Spacing, m	1982	1983	1984	1985	1986	1987
A	Plow-till	—	2.1	1.1	2.39	20.23	0.01	0.1
	No-till	—	0.8	0.2	0.18	0.55	2.3	0.7
B	*Leucaena*	4	2.1	1.7	0.08	2.10	0.2	0.6
	Leucaena	2	5.4	3.3	0.57	2.44	0.1	0.1
C	*Gliricidia*	4	1.5	1.1	0.32	0.82	0.3	0.4
	Gliricidia	2	0.4	0.2	0.71	4.67	3.1	0.4
Rainfall, mm			253	328	631	740	459	—
				LSD*				
				(0.05)			(0.10)	
(i)	Systems (S)			2.95			2.41	
(ii)	Treatments (T)			3.61			2.95	
(iii)	Years (Y)			4.67			4.82	
(iv)	$T \times Y$			8.09			6.61	
(v)	$T \times S$			5.11			4.48	

*LSD = least significant difference.
SOURCE: Lal, 1989.

1800 mm, *Leucaena* hedges are fully established in 2 years while *Gliricidia* grown from cuttings may take longer. Tables 13.22 to 13.25 give runoff and soil erosion data, respectively, from 1982 to 1985—immediately after planting shrubs to 4 years after. In 1982 when hedges were established, even though erosion was generally low because of good initial soil structure, there was no effect of hedges on reducing runoff or soil erosion. The effect was also negligible during 1983, the first year after planting hedges. Hedges became effective in reducing runoff and erosion in 1984, two years after their establishment. The no-till system for the maize-cowpea rotation, however, was effective against erosion right from the very beginning in 1982. Although erosion is not a serious problem for this soil and this slope, supplementary measures are required for erosion control during the first 2 or 3 years of hedge establishment on steeplands and when soils are susceptible to erosion. The time required for hedge establishment depends on the species, soil type, and the climate. It takes longer for hedges to establish in semiarid and arid climates than in subhumid and humid regions.

Spacing. There is an optimum spacing for erosion control for each slope and each species used as perennial hedges. This implies that hedgerow spacing depends on the soil, slope steepness, species, and management system. A spacing that is optimum for erosion control may not be optimum for yields of seasonal or annual crops. The data in Tables 13.22 to 13.25 show that a hedge spacing of 4 m was adequate for erosion control with *Leucaena* and 2 m for *Gliricidia*. A

TABLE 13.24 Soil Erosion from Maize in First Growing Seasons for the Period 1982 through 1987

Treatment			Soil erosion, t/ha					
Species		Spacing, m	1982	1983	1984	1985	1986	1987
A	Plow-till	—	0.02	2.50	14.16	3.64	3.80	1.48
	No-till	—	0.01	0.004	0.026	0.23	0.20	0.16
B	*Leucaena*	4	0.69	1.38	0.17	0.07	0.63	0.49
	Leucaena	2	0.25	0.18	0.07	0.03	0.03	0.02
C	*Gliricidia*	4	0.01	0.43	1.62	1.40	0.26	0.12
	Gliricidia	2	0.02	0.10	2.05	0.20	1.11	0.06
					LSD*			
				(0.05)			(0.10)	
(i)	Systems (*S*)			1.48			1.22	
(ii)	Treatments (*T*)			1.82			1.49	
(iii)	Years (*Y*)			2.57			2.12	
(iv)	$T \times Y$			4.46			3.67	
(v)	$T \times S$			2.57			2.12	

*LSD = least significant difference.
SOURCE: Lal, 1989.

TABLE 13.25 Soil Erosion from Cowpea in Second Growing Seasons for the Period 1982 through 1987

Treatment			Soil erosion, t/ha					
Species		Spacing, m	1982	1983	1984	1985	1986	1987
A	Plow-till	—	0.02	0.013	0.74	3.00	0.007	0.010
	No-till	—	0.02	0.003	0.006	0.03	0.036	0.066
B	*Leucaena*	4	0.24	0.06	0.02	0.13	0.008	0.325
	Leucaena	2	0.15	0.10	0.04	0.03	0.003	0.018
C	*Gliricidia*	4	0.04	0.06	0.05	0.21	0.023	0.040
	Gliricidia	2	0.02	0.02	0.27	0.70	0.034	0.048
				LSD*				
				(0.05)			(0.10)	
(i)	Systems (S)			0.35			0.29	
(ii)	Treatments (T)			0.43			0.35	
(iii)	Years (Y)			0.61			0.5	
(iv)	$T \times Y$			1.06			0.87	
(v)	$T \times S$			0.61			0.5	

*LSD = least significant difference.
SOURCE: Lal, 1989.

wider spacing may be acceptable if crops are to be grown with a no-tillage system or mulch farming than with conventional tillage practices. Hedge spacing also has a significant effect on the amount and velocity of water runoff and the sediment-carrying capacity of the overland flow. A strip of perennial shrubs is likely to be more effective in erosion control than a single row of widely spaced trees and shrubs.

Species. Some perennial species are more suited for erosion control in a given agroecological environment than others. The data in Tables 13.22 to 13.25 show

TABLE 13.26 Runoff Loss from Alley Cropping Maize with *Leucaena* in a Mechanized System for a 4-ha Watershed versus Manually Cultivated Plot of 0.07 ha (First Season 1985, IITA, Ibadan)

Treatment	Runoff, mm	Relative runoff*
Mechanized		
With alley cropping	38.4	662
Without alley cropping	109.8	1893
Manual		
With alley cropping	8.7	150
Without alley cropping	5.8	100

*Relative to manual treatment without alley cropping.

TABLE 13.27 Soil Erosion from Alley Cropping Maize with *Leucaena* in a Mechanized System for a 4-ha Watershed versus Manually Cultivated Plot of 0.07 ha (First Season 1985, IITA, Ibadan)

Treatment	Soil erosion, kg/ha	Relative soil erosion*
Mechanized		
With alley cropping	76	33
Without alley cropping	10,707	4655
Manual		
With alley cropping	70	30
Without alley cropping	230	100

*Relative to manual treatment without alley cropping.

that *Leucaena* is better adapted for alfisols in the subhumid regions of western Nigeria than *Gliricidia*. For example, *Leucaena leucocephala* does not grow well in acidic or alkaline soils. It has a favorable pH range of 5.5 to 7.0. Some commonly grown leguminous trees for acidic soils are *Calliandra callothyrsus, Sesbania grandiflora, Leucaena diversifolia, Acacia mangium, Albizzia falcataria, Albizzia procera, Dalbergia sissoo, Mimosa scarbella, Samanea saman, Cassia siamea, Cassia fistula,* and *Cassia orraria* (Sudjadi et al., 1985). The adaptability of different species needs to be validated for different agroecological regions.

Hedgerow management. The frequency and methods of pruning hedges also affect their ability to control erosion. If hedges are pruned mechanically, their effectiveness in erosion control is less than if they are pruned manually. The wheel ruts created by tractor tires traversing the same site 3 to 4 times during the cropping cycle are generally the sites of runoff initiation and sediment entrainment. The data in Tables 13.26 and 13.27 compare runoff and erosion from runoff plots pruned manually with watershed pruned mechanically. Both runoff and erosion were greater from mechanically than manually managed hedges.

The method of seedbed preparation within the hedges is another important

TABLE 13.28 Surface Runoff from Maize Grown in Alley Cropping System soon after Planting in April 1986 (April–May 1986, IITA, Ibadan, Nigeria)

Treatment	Total runoff, mm	Relative runoffs
Plow-till	99.5	100
No-till	10.3	10
Leucaena, 4 m	26.6	27
Leucaena, 2 m	9.0	9
Gliricidia, 4 m	30.5	31
Gliricidia, 2 m	27.8	28

SOURCE: Unpublished data of R. Lal.

factor. Erosion is better controlled if soil within alleys is managed with no-till and mulch than with plowed seedbed.

Crop growth. Stage of crop growth within alleys also affects the magnitude of runoff and erosion. The positive effect of alley cropping is enhanced when the annual crop has attained the maximum ground cover. Runoff and erosion measurements made during the first planting season of 1986 in southern Nigeria indicate that runoff during the first 3 weeks was in the order of no-till < 2-m *Leucaena* < 4-m *Leucaena* < 2-m *Gliricidia* < 4-m *Gliricidia* < plowed plots (Fig. 13.17). Soil erosion followed a similar pattern (Fig. 13.18). The stand of 2-m *Gliricidia* hedges was not as good as that of 4-m *Gliricidia*, hence the unexpected trends in erosion. Similar measurements made during 1987 showed that soon after the maize was planted, runoff and erosion were greater in alley-

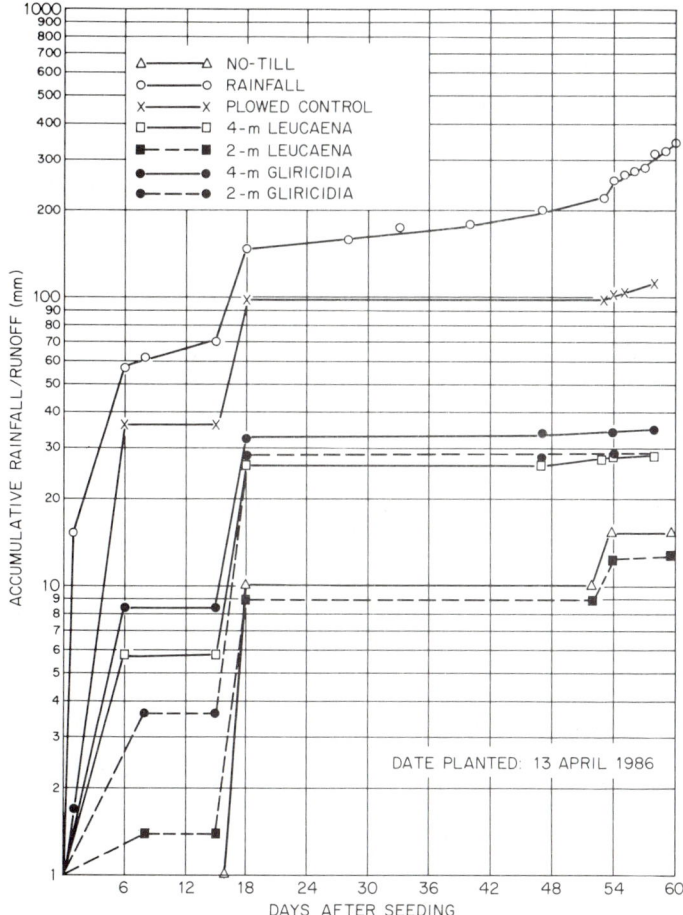

Figure 13.17 Effects of stage of maize growth within the alleys on water runoff from a tropical alfisol.

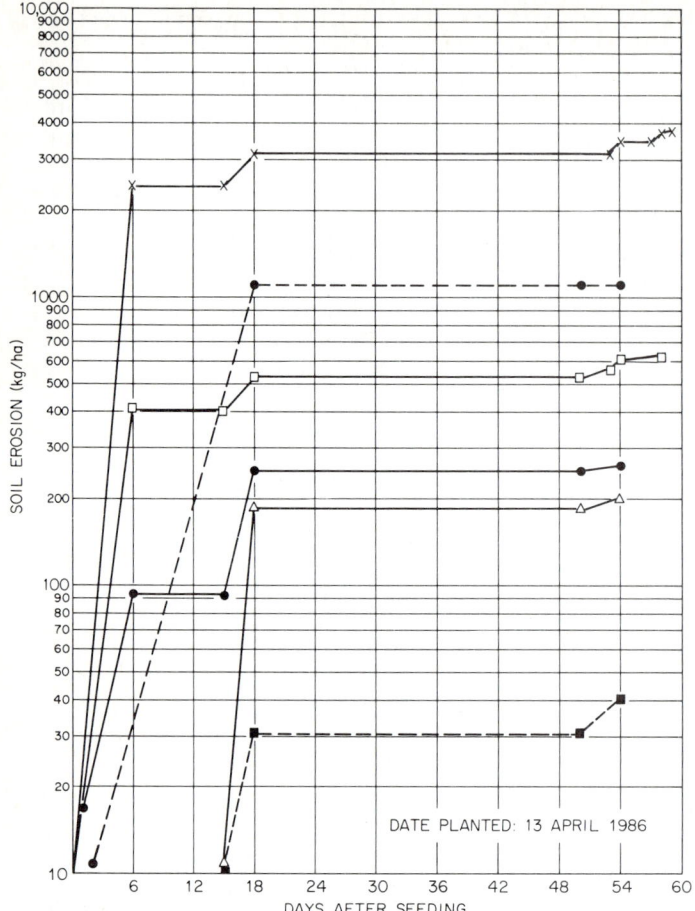

Figure 13.18 Effects of stage of maize growth within the alleys on soil erosion from a tropical alfisol. See Fig. 13.17 for meanings of symbols.

cropped maize than in no-till maize (Tables 13.28 and 13.29) or in maize sown with no-till after *Mucuna* fallowing (see Fig. 12.15).

Effects of Alley Cropping on Crop Yield

As for any mixed cropping system, the yield of an annual crop grown in association with perennials is subject to many factors. The interaction between two species may be of three types: symbiotic, neutral, and antagonistic. In the symbiotic interaction, both species are compatible and support each other. For example, a nitrophilic crop such as maize may benefit from the leguminous woody perennial if the mulch prunings are applied regularly as manure. Some shade-loving plants, such as coffee and cocoa, may also benefit from tall trees (Plate 13.15). The neutral effect means that there is no interaction among species, and

TABLE 13.29 Soil Erosion from Maize Grown in Alley Cropping System soon after Planting in April 1986 (April–May 1986, IITA, Ibadan, Nigeria)

Treatment	Soil erosion, kg/ha	Relative erosion
Plow-till	3127	100
No-till	186	6
Leucaena, 4 m	532	17
Leucaena, 2 m	31	1
Gliricidia, 4 m	234	7
Gliricidia, 2 m	1105	35

the crop yield of each species is proportional to the area allocated. Two species grow independently of each other in the space allocated. The antagonistic effect implies that one species suppresses the growth and yield of another through competition (for light, water, and nutrients), the allelopathic effect, or hibernating pests and diseases.

For alley cropping system to be compatible, the choice of suitable species is important. Usually the nitrophilic crops do better when grown in association with leguminous shrubs. An appropriate example is the growing of maize within the alley space formed by *Leucaena* hedgerows. Experiments conducted in different agroecological regions have shown some yield benefits to maize grown in a region of adequate moisture supply. Experiments conducted at IITA, Ibadan, Nigeria, have also demonstrated some yield increase of maize grown in association with *Leucaena* (Table 13.30).

The allelopathic effect may be caused by interactions in soil or in atmo-

Plate 13.15 Growing coffee under the shade of banana trees in Cameroon.

sphere. The soil interaction may lead to an inhibitory effect on seed germination or on root system establishment. Experiments conducted at IITA have shown that germination of cowpea is severely suppressed by the *Leucaena* mulch. The adverse effect is especially severe when cowpea is sown soon after the fresh *Leucaena* prunings are incorporated into the soil. Grain yield of cowpea is suppressed as a result of poor crop stand establishment (Table 13.31).

Another mechanism of allelopathic effect may be the suppression of root growth of one specie by another. Observations at Hyderabad, India, have shown that growth of castor beans is suppressed by *Leucaena*. Whereas the germination of castor beans was satisfactory, the growth of young seedlings was drastically suppressed.

Alley cropping system may promote crop growth through some beneficial effects especially on soil physical and nutritional properties. Experiments conducted at IITA showed that although physical and nutritional properties declined with cultivation, the rate of decline was less with alley cropping than continuous corn-cowpea cultivation with no-till or plow-till systems. In Indonesia, Sukmana et al. (1985) observed that growing *Flamingia congesta* improved soil organic matter content (Fig. 13.19) and soil moisture retention characteristics (Fig. 13.20).

The adverse effects of perennial shrubs on the growth of seasonal/annual crops may simply be related to competition for natural resources. There is an optimum space ratio for satisfactory growth of both species. A schematic of the relation between the yield of an annual crop grown at a different space alloca-

TABLE 13.30 Effects of Species, Spacing, and Tillage Methods on Grain Yields of Maize

System	Treatment Perennial species	Spacing, m	1982	1983	1984	1985	1986	1987
A	Plow-till	—	4.1	4.9	3.6	4.3	2.7	2.3
	No-till	—	4.0	4.1	4.0	5.0	2.4	2.7
B	*Leucaena*	4	3.7	3.3	3.7	4.8	2.1	2.0
	Leucaena	2	4.4	3.6	3.8	4.2	1.7	2.5
C	*Gliricidia*	4	3.9	3.9	3.6	4.5	2.6	2.2
	Gliricidia	2	3.8	3.2	3.3	4.8	1.6	2.8
Mean			4.0	3.8	3.7	4.6	2.2	2.4

		LSD*	
		(0.05)	(0.10)
(i)	Systems (S)	0.27	0.22
(ii)	Treatments (T)	0.34	0.28
(iii)	Years (Y)	0.48	0.39
(iv)	$S \times T$	0.48	0.39
(v)	$T \times Y$	0.83	0.68

*LSD = least significant difference.
SOURCE: Lal, 1989.

tion for the perennial crop is shown in Fig. 13.21. The growth of a nitrophilic crop may be enhanced at about 20 percent space allocation to the perennial shrubs in humid environments, provided that species chosen are compatible. Yield in annual crops may decline when about 25 percent of the land area is allocated to shrubs. In subhumid climate, the growth of an annual crop may not be affected for a compatible mixture up to about 10 or 15 percent of the area allocated to woody perennials. Yield suppression will occur when the land area allocated to perennial shrubs exceeds 15 percent. In semiarid and arid regions with severe limitations for plant-available water reserves, the yield of annuals is drastically reduced by perennials. The yields of annual crops may decline to as low as 50 percent when the land area allocated to perennials is about 25 percent.

In addition to compatibility among different species, the yield of annuals depends on the management, e.g., time and frequency of pruning, use of pruning as mulch, height of pruning, application of balanced fertilizers, tillage methods, etc.

Conclusions

Tree crops and tree crop-based cropping systems cause less erosion than systems based on annual crops. It would be dangerous to generalize, however, that erosion is not a severe problem in tree-based systems. Planned forests and crop-

TABLE 13.31 Effects of Species, Spacing, and Tillage Methods on Grain Yields of Cowpea

System	Treatment		Cowpea grain yield, t/ha					
	Perennial species	Spacing, m	1982	1983	1984	1985	1986	1987
A	Plow-till	—	720	442	447	435	992	369
	No-till	—	1520	829	1193	784	1000	213
B	Leucaena	4	1000	514	581	409	285	222
	Leucaena	2	730	319	503	159	146	236
C	Gliricidia	4	950	600	670	590	452	207
	Gliricidia	2	700	533	678	405	233	223
Mean			937	540	679	464	518	319
			LSD*					
			(0.05)			(0.10)		
(i)	Systems (S)		120			99		
(ii)	Treatments (T)		147			121		
(iii)	Years (Y)		208			171		
(iv)	$S \times T$		208			171		
(v)	$T \times Y$		361			297		

*LSD = least significant difference.
SOURCE: Lal, 1989.

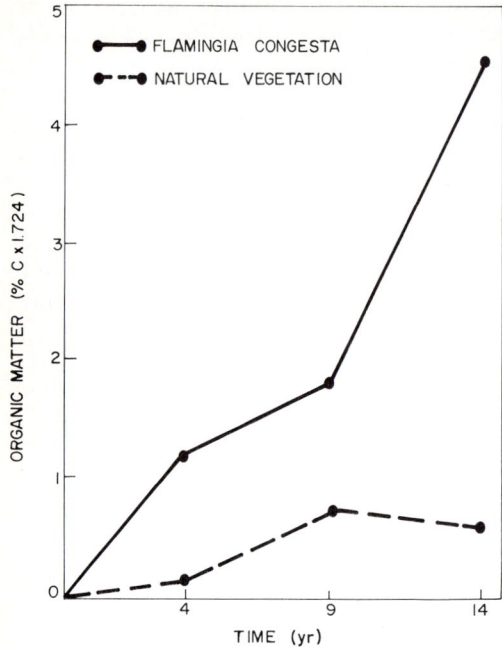

Figure 13.19 Effects of *Flamingia congesta* on soil organic matter content in Sumatra, Indonesia (Sukmana et al., 1985).

ping systems involving tree crops also require supplementary erosion control measures. Adopting suitable soil and crop management practices is as important in controlling erosion in perennials as in seasonal crops. Alley cropping can be a promising system for some soils and ecological regions provided that compatible mixtures are chosen and that the agronomic packages are developed for

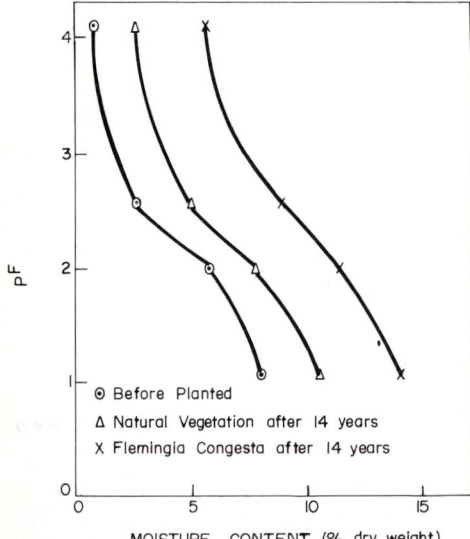

Figure 13.20 Effects of *Flamingia congesta* on soil moisture retention characteristics (Sukmana et al., 1985).

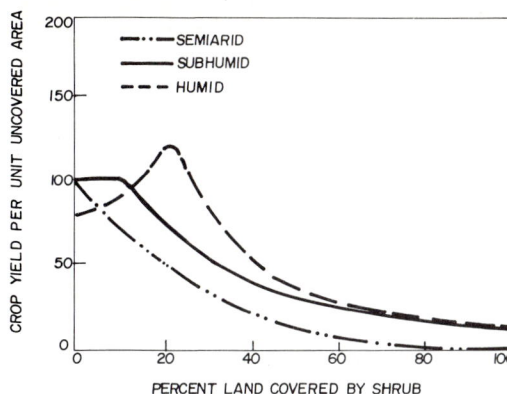

Figure 13.21 Schematic of relation between the yield of an annual crop grown at a different space allocation for the perennial crop.

appropriate management of annuals and perennials. Antierosive effects of alley cropping are observed from 2 to 3 years after establishment of perennial hedges. Soil surface management within the alleys, by mulching and no-till, is very important for erosion control and fertility maintenance. A strip of perennial crop may be more effective in controlling erosion on steeplands than a single row of tree cuttings.

References

Agpaoa, M., Endangan, D., Festin, S., Gumayagay, J., Hoenninger, Th., Seeber, G., Unkel, K., and Weidelt, H. J. 1975. *Manual of Reforestation and Erosion Control for the Philippines,* German Agency for Technical Cooperation (GTZ), Eschborn, West Germany.

Ahuja, L. R., and El-Swaify, S. A. 1979. Determining soil hydrologic characteristics on a remote forest watershed by continuous monitoring of soil-water pressures, rainfall and runoff. *J. Hydrol. (Amster.):* 44(1/2):135–147.

Alexander, T. G., Sobhaua, K., Balagopalan, M., and Mary, M. V. 1980. *Tamugya in Relation to Soil Properties, Soil Erosion and Soil Management,* KRFI Res. Rep. 4, Kerala, Forest Research Institute.

Alleyne, E. P., and Percy, M. J. 1966. Runoff and soil loss on two small watersheds in the Northern Range, Trinidad. *Trop. Agric. (Trin.)* 43:323–326.

Amphlett, M. B., and Tucker, S. M. 1984. *Soil Erosion Research Project, Bvumbwe, Malawi.* Report O.D. 62, 1982–1983, Hydraulics Research, Wallingford, U.K.

Arap Konuche, P. K. 1982. Effects of forest management practices on soil and water conservation in Kenya forests, National Soil Conservation Workshop, Kenya.

Atkinson, G. 1984. Soil erosion following wildfire in a sandstone catchment. *Proc. National Soils Conference,* May 13–18, 1984, Brisbane, Australia, CSIRO.

Attaviroj, P. (ed.). 1986. *Soil Erosion and Degradation in Northern Thai Uplands: An Economic Study.* Northern Thailand Upland Agricultural Project. Annexes of the paper presented to the International Conference on the Economics of Dry Land Degradation and Rehabilitation, 10–14 March 1986, Canberra, Australia.

Attaviroj, P. 1987. Northern Agricultural Land Development Project. In: S. Jantawat (ed.), *Soil Erosion and Its Counter Measures,* Soil and Water Conservation Society of Thailand, Bangkok, pp. 227–244.

Baconguis, S. R., and Punzalan, A. V. 1983. Geomorphological characteristics and infiltration capacities of different landuses at Buhisan watershed. *Proc. 33d Annual Convention Philippine Society of Agricultural Engineers,* Baybay, Leyte, Visaya State College Agriculture.

Baconguis, S. R., and Rondilla, C. S. 1979. Infiltration studies on giant ipil-ipil (*Leucaena leucocephala*) plantation sites at the Buhisan watershed. *Sylvatrop Phil. For. Res. J.* 4:23–29.

Balek, J. 1977. *Hydrology and Water Resources in Tropical Africa.* Elsevier, Amsterdam.

Barus, A., Suwardjo, H., and Suhardjo, H. 1986. Effect of land clearing methods on soil compaction, soil erosion and yield of cover crops and soybeans. *Proyek Penelitian Tanah,* Soil Research Institute, Bogor, Indonesia.

Benge, M. 1979. *Use of Leucaena for Soil Erosion Control and Fertilization,* USAID, Washington.

Birot, P. 1968. *The Cycle of Erosion in Different Climates,* Batsford, London.

Bonell, M., and Gilmour, D. A. 1978. The development of overland flow in a tropical rainforest catchment. *J. Hydrol.* 39:365–382.

Bonell, M., Gilmour, D. A., and Sinclair, D. F. 1979. A statistical method for modelling the fate of rainfall in a tropical rainforest catchment. *J. Hydrol.* 42:251–267.

Brinkman, W. L. F. 1972. Weekly water-loss from spherical water-loss integrators on a clearing and below secondary growth in central Amazonia. *Acta Amazonia* 2(1):33–36.

Celestino, A. F. 1984. *Establishment of Ipil-Ipil Hedges for Soil Erosion and Degradation Control in Hilly Land,* University of Philippines, Los Banos.

Celestino, A. F. 1985. Farming systems approach to soil erosion control and management. In: *Soil Erosion Management* (E. T. Craswell, J. M. Remenyi, and L. G. Nallana, eds.), pp. 64–70, ACIAR, Canberra, Australia.

Chadhokar, P. A. 1986. Principles of revegetation in soil and water conservation, FAO Regional Office, Addis Ababa, Ethiopia.

Chanphaka, U., Vajirajutiponk, T., and Vattanaphatheep, P. 1976. *The Study of Interception Stemflow and Throughfall in the M.SL. 350–500 Meters of the Evergreen Forest at Chinang Dao Watershed Research Station,* Chiang Mai.

Chinnamani, S., and Singh, G. 1971. Land management in red (Chalka) soils of Telencha. *Am. Arid Zone* 10(4):217–247.

Clegg, A. C. 1963. Rainfall interception in a tropical forest. *Caribbean Forester* 24(2):75–79.

Clinnick, P. F. 1984. Accelerated erosion resulting from forest operations including the effects of fire. *Proc. National Soils Conference,* May 13–18, 1984, Brisbane, Australia, CSIRO.

Cole, D. W. 1981. Nutrient cycling in world forests. In: *Proc. 17th IUFRO World Congress* (Div. 1), Ibaraki, Japan, pp. 139–157.

Commonwealth Science Council. 1986. *Amelioration of Soil by Trees,* Commonwealth Secretariat, Marlborough House, London.

Coster, C. 1983. Superficial runoff and erosion on Java. *Tectona* 31:613–728.

CTFT. 1979. *Conservation des Sols Sud du Sahara,* 2d ed., Centre Technique Forestier Tropical, Nogent, France.

Cunningham, R. K. 1963. The effect of clearing a tropical forest soil. *J. Soil Sci.* 14:334–345.

Dabral, B. G., and Pande, S. K. 1979. Soil moisture regime under forest plantations. *Indian Forest Records* 3(1):1–45.

Daubenmire, R. 1972. Some ecologic consequences of converting forest to savanna in northwestern Costa Rica. *Trop. Ecol.* 13(1):31–51.

De La Rosa, M. M. 1984. A study on the growth and yield of corn intercropped with varying population of giant ipil-ipil (*Leucaena leucocephala*) on a hillside. Thesis, Visayas State College Agriculture, Philippines.

Doty, R. D., Wood, H. B., and Merriam, R. A. 1981. Suspended sediment production from forested watersheds on Oahu, Hawaii. *Water Resources Bull.* 17:399–405.

Douglas, I. 1967. Erosion of granite terrain under tropical rainforest in Australia, Malaysia and Singapore. In: *Symposium on River Morphology,* General Assembly of Berne, Switzerland, pp. 32–39.

Dressler, J., and Neumann, I. 1982. Agriculture da couverture du sol. Un imperatif pour la lutte contre l'erosion au Rwanda. *Null. Agric. Rwanda* 4:215–222.

Ducreux, A., Godefroy, J., and LaCoenilhe, J. J. 1980. Some problems with working the soil in pineapple plantations in Martinique. *GERDAT* 35(10):595–604, 657–660.

Dyrness, C. T. 1967. Erodibility and erosion potential of forest watersheds. In: *International Symposium on Forest Hydrology* (W. E. Sopper and H. W. Lull, eds.), pp. 599–610, Pergamon Press, Oxford.

Eyles, R. J., and Ho, R. 1970. Soil creep on a humid tropical slope. *J. Trop. Geogr.* 31:40–42.

Flenley, J. R. (ed.). 1971. The water relations of Malesian Forests. *Transactions of the First Aberdeen-Hull Symposium on Malesian Ecology, Hull, 1970,* Miscellaneous series, Dept. of Geography, University of Hull, U.K.

Fonzen, P. F., and Oberholzer, E. 1984. Use of multipurpose trees in hill farming systems in western Nepal. *Agrofor. Systems* 2:187–197.

Fournier, F. 1967. Research on soil erosion and soil conservation in Africa. *Afr. Soils* 12:53–96.

Gananpin, D. J. 1978. Quo vadia, Philippine forests? *For. Dig.* 5(2):1–16.

Garcia, A. S. 1981. Initial effect of clear cutting on surface runoff, sedimentation and nutrient losses in ipil-ipil (*Leucaena leucocephala*) plantation. BIOTROP Special Publ. 13, pp. 83–92.

Garcia, A. S., and Bernollia, E. F. 1979. Closed nutrient cycle in tropical rainforest. *Canopy* 5:7.

Gilmour, D. A. 1971. The effects of logging on streamflow and sedimentation in a north Queensland rainforest catchment. *Comm. For. Rev.* 50:38–48.

Gilmour, D. A. 1977. Logging and the environment, with particular reference to soil and stream protection in tropical rainforest situations. In: *FAO Conservation Guide I: Guidelines for Watershed Management,* FAO, Rome.

Gilmour, D. A., and Bonell, M. 1977. *Stream Flow Generation Process in a Tropical Rainforest Catchment.* Institute Engineering Australia, Brisbane, Queensland, Publ. 77/5, pp. 178–179.

Gilmour, D. A., Bonell, M., and Sinclair, D. F. 1980. *An Investigation of Storm Drainage Processes in a Tropical Rainforest Catchment,* Department of National Development and Energy, Tech. Paper 56, Australian Government Publications Service, Canberra.

Greenfield, J. C. 1986. Narmada Sagar Area Development Project. *World Bank Report,* New Delhi, India.

Gwyer, G. 1978. Developing hillside farming systems for the humid tropics: The case of the Philippines. *Oxford Agrarian Stud.* 7:1–37.

Hamilton, L. S., and King, P. N. 1983. *Tropical Forested Watersheds, Hydrological and Soils Response to Major Uses or Conversions,* Westview Press, Boulder, Colo.

Hanif, M. 1979. Observation on Missa comparative watershed study. *Pakistan J. For.* 29(4):209–237. Watershed Management Branch, Pakistan Forestry Inst., Peshawar, Pakistan.

Hatch, T. 1981. Preliminary results of soil erosion and conservation trials under pepper (*Piper nigrum*) in Sarawak Malaysia. In: *Soil Conservation: Problems and Prospects* (R. P. C. Morgan, ed.), J. Wiley and Sons, Chichester, U.K., pp. 255–262.

Hibbert, A. R. 1967. Forest treatment effects on water yield. In: *Proc. International Symposium on Forest Hydrology, Pennsylvania State University,* pp. 527–543, Pergamon Press, N.Y.

Hopkins, B. 1960. Rainfall interception by tropical forest in Uganda. *E. Afr. Agric. J.* 25(4):1360.

Hudson, N. 1977. Research needs for soil conservation in developing countries. In: *Soil Conservation and Management in Developing Countries,* FAO Soils Bull. 33, pp. 169–184, FAO, Rome.

Hulugalle, N. R., Lal, R., and Ter Kuile, C. H. H. 1984. Soil physical changes and crop root growth following different methods of land clearing in western Nigeria. *Soil Sci.* 138:172–179.

Hulugalle, N. R., Lal, R., and Ter Kuile, C. H. H. 1986. Amelioration of soil physical properties by *Mucuna* after mechanized land clearing of a tropical rainforest. *Soil Sci.* 141:219–224.

IDRC. 1982. *Leucaena Research in the Asian-Pacific Region,* Ottawa, Canada.

IITA-UNU. 1985. Effects of deforestation and landuse on soil, hydrology, microclimate and productivity, progress report, December 1985, IITA, Ibadan, Nigeria.

Jackson, I. J. 1975. Relationships between rainfall parameters and interception by tropical forest. *J. Hydrol.* 24(3/4):215–238.
Jones, P. H. 1983. Lamtoro and the Amarasi model from Timor. *Bull. Indones. Econ. Stud.* 19:106–112.
Jurion, F., and Henry, J. 1969. *Can Primitive Farming Be Modernized?* I.N.E.A.C. Hors Eries, Yangambi, Zaire.
Kang, B. T., Wilson, G. F., and Lawson, T. L. 1984. *Alley Cropping: A Stable Alternative to Shifting Cultivation,* IITA, Ibadan, Nigeria.
Kang, B. T., Grimme, H., and Lawson, T. L. 1985. Alley cropping sequentially cropped maize and cowpea with *Leucaena* on a sandy soil in southern Nigeria. *Plant Soil* 85:267–277.
Kellman, M. C. 1969. Some environmental components of shifting cultivation in upland Mindanao. *J. Trop. Geogr.* 28:40–56.
Kerfoot, O., and McCulloch, S. S. G. 1967. The interception and condensation of atmospheric moisture by forest canopies. In: *8th British Commonwealth Forestry Conference, East Africa,* East African Agriculture and Forestry Research Organization, Muguga, p. 7.
Krishnarajah, P. 1984. The magnitude and extent of soil erosion under different landuses in Sri Lanka. *Proc. Malama Aina Conference,* Honolulu, Hawaii, ISCO and Soil and Water Conservation Society, Ankeny, Iowa.
Lal, R. 1981. Deforestation of tropical rainforest and hydrological problems. In: *Tropical Agricultural Hydrology* (R. Lal and E. W. Russell, eds.), pp. 138–140, Wiley, Chichester, U.K.
Lal, R., 1986. Deforestation and soil erosion. In: *Land Clearing and Development in the Tropics* (R. Lal, P. A. Sanchez, and R. W. Cummings, Jr., eds.), pp. 299–312, A. A. Balkema, Rotterdam, Holland.
Lal, R. 1987. *Tropical Ecology and Physical Edaphology,* Wiley, Chichester, U.K.
Lal, R. 1989. Agroforestry systems and soil surface management of a tropical Alfisol. Parts I–V. *Agrofor. Sys.* Vol. 8, pp. 1–6, 7–29, 97–111, 113–132, 197–215, 217–238.
Lal, R., and Cummings, D. J. 1979. Clearing a tropical forest. I. Effects on soil and microclimate. *Field Crops Res.* 2:91–107.
Lal, R., Sanchez, D. A., and Cummings, Jr., R. W. 1986. *Land Clearing and Development in the Tropics.* A. A. Balkema, Rotterdam, Holland.
Lal, R., Wilson, G. F., and Okigbo, B. N. 1979. Changes in properties of an Alfisol produced by various crop covers. *Soil Sci.* 127:377–382.
Lawson, T. L., Lal, R., and Oduro-Afriyi, K. 1981. Rainfall distribution and microclimatic changes over a cleared watershed. In: *Tropical Agricultural Hydrology* (R. Lal and E. W. Russell, eds.), pp. 141–151, Wiley, Chichester, U.K.
Leigh, C. H. 1982. *Development and Environment in Peninsular Malaysia,* McGraw-Hill, New York.
Lembaga Ekologi. 1980. *Report on Study of Vegetation and Erosion in the Jatiluhur Catchment 1980,* Institute of Economics, Padjadjaran University, Bandung, Indonesia.
Lewis, L. A. 1974. Slow movement of earth under tropical rainforest conditions. *Geology* 2:9–10.
Liao, M. C., and Chang H. M. 1975. Study on soil conservation practices for juvenile orange plantation. *J. Agric. Assoc. China (Taiwan)* no. 88, pp. 74–82.
Liao, M. C., and Wu, H. L. 1987. *Soil Conservation on Steep Lands in Taiwan,* The Chinese Soil and Water Conservation Society, Taipei, Taiwan.
Lieu, T. C. 1974. A note on soil erosion at Tawan Hills forest reserve. *Malaysia Natl. J.* 27:20–26.
Lipman, E. 1986. *Etat de la recherche agroforestière en Twanda, Ruhande, Butare,* Dept. For., Inst. Sci. Agron., Bature, Rwanda.
Loch, R. J. 1985. *Soil Erosion in the Philippine Uplands: Observations of the Problem and Recommendations for Research,* Queensland Dept. of Primary Industries, Brisbane, Australia.
Lockwood, J. G. 1980. Problems of humid equatorial climates. *Malaysian J. Trop. Geogr.* 1:12–20.

Lombardi, F. Neto, Bertoni, J., and Benatti, Jr., R. 1975a. Praticas conservacionistas en cafezal e as perdas por erosao em solos podzolizados. *An. Do XV Congresso Bras. Cienc. Solo,* pp. 559–562.

Lombardi, F. Neto, Bertoni, J., and Benatti, Jr., R. 1975b. Praticas conservacionistas en cafezal e perdes por exosao en Latossolo Roxo. *An. Do XV Congresso Bras. Cienc. Solo,* pp. 581–583.

Low, K. S. 1972. Interception loss in the humid forested areas (with special reference to Sungai Lin catchment, West Malaysia). *Malaysia Natl. J.* 25:104–111.

Lundgren, B. O., and Raintree, J. B. 1983. Sustained agroforestry. In: *Agricultural Research for Development: Potentials and Challenges in Asia* (B. Nestel, ed.), pp. 37–49, ISNAR, The Hague.

Lundgren, L. 1980. Comparison of surface runoff and soil loss from runoff plots in forested and small-scale agriculture in the Usambara Mts. Tanzania. *Geograf. Ann.* 62A:113–148.

Lundgren, L. 1981. Soil and vegetation development on fresh landslide scars in the Mgeta Valley, Western Uluguru Mountains, Tanzania. In: *Tropical Agricultural Hydrology* (R. Lal and E. W. Russell, eds.), pp. 227–236, Wiley, Chichester, U.K.

Management of soils under hevea in peninsular Malaysia. 1974. *Planters Bull. (Malaysia)* no. 134, pp. 147–156.

Manokarau, N. 1980. Stemflow, throughfall and rainfall interception in a lowland tropical rainforest in peninsular Malaysia. In: *Proc. Fifth International Symposium of Tropical Economics and Development,* Kuala Lumpur, Malaysia, pp. 91–94.

McGregor, D. F. M. 1980. An investigation of soil erosion in the Colombian rainforest zone. *Catena* 7:265–273.

Millington, A. C. 1981. *Relationship between Three Scales of Erosion Measurement on Two Small Basins in Sierra Leone.* IAHS Publication 133, pp. 485–492.

Mishra, B. K., and Ramakrishnan, P. S. 1983. Slash and burn agriculture at higher elevations in North-eastern India. 1. Sediment, water and nutrient losses. *Agric. Ecosystem Environ.* 9:69–82.

Mitchell, H. W. 1965. Soil erosion losses in coffee. *Tanganyika Coffee News,* April/June, pp. 135–155.

Molchanov, A. A. 1960. The hydrological role of forests. *Acad. Sci. USSR. Inst. For.,* Moscow (trans. by A. Govrevitch), Israel, 1963.

Nair, P. K. R. 1987. Agroforestry in the context of land clearing and development in the tropics. In: *Tropical Land Clearing for Sustainable Agriculture* (R. Lal, M. Nelson, H. W. Scharpenseel, and M. Sudjadi, eds.), IBSRAM, Bangkok, Thailand.

Naprakob, B., Lapudomlers, P., and Witchirjutipong, Y. 1975. *Sediment Yield from Shifting Cultivation at Chiang Dao.* Watershed Management Division, Royal Thai Forest Dept., Bangkok, Thailand.

Nieuwolt, S. 1965. Evaporation and water balance in Malaysia. *J. Trop. Geogr.* 20:34–53.

Noer, Z. M. 1981. Effects of different vegetation on soil protection against rainsplash. Ph.D. dissertation, Department of Biology, Padjadjaran University, Bandung, Indonesia.

Nortcliff, S., Thomes, J. B., and Waylen, M. J. 1979. Tropical forest systems: a hydrological approach. *Amazonia* 6(4):557–568.

Nye, P. H., and Greenland, D. J. 1964. Changes in soil after clearing tropical forests. *Plant Soil* 21:101–112.

Odum, H. T. 1970. Rainforest structure and mineral cycling homeostasis. In: *A Tropical Rain Forest* (H. T. Odum and R. F. Pigeon, eds.), pp. H3–H52, U.S. Atomic Energy Commission, Washington.

Ojeniyi, S. O., and Agbede, O. O. 1980. Effect of inter-planting *Gmelina arborea* with food crops on soil conditions. *Turrialba* 30:268–271.

Othieno, C. O. 1974. *First Year Studies on Rainfall, Runoff and Soil Erosion in a Field of Young Tea,* Tea Research Institute of East Africa, Kericho, Kenya.

Othieno, C. O. 1975. Surface runoff and soil erosion on fields of young tea. *Trop. Agric.* 52(4):299–308.

Othieno, C. O. 1982. Diurnal variations in soil temperature under tea plants. *Exp. Agric. (U.K.)* 18(2):195–202.

Pacardo, E. P. 1985. Soil erosion and ecological stability. In: *Soil Erosion Management* (E. T. Craswell, J. V. Remenyi, and L. G. Nallana, eds.), ACIAR Proc. Series, no. 6, Canberra, Australia.

Pacardo, E. P., and Montecillo, L. 1983. Effect of corn/ipil-ipil cropping system on productivity and stability of upland agroecosystem. Annual Report, UPLB-PCARRD Research Project.

Pandey, A. N., Pathak, P. C., and Singh, J. S. 1983. Water sediment and nutrient movement in forested and non-forested catchments in Kumaun Himalaya. *For. Ecol. Manage.* 7(1):19–29.

Paningbatan, E. P. 1987. Alley cropping in the Philippines. In: *Soil Management under Humid Conditions in Asia and the Pacific,* pp. 386–395, IBSRAM Proc. 5, IBSRAM, Bangkok, Thailand.

Parera, V. 1983. *Leucaena* for erosion control and green manure in Sikka. In: *Leucaena Research in the Asia-Pacific Region,* pp. 169–172, IDRC, Ottawa, Canada.

Parfait, J. A., and Lallmahomed, H. 1980. The effects of change in land use on the hydrological regimes of three small basins in Mauritius. *IASH* 130:351–358.

Pereira, C. 1979. Hydrological and soil conservation aspects of agroforestry. In: *Soil Research in Agroforestry* (H. A. Mongi and P. A. Huxley, eds.), pp. 315–326, ICRAF, Nairobi, Kenya.

Pereira, H. C. 1965. Landuse and streamflow. *E. Afr. Agric. For. J.* 30:395–397.

Pereira, H. C. 1973. *Landuse and Water Resources in Temperate and Tropical Climates,* Cambridge University Press, Cambridge, England.

Pereira, H. C., and Hosegood, P. H. 1962. Comparative water use of softwood plantations and bamboo forest. *J. Soil Sci.* 13:299–314.

Roche, M. A. 1981. Watershed investigations for development of forest resources of the Amazon region in French Guyana. In: *Tropical Agricultural Hydrology* (R. Lal and E. W. Russell, eds.), Wiley, Chichester, U.K.

Rodriguez, G. 1958. Sistemas de conservacion de suelos en plantaciones do cafe al sol. *Cenicofe (Colombia)* 9:277–290.

Roose, E. J. 1979. Present dynamics of a highly desaturated ferrallitic soil derived from sandy clay deposits under cultivation and under dense subequatorial rainforest in southern Ivory Coast. Adiopodoume 1964 to 1976. *Cah. ORSTOM Pedol.* 17(4):259–281.

Rubber Research Institute (RRI). 1974. Management of soils under Hevea in Peninsular Malaysia. *Planters Bull.* 134:147–154.

Sarkar, S. K. 1974. Problems of tea in Darjeeling two and a Bud. *J. Soil Water Cons. (India)* 21(2):45–49.

SCSA. 1982. *Resource Conservation Glossary,* SCSA, Ankeny, Iowa.

Seubert, C. E., Sanchez, P. A., and Valverde, C. 1977. Effects of land clearing methods on soil properties of an Ultisol and crop performance in the Amazon jungle of Peru. *Trop. Agric. (Trin.)* 56:325–331.

Sheng, T. C., and Koh, C. C. 1965. Forest hydrology research in Taiwan, Republic of China. In: *International Symposium on Forestry Hydrology,* Taipei, Taiwan, pp. 89–94.

Sheppard, J. S. 1978. Woody plants for soil stabilisation and erosion control. *R. NZ Inst. Hort. Annu. J.* 6:46–53.

Sim, L. K., and Hock, P. C. 1986. Conversion of forest to tree crops: General Malaysian experience. In: *Landuse, Watersheds, and Planning the Asia-Pacific Region,* FAO Regional Office for Asia and the Pacific, Bangkok, Thailand, RAPA Rep. 1986/3, pp. 60–71.

Ssekabembe, C. K. 1985. Perspectives on hedgerow intercropping. *Agrofor. Systems* 3:339–356.

Stockdale, F. A. 1937. Soil erosion in the Colonial Empire. *Emp. J. Exp. Agric.* 5:281–297.

Sudjadi, M., Manuelpillai, R. G., and Caguan, B. 1985. Performance of some nitrogen fixing trees on typic Paleudult. *Pembr. Pen. Tanah Dan Pupuk* 4:10–41.

Sukmana, S., Suwardjo, H., Abdurachman, A., and Dai, J. 1985. Prospect of *Flemingia congesta* Roxb. for reclamation and conservation of volcanic skeletal soils. *Pembr. Pen. Tanah Dan Pupuk* 4:50–54.

Takahashi, T., Nagahori, K., Mangkolsawat, C., and Losirkul, M. 1983. Runoff and soil loss. In: *Shifting Cultivation* (K. Kyuma and C. Pairintra, eds.), pp. 84–109, Ministry of Science, Technology, and Energy, Bangkok, Thailand.

Thaiutsa, B., and Granger, O. 1980. Climate and the decomposition rate of tropical forest litter. *Unasylva* 31:28–35.

Toky, O. P., and Ramakrishnan, P. S. 1981. Runoff and infiltration losses related to shifting agriculture (Jhum) in northeastern India. *Environ. Conserv.* 8:313–321.

Torquebian, E. 1984. Man-made dipterocarp forest in Sumatra. *Agrofor. Systems* 2:103–127.

Trimble, S. W., and Weirich, F. H. 1987. Stream flow decreased by reforestation in the southeastern USA. *J. Soil Water Conserv.* 42:274–276.

Turvey, N. D. 1975. Water quality in a tropical rainforest catchment. *J. Hydrol.* 27:111–125.

TVA (Tennessee Valley Authority). 1962. *Reforestation and Erosion Control Influences upon the Hydrology of the Pine Tree Branch Watershed, 1941–1960,* Knoxville, Tenn.

Van der Weert, R. 1974. Influence of mechanical forest clearing on soil conditions and the resulting effects on root growth. *Trop. Agric. (Trin.)* 51:325–331.

Van Kraayenoord, C. W. S. 1976. Plant materials for erosion control. *NZ Agric. Sci.* 10:29–33.

Vergara, N. T. 1982. *New Directions in Agroforestry: The Potential of Tropical Legume Trees.* EPI–East West Centre, Honolulu, Hawaii.

Vergara, N. T. 1987. Agroforestry: A sustainable landuse for fragile ecosystems in the humid tropics. In: *Agroforestry: Realities, Possibilities and Potentials* (H. L. Gholz, ed.), pp. 7–20, Martinus Nijhoff Publishers, Dordrecht.

Von Uexkull, H. R. 1984. Managing Acrisols in the humid tropics. In: *Ecology and Management of Problem Soils in Asia,* FFTC Book Series, no. 27, pp. 382–397, FFTC, Taipei, Taiwan.

Watnaprateep, P. 1980. First year study on soil loss and runoff from cropping under a forest plantation, In: *Proc. ASEAN Seminar on Management of Tropical Forests,* Chiang Mai, Thailand.

Weber, F. R. 1977. Reforestation in arid lands. *VITA Manual,* Series 37E, Maryland.

Weidelt, H. J. 1976. *Manual of Reforestation and Erosion Control for the Philippines,* GTZ, Eschborn, Germany.

Weinstabel, P. E., and Zech, W. 1982. Forestry as a possible solution to the environmental and energy difficulties facing semiarid Africa. In: *Forestry in Semiarid Africa,* pp. 71–87, Institut fur Wissenschaftliche Zusammenarbeit Landhausstrasse 18, Tubingen, Germany.

Wiersum, K. F. 1981. Introduction to agroforestry. In: *Observation of Agroforestry in Java, Indonesia* (K. F. Wiersum, ed.), pp. 3–29, Forestry Faculty Gadjah, Mada University, Yogyakarta.

Wiersum, K. F. 1983. Effects of various vegetation layers of an *Acacia auriculiformis* forest plantation on surface erosion at Java, Indonesia. *Proc. Malama Aina Conference,* Honolulu, Hawaii. ISCO and Soil and Water Conservation Society, Ankeny, Iowa.

Wiersum, K. F. 1984. Surface erosion under various tropical agroforestry systems. In: *Symposium on Effects of Forest Landuse on Erosion and Slope Stability* (C. L. O'Loughlin and A. J. Pearce, eds.), East-West Center, Honolulu, Hawaii.

Wiersum, K. F. 1985. Effects of various vegetation layers in an *Acacia auriculiformis* forest plantation on surface erosion in Java, Indonesia. In: *Soil Erosion and Conservation* (S. A. El-Swaify, W. C. Moldenhauer, and A. Lo, eds.), pp. 79–89, SCSA, Ankeny, Iowa.

Wijewardene, R. 1982. Shift to "conservation farming." *Int. Agric. Dev.* 2(2):11.

Williams, M. A. J. 1973. The efficacy of creep and slope wash in tropical and temperate Australia. *Austr. Geogr. Stud.* 11:62–78.

Yen, C. P. 1984. Tree root patterns and erosion control. *Proc. International Workshop on Soil Erosion and Its Countermeasures,* November 11–19, 1984, Chiangmai, Thailand, Soil and Water Conservation Association of Thailand.

Young, A. 1986. *The Potential of Agroforestry for Conservation,* ICRAF Working Paper 42, ICRAF, Nairobi, Kenya.

Part 5

Special Topics

Chapter

14

Erosion Hazard on Steeplands

Introduction

Steeplands, in the context of this book, are defined as those lands with a slope gradient of 20 percent or more. Lands with a slope gradient of less than 20 percent are considered as flat to undulating or rolling lands. Because of erosion hazard and the constraint to mechanization, steeplands are considered marginal for arable use. Some countries, with pressure of high human and animal populations, however, have no choice but to expand their land bases to include steeplands for food production. Countries in that category include those in the Himalayan-Tibetan ecosystem, Thailand, Philippines, Indonesia, the Andean region, eastern Africa (Plate 14.1), the Pacific, and the Caribbean regions. It is ironic that people in some of the most densely populated regions, where land pressure is too great, have no choice but to exploit steeplands. For example, in Nepal the hillslopes up to about 45 percent slope are cultivated. The topographic map of Nepal shows that the zone of middle hills and high mountains is steepland and is being intensively cultivated. As much as 83 percent of the population in these hills are resource-poor small farmers. Cultivation of steep hills in these regions has caused severe soil erosion problems.

Use of steepland is not confined to any specific region. Steeplands are used in diverse climates and ecological regions. In some regions, flatlands are relatively less available. In others, landless farmers have been forced to move uphill to carve out a subsistence living. In Ethiopia, 54 percent of the total land area has a slope gradient exceeding 8 percent, and 29 percent has a slope gradient exceeding 30 percent. In comparison, only 6 percent of Sudan's land area exceeds a slope gradient of 30 percent (Purnell, 1986). Extremely steep slopes also

490 Special Topics

(a)

(b)

Plate 14.1 Steeplands are being used intensively for food crop production in (a), (b) Ethiopia and (c), (d) Rwanda.

Erosion Hazard on Steeplands   491

(c)

(d)

Plate 14.1 (*Continued*)

are being used in the Andes and the Caribbean, and steeplands are cultivated in temperate regions, e.g., New Zealand and western Europe. In New Zealand, primary production from pastoral steeplands contributes nearly one-third of the nation's agricultural products, and these in turn account for 60 to 70 percent of the export income (Thomas and Trustrum, 1984). Steeplands in New Zealand are, however, subject to severe erosion and erosion-caused degradation. Trustrum and Hawley (1986) estimated that in the Wairarapa, percentage of hillslope area eroded has increased from about 10 percent in 1932 to 55 percent in 1980. An increasing level of erosion has also adversely affected pasture production. In North America, Purnell (1986) estimated percentages of land area in different slope classes as 36 percent in 0 to 8 percent, 50 in 8 to 30 percent, and 14 in slope steepness exceeding 30 percent. The corresponding percentages for land areas in Europe and north Asia are 43, 38, and 19, respectively.

Factors Affecting Soil Erosion

Cultivation of steeplands can cause severe soil erosion. The extent and severity of erosion depend on the soil, climate, land use, and farming systems. Most steeplands in the densely populated regions are used intensively, and the erosion problem is severe regardless of the soil and climatic conditions. The erosion is highly accentuated if the soil and climatic conditions are also prone to erosion. Soil erosion from steeplands can be extremely severe even with low cropping intensity. Sabhasri (1978) reported extremely high erosion rates from steeplands used by shifting cultivators in Thailand. Estimates of soil erosion on the basis of the change in level of soil surface indicated an annual soil erosion rate of 10 to 12 cm (Table 14.1). For such a steepland, high erosion rates estimated even under forest cover may be attributed to selective logging, footpaths, and landslides. The change in the level of soil surface, however, can be due to many factors, e.g., subsidence, compaction, and erosion. More conservative estimates indicated erosion rates of 0.1 to 0.8 cm/yr. Soil erosion (type and magnitude) from steeplands is affected by many factors, some of which are explained now.

TABLE 14.1 Soil Erosion under Shifting Cultivation from Steeplands in Thailand during 1968–1969

Station	Slope, %	Exposure	Average soil loss, cm/yr
1	29	Northeast	10.33
2	44	East	12.02
3	42	Southeast	13.00
4	33	South	12.65
5	39	Southwest	12.62
6	40	Northwest	11.2
7 Forest control	44	North	8.6

SOURCE: Sabhasri, 1978.

Land use

The severity of erosion on steeplands is governed by the land use. The unprotected arable lands and rangelands with uncontrolled grazing and a high stocking rate often suffer from severe soil erosion. The forest-covered steeplands generally have less severe soil erosion, especially if the soil surface is protected by leaf litter and regrowth. The data of Fleming (1983) and Pandey et al. (1983) from Nepal showed that the maximum erosion occurred on grazed lands (Table 14.2). Some researchers estimate that the denudation rates in Nepal range from 0.5 to 14 mm/yr (Starkel, 1972; Caine and Mool, 1982; Gilmour, 1986). In the Rif Mountains of northern Morocco, erosion rates of 43 to 285 t/(ha · yr) were measured from steepland watersheds. Erosion rates of 20 to 39 t/(ha · yr) have been reported in steeplands of northern Thailand. Severe erosion is also reported from Australia. In Queensland, soil losses of 200 to 328 t/(ha · yr) have been reported from sugarcane fields on sloping lands in central and north Queensland (Sallaway, 1979; Matthews and Makepeace, 1981). Severe erosion is caused by the misuse of these fragile lands, e.g., overgrazing, deforestation, removal of leaf litter, and fire.

Steeplands are equally vulnerable to forces causing soil erosion even in the temperate-zone climate. Following is a pertinent example of the catchment of Lake Balatan in Hungary. Balatan is the largest lake (area 600 km^2) in central Europe. The hilly regions situated around the lake have slopes of up to 50 percent. The catchment surrounding Lake Balatan has been used intensively, and this has caused severe erosion and pollution problems in the lake. These slopes are extensively used for animal husbandry farms, vineyards, and croplands. The data in Table 14.3 show extremely high losses of soil and humus content, major pollutants of the lake water.

In New Zealand, severe debris slides and mass movement are observed in the Wirapa hill country, intensively used for sheep and cattle grazing. The original native forest (podocarp broad leaf) was removed and converted to pastures in the latter part of the nineteenth century (Thomas and Trustrum, 1984). Intense rains falling on denuded hills triggered widespread soil slipping. The resistance of the soil to slipping decreased as the tree roots progressively deteriorated. The prolonged winter wetness also increased the incidence of slipping.

In fact, erosion from steeplands can be severe in any ecological region, especially if steeplands are put to an intensive arable use. This conclusion is sup-

TABLE 14.2 Estimated Annual Erosion Rates from Land Use Categories in the Phewa Tal Catchment, Nepal

Land	Erosion rate, t/(ha · yr)
Forest	8
Scrub land	15
Grazing land	34.7
Terrace land	10

SOURCE: Fleming, 1983.

TABLE 14.3 Effects of Slope Steepness on Soil Erosion and Loss of Humus from the Catchment of Lake Balatan in Hungary

Slope category, %	Area, ha	Soil erosion, t/yr	Humus loss, t/yr
0–5	175	1030	23.7
0–12	164	2475	46.3
12–17	129	3603	86.2
17–25	169	8672	122.4
25–50	31	1764	21.2

SOURCE: Toth and Fekete, 1974.

ported by the literature survey compiled by Liao and Wu (1987) from cultivated steeplands around the world (Table 14.4). Appropriate land use, however, can drastically reduce the erosion risks.

Rainfall patterns

High-intensity monsoon rains, concentrated in a short period and falling on denuded hills, cause severe soil erosion. The kinetic energy, drop size distribution, and effective rainfall on highlands are not studied for a wide range of ecological regions. Gilmour (1986) reported the depth-duration-frequency curves of rainfall at Kathmandu, Nepal. The data in Fig. 14.1 show that intense rains are a common occurrence in the Himalayan-Tibetan ecosystem. The "catastrophic" rains are considered as a major factor in the shaping of the mountain relief in many regions (Starkel, 1972). High-intensity rains are also reported in steepland regions in East Africa and the Caribbean.

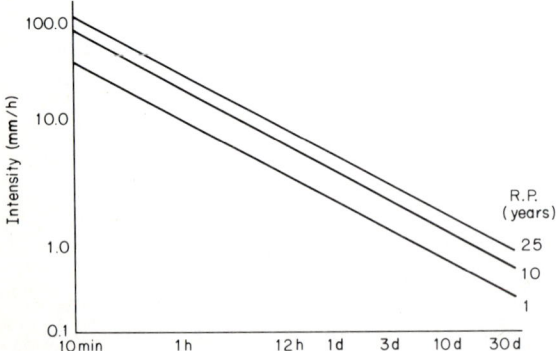

Figure 14.1 The depth-duration-frequency curves of rainfall at Kathmandu, Nepal. R.P. = return period. (Gilmour, 1986.)

Soils

Soils in the steepland regions of volcanic origin generally have a favorable structure and thus low susceptibility to erosion. For example, intensively cultivated steeplands in Rwanda (Plate 14.2), Burundi, and the Kivu region of Zaire suffer less from accelerated erosion than steeplands in the Machakos region of Kenya or the highlands in Tanzania. In addition to the difference in the inherent properties of the soil, soil surface management systems have an important influence on the magnitude of soil erosion. Steeplands in Rwanda are used with extreme care. Some of these useful management techniques are discussed in another section.

TABLE 14.4 Runoff and Erosion from Steep Orchards and Crops in Different Regions

Place	Soil texture	Crops	Rainfall, mm	Slope, %	Runoff, %	Erosion, t/(ha · yr)
United States						
Wisconsin	Silt loam	Corn	867	16	20	159.0
		Fallow			20	151.9
Washington	Silt loam	Wheat	552	30	10	7.7
		Fallow			25	27.8
Ohio	Silt loam	Corn	837	12	17	11.4
Puerto Rico (Mayaguez)	Clay loam	Coffee with natural cover and terrace	1883	62	3	1.8
		Coffee without cover and terrace	1883	62	8	15.3
		Beans, yams, and corn sequences	1899	40	7	39.5
Jamaica (Smithfield)	Clay loam	Banana		30		132.8
El Salvador (Metapan)	Loam to clay loan	Banana		30		182.9
		Yams				132.8
Indonesia		Coffee		59–63		3.1
India		Pomegranate		11	7	1.4
		Pineapple			6	1.7
		Fallow			16	17.5
Nigeria		Bare soil		10	37	142.0
		Mulch 6 t/ha			3	0.2
Tanzania		Coffee		33	5	30–35

SOURCE: Liao and Wu, 1987.

Plate 14.2 Intensively used steeplands in Rwanda.

Predominant Soil Erosion Processes

Processes governing erosion in steeplands are the same as those on undulating and rolling terrains. However, subtle differences exist in the relative importance of different processes involved on steep versus flatlands. Predominant types of erosion on steeplands include rills, gullies, land slips, and mass movement. On flatlands, the predominant type of erosion comprises sheet wash, directional wash, and reticular wash (Bergsma, 1986).

Gully erosion

Because of the steep terrains involved, gully formation is a common mode of erosion on steeplands (Plate 14.3).

Mass movement

Landslides, debris flow, soil slips, and other mechanisms of mass movement are common on steeplands (Plate 14.4). The net positive effect of trees in reducing mass movement is partly due to the mechanical reinforcement provided by the tree root system. Deforestation on hillslopes increases mass movement. In Nepal, Laban (1979) estimated that in recent years deforestation has increased mass movement by 26 percent.

Sediment deposition

The sediment-carrying capacity of high-velocity streams draining the steeplands is generally high. For example, the sediment load in some Himalayan rivers is reported to be 1830 to 4730 $m^3/(km^2 \cdot yr)$. Similarly, in Morocco sedimentation rates of 0.4 to 5.6 million m^3/yr are reported for Loukos, Tieta, and Nekor watersheds. Sediments are readily deposited, however, even with the slightest change in velocity. Claims are also made that because of high sediment load, the streambeds in many hilly regions have been raised. Although little, if any, quantitative data exist in support of these claims, high sedimentation rates are a severe menace in rivers, lakes, and reservoirs of steeplands. The deposition usually occurs as fans or calluvium slope-wash deposits.

Riverbank erosion

The riverbank erosion is also severe in streams draining steeplands. Undercutting and slumping are the principal mechanisms involved.

Plate 14.3 Gully formation in eastern Nigeria.

Plate 14.4 Landslides on road cuts in Malaysia.

Magnitude and Extent of the Problem

The extent of steeplands in different slope classes is not precisely known. It is difficult to develop effective solutions to soil erosion and erosion-caused degradation on steeplands if we do not precisely know their extent and the regional and geographic distributions of the different slope classes. Planning for proper resource management also requires knowledge of major soils being cultivated, or likely to be cultivated, in different slope classes, along with their physical, nutritional, and biologic properties, their productive potentials, and major constraints. We should also know the current or existing land uses or farming systems and what are likely trends in farming systems due to changing needs within the next 20, 50, or 100 years. In the long run, it is also useful to know the productivities of existing farming systems and what act as soil, climate, and sociopolitical constraints to increasing production of the existing farming systems.

Some useful data are available. For example, Purnell (1986), using a soil map of the world, estimated percentages of areas of sloping land for different regions or continents (Table 14.5). Although this is useful information for detailed, effective land use planning, the information derived from soil and topographic maps based on reconnaissance surveys at 1:5,000,000 scale have limited utility. Attempts should be made to produce topographic and soil maps by using semi-detailed surveys at 1:25,000 to 1:100,000 scale. Most information now available is scanty, sketchy, often unreliable, obtained with nonstandardized methodology, and difficult to use in comparisons; so valid generalizations and conclusions are not possible. It will be an important step forward if reliable estimates are obtained of (1) the soil types in different slope classes, (2) current and projected farming systems, and (3) important constraints to increasing production.

TABLE 14.5 Area (%) of Sloping Lands in Tropical Regions

Slope, %	Africa	Southwest Asia	South America	Central America	Southeast Asia	Total area	
						10^6 ha	%
0–8	58	45	52	35	40	3340	51
8–30	34	31	30	40	31	2107	33
>30	8	24	18	25	29	1048	16

SOURCE: Purnell, 1986.

Available information must be collated and new data obtained to build and strengthen the needed data base. An example of the kind of information needed is shown in Table 14.6. Although the data shown in Table 14.7 are based on reconnaissance surveys, this type of information obtained for regions of a few hundred to a few thousand square kilometers would be extremely useful for planning purposes.

Steep Agricultural Land Technologies

Steeplands have long been used for food production. In many regions, e.g., southeast Asia and South America, technology has long been developed to permit an intensive land use. In fact, terraced agriculture has been a cultural tradition of many ancient civilizations that farmed steeplands. Terraced farming, however, is a labor-intensive technology and requires continuous maintenance. Failure of a terraced system, often due to poor maintenance, causes severe gullying and soil degradation. Gully erosion of once terraced land is a widespread problem in the Andes, the Himalayas, and elsewhere.

Because terraced and contour ridge farming have been widely practiced in traditional agriculture, their failure is attributed to the lack of proper attention given to the interterrace soil management. In fact, soil surface management is the second most important consideration in erosion control after the ecologically suitable land use system has been chosen. On the basis of the research in-

TABLE 14.6 Desirable Land Resource Inventory for a Region Based on the Semidetailed Soil and Topographical Surveys

	Area in west Africa, %					
Slope, %	Alfisol	Ultisol	Oxisol	Inceptisol	Vertisol	Aridisol
0–8						
8–20						
20–30						
>30						

The land inventory should be complemented with surveys of the vegetation and land resources.

TABLE 14.7 Distribution of Steeplands (10^6 ha) in the Amazon Basin

Soil grouping	Flat, poorly drained	Well drained			Total (%)
		Slope 0–8%	Slope 8–30%	Slope >30%	
Oxisols, ultisols (acid soils)	43	207	85	23	361 (75)
Aquepts, aquents, gleysols (poorly drained, alluvial soils)	56	13	1	—	70 (14)
Alfisols, mollisols, tropeps, fluvent (moderately fertile, well drained)	0	17	13	7	37 (8)
Psamments, spodosols, podzols (sandy, infertile soils)	10	5	1	—	16 (3)
Total	109	242	103	30	484

SOURCE: Sanchez et al., 1982.

formation available, the following technological options are relevant for intensive use of of steeplands:

Appropriate land use

Erosion risks are accentuated by land misuse. The suitable land use should be chosen on the basis of a detailed soil survey and assessment of its physical, chemical, nutritional, and biologic properties. It is on the basis of these properties that a rational decision should be made regarding which part of the catena should be used for seasonal crops, perennial crops, pastures, or their appropriate combinations. The most suitable land use system can only be chosen on the basis of a detailed resource inventory. Once the land use system has been selected, appropriate resource management strategy involves the following steps (Fig. 14.2):

1. Decide on the cropping systems (crop combination and rotation), agroforestry system that involves growing annuals in association with perennials during the initial stages of tree crop establishment, and appropriate pasture species to be planted.

2. Soil surface management is the crucial point in runoff and erosion control and in curtailing soil degradation. For arable land use, soil surface management may involve no-till farming, conservation tillage, and use of crop residue mulch. Even in the case of perennial crops, providing an effective ground cover by seeding an appropriate cover crop is equally important. The cover crop should be managed so as to minimize the competition. In the case of pasture, an appropriate lea farming system can be developed to grow sea-

sonal crops in rotation with pasture (McCown et al., 1985). Grain crops can be grown through the lightly grazed pasture in a no-till system. Equally important for soil and water conservation is the establishment of an adequate crop stand. Early sowing and ensuring an optimum plant population are useful measures to establish a quick ground cover.

3. Soil fertility and its management are important conservation measures. The soil with imbalanced plant nutrients and low soil fertility is also prone to erosion (Hudson and Jackson, 1959). Soil fertility management is more than just the addition of chemical fertilizers. Supplying a balanced amount of all essential nutrients involves a combination of using organic manures, growing leguminous cover crops, adding crop residue mulch and prunings from perennial crops, and supplementing with chemical fertilizers. Recycling animal waste is an important step in fertility maintenance and in minimizing pollution of natural waters (Plate 14.5).

4. The last consideration is the use of engineering techniques for water management. If the previous three steps are carefully carried out, installing water management devices is not so critical. Under special conditions, how-

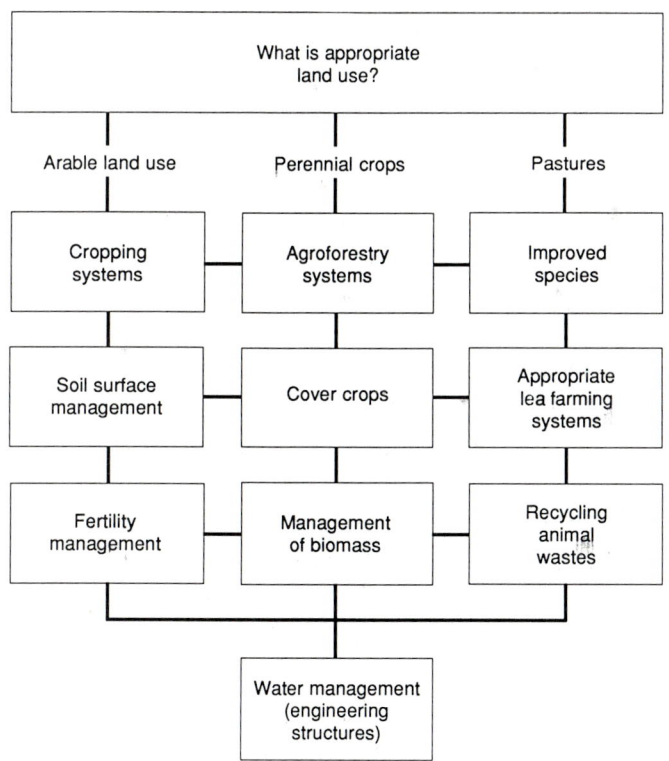

Figure 14.2 Flowchart outlining different steps in resource use strategy.

Plate 14.5 Recycling animal waste to replace depleted plant nutrients and restore soil organic matter content but with low stocking rate.

ever, construction of diversion channels, contour terraces, and farm pond may be necessary measures for the integrated development and use of natural resources. Developing a farm pond is an important input in water management on steeplands. If the pond is to be constructed, the site selection is crucial; otherwise, landslides and mass movement can be caused by the seepage water from the pond.

TABLE 14.8 Farming Systems and Soil and Water Conservation Packages Tested at Two Sites—Jratunseluna Watershed in Central Java and Brantas Watershed in East Java, 1985/86–1986/87

	Effective soil depth					
	>90 cm		90–40 cm		<40 cm	
Slope, %	Low	High	Low	High	Low	High
<15	B	B	B	B	C	C
15–30	B	B	B	C	C	C
30–45	B	C	C	C	C	D
>45	D	D	D	D	D	D

Package A: Traditional conservation techniques and crop management for comparison
    B: Bench terrace, 75% food crops, 25% perennial crops plus livestock
    C: Ridge terrace, 25% food crops, 75% perennial crops plus livestock
    D: Contour alley cropping, nearly 100% perennial crops
SOURCE: Fagi and Mackie, 1987.

An example of using such a scheme as outlined above for watershed management in Java is shown in Table 14.8. The authors have, however, given more importance to the water management and terracing for slopes of up to 30 percent. It would have been more appropriate to select the most suitable cropping systems, techniques of soil surface management, and fertility maintenance prior to deciding upon the water management techniques.

Cropping systems

The great importance of cropping systems in soil and water conservation on steeplands has already been indicated. Soil erosion is always less with diversified cropping systems that involve crop rotations and sequential cropping or mixed and relay cropping than with simplified monoculture systems. Those crops that provide an early and a quick ground cover are more effective in preventing raindrop impact than slow-growing crops.

Examples of the effects of cropping systems on runoff and erosion from steeplands of Rwanda, Trinidad, Peru, and Taiwan are shown in Tables 14.9, 14.10, 14.11, and 14.12, respectively. The data from Rwanda show that the maximum erosion potential of 880 t/(ha · yr) from bare soil on a 40 percent slope can be reduced to a negligible 0.4 t/(ha · yr) or less by choosing an appropriate cropping system. The most conservation-effective systems for steeplands in Rwanda were perennial crops, e.g., coffee and pine kesiya. Cassava grown with two strips of mulch was also effective. Growing seasonal crops such as maize and beans, however, and growing cassava with bench terracing caused severe erosion losses of up to 150 t/(ha · yr).

The data from Trinidad in Table 14.10 show that providing any type of ground cover is better than leaving the soil unprotected. In comparison with bare uncultivated surface, growing maize significantly reduced soil erosion.

The data in Table 14.11 from San Ramon, Peru, are another example high-

TABLE 14.9 Effects of Cropping Systems on Runoff and Erosion in Rwanda

Cropping system	Conservation treatment	Slope, %	Surface area, m^2	Runoff rate, %	Erosion, t/(ha · yr)
Bare	None	40	100	7.2	440–880
Pine kesiya	Weeding	50	750	0.2	0–1
Coffee	Mulch	50	750	0.1	0–0.4
Cassava on bunds	Mulch + 2 strips	49	250	0.4	0–0.4
Cassava on bunds	2 strips of tripsacum	49	226	8–11	29–55
Cassava on bunds	Bench terracing with stone wall	49	358	2.4	5–11
Cassava on bunds	Traditional bunds	49	252	10–17	72–87
Maize + beans	Traditional bunds	49	250	—	150

SOURCE: Roose, 1987.

TABLE 14.10 Soil Erosion from Bare Soil and in Maize Grown on Steeplands in Trinidad

Slope, %	Soil erosion, t/ha	
	Bare uncultivated	Maize
11	27.9	8.3
22	14.7	1.6
52	42.1	4.3

SOURCE: Gumbs and Lindsay, 1982.

lighting the important effect of cropping systems and of crop rotations on runoff and erosion. The soil erosion under different cropping systems varied by several orders of magnitude. Soil erosion under pasture (*Centrosema pubescens*) was 1.3 t/(ha · yr) compared with 148 t/(ha · yr) on bare soil and 119 t/(ha · yr) under potato-fallow and burning potato treatments. Such steep slopes are obviously suited more for pastures than for seasonal crops. If seasonal crops must be grown, they should be grown with the most effective soil surface management practices. The data from Taiwan in Table 14.12, though difficult to compare among different crops, show that on steeplands erosion is more severe from clean cultivated plots when sown to banana, mulberry, and pineapple than from sugarcane, tea, or mango. Under each crop, the erosion rates were drastically reduced with mulching, growing cover crops, and installing reverse-slope terraces.

If the intended land use is pasture, it is important to keep the stocking rate even below that recommended for the region. Excessive, uncontrolled grazing is the major cause of accelerated erosion on steeplands. In Kenya, Dunne (1979) observed maximum sediment yield from those catchments with heavy grazing pressure (Table 14.13). In the steeplands of Machakos, Kenya, Thomas et al.

TABLE 14.11 Effects of Rotation and Cropping Systems on Yields, Runoff, and Erosion at Tupac Amaru, San Ramon, Peru

Treatment	Corn, t/ha	Cowpea, t/ha	Potato, t/ha	Pasture,* t/ha	Pineapple, t/ha	Runoff, mm	Erosion, t/ha
1	—	—	—	—	—	222	148
2	8.3	1.1	13.1	—	—	198	119
3	—	—	—	28	—	73	1.3
4	9.2	1.2	13.0	—	—	136	46
5	—	—	—	—	16.3	228	72

*Four cuts.
1 = Soil continuous fallow
2 = Potato-fallow and burning potato (rows following the maximum slope)
3 = Potato-pasture (*Centrosema pubescons*)
4 = Potato-cowpea (*Vigna unguiculata*) contour rows
5 = Potato-cowpea-potato (rows following the slope with mulch)
SOURCE: Alegre et al., 1987.

TABLE 14.12 Effects of Soil and Crop Management on Soil Erosion and Crop Yields from Steeplands in Taiwan

Crops and treatments	Soil texture	Slope, %	Erosion, t/(ha · yr)	Yield, t/(ha · yr)	Index, %
Sugarcane (1959–1961)	Clay loam	16			
Reverse-slope terrace			1.07	80.4	96
Contour planting			6.47	91.7	110
Up and down			34.98	83.5	100
Banana (1966–1968)	Clay loam	23			
Reverse-slope terrace			0.00	19.4	88
Individual basin			4.06	29.2	133
Up and down			75.90	22.0	100
Tea (1956–1965)	Clay loam	17, 32			
Reverse-slope terrace			0.07	0.61	82
Level bench terrace			0.15	0.65	88
Outward-slope terrace			1.32	0.71	96
Contouring and mulching			0.61	0.85	115
Clean cultivation			6.39	0.74	100
Pineapple (1956–1964)	Sandy clay loam	20			
Reverse-slope terrace			0.03	47.2	158
Contour, close planting with mulching			2.91	52.6	177
Contour, close planting			9.46	47.2	158
Up and down			38.94	29.8	100
Pineapple (1976–1978)	Silt loam	36			
Pineapple stubble mulching			1.39	42.4	123
Plastic sheet mulching			9.04	40.5	117
Clean cultivation			18.69	34.5	100
Mango (1971–1976)	Loam	14			
Bahia grass covering and mulching			0.52	6.1	82
Bahia grass stripping and mulching			1.30	7.9	107
Volunteer grass covering			3.30	6.7	91
Indigofera covering			1.50	5.8	78
Weeping love grass mulching			1.38	7.9	107
Clean cultivation			11.68	7.4	100
Mulberry (1967–1977)	Clay	26			
Bahia grass stripping and mulching			1.97	18.3	110
Desmodium covering			9.95	17.3	104
Straw mulching			6.03	19.9	119
Clean cultivation			66.78	16.7	100

SOURCE: Liao and Wu, 1987.

TABLE 14.13 Land Use and Sediment Yield in Kenya

	R^2
Forest	
$SY = 1.56Q^{0.46}S^{-0.03}$	0.98
$SY = 2.67Q^{0.38}$	0.98
Forest > agriculture	
$SY = 1.10Q^{1.28}S^{0.05}$	0.76
Agriculture > forest	
$SY = 0.14Q^{1.48}S^{0.51}$	0.74
Rangeland	
$SY = 4.26Q^{2.17}S^{1.12}$	0.87

SY = sediment yield [t/(km$^2 \cdot$ yr)].
Q = mean annual runoff (mm).
S = relief.
SOURCE: Dunne, 1979.

(1980) also observed that well-maintained pastures had only a slight and tolerable level of soil erosion. With uncontrolled grazing on degraded lands, however, erosion increased by 50 times (Table 14.14).

Soil surface management

Soil surface management is a term used to encompass a wide range of practices including methods of seedbed preparation and crop residue management. Whereas seedbed preparation includes practices such as the frequency and intensity of plowing and the type of equipment used, crop residue management refers to methods of using crop residue, e.g., mulching, burning, or residue incorporation. Crop residue is an effective measure to improve water infiltration into the soil and to reduce the velocity and shearing and transport capacity of water runoff. A successful example of the effects of crop residue mulch on erosion control on a pineapple plantation for a 20 percent slope in the Ivory Coast is shown by the data of Roose and Asseline (1978) in Table 14.15. Soil erosion with residue burning and residue incorporation treatment was several hundred times greater than when residue was left on the soil surface as mulch. Residue mulch placed between tea bushes has also proved an effective erosion control

TABLE 14.14 Effects of Grazing on Soil Erosion in Machakos, Kenya

Land use	Soil loss, mm/yr	Ratio
Degraded grazing land	4.5	50
Cultivated land	1.3	15
Good grazing land (bush and woodland)	0.07	1

SOURCE: Thomas et al., 1980.

TABLE 14.15 **Effects of Residue Management on Soil Erosion in Pineapple Plantations in the Ivory Coast on a 24 Percent Slope**

Treatment	Soil erosion, t/(ha · yr)	Soil erodibility K
Bare soil	253	0.07
Residue burned	16.7	—
Residue incorporated	9.7	—
Residue mulch	0.007	—

SOURCE: Roose and Asseline, 1978.

measure on steeplands in Sri Lanka (Manipura, 1972) and east Africa (Shaxson, 1975, 1981). The data of Othieno (1975) and Othieno and Laycock (1977) depicting the effects of mulch on coffee are another example of successful erosion control with residue mulch (Table 14.16). Crop residue mulch is also recommended for erosion control in orange plantations on steeplands in Taiwan (Liao and Chang, 1974; Liao and Wu, 1987; Table 14.17). The data in Table 14.17 show that mulching reduced erosion to practically zero on all slopes ranging from 14 to 46 percent and in all fruit orchards.

The effects of mulching on runoff and erosion control from seasonal food crops grown at San Ramon in the Andean region of Peru are shown in Tables 14.18 and 14.19. Mulching significantly reduced soil erosion under corn-cowpea-potato and peanut-cassava rotations. In addition to erosion control, mulching significantly reduced water runoff. In the case of the corn-cowpea-potato ro-

TABLE 14.16 **Effects of Mulching on Runoff and Erosion from a Young Tea Plantation in Kenya**

Variable/treatments	Years after planting		
	1	2	3
1. Ground cover, %	1–30	30–60	60–70
2. Rainfall, mm	2083	2045	1985
3. Runoff, mm			
a. Manual weeding	181	127	32
b. Herbicides	160	162	90
c. Interplanted with oats	65	80	39
d. Mulch	54	27	22
4. Erosion, t/ha			
a. Manual weeding	161	48	1.2
b. Herbicides	168	81	6
c. Interplanted with oats	35	4	0.4
d. Mulch	0.5	0.1	0.08

SOURCE: Othieno, 1975.

TABLE 14.17 Effects of Different Soil Conservation Measures, Including Various Cover Crops and Mulching, on Runoff and Erosion from Orchards Established on Steeplands in Taiwan

Crop and treatment	Period, yr	Rainfall, mm/yr	Slope, %	Runoff, %	Erosion, t/(ha · yr)	Index, %
Citrus	4	1634	28			
Bench terrace				19	5.0	3
Bahia grass covering and mulching				6	0.9	1
Bahia grass stripping and mulching				6	2.8	2
Straw mulching				6	0.0	0
Control				74	156.4	100
Pineapple	6	1501	20			
Bench terrace				3	0.0	0
Contour, close planting and mulching				4	5.5	14
Up and down				32	38.9	100
Banana	3	2349	23			
Bench terrace				28	0.7	1
Guinea grass barrier with mulching				39	1.8	3
Weeping love grass barrier with mulching				37		
Control				53	63.7	100
Litchi	4	1743	46			
Bahia grass covering				1	0.2	1
Desmodium buerger covering				16	2.6	5
Weeping love grass mulching				17	2.0	4
Control				54	54.8	100
Mango	6		14			
Bahia grass covering and mulching					0.5	4
Bahia grass stripping and mulching					1.3	11
Indigofera covering					1.5	13
Weeping love grass mulching					1.4	12
Control					11.7	100
Apple	3	2077	44			
Bahia grass stripping				2	0.02	3
White clover stripping				2	0.02	3
Control				5	0.70	100
Mulberry	2	1810	26			
Bahia grass stripping and mulching					2.0	3
Desmodium heteropyllum covering					10.0	15
Straw mulching					6.0	9
Control					66.8	100

SOURCE: Liao and Wu, 1987.

TABLE 14.18 Effects of Crop Rotation, Mulching, and Fertilizer Application on Yields, Runoff, and Erosion at San Ramon, Peru

Treatment	Corn, t/ha	Cowpea, t/ha	Potato, t/ha	Runoff, mm	Erosion, t/ha
1. Bare soil	—	—	—	205	45.3
2. Corn-cowpea-potato without fertilizer	3.2	1.30	0.3	133	32.0
3. Corn-cowpea-potato with fertilizer	5.7	1.20	2.2	149	33.7
4. Corn-cowpea-potato (mulch and fertilizer)	6.2	1.60	2.9	34	3.8

Amount of rain: 1636 mm, 1978–1979 cropping season.
SOURCE: Alegre et al., 1987.

tation, mulching decreased runoff from 149 to 34 mm and soil erosion from 34 to 4 t/ha (Table 14.18). In the case of the peanut-cassava rotation, mulching decreased runoff from 156 to 56 mm and erosion from 70 to 12 t/ha (Table 14.19). The runoff from mulched plots was merely 2 and 4 percent of the rainfall corresponding to a mean annual rainfall of 1636 and 1484 mm, respectively.

While mulching is an effective technique to control erosion on steeplands, the practical problem lies in procuring an adequate amount of mulch material. Mulch is normally gotten by using the previous crop residue as mulch by using no-till in a reduced tillage system. Lal (1976) observed that no-till farming was an effective erosion control measure on slopes of up to 15 to 19 percent. Soil erosion and water runoff decreased exponentially with an increasing mulch rate (Lal, 1976). Furthermore, residue from the previous maize crop was adequate to control erosion in the following cowpea and soybean crop. The mulch can also be produced in situ by growing an appropriate cover crop in rotation with seasonal crops. Grass and legume covers reduce erosion even more effectively than plantation or tree crops (Virgo and Ysselmuiden, 1979). Cover crops are also grown for erosion control in between the plantation crops, e.g., oil palm. A range

TABLE 14.19 Effects of Fertilizer and Mulching on Crop Yields, Runoff, and Soil Erosion from Peanut-Cassava Rotation at Santo Atahualpa, San Ramon, Peru

Treatment	Peanut,* t/ha	Cassava, t/ha	Runoff, mm	Erosion, t/ha
1. Bare soil	—	—	166	53.4
2. Peanut-cassava without fertilizer	4.2	14.7	138	55.8
3. Peanut-cassava with fertilizer	3.7	15.3	156	70.4
4. Peanut-cassava	4.0	20.8	56.4	12.1

Amount of rain: 1484 mm, 1978–1979 cropping season.
*With shell.
SOURCE: Alegre et al., 1987.

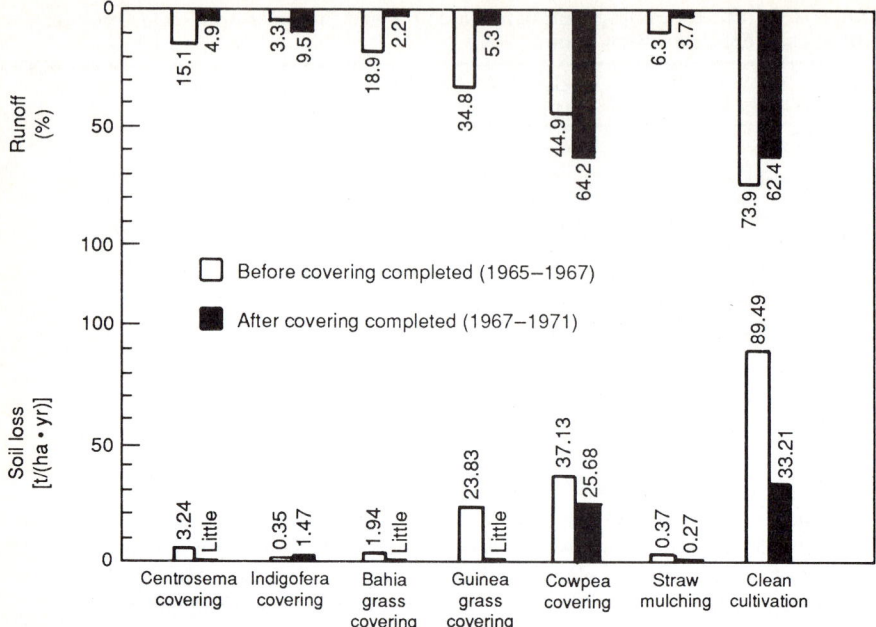

Figure 14.3 Effects of different cover crops on runoff and soil erosion from steeplands in Taiwan (Liao and Wu, 1987).

of cover crops are available, although the choice of an appropriate one depends on the soil, climate, and land use. In Taiwan, cover crops commonly used include Bahia grass, weeping love grass, guinea grass, centrosema, indigofera, desmodium, and white clover. Material mowed from different cover crops is used to mulch around the fruit trees. The results shown in Fig. 14.3 show a very high effectiveness of centrosema, indigofera, Bahia grass, and straw mulch in reducing runoff and soil erosion. For growing a high-value cash crop, mulch material can be grown on a contiguous area and brought in as mulch whenever needed.

Agroforestry and grass strips

Establishing deep-rooted shrubs or trees is an important strategy in stabilizing steep slopes. Trees can be established on the field boundary, on gullies, or on the contour. Trees are the most effective in reducing runoff and controlling erosion, when they are established on the contour. Establishing trees and shrubs on degraded steeplands, however, requires special efforts to provide a strong root bed for the young seedlings. Some practical ways to establish seedlings in shallow soils on steep terrain are shown in Fig. 14.4, if trees or shrubs are established as hedgerows or strips on the contour at regular intervals. This system lends itself to adoption by agroforestry systems.

Seasonal food crops are also grown in association with deep-rooted perennials to provide ecological diversity and to reduce risks of soil erosion. Leguminous shrubs and woody perennials are often grown in one row or multiple rows

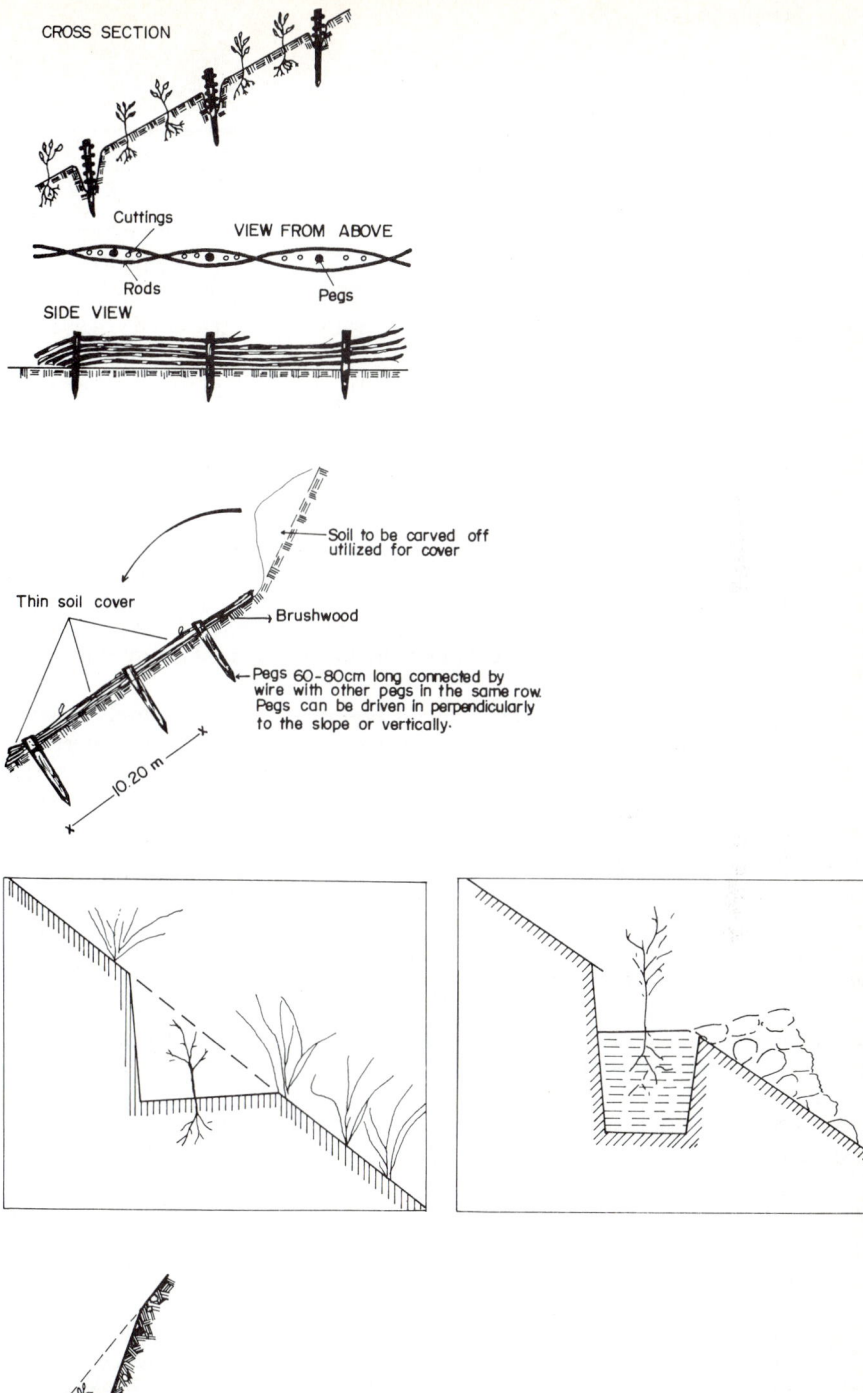

Figure 14.4 Schematics of techniques for establishing tree seedlings on steep terrain.

512   Special Topics

TABLE 14.20 Runoff and Soil Loss with *Leucaena* Hedges for Growing Maize on Steeplands in the Philippines

	Dapdap, Carcar, Cebu		Patupat, Barili, Cebu	
Treatment	Runoff, mm	Erosion, g/plot	Runoff, mm	Erosion, g/plot
Bare fallow				
Corn, stubble removed	87	3156	56	5667
Corn with *Leucaena* hedges	30	680	33	2214
Corn with *Leucaena* hedges, stubble retained	2	14	13	712
Corn with *Leucaena* hedges, stubble removed	4	8	16	820

SOURCE: Pacardo and Montecillo, 1983.

as contour hedges. Hedges are spaced 2 to 10 m apart, depending on the slope and cropping systems. The space between the hedges is used for growing seasonal food crops.

Depending on the interterrace soil surface management, closely spaced hedgerows of perennial shrubs can be very effective in reducing the velocity of water runoff and in controlling erosion. The data shown in Table 14.20 from the Philippines indicate the effectiveness of these hedges in reducing soil erosion. Sediments originating from the interterrace region are trapped by the hedgerows and over time form natural terraces (Fig. 14.5). Maintaining hedgerows, including regular replanting of the missing shrubs, is necessary. Care must also be taken to minimize the risk of terrace breakage through channels created by rodents and burrowing animals. Woody species found useful in such a system for the tropical regions are *Leucaena leucocephala, Gliricidia sepium, Sesbania grandiflora, Tephrosia candida, Acio bartehi, Cajanus cajan, Flamingia congesta,* and others. The choice of an appropriate woody species, however, depends on soil properties, crops to be grown, and alternate uses, e.g., fuel, fodder, and mulch material.

An alternative to perennial shrubs is the use of narrow grass strips. Wide grass strips are used for flatlands, as in strip cropping practiced in the United States. Strips of perennial grasses are grown on the contour as a source of fodder and mulch material and to retard the velocity of water runoff and trap sediments. Roose (1977) observed that grass strips are extremely effective in erosion control from steeplands in the Ivory Coast. Absorptive grass strips 2 to 4 m wide reduce runoff and erosion to 30 to 10 percent of that of the control. Grass strips have also proved effective in erosion control from steeplands in Kenya. Thomas and Senga (1982) and Fissiha (1983) recommended the use of narrow grass strips 2 m wide as an alternative to terraces for erosion control on steeplands in east Africa.

The data from the University of Nairobi, shown in Table 14.21, indicate that grass strips become effective in controlling erosion in the second and third sea-

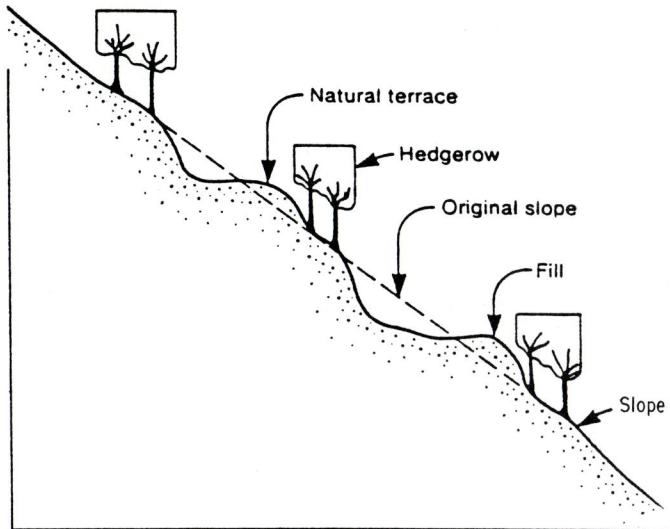

Figure 14.5 Schematic of the natural terraces formed by deposition of sediments by contour hedges.

sons after planting. Furthermore, a 1.5-m-wide grass strip is more effective than a 0.5-m-wide strip. The cumulative soil erosion for three seasons (from November 1982 to April 1983) was 97.8, 35.3, 35.6, and 17.8 t/ha for the control for 0.5-, 1.0-, and 1.5-m-wide strips, respectively. The reductions in soil loss and water loss for a 1.5-m-wide strip were 82 and 76 percent, respectively. The data in Table 14.21 also show that interstrip soil surface management could have been useful in further reducing runoff and soil erosion not only during the initial stages of grass strip establishment but also at later stages when the grass was fully established. Soil surface management is a crucial factor in all erosion control measures.

Use of grass strips in controlling erosion has also proved effective in Uganda (Stephens, 1970), Tanzania (Temple, 1972a, 1972b), Kenya (Kimutai, 1979;

TABLE 14.21 Effects of Width of Grass Strip on Soil Erosion and Water Runoff

Season	Rainfall, mm	Soil erosion, t/(ha · season)				Relative runoff			
		0	0.5 m	1.0 m	1.5 m	0	0.5 m	1.0 m	1.5 m
1982 April	328	75.3	61.8	59.0	—	100	86	91	—
1982 November	175	7.0	2.1	0.1	0.0	100	35	15	0
1983 February	113	17.5	5.3	1.5	0.5	100	35	15	8
1983 April	212	73.3	27.9	34.0	17.3	100	61	34	30

SOURCE: Thomas, 1983.

Niebling and Alberts, 1979), and the Ivory Coast (Roose and Bertrand, 1971). Some multipurpose grasses recommended for grass strips include Nandi Setaria (*Setaria anceps*), Guatemala grass (*Tripsacum laxum*), elephant grass (*Pennisetum purpureum*), and bana grass (*P. purpureum* × *P. typhoides*). Like hedgerows of perennial shrubs, however, these grass strips also facilitate formation of natural terraces over time. Grass strips are effective in trapping sediments, and banks 50 to 100 cm high can be developed in 3 to 5 years (Thomas and Barber, 1982).

In Taiwan, strips of Bahia grass are used to control erosion under season crops (Table 14.22). Compared with conventionally cultivated plots without grass barriers, erosion is drastically reduced by Bahia grass barriers for closely planted sorghum, corn, and peanuts. Because the soil within the strips is cultivated, erosion can still be severe in many slow-growing and open-row crops, e.g., yam and soybean. Grass strips are more effective in erosion control for orchards, especially when the interstrip areas are mulched. The data in Table 14.17 show that grass strips and mulching effectively controlled erosion in citrus, banana, litchi, mango, and apple.

TABLE 14.22 Effects of Grass Strip Barriers on Runoff and Erosion from Different Crops Grown on Steeplands in Taiwan

Treatments	Crops	Period	Rainfall, mm	Runoff, %	Erosion, t/ha Period	Erosion, t/ha 100 mm rain	Index, %
Close planting with Bahia grass barriers	Sorghum	Mar–Jun	1137	39	54.8	2.13	
	Soybean	Jul–Oct	1160	48	122.1	10.52	
	Corn	Nov–Feb	76	2	0.0	0.00	
	Peanut	Mar–Jul	891	20	24.7	2.13	
	Yam	Aug–Dec	649	36	36.9	5.69	
	Subtotal	(2 yr)	3913	36	238.5	6.10	53
Conventional cultivation with Bahia grass barriers	Peanut	Mar–Jun	1755	45	144.0	8.20	
	Yam	Aug–Dec	754	36	33.7	4.47	
	Corn	Feb–Jun	498	21	14.6	2.92	
	Soybean	Jul–Oct	1202	50	90.8	7.55	
	Sorghum	Oct–Jan	29	0	0.0	0.00	
	Subtotal	(2 yr)	4238	42	283.1	6.68	63
Check (conventional cultivation without barriers)	Sorghum	Mar–Jun	1137	53	119.5	10.51	
	Peanut	Jul–Oct	1285	73	199.2	15.50	
	Yam	Nov–May	233	2	0.4	0.17	
	Soybean	May–Sept	1526	40	116.7	7.65	
	Corn	Sept–Jan	117	15	16.5	14.06	
	Subtotal	(2 yr)	4298	51	452.3	10.52	100

SOURCE: Liao and Wu, 1987.

A grass widely adapted for erosion control is the vetiver grass (*Vetiveria zizanioides*). It has been observed that in arid and semiarid climates, such as in India, strips of vetiver grass planted on the contour are more effective in erosion control than conventional earth bunding systems. Of the 10 species of coarse perennial grasses in the tropics of the Old World belonging to the family Andropogoneae, *V. zizanioides* has proved most versatile and highly adaptive to many diverse soils and climate regimes. (See p. 347.)

For maximum protection against erosion on steeplands, grass strips can be combined with perennial shrubs. A suggested scheme for such a mixed system is such that grass is grown in between the trees. Grass, however, must be shade-tolerant.

Water management

A range of water management systems have evolved over the centuries for erosion control on steeplands. Steeplands in Asia have been used intensively for thousands of years by elaborate construction of terraces and other devices for water management and safe disposal of access water. Terraces are constructed for cultivation of rice and fruit crops. Terraces have also been used for traditional farming in the highlands of Kenya (Thomas et al., 1980) (Plate 14.6).

Terraces and other mechanical structures are secondary to more effective biologic control measures. When constructed and maintained properly, however, terraces can reduce both runoff and soil erosion. Terraces are constructed primarily to reduce the velocity of water runoff and for water distribution rather than to decrease the runoff amount. The objective is more to safely conduct the surplus water out of the farmland than to increase water infiltration. Examples

Plate 14.6 Effectiveness of different types of terraces is evaluated on a hillslope at the research farm of the University of Nairobi, Kenya. Note the different types of grasses established to stabilize the terraces.

TABLE 14.23 Comparison of Soil Erosion Losses from Various Conservation Techniques in Sierra Leone

Terraces	Soil loss, t/ha	
	Rice	Cassava
Bench terraces	7.5	—
Stone bunding	29.5	4.4
Stick bunding	27.3	27.3
Contour bunding	18.0	16.8
No conservation	40.7–54.5	11.2–55.1

SOURCE: Millington, 1982.

of using terraces for water management and erosion control are shown in Tables 14.23 to 14.26. In Sierra Leone, soil erosion was reduced under both upland rice and cassava cultivation by bench terraces. Bench terraces were more effective than contour bunds constructed from soil, stones, or sticks (Table 14.23). The labor costs for constructing bench terraces were, however, prohibitive. In spite of the high inputs required for terrace construction, soil erosion even with bench terraces may have been greater than the acceptable level. In South Africa, Whitmore (1969) developed runoff rainfall relationships for terraced and unterraced watersheds. The runoff was found to be related to rainfall according

TABLE 14.24 Effects of Soil and Water Conservation Techniques and Cropping Patterns on Soil Erosion and Runoff from Oxisols in Citanduy Watershed, West Java (1984–1986)

Soil and water conservation techniques	Cropping patterns	Erosion, t/(ha · yr)	Runoff, m³/(ha · yr)
Bench terrace:			
Cultivated areas	Upland rice + cowpea + corn – cowpea + soybean – peanut + corn	1.5	21,646
Risers	*Brachiaria* grass		
Ridge terrace:			
Cultivated areas	Cowpea + corn + *Brachiaria* – cowpea + corn + *Leucaena*	5.7	49,635
Ridges	*Brachiaria*		
Individual terrace with *Gliricidea* strip	Centrocema + corn – cowpea/corn + *Brachiaria*	9.6	48,732
Farmer's technique:			
Cultivated areas	Upland rice + corn + cassava	12.6	42,634

SOURCE: Fagi and Mackie, 1987.

to a power function $Y = ax^b$. Both constants a and b were lower in terraced than in unterraced land.

The data of Fagi and Mackie in Tables 14.24 and 14.25 from Java and Indonesia are also examples of use of terraces in managing steeplands. In contrast to the data from Sierra Leone, however, the reduced rate of erosion by terracing was within the permissible limit of soil loss for the region (Table 14.25). In some cases, erosion from newly terraced land with severe soil disturbance may, however, exceed that from the unterraced land. The data from Taiwan in Table 14.26 show that the reverse-slope and level bench terraces are more effective in erosion control in tea plantations and orchards than outward-slope terraces. Individual basins, commonly used for banana, are also not as effective as the reverse-slope terraces. Tables 14.27 and 14.28 give data from Java for erosion and runoff under various vegetative covers and as the result of various soil and water conservation techniques and cropping patterns.

The effectiveness of terraces can be increased and their maintenance costs considerably reduced by stabilizing the terrace walls. Although stone pitching and rock rip-rap are commonly used to strengthen terrace walls, biologic methods of wall stabilization are usually more effective than rock pitching. Grasses are usually more effective in stabilizing terrace walls than legumes. In Kenya, Thomas (1983) recommended the use of the following grasses for stabilizing terrace walls: Donkey grass (*Panicum trichoiladum*), Makarikari grass (*Panicum coloratum*), signal grass (*Brachiaria brizantha*), elephant grass (*Pennisetum purpureum*), and others.

These examples, indicating relatively low effectiveness of terraces in erosion control, support the argument that the interterrace soil management and appropriate land use systems are more important in erosion control on steeplands than terraces and other engineering techniques. The effectiveness of terraces increases with the adoption of appropriate soil surface management and cropping systems. Terraces may, however, be more effective for erosion control in fruit orchards and perennial crops than in seasonal crops and arable land use.

TABLE 14.25 Effects of Bench Terraces and Alley Cropping on Erosion Rates for Two Soils in Indonesia

Soil and water conservation technique	Srimulyo site (Aquic Tropudalfs)		Sumberkembar site (Vertic Eutropepts)	
	Erosion, t/(ha · 3 mo)*	Permissible limit, t/(ha · yr)	Erosion, t/(ha · 3 mo)*	Permissible limit, t/(ha · yr)
Traditional	29.4 (59)	10.8	15.4 (26)	7.3
Bench terrace	2.7 (53)	9.4	7.5 (23)	9.3
Ridge terrace	5.8 (87)	5.5	14.5 (66)	5.3
Contour alley cropping	1.0 (41)	3.3	6.5 (8)	3.0

*Values in parentheses are the percentage of runoff water over rainfall.
SOURCE: Fagi and Mackie, 1987.

The successful use of terraces for erosion control in orange plantations in Taiwan is shown by the data in Table 14.26.

One of the important factors responsible for gully erosion from steeplands is the surplus water from the access roads and footpaths (Plate 14.7). So provisions must be made for safe disposal of runoff from access roads. It is for this purpose that some engineering devices, e.g., gabiens (Plate 14.8) and check dams (Plate 14.9), may be useful. Establishing vegetative cover on access roads and waterways is the first priority. In Taiwan, Wang et al. (1975) recommended the use of *Paspalum notatum* (Plate 14.10), *Desmodium buergeri,* and weeping love grass for steep slopes of 25°.

TABLE 14.26 Comparison of Runoff and Erosion Losses from Various Types of Bench Terraces Used for Tea and Fruit Orchards in Taiwan

Crop and treatment	Period	Rainfall, mm	Slope, %	Runoff, %	Erosion, t/(ha · yr)	Index, %
Tea	1955–1961	1472	32			
Reverse-slope terrace				7.3	0.9	4
Level bench terrace				9.0	3.2	12
Outward-slope terrace				15.5	17.4	70
Control				18.4	24.6	100
Tea	1956–1965	1757	17			
Reverse-slope terrace				2.1	0.0	0
Level bench terrace				5.5	0.2	2
Outward-slope terrace				17.9	1.9	27
Control				16.4	7.1	100
Tea	1956–1965	1457	27			
Reverse-slope terrace				6.8	0.4	13
Level bench terrace				8.9	0.9	28
Outward-slope terrace				10.7	3.5	104
Control				12.0	3.3	100
Banana	1966–1968	2349	28			
Reverse-slope terrace				37.5	1.6	2
Level bench terrace				10.2	1.6	1
Individual basin				38.2	6.8	7
Control				65.3	92.5	100
Citrus	1971–1974	1743	28			
Reverse-slope terrace				25.0	8.1	5
Level bench terrace				1.8	0.0	0
Outward-slope terrace				29.6	6.5	4
Control				73.8	156.4	100

SOURCE: Liao and Wu, 1987.

TABLE 14.27 Erosion Levels and Runoff under Various Vegetative Covers on Regosol Soil, 10 Percent Slope, Ciparay, West Java

Crop cover	Erosion, t/(ha · yr)	Runoff, %
Potatoes planted up- and downslope	136.1	17.3
Grasses	0.2	0.7
Contour potatoes planting	43.5	14.3
Onion	11.0	6.2
Onion planted on terrace	3.1	4.7
Forest	0	2.0
Trees without shrub	29.1	33.0
Trees without shrub, mulched	1.0	9.0
Shrub, mulched	0.4	7.3
Trees, cultivated underneath	27.1	36.8
Trees, cultivated and mulched	6.8	13.1

SOURCE: Fagi and Mackie, 1987.

Research Needs in Soil Erosion and Its Control

Most research on soil erosion and its control has been done on relatively flat to undulating or rolling lands. Research on erosion problems on hilly lands and steeplands, with slopes exceeding 20 percent, have been neglected because steeplands traditionally have been considered marginal for arable land use.

TABLE 14.28 Effects of Soil and Water Conservation Techniques and Cropping Patterns on Soil Erosion and Runoff from Oxisols in Citanduy Watershed, West Java (1984–1986)

Soil and water conservation techniques	Cropping patterns	Erosion, t/(ha · yr)	Runoff, m³/(ha · yr)
Bench terrace:			
Cultivated areas	Upland rice + cowpea + corn – cowpea + soybean – peanut + corn	1.5	21,646
Risers	*Brachiaria* grass		
Ridge terrace:			
Cultivated areas	Cowpea + corn + *Brachiaria* – cowpea + corn + *Leucaena*	5.7	49,635
Ridges	*Brachiaria*		
Individual terrace with *Gliricidea* strip	Centrocema + corn – cowpea/corn + *Brachiaria*	9.6	48,732
Farmer's technique:			
Cultivated areas	Upland rice + corn + cassava	12.6	42,634

SOURCE: Fagi and Mackie, 1987.

Plate 14.7 Gully erosion caused by surplus water from footpaths and access roads in eastern Nigeria.

Limited research data, therefore, are available to help plan effective resource management strategies for such lands or to develop sustainable land use systems. This does not imply, however, that steeplands are not being used or cultivated. On the contrary, some countries, with the pressure of high human and animal populations, have no choice but to expand their land bases to include steeplands for food production.

For an effective resource management, additional research information needed includes the following:

Soil and water loss

Quantitative, reliable measurements of soil erosion in relation to different factors and causes of erosion are few. Research information is needed for the following:

Soil erosion and land use. Quantifying soil erosion and water runoff losses in relation to existing and new farming systems for different slope gradients and slope lengths, and for different soils and rainfall regimes, is important. And all that is only the first step in developing effective soil conservation measures. Although the literature on soil erosion is voluminous, still needed are accurate, reproducible, and reliable field data. Such data should provide the basis for developing conservation-effective land use systems.

The term *conservation-effective* needs further clarification. Attempts have been made to estimate the erosion potential by using empirical models such as the USLE, MUSLE, and other variations of parametric models. The results obtained by using a simulation model, however, are useful only when they can be validated against actual field data. The field data must be obtained by uniform, standardized methodologies so that results from different soils and ecological regions are comparable. The data published in tables with an added footnote "the storage tank overflowed for x percent of rains" are useless because the heavy rains cause the greatest damage. Runoff and soil loss equipment must therefore be properly designed.

Erosion can be measured on hillslopes by using field runoff plots, agricultural watersheds, or small river catchments. Although sediment yields from river catchments can provide useful information on denudation rates over the catchment, knowing the delivery ratio for different parent materials and land uses is a major bottleneck. To know the erosion potential of agricultural lands, it is important that soil losses be measured on field runoff plots and/or on small agricultural watersheds.

Erosion-induced productivity decline. Erosion-caused alterations in soil properties and crop yields should be assessed to establish levels of tolerable soil losses for different soils, crops, and management systems. For many shallow soils of the tropics and subtropics, the presumed levels of tolerable soil losses used in land use planning are too high. For example, Laban (1978) suggested tolerable soil loss limits for hilly regions of Nepal to be between 10 and 20 t/(ha · yr). These limits are too high, even for the relatively young landscape. Tolerable levels of soil loss should be evaluated on the basis of the erosion-caused productivity decline and the rate at which new soil is formed. The magnitude of either of those determinants of soil loss tolerance is not known for a majority of regions where steeplands are intensively used. Also unknown are short- and

Plate 14.8 Gabiens and drop structures installed to control erosion.

Plate 14.9 Check dams to control erosion may be composed of any feasible material. These photographs were taken on farmland in Burkina Faso.

long-term economic losses caused by on-site and off-site effects of accelerated erosion.

Erosion processes. It is not known whether the relative importance of the physical processes governing erosion on steeplands is the same as that on flatlands or gently sloping lands. Subtle differences are likely.

Slope gradient and length. The effects of slope gradient and slope length for steep slopes presumably differ from those for gentle slopes. Some researchers believe that these effects can be easily computed from physical principles. That may be the case for regular slopes of uniform soil characteristics. In nature, however, regular slopes are the exception rather than the rule. The effects of slope gradient on soil erosion and runoff should therefore be assessed for different shapes, lengths, and slope aspects. How does overland flow originate? What are the roles of interflow in runoff and erosion? What are the threshold slopes for different soil types at which a rill is transformed to gully?

Directional rainstorms. There is another important but less understood aspect of slopes—the interaction between directional storm and the slope aspect in relation to soil detachment and splash downslope for different slope gradients. Researchable questions include: What is the maximum effective rainfall on a given steep slope? What effects does a directional storm have on interrill erosion (Fig. 14.6)? Is interrill erosion as effective on steeplands as on gentle to rolling landscape? Some researchers argue that it is not (Bryan, 1979).

Rain splash and overland flow. What is the interaction between sheet wash and rain splash on high slope angles? The maximum soil detachment occurs at a specific combination of drop diameter and thickness of overland flow. How does this interaction change, if any, by steep slope gradients? It is argued by some that this interaction differs drastically from that on gentle slopes (Bryan, 1979). However, the interaction has not been widely studied for different soils, rainfall regimes, and slope angles.

Critical areas. What are the critical areas of soil loss from catchment of steep terrain, and what are the factors that determine the critical areas? How does the run-on of a steep slope compare with that of a gentle slope?

Plate 14.10 *Paspalum notatum* or Bahia grass used in waterways and roadsides.

Mass movement. Processes governing mass movement, a common problem on cultivated and uncultivated steeplands alike, are not yet adequately understood. They should be assessed in relation to surface soil thickness and the hydrologic and edaphologic properties of the subsoil (Fig. 14.7) (Yoshinori and Osamu, 1984). What is the critical shear strength below which mass wasting is common? What effect does the antecedent moisture index have on mass move-

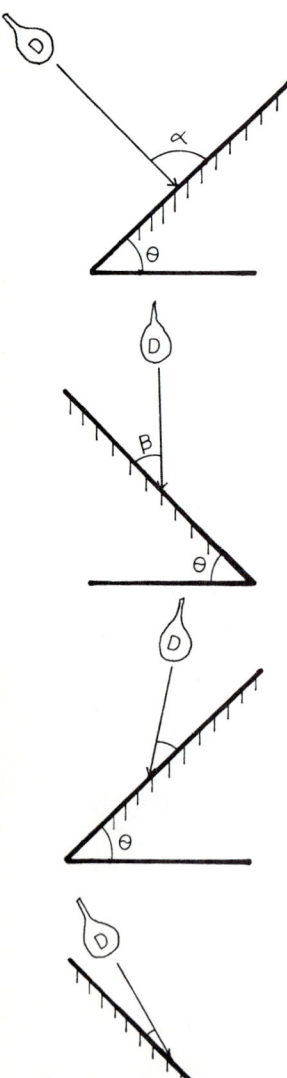

Figure 14.6 Effect of directional storm on runoff and erosion depends on slope aspect. D is an impacting raindrop, α and β are angles of impact, and θ is slope.

ment? Can mass movement be predicted with some index of antecedent precipitation and soil profile characteristics?

Land capability evaluation. One approach to developing appropriate land use systems for conserving soil and water resources is to use the land for whatever it is capable of, by management that is ecologically compatible. In this regard, land evaluation is an important step for conservation and land use planning. The FAO framework for land evaluation (FAO, 1976) is a system of land evaluation that can be applied to sloping lands.

Erosion is caused by bad farm practices. Packages of cultural practices and management systems that have proved successful in similar conditions elsewhere should be validated and adapted for different soils and environments. The sloping agricultural land technologies (SALT) (Watson and Laguihon, 1985) should be assessed and adapted in different regions. Appropriate SALT technologies to be evaluated include conservation tillage, strip cropping, cover crops, contour terraces, alley cropping and agroforestry, and mixed cropping. Suitable combinations of different technological components may be appropriate for different soils and climatic regimes.

Erosion control. Some countries and regions have no choice but to use marginal lands to produce food, fuel, fodder, and fiber. The scientific community cannot afford an attitude of indifference that says not to use lands with a slope gradient exceeding, say, 8 percent. We must test and evaluate land uses and soil and crop management systems that permit ecologically sustainable production from these lands. Even on flatlands, soil and water conservation is a continuous, ongoing, never-ending endeavor. For steeplands, one has to be extra careful.

Soil and water conservation systems are not synonymous with engineering techniques designed to alter the slope length or its gradients. Biologic measures that provide an effective cover close to the ground surface are likely to be more effective, more ecologically compatible, and more durable.

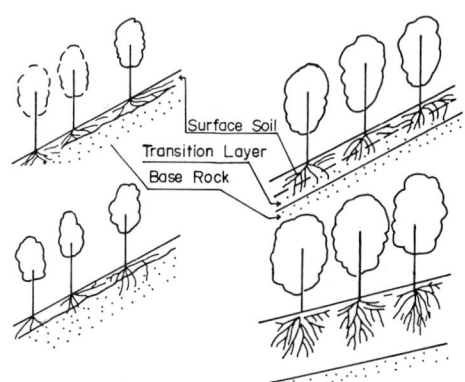

Figure 14.7 Surface and subsoil characteristics in relation to mass movement on steep slopes; four examples.

Erosion-induced soil degradation

The term *soil degradation* is used vaguely to qualitatively describe the decline in soil quality through its use by humans. Erosion-caused degradation includes a decrease in effective rooting depth, reduction in plant-available water and nutrient reserves, and decline in organic matter content and structural properties. To avoid ambiguity, it is important that erosion-caused degradation be measured quantitatively.

Research data should be obtained to define and delineate critical limits of those soil properties that affect crop production. The critical limits may differ with soil, prevailing climate, land use, crops, and ecological regions. If the critical limits of organic matter content, water and nutrient status, porosity, and compaction are not known for major soils and crops, it is difficult to judge whether a soil is degraded or, if so, to what degree.

There is a paucity of basic information delineating certain physical processes of erosion-caused soil degradation, e.g., compaction, porosity, critical rooting depth, and plant-available water reserves. The quantity and quality of soil organic matter necessary to maintain adequate structural condition vary for different soils and environments and are not known.

Restoration of degraded lands

It is estimated that we currently have 1.5 billion ha of land cultivated in the world and that an additional 2 billion ha of land that was once biologically productive has gone out of production. It is also feared that some 5 to 7 million ha of cultivated land now is completely lost for agricultural production every year through soil degradation (UNEP, 1986). One wonders how reliable these estimates are. Since neither the critical levels of degradation nor responses of different crops to the properties for different soils and management conditions are known, are we simply "crying wolf" regarding the magnitude and trends in soil degradation, or is soil degradation a genuine threat to humans? If the estimates are somewhere near correct, the consequences are indeed alarming and are the basis for one of the greatest challenges facing the scientific community.

Regardless of the reliability of data on degraded lands, clearly the world is running out of good, arable land. As the population increases, the greatest opportunity for increasing the world food supply lies in restoring productivity where it has been lost through misuse. To improve our food security, we must develop systems of recharging our soil and water resources on a continuous basis. Soil restoration involves more than just physically saving the fragile soil. It involves restoring, and even enhancing, its productive capacity by improving its organic matter content, porosity, infiltration rate, available water capacity, and biotic activity.

Systematic, long-term research aimed at restoring degraded steeplands should be initiated for different soils and ecological regions. Degraded ecosystems can be restored through judicious land use and by adopting management systems that do not cause gross imbalances in the soil-water-climate equilibrium. Because soil is a finite, nonrenewable resource, there is no choice but to restore the productivity of degraded lands. Through technological innovations, we have the capacity to do it.

Conclusions

Sloping lands, which form a substantial portion of the land resources of the world, are intensively cultivated in many of the world's regions. However, a precise resource inventory based on semidetailed soil and topographic surveys is not available for many regions. Sloping lands are a fragile ecosystem and are susceptible to rapid soil degradation due to physical, chemical, and biologic processes.

Sloping agricultural land technologies for erosion control comprise appropriate land use based on the resource evaluation, suitable cropping system, ecologically compatible soil surface management, and engineering devices for the safe disposal of excess runoff. Proved technologies include mulch farming, use of grass strips and perennial hedgerows, cover crops, and reverse-slope and bench terraces. Terraces are effective only if combined with appropriate soil surface management, e.g., mulching and cover crops. In general, terraces are more effective for erosion control in orchards and plantation crops than in seasonal or annual crops.

Research needs on soil erosion include reliable measurements of soil and water losses under different land use systems. It is also important to evaluate the economic consequences of soil erosion, including knowledge of erosion-caused productivity decline, because research is also needed on the relative importance of different processes involved in sediment origin and transport in relation to slope gradient and length, direction of rainstorms and effective rains, rain splash and its interactions with overland flow, critical areas contributing to sediments and water runoff, and mass movement. Assessing the degree of soil degradation and developing methods of restoring productivity of degraded lands are important research priorities. We need proved "sloping agricultural land technologies" so these lands can be properly managed.

References

Alegre, J. C., Felipe-Morales, C., La Torre, B., and Elserbeer, H. 1987. Soil erosion studies in Peru. In: *Proc. Soil and Water Conservation on Steeplands,* March 22–27, 1987, San Juan, Puerto Rico. SCSA/WAWSC, Ankeny, Iowa.

Anonymous. 1984. *Resultats des experimentations sur l'erosion des sols.* Million Forestiere Crete Zaire-NIL, Bujumbura.

Bergsma, E. 1986. *Aspects of Mapping Units: The Rain Erosion Hazard Catchment Survey.* International Institute for Land Reclamation and Improvement (IILRI), Publ. No. 40, pp. 84–104, Enschede, Netherlands.

Bryan, R. B. 1979. The influence of slope angle on soil entrainment by sheetwash and rainsplash. *Earth Surf. Processes* 4:43–58.

Caine, N., and Mool, P. K. 1982. Landslides in the Kalpu Khola drainage, middle mountains, Nepal. *Mountain Res. Dev.* 2(2):157–173.

Dunne, T. 1979. Sediment yield and landuse in tropical catchments. *J. Hydrol.* 42:281–300.

Fagi, A. M., and Mackie, C. 1987. Watershed management in upland Java: Past experience and future directions. In: *Proc. Soil and Water Conservation on Steeplands,* March 22–27, 1987, San Juan, Puerto Rico, SCSA/WAWSC, Ankeny Iowa.

FAO. 1976. *A Framework for Land Evaluation, Soils Bull.* 32, FAO, Rome, Italy.

Fissiha, T. W. 1983. The effects of narrow grass strips in controlling soil erosion and runoff on sloping lands. M.Sc. Thesis, Department of Agricultural Engineering, University of Nairobi, Kenya.

Fleming, W. M. 1983. Phewa Tal catchment management program: Benefits and costs of forestry and soil conservation in Nepal. In: *Forest and Watershed Development and Conservation in Asia and the Pacific* (L. S. Hamilton, ed.), pp. 217–288, Westview Press, Boulder, Colo.

Gilmour, D. A. 1986. Reforestation or afforestation of open lands: A Nepal perspective. In: *Land Use, Watersheds and Planning in the Asia Pacific Region,* pp. 158–169, FAO Regional Office for Asia and the Pacific, Bangkok, Thailand.

Gumbs, F. A., and Lindsay, J. I. 1982. *Soil Sci. Soc. Am. J.* 46:1264–1266.

Hudson, N. W., and Jackson, D. C. 1959. Results achieved in the measurement of erosion and runoff in South Rhodesia. *Third Inter-African Soils Conference,* Dalaba, pp. 575–583, Africa Soil Bureau, CCTA.

Kimutai, J. N. 1979. A survey of the effectiveness of cutoff chains and grass strips as soil conservation measures in Tuloi, Kapkangani location, Nandi Oislrid. Dip. Thesis, University of Nairobi, Kenya.

Laban, P. 1978. *Field Measurements on Erosion and Sedimentation in Nepal.* Integrated Watershed Management, Torrent Control and Land Use Development Project, Department of Soil and Water Conservation, Ministry of Forestry, Kathmandu.

Laban, P. 1979. Landslide occurrence in Nepal. IWM/WP 13. HMG/FAO, Kathmandu.

Lal, R. 1976. *Soil Erosion Problems on an Alfisol in Western Nigeria and Their Control,* IITA Monogr. 1, IITA, Ibadan, Nigeria.

Liao, M. C., and Chang, M. H. 1974. Soil conservation practices for a young orange plantation. *J. Agric. Assoc. China* 88: 74–82.

Liao, M-C., and Wu, W-L. 1987. *Soil Conservation on Steeplands in Taiwan,* Chinese Soil and Water Conservation Society, Taipei, Taiwan.

Manipura, W. B. 1972. Influence of mulch and cover crops on surface runoff and soil erosion on tea lands during the early growth of replanted tea. *Tea Q.* 43(3):95–102.

Matthews, A. A., and Makepeace, P. K. 1981. A new slant on soil erosion control. *Cane Grow. Q. Bull.* 45:43–47.

McCown, R. L., Jones, R. K., and Peak, D. C. I. 1985. Evaluation of a no-till, tropical legume ley-farming strategy. In: *Agro-Research for the Semi-arid Tropics: Northwest Australia* (R. C. Muchow, ed.), pp. 450–469, University Old Press, St. Lucia, Australia.

Millington, T. 1982. Soil conservation techniques for the humid tropics. *Approp. Tech.* 9(2):17–18.

Niebling, W. H., Alberts, E. E. 1979. Composition and yield of soil particles transported through sod strips. In: *Symp. Proc. 2d National Workshop on Soil and Water Conservation in Kenya,* March 10–13, 1982, University of Nairobi.

Othieno, C. O. 1975. Surface runoff and soil erosion on fields of young tea. *Trop. Agric.* 52:299–308.

Othieno, C. O., and Laycock, D. H. 1977. Factors affecting soil erosion within tea fields. *Trop. Agric.* 54(4):323–330.

Pacardo, I. P., and Montecillo, L. 1983. *Effect of Corn/Ipil-Ipil Cropping System on Productivity and Stability of Upland Agroecosystems.* Annual Report, University of the Philippines at Los Banos–Philippine Council for Agriculture and Resources Research and Development Research Project.

Pandey, A. N., Pathak, P. C., and Singh, J. S. 1983. Water, sediment and nutrient movement in forested and non-forested catchments in Kumaun Himalaya. *For. Ecol. Manage.* 7:29.

Purnell, M. F. 1986. Application of the FAO framework for land evaluation for conservation and land-use planning in sloping areas; potentials and constraints. In: *Land Evaluation for Land-Use Planning and Conservation in Sloping Areas* (W. Siderius, ed.), pp. 17–31, ILRI, Wogeningen, Holland.

Roose, E. J. 1977. Use of the universal soil loss equation to predict erosion in West Africa. In: *Soil Erosion Prediction and Control,* Spec. Publ. 21, pp. 60–74, SCSA, Ankeny, Iowa.

Roose, E. J. 1987. Soil conservation on steeplands in Africa. *Proc. Soil and Water Conservation on Steeplands,* March 22–27, 1987, San Juan, Puerto Rico, SCSA/WAWSC, Ankeny, Iowa.

Roose, E. J., and Asseline, J. 1978. Measurement of erosion phenomena under simulated rainfall at Adiopodoume: Soluble and solid load, in runoff water from soil and pineapple plantations. *Cah. ORSTOM Pedol.* 16:43–72.

Roose, E. J., and Bertrand, R. 1971. Contribution to the study of the method of buffer strips for controlling water erosion in West Africa: Experimental results and observations on the ground. *Agron. Trop.* 26:1270–1283.

Sabhasri, S. 1978. Effects of forest fallow cultivation on forest production and soil. In: *Farmers in the Forest* (P. Kunstadter, E. C. Chapman, and S. Sabhasri, eds.), pp. 160–184, East-West Center, Honolulu, Hawaii.

Sallaway, M. M. 1979. Soil erosion studies in the Mackay district. *Proc. Australian Society Sugar Cane Technology 1979 Congress,* Brisbane, pp. 125–132.

Sanchez, P. A., Bandy, D. E., Villadrica, J. H., and Nicholaides, J. J. 1982. Amazon basin soils: Management for continuous crop production. *Science* 216:821–827.

Shaxson, T. F. 1975. Soil erosion, water conservation and organic matter. *World Crops* 21:6–10.

Shaxson, T. F. 1981. Determining erosion hazard and land use incapability: A rapid subtractive survey method. *Soil Surv. Land Eval.* 1(3):44–50.

Starkel, L. 1972. The role of catastrophic rainfall in shaping of the relief of the lower Himalaya (Darjeeling Hills). *Geogr. Pol.* 21:103–147.

Stephens, D. 1970. Soil fertility. In: *Soil and Water Conservation and Erosion in Agriculture in Uganda* (J. D. Jameson, ed.), pp. 66–88, Oxford University Press, London.

Swanston, D. N., and Dyrness, C. T. 1971. Stability of steepland. *J. For.* 71:264–269.

Temple, P. H. 1972a. Measurement of runoff and soil erosion at an erosion plot scale with particular reference to Tanzania. In: *Studies of Soil Erosion and Sedimentation in Tanzania* (A. Rapp, L. Berry, and P. Temple, eds.), pp. 203–219, Bureau of Resource Assessment and Landuse Planning, University of Dar Es-Salaam, Tanzania.

Temple, P. H. 1972b. Soil and water conservation policies in the Ulugwru Mountains, Tanzania. In: *Studies of Soil Erosion and Sedimentation in Tanzania* (A. Rapp, L. Berry, and P. Temple, eds.), pp. 110–123, Bureau of Resource Assessment and Landuse Planning, University of Dar Es-Salaam, Tanzania.

Thomas, D. B. 1983. *Erosion Control Research at Kabete, Progress Report,* University of Nairobi, Department of Agricultural Engineering.

Thomas, D. B., Barber, R. G., and Moore, T. R. 1980. Terracing of cropland in low rainfall areas of Machokos District, Kenya. *J. Agric. Eng. Res.* 25:57–65.

Thomas, D. B., and Senga, W. M. 1982. Soil and water conservation in Kenya. *Proceedings of the Second National Workshop,* 10–13 March 1982, University of Nairobi, Kenya.

Thomas, V. J., and Trustrum, N. A. 1984. A simulation model of soil slip erosion. In: *Proc. Symp. Effects of Forest Land Use on Erosion and Slope Stability,* East-West Center, Honolulu, Hawaii.

Toth, A., and Fekete, Z. 1974. *The Effects of Agriculture on the Surrounding of Lake Balatan.* Wallingford, U.K., IAHS Publ. 13, pp. 36–39.

Trustrum, N. A., and Hawley, J. G. 1986. Conversion of forest land in grazing: A New Zealand perspective on the effects of land slide erosion on hill country productivity. In: *Landuse, Watersheds and Planning in the Asia-Pacific Region,* pp. 73–93, FAO Regional Office for Asia and the Pacific, Bangkok, Thailand, RAPA Report 1986/3.

UNEP (United Nations Environment Program). 1986. *Farming Systems Principles: For Improved Food Production and the Control of Soil Degradation in the Arid, Semiarid and Humid Tropics,* Experts Meeting, June 20–30, 1983, ICRISAT, Hyderabad, India.

Virgo, K. J., and Ysselmuiden, I. L. 1979. Cultivating soils of tropical steeplands. *World Crops* 31:216–221.

Wang, S. T. 1984. Management of problem soils in Taiwan. ROC. In: *Ecology and Management of Problem Soils in Asia,* pp. 74–88, FFTC Book Series 27, Taipei, Taiwan.

Wang, S. T., Chiang, S. M., and Cheng, C. W. 1975. Observation on cover crops and mulching on steep slopeland orchard. *J. Agric. Assoc. China.* 91:69–76.
Watson, H. R., and Laguihon, W. Q. 1985. *Sloping Agricultural Land Technology. Field Manual for Philippines.* ECHO, Ft. Meyers, Florida.
Yoshinori, T., and Osamu, K. 1984. Vegetative influences on debris slide occurrences on steep slopes in Japan. In: *Symposium on Effects of Forest Land Use on Erosion and Slope Stability* (C. O. O'Loughlin and A. J. Pearce, eds.), pp. 63–72, EPI, East-West Center, Honolulu, Hawaii.

Chapter 15

Wind Erosion and Its Control

Introduction

Wind erosion means soil detachment and its transport by forces generated by wind. Wind erosion is caused by wind blowing across long, bare fields, where soil tilth is in the form of fine, single-grain particles. It occurs in regions of low precipitation and high temperatures and wind velocity. These conditions predominate in arid and semiarid climates (Fig. 15.1). Arid and semiarid lands are extensive and comprise about one-third of the world's land area. About one-sixth of the world's population lives in these climates (Dregne, 1976; Gore, 1979). Areas most susceptible to wind erosion include much of north Africa and the Near East, parts of southeastern Africa, some parts of southern and eastern Asia, Siberian Plain (Baraev, 1974), South America, semiarid and arid parts of North America, and the drylands of Australia (Lorimer, 1984).

Similar to water erosion, wind erosion is accentuated by human perturbations, e.g., intensive farming, clean cultivation, overgrazing, fire. Desertification has become an important environmental issue in recent years and should not be confused with wind erosion. Although they are interrelated, desertification and wind erosion are caused by distinctly different processes. Desertification, also a problem of the semiarid and arid regions, is caused by several factors, including a severe wind erosion. Desertification refers to the increasing aridity of the environments whereby plant and animal life is adversely affected. In simple terms, *desertification* means "the spread of desert-like conditions in and around semiarid areas" (Rapp, 1974) or a change in character of the land to more desert conditions. The United Nations Environment Program defines desertification as the "impoverishment of arid, semiarid and subhumid ecosys-

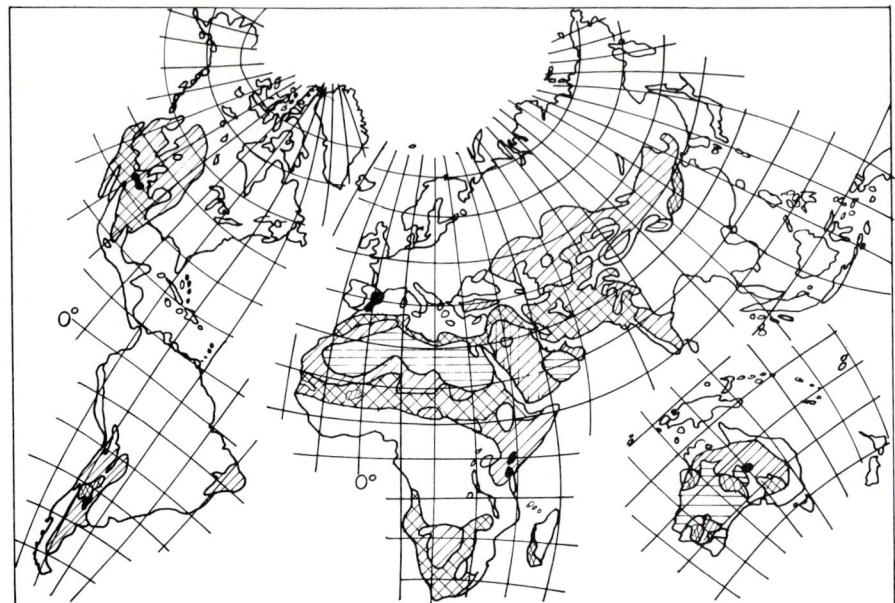

Figure 15.1 Wind erosion is a serious problem in the arid and semiarid regions of the world (Dregne, 1976).

tems by the impact of man's activities" (UNEP, 1977). The spread of desertlike conditions may reduce the productivity of desirable plants, reduce the biomass production, decrease the diversity of life forms, accelerate soil degradation, and increase the hazard for human occupancy.

Desertification is one cause of soil degradation in arid and semiarid ecosystems. Similar to the concept of degradation, however, desertification is also vaguely used. To be quantitative and precise is to define the critical limits of parameters important for plant and animal production, e.g., the plant-available water reserve in the soil, heat and water balance over a watershed, water deficit, etc.

Unlike with water erosion, there are few, if any, measurements of the magnitude of wind erosion in different geographic locations. Erosion rates can be as high as 0.5 mm/yr, as were estimated during the dustbowl period around Kansas. In general, the rate of water erosion is not as high as that of wind erosion. Further, water erosion may occur in localized areas within the watershed. The effect of wind-blown erosion, however, is observed over a large area. This is particularly true for relatively smooth, denuded soil surface, as is commonly observed in large deserts and along broad floodplains and coastlines.

Mechanisms of Wind Erosion

The magnitude of wind erosion is affected by four factors, wind, meteorological factors, soil surface, and land use:

$$A = f(V,S,M,U)$$

Here V is wind speed, S is soil surface, M is meteorological factors, and U is land use including the vegetation cover. Important parameters of these factors that affect wind erosion are shown in Fig. 15.2.

Wind erosion is the resultant of six forces. Three forces involved in sediment entrainment are lift, shear (drag), and ballistic impact. Three forces opposing sediment movement are gravity, friction, and cohesion. Wind erosion occurs if three forces causing entrainment exceed three resisting it.

Sediment entrainment

Three forces involved in sediment entrainment are lift, drag, and ballistic impact.

Lift. Moving wind provides energy through two mechanisms, e.g., lift and drag, or shear. Lift of soil particles is caused by the Bernoulli effect—the air moving at high velocity has low pressure which causes bodies to move in the direction of the pressure differential. Chepil (1945a, b, c) observed that differences in wind velocity around soil particles on the ground may be caused by the acceleration of the flow over the obstacle and/or by the differences in pressure between the slowly moving air in the pores within the upper soil layer and that in the space just above the soil surface. There is in fact a steep velocity gradient near the ground.

Drag. The major part of energy required for work to be performed in the detachment and entrainment of soil particles is provided by the fluid drag τ_w imposed on a soil surface by the wind velocity. The higher the velocity, the greater the drag (see Chap. 2). Soil erosion is almost proportional to the magnitude of the drag.

There exists a critical or threshold shear velocity or incipient velocity necessary to initiate particle movement. Bagnold (1941) related the threshold velocity to the properties of air and soil by

$$V_{*t} = K[(\sigma - \rho)gD]^{1/2} \qquad (15.1)$$

where σ and ρ are the grain and air densities, respectively, g is the acceleration due to gravity, D is the mean grain diameter, and K is an empirical constant

Wind	Meteorological factors	Soil	Land use
• Velocity	• Aridity	• Structure	• Roughness
• Density	• Air and soil temperature	• Texture	• Cover
• Viscosity	• Rainfall amount and distribution	• Moisture regime	• Obstructions
	• Relative humidity	• Organic matter content	• Topography
			• Temperature
			• Length

Figure 15.2 Factors affecting wind erosion.

equal to 0.1 for the Reynolds number V_*D/j greater than 3.5 with V_* the shear velocity and j the kinematic viscosity of air. Chepil (1945a, b, c) reported that this threshold velocity ranges from about 17 to 48 km/h at 30-cm height depending on the previous history of the field.

Once the particle movement is initiated, particles can stay in motion due to the collision or impact between particles. This implies that the velocity needed to start the movement (the fluid threshold velocity) is much greater than that needed to maintain it (the impact threshold velocity). Once the fluid threshold velocity is attained for specific soil and surface conditions, the rate of soil movement is related to the power function of the wind velocity. Malina (1941), Cooke and Warren (1973), and others have observed that the amount of soil transported is proportional to the wind speed cubed after a wind speed reaches a critical velocity of 4.08 m/s:

$$Q_1 = C_1 V_{*t}^3 \tag{15.2}$$

Here Q_1 is the amount of soil moved per unit area, V_{*t} is the wind critical shear velocity above the threshold level, and C_1 is the constant of proportionality which depends on soil conditions and the ground cover. Hsu (1973) developed this formula to estimate the weight of soil movement

$$Q = (4.97D - 0.47)\left[\frac{0.4(\bar{V}_z - 275)}{\ln z(gD)^{1/2}}\right]^3 \tag{15.3}$$

in which Q is the weight of sand moved in tons per meter of width per year, z is the height of the wind speed measurement in meters, D is the mean grain diameter in millimeters, and $\bar{V}_z$ in centimeters per second is the hourly averaged annual wind velocity from a particular direction.

Effect of air turbulence. Just as soil particles are not entrained in laminar-type overland flow or rill flow, a wind strong enough to produce soil particle movement is always turbulent. Little or no geomorphologic work is accomplished when the wind flow is laminar. Chepil and Woodruff (1963) suggested that a turbulence factor T be included in an equation to compute the critical drag τ_c. The turbulence factor T is computed

$$T = \frac{3\sigma_p + \bar{P}}{\bar{P}} \tag{15.4}$$

where σ_p is the root mean square of pressure fluctuation and $\bar{P}$ is the mean pressure. An average value of T is about 2.5 and ranges between 2.1 and 3.0. Lyles and Krauss (1971) studied threshold conditions for particle motion over three surfaces with known levels of local turbulence intensity. They observed that the threshold mean wind speed for a given particle size range decreased as the longitudinal turbulence intensity increased. However, Lyles and Krauss observed that the threshold friction velocities were approximately the same regardless of turbulence intensity.

Effect of temperature. Wind erosion, just like water erosion, occurs in all latitudes. However, because of the temperature-induced differences in density and viscosity, the threshold velocity for a given particle size is less in the Antarctic winter than in the subtropical deserts. For example, Selby et al. (1974) reported a linear increase in the threshold wind velocity required to lift a 2-mm particle to a height of 2 m with increasing latitude.

Ballistic impact. Considerable particle movement occurs due to the collisions among particles. Once a few particles are lifted, the collisions among particles play an important role in sediment entrainment. The collisions or bombardment among particles begins a chain reaction that initiates a "zone of particle movement" that spreads downwind. This process is called *avalanching*. The impact of colliding particles and the resultant abrasion can break up even relatively large clods (Chepil, 1946).

Particle Movement

Bagnold (1941), Malina (1941), and Chepil (1945a, b, c) were the pioneers in explaining an initial motion of the first particle moved by wind or any other fluid. Particle movement is caused by at least three mechanisms.

Surface creep

The rolling or sliding of particles along the ground is called *creep*. Large and heavy particles (greater than 0.5 mm in diameter) move predominantly by surface creep. Particles are rolled along the surface by the direct pressure of wind or due to the bombardment by saltating particles.

Saltation

Once the rolling particles have gathered sufficient speed, they start to hop, leap, or bounce off the ground (the *Latin* word *saltare* means to leap). The bouncing motion of saltating grains is also initiated by the impact of one particle on another. Bouncing particles are lifted steeply from the surface and transported downwind along trajectories. Soil particles have a low parabolic trajectory because their specific gravity is about 2000 times greater than that of the air. Particles are ejected almost vertically into a faster and faster wind, and then particles momentarily travel parallel to the flow when the upward acceleration is balanced by gravity. The final descent of the particles is a resolution of the gravitational and drag forces. The process of saltation over a stony bed is expectedly faster than that over an eroding sand patch. The cloud of sand formed over an eroding patch removes energy from the boundary layer and slows it down. Surfaces that release large amounts of dust to the wind also slow it down. This implies that a sand patch across the path of a dust storm will retard wind velocity and may even cause deposition of the suspended material.

The process of saltation can carry a particle to a maximum height of about 1 m. The actual height attained depends on the wind velocity, particle size, and ground cover. Sharp (1964) reported that the average grain size of the particles in high winds increases with height in the lower part of the dust cloud. An impacting soil particle can bounce another particle 6 times its size; e.g., a saltating particle 0.2 mm in diameter can induce surface creep in a particle 1.2 mm in diameter. Particles between 0.5 and 0.1 mm tend to move by saltation.

Suspension

Particles less than 0.1 mm in diameter are carried into suspension. The tendency to settle out of the airstream is counterbalanced by the lifting force of eddies in the fluid. Once fine silt and clay particles are lifted, these fall back to earth slowly. The Saharan dust, moving by wind called *harmattan,* has been collected as far north as northwestern Europe (Stevenson, 1969) and as far southwest as the West Indies (Morales, 1979). The fine harmattan dust can move at a height of about 1.5 km and drastically reduces visibility (Plate 15.1).

The size range of particles that travel by each of these modes increases with an increase in wind velocity. These simple explanations, put forward by Bagnold (1941) and Chepil (1945a, b, c), were not questioned until the work of Bisal and Nielsen (1962) and Lyles and Krauss (1971). Bisal and Nielsen hypothesized that erosive particles vibrate as the wind velocity approaches the threshold value and then bounce off the surface as the velocity is increased further. They attributed the motion to impulse forces caused by pressure fluctuations. Oscillations in soil particles may be caused by an intensive eddying, e.g., wild fluctuations of velocity and pressure. Observations by Bisal and Nielsen were confirmed by Lyles and Krauss, who reported that vibrations were seldom steady. After 3 to 5 vibrations, the particles ceased vibrating momentarily before suddenly bouncing off the ground. The sudden uplift may be caused by a rare pocket of very low pressure. When the wind velocity was increased considerably above the threshold velocity, the vibrations were not observed, but particles bounced off the ground immediately because of the steeply rising pressure gradient. The average vibration frequency for particles of 0.59 to 0.84 mm is

Plate 15.1 Fine harmattan dust originating in Sahara and photographed at Ibadan, Nigeria, in January 1987.

1.8 ± 0.3 hertz (Hz). It is believed that the particle vibration frequency is related to the frequency band containing the maximum energy of the energy of the turbulent motion.

Forces Opposing Soil Uplift and Entrainment

Forces opposing particle detachment and transport are gravity, friction, and cohesion.

Gravity

Soil particles are 2000 times heavier than air. The soil organic matter fraction, being the lightest, is most easily eroded. Since σ is much larger than ρ, for practical purposes Eq. (15.1) can be reduced to

$$V_{*t}^2 = K^2 C \tag{15.5}$$

Equation (15.5) implies a linear relationship between the threshold velocity and the particle size diameter exceeding 0.1 mm. The following conclusions can be drawn: The wind velocity gradient is zero near the surface, the threshold velocity is proportional to the square root of the diameter, and the particles smaller than about 80 to 100 μm have higher threshold velocity than larger particles. Because of their size, particles greater than 1 mm are moved only by very strong winds.

Friction

The friction between particles is influenced by their shape. Particle shape has a marked effect on the bouncing properties and on the bombardment during saltation. The angle of trajectory during saltation is also influenced by the shape. The abrasive action of wind produces smooth aerodynamic shapes. Smooth shapes called *ventifacts* cause a minimum resistance to winds moving around them.

The surface conditions and vegetation cover increase resistance to wind movement (see Chap. 2). The resistance effect of vegetation cover is an important aspect of effective erosion control measures.

Cohesion

Next to gravity, cohesive forces between particles play an important role in wind erosion. Cohesive forces are high in fine rather than coarse soil fraction. Soil properties that influence wind erosion are the particle size distribution, clod size and surface roughness, and soil moisture (Chepil and Woodruff, 1963). The force required to break a clod is inversely proportional to the third power of its diameter ($F \propto 1/D^3$) (Smalley, 1970). Greater threshold velocity is required for the finer particles because of the greater cohesion and moisture retention. Clay particles retain more water than sand particles, and a soil at a moisture potential of −1.5 megapascals (MPa) or less is practically resistant to wind erosion.

Surface cloddiness is an important factor in susceptibility of a field to wind erosion. The cloddier the surface, the less the erosion potential. The critical minimum diameter for wind erosion is 0.84 mm. The ideal range of clod size to resist erosion is 0.84 to 6.40 mm. Clods larger than 6.4 mm cover less area in proportion to their weight.

Factors Affecting Wind Erosion

Important factors affecting wind erosion and the mechanism of their action have already been discussed. Similar to water erosion, factors affecting the magnitude of wind erosion are the wind erosivity, soil erodibility, landform, and cover.

Wind erosivity

Wind erosion is affected by the aggressivity of the wind and the climate to cause soil detachment and entrainment. Important parameters of wind are the velocity (speed and prevailing dominant wind), frequency and duration, and seasonality of strong winds.

Soil erodibility

Soil particles are cemented together due to the forces of cohesion and adhesion and offer a shearing resistance. The magnitude of the shearing resistance depends on soil properties. The susceptibility of a soil to wind erosion is influenced by its properties, e.g., texture, structure, crusting, moisture content, and tilth. Some soil particles, because of their size and shape, are more erodible than others. Many empirical predictive equations have been developed on the basis of particle size distribution (Lorimer, 1984). Wind erosion occurs if the shear stress exerted by the wind exceeds the shearing resistance offered by the soil. Little, if any, soil movement occurs if the shearing resistance is high.

Landform

Features of the landscape that affect wind erosion are the field length, exposed topographic locations, field orientation, and shelter relative to erosive wind.

Cover

Vegetation cover is an important factor in resisting the forces generated by blowing wind. The vegetation cover decreases the wind velocity close to the ground surface.

Causes of Wind Erosion

The effectiveness of agents and factors in causing wind erosion is grossly influenced by human activities. Soils and landscape that are otherwise less susceptible to the forces of wind erosion become highly erodible due to perturbations caused by people. Even soils resistant to wind erosion can be blown away due to the trampling by animals, loosening of soil by plowing and tillage operations, pulverization of dry soil by excessive traffic (humans, animals, and vehicles), and denudation of vegetation cover by expansion of agriculture or by fire. The interaction of these causes with the agents and factors led to the dust bowl era in the 1930s in the United States and to holocaust over urban centers in Australia in 1983.

Wind Erosion Prediction

Predictive models have been developed to estimate the wind erosion potential. Both empirical and deterministic models have been developed.

Wind erosion equation

Based on about 30 years of research on primary factors that cause soil erosion by wind, Woodruff and Siddoway (1965) proposed a parametric equation to predict wind erosion

$$E = f(I',K',C',L',V) \qquad (15.6)$$

where E is the potential average annual erosion, I' is the soil erodibility index, K' is the soil-ridge roughness factor, C' is the climatic factor, L' is the median unsheltered travel distance across a field, and V is the equivalent quantity of vegetative cover. This equation was designed to determine the potential average annual soil loss or the condition of any one of the variables (I', K', L', or V) needed to control erosion. The variables involved are briefly described.

Soil erodibility index. The *soil erodibility index* I' is defined as the potential soil loss in tons per acre per annum from a wide, unsheltered, isolated field with a bare, smooth, noncrusted surface. The index is related to cloddiness, and its value increases as the percentage of soil fraction greater than 0.84 mm in diameter decreases. The index is related to climatic conditions, such as at Garden City, Kansas.

The soil erodibility index I' is computed as

$$I' = \frac{X_2}{X_1}$$

where X_1 is the quantity eroded from soil containing 60 percent of clods exceeding 0.84 mm and X_2 is the quantity eroded under the same set of conditions from soil containing any other proportion of clods exceeding 0.84 mm. Percentages of dry fractions greater than 0.84 mm can be obtained by standard dry sieving in the field or in the laboratory (Chepil and Woodruff, 1959). Woodruff and Siddoway have computed tables of factor I' for different soils.

Soil-ridge roughness. The factor K' is a measure of soil surface roughness other than that caused by clods and vegetation; i.e., it is the natural or artificial roughness of the soil surface in the form of ridges or small undulations. It is determined from a linear measure of surface roughness.

Wind erosivity. The wind erosion climatic factor C' is computed according to (Chepil et al., 1963)

$$C' = 34.483 \frac{V^3}{(P\text{-}E)^2} \tag{15.7}$$

where V is mean annual wind velocity for a particular geographic location, corrected to a standard height of 30 ft (9.14 m), and $P\text{-}E$ is Thornthwaite's $P\text{-}E$ ratio (Thornthwaite, 1931), given by

$$P\text{-}E \text{ ratio} = 10\frac{P}{E} = 115\left(\frac{P}{T-10}\right)^{1.111} \tag{15.8}$$

The factor C' has been computed for many locations in the United States, and an isoerodent map has been compiled (Chepil et al., 1962).

The original climatic index proposed by Chepil et al. (1963a, b) was expressed somewhat differently:

$$C_1 = \frac{100}{2.9} \frac{V^3}{(P\text{-}E)^2} \tag{15.9}$$

As the $P\text{-}E$ index gets small when precipitation is slight, as in arid regions, the climatic factor approaches infinity. To prevent C' from approaching infinity, a restriction of 12.7 mm for minimum monthly precipitation was placed on P for calculating the $P\text{-}E$ index (Lyles, 1983). (A similar restriction exists for calculating the R factor of the USLE.) In addition, Woodruff and Armbrust (1968) proposed that monthly climatic factors be calculated by using an annual $P\text{-}E$ index with the monthly mean wind speed.

The average annual value of $C = V^3/(P\text{-}E)^2$ for Garden City, Kansas, is equal to 2.9. Hence C_1 for any location and any year is expressed as a percentage of the average annual wind erosion climatic index C for Garden City, Kansas. The climatic index C' ranges from 0 to 200.

Field length L'. The equivalent field length is the unsheltered distance across the field along the prevailing wind erosion direction

$$L' = D_f - D_b \tag{15.10}$$

where D_f is the distance across the field and D_b is the sheltered distance. Cole et al. (1983) have simplified the procedure by ignoring the wind direction distribution. Williams et al. (1984) used the following method of computing factor L:

$$L = \frac{lw}{l_1 \cos(\pi/2 + \theta - \Phi) + w_1 \sin(\pi/2 + \theta - \Phi)} \quad (15.11)$$

Here l and w are large (length) and small (width) dimensions, respectively, of a rectangular field, θ is the wind direction clockwise from north, in radians, and Φ is the clockwise angle between field length and north, in radians.

Vegetative cover. Crop cover is an important factor and is related to the presence of stubble and growing crops. The vegetative cover variable V can be obtained by multiplying the variables, R', S, and K_0. Values of V have been computed for various kinds and amounts of residue mulch (Woodruff and Siddoway, 1965). The factor R' refers to the quantity of vegetative cover. The surface residue amounts are determined by weighing the amount of cover present per unit area. The factor R' is based on the quantity of washed oven-dry residue, multiplied by 1.2 to make it comparable to the usual field measurements where samples are dry-cleaned and air-dried.

The variable S denotes the kind of vegetative cover and is related to the total cross-sectional area of the vegetative material. A fine material with greater surface area is more effective in wind erosion control than a coarse material. The value of S is assigned as 1.00 for small grain stubble and stover, 0.25 for sorghum stubble and stover, 0.20 for corn stubble and stover, and 2.50 for small grain in seedling and stooling stages.

The variable K_0 refers to the orientation of the vegetative cover and is a measure of the vegetative surface roughness. The erect vegetative matter is an effective barrier in decreasing the wind velocity near the ground. This variable also reflects the width and direction of rows, uniformity of distribution, and whether the vegetation is in a furrow or on a ridge. The numerical value of K_0 is 1.0 for absolutely flat, small grain stubble for straw aligned parallel with wind direction on smooth ground in rows 25 cm apart at right angles to the wind direction. The variable K_0 is a function of the amount of residue [$K_0 = f(R')$] and varies as a power function of the residue amount. The numerical value of the exponent ranges from 0.5 for flattened small grain or sorghum to 0.25 for standing small grain and 50-cm-high sorghum. The cover factor has also been computed for many additional crops (Lyles and Allison, 1980, 1981).

The use of the wind erosion equation to predict soil loss by wind erosion is more complex than use of the USLE to estimate the upland erosion by water. Numerical values of different variables have been computed for a range of soils, crop covers, and geographic locations in the United States (Skidmore and Woodruff, 1968). Many users have pointed out the need for simplifying the equation. Skidmore et al. (1970) have developed a computer solution of the equation, and a slide rule calculation has also been developed (Skidmore, 1983).

Another parametric equation to estimate the wind erosion potential is that by Gillette et al. (1972). This equation includes the factor denoting the surface roughness, an important consideration in wind erosion:

$$g_s = \frac{0.295 g_m L I r^{1.5}}{(RK)^{1.26} r_r^{1.5}}$$ (15.12)

where g_s = soil erosion, g/(cm · s)
g_m = maximum soil flux, equal to 1 g/(cm · s)
L = length of field exposed to wind erosion, m
I = erodibility index
r = wind drag corrected for soil moisture
r_r = reference wind drag of 3300 kg/ha
R = surface residue, kg/ha
K = soil surface roughness, cm

The factor RK is very important in developing wind erosion control strategies through appropriate soil and crop management practices.

Another wind erosion equation was developed by Pasak (1974) for estimating the potential wind erosion in Czechoslovakia. Pasak related erosion to soil properties such as clay content, aggregation, and moisture content by

$$E = 2.28 \times 10^2 \times 10^{-0.0787Z}$$ (15.13)

where E = amount of soil transported by wind (g/m²) and Z = clay content (%). Equation (15.13) apparently indicates that erosion decreases with an increase in clay content. This similar empirical equation was developed on the basis of surface aggregation

$$E = 2.105 \times 10^2 \times 10^{-0.0358A}$$ (15.14)

where A = percentage of soil aggregates greater than 0.8 mm. Because the soil moisture content increases the soil resistance to erosion, Pasak related potential erosion to soil moisture content for sandy soil and loamy sand soil, respectively:

$$E = 4.65 \times 10^2 \times 10^{-0.1229 \theta_g}$$ (15.15)

$$E = 6.98 \times 10^{-0.0708 \theta_g}$$ (15.16)

Here θ_g is the percentage of soil moisture by weight. Subsequently, Pasak combined these variables into one equation

$$E = 22.02 - 0.72P - 1.69V + 2.64R$$ (15.17)

where P = percentage of nonerodible particles in soil
V = proportion of soil moisture to nonavailable moisture in soil
R = wind velocity near soil surface, m/s

Pasak also developed wind erosion equations for soils of different texture. Sandy soils:

$$E = 269.244 + 24.506K - 1.652M + 20.919R$$ (15.18)

Loamy sand:

$$E = 8.95 - 0.63K - 0.51M + 1.22R \quad (15.19)$$

Sandy loam:

$$E = 16.091 - 0.584K - 0.177M + 0.422R \quad (15.20)$$

where E = wind erosion, g/m^2
K = clay content, % < 0.01 mm
M = antecedent moisture content, %
R = wind velocity at soil surface, m/s

Recent advances in wind erosion prediction. At the same time it is developing WEPP (see Chap. 8), the United States Department of Agriculture (USDA) Agricultural Research Service (ARS) is also developing the wind erosion research model (WERM), based on basic principles. Details on the wind erosion prediction system (WEPS), a variation of WERM, are given by Hagen (1989).

WERM simulates erosion in a field or a few adjacent fields by defining boundary conditions. It is based on many submodels, including weather, crop growth, decomposition, soil, hydrology, and tillage. The weather submodel generates meteorological variables needed to drive the other submodels. The weather generator used in WEPP (see p. 270) can be used as part of the weather submodel. Submodels dealing with crop growth, decomposition, soil, hydrology, and tillage are used to assess the temporal soil and vegetative cover variables that control soil erodibility in response to inputs generated by the weather submodel. The erosion submodel is used to compute soil loss or deposition in relation to wind speed.

Application of wind erosion equation in the tropics

Unlike the USLE, the wind erosion equation has not been widely used. The wind erosion equation is not designed to estimate the soil loss from a single dust storm. This equation can be used to estimate the annual cumulative soil loss.

This equation has been used in India to estimate the wind erosion hazard in Rajsthan (Singh, 1977). FAO (1979) used the wind erosion equation to estimate the erosion potential of Africa north of the equator and of the Middle East. Some modifications were made in computing the variables (climatic factor C', soil erodibility index I', soil-ridge roughness factor K', field length factor L') and the vegetative cover V. FAO used this modified form of the climatic index

$$C' = \frac{1}{100}\sum_{i=1}^{12} V^3 \left(\frac{\text{PET} - P_{xn}}{\text{PET}}\right) \quad (15.21)$$

where V is the mean monthly wind speed at 2-m height (m/s), P is the precipitation (mm), PET is the potential evapotranspiration (mm), and the factor $(\text{PET} - P_{xn})/\text{PET}$ is the number of erosive days per month. This factor was further modified for irrigated lands as

$$\frac{\text{PET} - P_{xn} + Q}{\text{PET}} \tag{15.22}$$

where Q is the amount of water supplied in irrigation (mm). The climatic factor C' approaches zero if $P + Q$ equals PET. The wind erosivity map of Africa and the Middle East computed according to Eqs. (15.10) and (15.11) is shown in Fig. 15.3. Some additional modifications in the wind erosivity factor have been proposed by Skidmore (1987).

Wind Erosion Hazard

Wind erosion is a serious problem in all continents in semiarid regions. The deserts, semideserts, and adjoining wind-threatened regions exist in Asia (12,500,000 km^2), Africa (12,000,000 km^2), Australia, North America, South America, and Europe. The total area in the world prone to wind erosion is estimated to be 30,000,000 km^2 or approximately 20 percent of the land area (Zachar, 1982). The wind-erosion-caused world shown in Fig. 15.1 indicates the extent and severity of this problem.

The intensity of wind erosion can be as great as or greater than the water erosion. A single dust storm may carry as much as 1000 t/ha of soil. A vast amount of literature is available on this subject (Dregne, 1982). Therefore, a few examples are cited here from some tropical regions.

The wind erosion hazard has been assessed for the Rajsthan desert in India (Mann, 1980; Mann and Rama Krishna, 1980). The Indian desert occupies

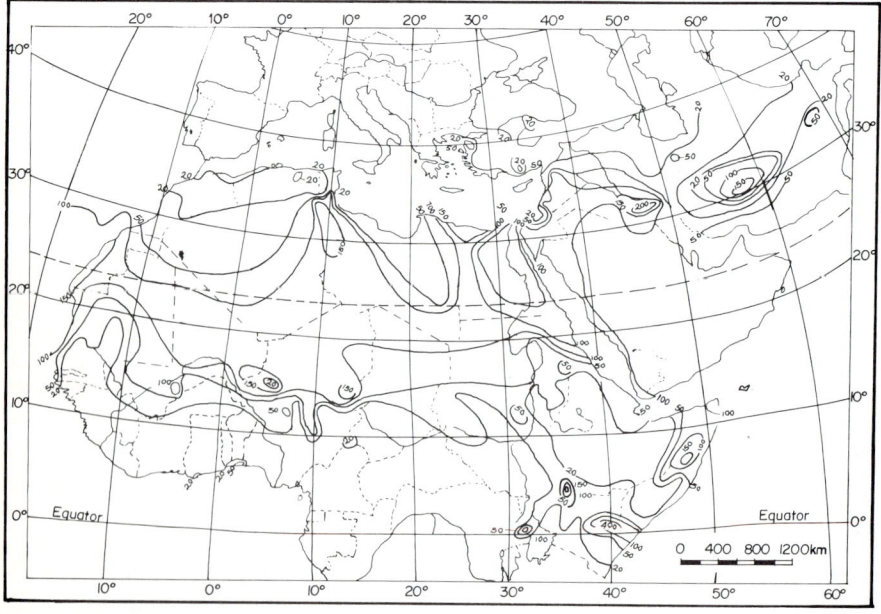

Figure 15.3 Wind erosivity map of Africa north of the equator and the Middle East up to Pakistan (FAO, 1979).

TABLE 15.1 Frequency Distribution of Mean Daily Wind Speeds during Summer Season for 1979 and 1980

Mean daily wind speed, km/h	Number of days during year, %	
	1979	1980
0.0–5.0	12.1	4.4
5.1–10.0	44.0	41.8
>10.0	43.9	53.9

SOURCE: Gupta et al., 1983.

about 286,000 km² in the states of Rajsthan, Gujrat, Haryana, and Punjab, where wind erosion is a serious problem (Sidhu, 1977; Sen and Singh, 1977; Dhir, 1977; Goudie and Wilkinson, 1977). Excessive wind erosion is caused by strong and directional winds, sandy and structureless soils, and lack of protective vegetation cover. Due to these factors, the amount of measured wind-blown soil in Rajsthan ranges from 50 to 420 kg/(ha · day) in the Jodhpur area and exceeds 500 kg/(ha · day) for the Jaisalmer region (Rama Krishna and Sastri, 1980). In Chandan region about 50 km east of Jaisalmer, the removal of vegetation cover from an area of 2000 m² caused the accumulative loss of 53 cm of soil over 3 years. Using the wind erosion equation has led to the estimated erosion rate of 250 tons/(acre · yr) (Singh, 1977). Gupta et al. (1981) measured soil erosion by wind from bare sandy plains near Bikaner, India. They reported the loss of as much as 615 and 325 t/ha of soil from Bikaner and Chandan regions,

TABLE 15.2 Monthly Distribution of Wind Energy Potential (Wh/m²) for Jaipur and Jodhpur Regions of Rajsthan, India

Month	Jaipur (29°49′N, 75°48′E)		Jodhpur (26°18′N, 73°01′E)	
	A*	B	A	B
January	0.0	0.0	95.4	0.0
February	0.0	0.0	71.2	0.0
March	24.2	0.0	55.3	0.0
April	32.5	0.0	62.4	0.0
May	36.0	10.2	236.0	159.5
June	110.7	54.4	446.0	411.8
July	66.4	0.0	293.6	215.3
August	28.3	0.0	131.0	112.4
September	39.1	0.0	63.6	0.0
October	0.0	0.0	0.0	0.0
November	0.0	0.0	0.0	0.0
December	0.0	0.0	59.0	0.0

*A = wind speed of 2.2 m/s or more, B = wind speed of 3 m/s or more based on $P = 0.37V^3$, where P is energy flux (W/m²) and V is wind speed (m/s).
SOURCE: IMD, 1985.

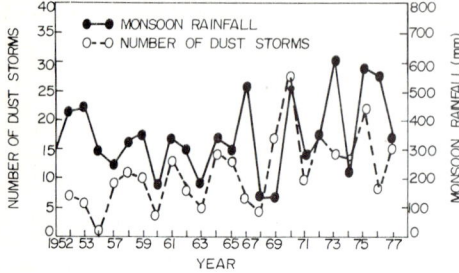

Figure 15.4 Relation between frequency of dust storms and monsoonal rainfall (CAZRI, 1985).

respectively, during a 75-day period from April to June. A very high exponential correlation between wind velocity and soil loss was observed at both sites. High wind energy in summer is responsible for this high erosion potential (Table 15.1). The data in Table 15.2 show the monthly distribution of wind energy potential for Jaipur and Jodhpur regions in Rajsthan. The wind energy is particularly high for the Jodhpur region during May through August. Wind erosion in India is also related to the monsoon rainfall. There is a phase difference between the amount of rainfall and the number of dust storms (Fig. 15.4). The number of dust storms was less in years following the year with good monsoon rains. Good vegetation cover following favorable rains controls wind erosion. The drought during the preceding years was found to have a multiplier effect on dust storm activity in the succeeding year (CAZRI, 1985).

Vast areas of Africa are subject to severe wind erosion. The arid and desert regions of north Africa include countries such as Morocco, Algeria, Tunisia, and Libya. An areal distribution of different ecological regions of these countries is shown in Table 15.3. Erosion in these regions is estimated to remove 0.5 to 1.5 mm/yr of soil (Le Houerou, 1970). General assessment of the erosion potential of this region has been made by FAO (Riquier, 1982). Desertification, caused by both wind and water erosion, is a serious problem in northern Africa. The total wind-blown dust originating from the Sahara is estimated at 260 million t/yr (Jenicks, 1979). Zachar (1982) reported that a sandstorm in the Sahara continuing from March 9 to 12, 1901, caused 1,960,420 t of dust to be carried to Europe. The dust was transported over 3000 to 4000 km, and if it evenly covered Europe, it would form a layer 0.25 mm thick. The underlying cause of this degradation of the environment is excessive use of the limited resources. The atmo-

TABLE 15.3 Arid and Desert Zones (10^3 km^2) of North Africa

Country	Total area	Nonarid, nondesert	Arid	Desert	Arid and desert
Morocco	447	197	120	130	250
Algeria	2381	181	200	2000	2200
Tunisia	155	37	55	63	118
Libya	1760	5	90	1665	1755
Total	4743	420	465	3858	4323

SOURCE: Le Houerou, 1970.

spheric transport of dust particles from the Sahara has been reported as far west as South America (Prospero et al., 1981). In summer, the suspended particles often reach 5- to 7-km altitude and spread over hundreds of kilometers in latitude and extend as far as the Caribbean sea and the southeast United States. The absolute maximum size of these aerosol particles is 0.1 µm (D'Almeida and Schutz, 1983).

Wind erosion is equally severe in the sub-Saharan Sahel comprising Niger, Burkina, Chad, Mali, and Senegal (Fig. 15.1). Maingnet and Chemin (1977) used aerial photographs in the Sahel region of Niger to estimate wind erosion. They observed that the eolian activity occurs up to the 625-mm isohyet. The Sahel suffers from both wind and water erosion (Talbot and Williams, 1978).

Since Australia is the driest continent, wind erosion is quite serious. Eolian influence on soil formation has been a major factor even for soils in Victoria, apparently developed on basalt (Jackson et al., 1972). Wind erosion is also severe in western New South Wales. Thompson (1981a,b) related wind erosion in western New South Wales (NSW) to wind velocity and erosivity. He observed that a velocity of 32 km/h was required at the anemometer height to initiate erosion at ground level. The wind velocity recorded between 1952 and 1976 at Broken Hill, NSW, showed that potentially erosive wind gusts occurred on more than 50 percent of recorded days, with the highest number in spring and the lowest in autumn. Extreme wind velocity exceeding 95 km/h was recorded for 31 events during the 24-yr period.

Wind erosion is also a severe problem in the arid region of the United States. In fact, most of the research information available on wind erosion is based on the data obtained from this region (Chepil and Woodruff, 1963; Woodruff and Siddoway, 1965). Rapid expansion of agriculture and rangelands in arid regions caused extensive soil erosion in the Great Plains of the United States, and in western Canada in the early part of the twentieth century. Hagen and Woodruff (1973) estimated that wind erosion in the Great Plains contributed some 244 and 77 million t/yr of dust to the atmosphere in the 1950s and 1960s, respectively.

Severe wind erosion is reported in some parts of Europe, including Germany, and the U.S.S.R. There are extensive sand dunes in France, Spain, Italy, and the Baltic and Scandinavian countries.

Measuring Wind Erosion

The methodology for wind erosion measurement is not as advanced as that for water erosion. Methods of measuring wind erosion can be broadly classified into two categories: indirect and direct.

Indirect methods

Theses methods are based on the properties of soil and climate, e.g., determining the wind resistance of soil aggregates, measuring a field's macro- and microrelief, estimating the ground cover, measuring the wind velocity and its energy potential, determining climatic aridity, etc. Because climatic data are frequently limited in arid regions, statistical techniques have been developed to

548 Special Topics

qualitatively assess the climatic factors related to wind erosion (Lynch and Edwards, 1980).

Comparing soil properties (texture and gravel content) of eroded and uneroded profiles can also provide an indirect measure of the amount of soil fraction blown away. The index property used may vary among soil types and may include organic matter content, nitrogen, phosphorus, and clay content.

Electronic devices based on acoustics can also be developed to measure the number of saltating particles impacting on a sensor per unit time. The principle of this technique is the same as that of measuring the drop size distribution of rain on the basis of the sound it makes on a sensor.

The empirical formulas, such as the wind erosion equation, are also the indirect methods of estimating the wind erosion potential. Laboratory techniques involving elaborate wind tunnels are used to evaluate the wind resistance of soil aggregates (El-Asswad et al., 1986).

Reconnaissance and seimdetailed surveys are also used to assess the extent of erosion over large areas. Hamza and El-Amani (1977) used satellite imagery in morphopedologic mapping of erosion hazard in Tunisia. Maps were prepared at 1:200,000 scale. Carter and Houghton (1982) reported that the LANDSAT data were accurate for sensing sandblasted areas in western Australia, especially when combined with good backup aerial and ground information.

Direct methods

Direct methods of measuring wind erosion are based on techniques involving measurement of drift and measurement of particles entrained by wind.

Plate 15.2 Plant roots exposed by wind erosion can be used as an indication of degree of erosion. This photograph was taken in Niger.

Measurement of drift. The amount of soil blown away by the wind can be estimated by measuring the microrelief of arable land before and after the windstorm. The depth of soil removed over a specific time can be estimated from the observations made on plants and exposed roots (Plate 15.2). These measurements are merely an approximate indication of the extent of erosion.

The amount of soil blown away by wind can be estimated from the rods or marker embedded in the soil. The technique is similar to that used for estimating the erosion by water, and the number of rods used should be sufficient to obtain a reliable estimate. The calibration of rods to a millimeter division is essential. This method is useful for estimating soil erosion over a long period because the margin of error over a short time is rather large. The sand deposited along the fence lines, shelterbelt, ridges, and dikes is often assessed to estimate erosion on the basis of area exposed to the wind (Plate 15.3).

The most direct measurement of wind-blown soil is gotten by trapping the solid particles entrained by moving wind. Different devices are used to trap particles moving by different mechanisms—suspension, saltation, rolling. The assessment of soil carried in air by any of these mechanisms may be made either by measuring the air density or by collecting the sediments and weighing them. Different types of instruments thus developed have been described in detail by Bocharov (1986).

Control Measures

Considerable research and extension-oriented information are available regarding agronomic measures to control wind erosion (FAO, 1960; Chepil et al., 1963a, b; Chepil and Woodruff, 1963; WMO, 1964; USDA, 1972; Lyles et al., 1973; Singh, 1982). It is not necessary, therefore, to repeat here the details of various methods of wind erosion control and their potentials and limitations for different soils and ecological environments. There are two basic principles to control wind erosion: reduce the wind velocity near the soil surface, and increase the resistance of the soil surface to wind drag.

Methods to reduce wind velocity near soil surface

These methods are based on the use of vegetation cover to decrease the wind velocity and its drag on the soil surface. Some of the cultural practices used include the following:

1. *Afforestation.* Establishing the vegetation cover is an important control measure. Suitable tree species for arid Africa comprise neem (*Azadirachta indica*), *Eucalyptus* spp., *Acacia albida, Acacia tortilis, Prosopis juliflora, Albizzia lebbek, Cassia siamea,* and others (Plate 15.4).
2. *Planting of dry-season cover crops.*
3. *Strip cropping of erosion-vulnerable and erosion-resistant crops.* Grass and small-grain crops are the erosion-resistant crops.
4. *Use of crop residue mulch and no-till system to provide an effective ground cover.* Any vegetation cover left on the soil surface protects the soil from being

blown away and traps the blowing sand that helps reestablish the native vegetation. In Niger, Chase and Boudouresque (1987) observed that the mulch of twigs and crop residue trapped wind- and water-entrained sand and stimulated vegetation establishment (Plate 15.5). The year-old mulched plots produced 5 times more biomass than the newly established plots.

(a)

(b)

Plate 15.3 The amount of wind erosion can also be estimated by measuring the quantity of sand deposited (a) along the fence line or the field boundary and (b) along the contour line or a deliberately erected barrier. These photographs were taken in Niger.

Plate 15.4 *Acacia* is a commonly used tree for wind erosion control in the tropics.

The beneficial effects of mulching on wind erosion control have been reported from India (Singh, 1982). In addition to controlling erosion, mulching increases crop yield through the improvement of soil moisture regime and microclimate. At the ICRISAT Sahelian Center near Niamey, Niger, it has been observed that millet yield is increased as much by crop residue mulch as by fertilizer (Fig. 15.5) (ISC, 1985). In general, the more the biomass covers the ground, the less the erosion.

5. *Planting windbreaks.* Installing effective windbreaks requires careful design because they should be as nearly perpendicular to the prevailing erosive winds as possible. The most effective windbreaks are the wide, permeable belts of multipurpose trees.

Windbreak or shelterbelt comprises a wind barrier of living trees and shrubs maintained for the purpose of protecting the farmland and associated structures (WMO, 1964). A belt of trees and shrubs constitutes an effective protec-

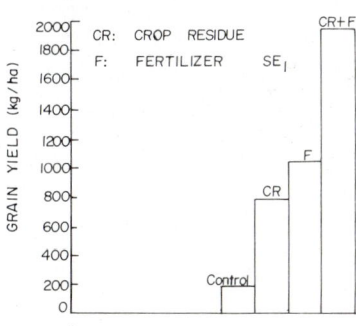

Figure 15.5 Comparative effects of fertilizer and crop residue mulch on yield of millet at ICRISAT, Niamey. SE = standard error. (ISC, 1985.)

tion against strong winds. Windbreaks reduce wind velocity within a certain area in the leeward side and also alter the microclimate, e.g., relative humidity, soil and air temperature, and soil moisture. The change in airflow by the wind barrier depends on the velocity, direction, and degree of turbulence of the wind as well as the length, height, thickness, and porosity of the barrier. If the wind

Plate 15.5 Mulch of the twigs and branches left on the soil surface traps sand and silt and facilitates revegetation. (Courtesy of FAO, Rome.)

conditions of an infinitely long belt are to be achieved, the ratio of length to height of the barrier should be at least 11.5 (Naegeli, 1953), 24 (Bates, 1937), or 30 (Staple and Lehane, 1955) depending on soil and climatic conditions. The shelter effect can be computed from the following equation by Frankenberger (1951):

$$S = 1 - \frac{V}{V_F} = \exp\frac{a-3}{a} \qquad (15.23)$$

where S is the shelter effect, V is the wind velocity at distance a from the belt, V_F is the wind speed in the open, and exp is the naperian logarithm.

In practice, the shelterbelts comprise rows of trees that also have some economic return. Trees of different canopy structure are often grown in alternate rows (Figs. 15.6 and 15.7). The shelterbelt may also comprise a tall crop or a hedge of crop residue. The tall crop generally consists of cereal, e.g., sorghum or millet, rows a few weeks ahead of the short-stature vegetable or any other crop (Black and Siddoway, 1971). In Rajsthan, India, Gupta et al. (1983) studied the performance of tree shelterbelts of *Acacia tortilis, Cassia siamea,* and *Prosopis juliflora.* Their data, shown in Fig. 15.8, indicate that *Cassia siamea* and *Acacia tortilis* have higher effectiveness indices in reducing wind speed than *Prosopis juliflora.* The effect of these shelterbelts on soil erosion and the loss of plant nutrients in eroded soils is shown in Table 15.2. The mean soil loss in plots protected by the shelterbelt of *Cassia siamea* was merely one-third that from the bare soil without a shelterbelt. Also in Rajsthan, Rama Krishna (1985) reported the effects of the microcrop shelterbelt system on reducing the wind speed. The data in Table 15.4 show that the microcrop shelterbelt system reduced wind

Plate 15.5 (*Continued*)

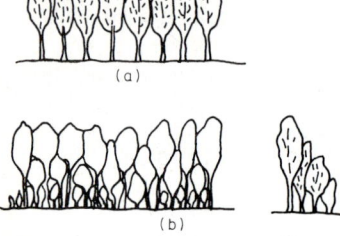

Figure 15.6 Creating a shelterbelt comprised of trees of different heights and canopies. (*a*) Single row: too open; wind blows through. (*b*) Multiple rows: various species form a solid wind wall.

speeds to about 60 to 70 percent in the sheltered vegetable plots. Ujah and Adeoye (1984) studied the effects of shelterbelts in the Sudan-savanna zone of northern Nigeria. The shelterbelt comprised lines of *Eucalyptus camaldulensis* of an average height of 10 m about 30 m long and spaced 300 m apart. The maximum reduction in wind velocity was at a distance of about twice the height of the shelterbelt (Table 15.5). However, the reduction in wind speed was measured at a distance 20 times the height of trees.

Because of the generally improved microclimate in the leeward side of the shelterbelt, crop growth and yield are often better with than without shelterbelts. For example, Ujah and Adeoye (1984) observed that millet yield increased by 7 to 21 percent compared with the yield from unsheltered plots. In India, Rama Krishna (1985) reported a significant increase in yields of sheltered okra (Table 15.6). Rama Krishna reported that sheltered plants developed larger leaves, better rooting depth and root biomass, more pods per plant, and more dry matter production than unsheltered plants. Crop growth in sheltered plots is also improved by low or negligible losses of plant nutrients in eroded soil (Table 15.7).

Crop yield in the immediate vicinity of the shelterbelt may, however, be decreased due to the competition effect of the tree belt. Actively growing trees can deplete soil moisture and nutrient reserves. In Niger, Long et al. (1987) showed that a shelterbelt of neem (*Azadirachta indica*) adversely affected the grain yield of millet. The grain and dry matter yields were significantly reduced near the windbreak rows (Table 15.8, Fig. 15.9). The yield reduction of cowpea and castor bean by hedges of *Leucaena leucocephala* have been reported from Hyderabad, India.

Methods to increase soil resistance

Tillage practices. No-till with crop residue mulch, where applicable, reduces wind erosion. If a plow-based system is required, the fine seedbed must be avoided. Rough, cloddy seedbed reduces wind erosion. Plowing should be done

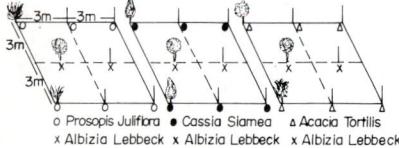

Figure 15.7 Trees of different canopy structure are grown in alternate rows.

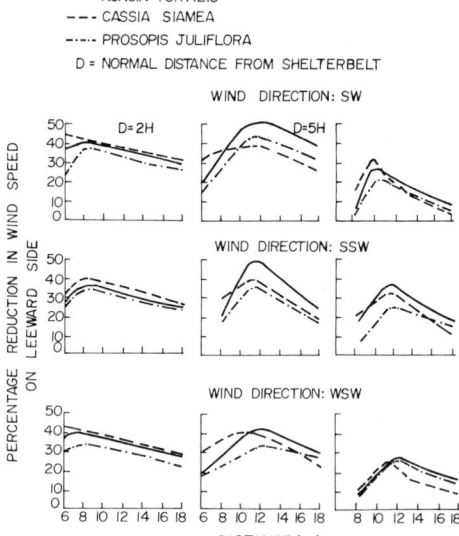

Figure 15.8 Performance of tree shelterbelts on wind reduction in Rajsthan, India. H = 10.54 m. (Gupta et al., 1983.)

perpendicular to the erosive wind direction. Emergency tillage involves ripping of rough strips at right angles to the wind direction. The rough surface slows the intensity of saltation and surface creep.

Tillage practices that leave crop residue on the surface reduce wind velocity and protect the soil from being blown away. The ridge-furrow seedbed is another effective method of wind erosion control. The practice of sowing in the fur-

TABLE 15.4 Influence of Microcrop Shelterbelt on Wind Speed Reduction in Rajsthan, India

	Total wind run (in 12 h), km			
	In unsheltered plot	In sheltered 6-m vegetable plot at:		
Days after sowing		1 m	3 m	5 m
40	116.1	1.0	6.3	16.5
45	68.5	2.0	5.6	20.5
50	134.2	2.3	47.5	74.8
62	85.8	1.0	13.1	45.2
80	15.9	0.2	8.7	11.4
Mean	84.1	1.3	16.2	33.7
Reduction in wind speed due to shelter, %		98	81	60
Mean reduction, %			80	

SOURCE: Rama Krishna, 1985.

TABLE 15.5 Mean Wind Speed Reduction (%) in Sheltered Farmland at Dambatta, Nigeria, 1980/81

Month	Distance from shelterbelt (in multiples of tree height, H*)				
	2H	4H	10H	15H	20H
August	19.30	18.7	14.4	12.5	7.0
September	20.1	19.6	15.6	12.6	7.5
October	23.7	20.4	14.6	12.4	8.5
November	19.5	18.4	13.8	8.4	—
December	—	—	—	—	—
January	22.1	20.7	15.4	12.3	6.9
February	19.2	18.7	14.9	9.3	6.5
March	21.5	21.9	14.3	9.9	7.1
April	21.5	21.2	11.1	9.4	6.2
May	21.3	20.9	11.2	9.4	7.7
June	20.9	19.6	16.9	11.6	7.0
Millet yield, % of open farmland	—	121	115	110	107

*H = 10 m. Mean wind velocity in the windward side = 7 to 16 km/h.
SOURCE: Ujah and Adeoye, 1984.

row protects the young seedlings from the sand blast, an important cause of seedling mortality.

A smooth seedbed is most susceptible to wind erosion. In Hungary, Bodolay (1974) reported that the least erosion was observed from a seedbed with stubble mulch and the maximum from a rolled and smooth seedbed. Similar results are observed from the experiment conducted in North America.

Improving soil structure. Soil amendments have been effectively used to improve soil structure. Improvement of soil aggregation by the buildup of organic matter content, mulching, and soil conditioners are useful erosion control measures.

TABLE 15.6 Influence of Microcrop Shelterbelt on Growth and Yield of Okra in Rajsthan, India

Year	Fingers per plant		Vegetable yield, t/ha	
	With shelter	Without shelter	With shelter	Without shelter
1976	19	10	4.8	2.4
1977	14	8	2.2	1.8
1978	15	11	2.9	2.9
Mean	16	10	3.3	2.4

SOURCE: Rama Krishna, 1985.

TABLE 15.7 Effect of Different Types of Shelterbelt on Soil Erosion and Nutrient Loss due to Wind in Rajsthan, India

Type of shelterbelt	Total soil loss 20 April–26 June, kg/ha			Nutrient loss due to wind erosion, g/ha								
				Nitrogen			Phosphorus			Potassium		
	1979	1980	Mean	1979	1980	Mean	1979	1980	Mean	1979	1980	Mean
Prosopis juliflora	93.2	609.3	351.2	32.6	213.3	123.0	17.2	112.7	65.0	177.1	1157.7	667.4
Cassia siamea	91.5	277.1	184.3	32.0	97.0	64.5	17.0	51.3	34.1	174.0	526.5	350.1
Acacia tortilis	106.0	494.1	300.0	37.1	173.0	105.0	19.6	91.4	55.5	201.2	938.8	570.0
Bare soil (without a shelterbelt)	262.7	831.0	546.8	92.0	290.8	191.4	48.6	153.7	101.2	499.1	1579.0	1039.0

SOURCE: Gupta et al., 1983.

TABLE 15.8 Effect of Distance from Windbreak Row on Millet Grain Yield and Dry Matter Production in Niger

Distance from shelterbelt ($H = 10.54$ m)	Grain yield, kg/ha	Dry matter yield, kg/ha
0.5H	327 a	2179 a
2.0H	593 b	4211 b
4.0H	572 b	3858 b
6.0H	566 b	3971 b
8.0H	447 ab	3661 bc
9.5H	424 ab	3183 c

Figures followed by the same letter are statistically identical.
SOURCE: Long et al., 1987.

Use of soil conditioners to control wind erosion is now receiving a great deal of attention. Earlier studies, however, were less successful. For example, the use of synthetic conditioners to control wind erosion was tested in the United States by Chepil (1951). Chepil observed that application of VAMA (a modified vinyl acetate maleic acid compound) and HPAN (a hydrolyzed polyacrylonitrile) substantially increased the proportion of water-stable aggregates and decreased the proportion of fine water-dispersible particles. However, the great majority of water-stable aggregates formed by VAMA were of the size erodible by wind. VAMA applied to the soil surface increased the erodibility by wind but decreased that by water.

Recent studies with conditioners, however, have proved it to be a promising method of controlling wind erosion. In Ontario, Canada, El-Asswad et al. (1986) evaluated the effects of PVA (polyvinyl alcohol) on the threshold shear velocity and soil loss in wind tunnel experiments. PVA was found to be effective in reducing or even eliminating wind erosion. The effects, however, varied among soil types.

The use of soil conditioners to control wind erosion has also been tested in Africa. In Egypt, Wakhba (1980) tested the sand surface treated with K-4 and K-9 polymers, polyacrylamide, polyvinyl alcohol, and bitumen emulsion for resistance to erosion. He observed that application of polymers at the rate of 10 kg/ha and of the bitumen emulsion at the rate of 100 kg/ha completely pre-

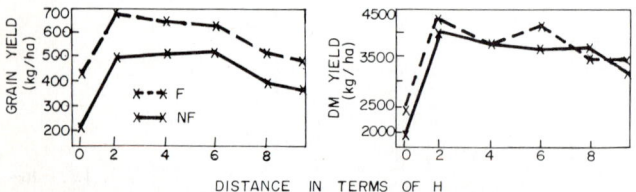

Figure 15.9 Effect of distance from the windward row and fertilizer on grain and dry matter yields. F = fertilizer applied, NF = no fertilizer applied, DM = dry matter, H = 10.54 m. (Long et al., 1987.)

vented erosion in a wind tunnel study at a wind speed of 21 m/s. In Tunisia, De Kesel and De Vleeschauwer (1981) observed that application of uresol stabilized sand dunes and prevented wind erosion.

The use of soil conditioners in controlling wind erosion has also been tested in the U.S.S.R. Kutsenko (1981) reported the results of field tests using polycomplexes to control sand drift. He tested different rates of application of polyethyleneimine (PEI) + polyacrylic acid (PAA) at the Dzhalykovo state farm. The results showed that the optimum dose of polycomplex to control wind erosion is 80 kg/ha. The soil-polymer crust formed by PEI + PAA reduced soil evaporation and protected the seedling from the sandblast. In Hungary, Bodolay (1974) evaluated the effects of unisol and curasol A in improving soil structure and controlling wind erosion. These conditioners were most effective on the ridged seedbed. In addition to stabilizing soil structure, these conditioners improved soil moisture regime and increased soil resistance to erosion. The use of bitumen emulsion seal for wind erosion in New South Wales, Australia, has been recommended by Junor (1978).

Conserving soil moisture. Moist soil cannot be blown away by winds. Cultural practices that conserve soil water also reduce the risk of wind erosion. These practices include mulches, conservation tillage, windbreaks, and cover crops.

One method of increasing the soil moisture content is supplementary irrigation. Irrigation reduces wind erosion by increasing soil resistance and by promoting vegetative growth. Proper regulation of soil moisture content is an effective technique to prevent wind erosion. The difficulty, of course, lies in the practicality of this method in arid regions where water is a scarce commodity.

References

Bagnold, R. A. 1941. *The Physics of Blown Sand and Desert Dunes,* Chapman and Hall, London.
Bagnold, R. A. 1954. Experiments on a gravity dispersion of large solid spheres in a Newtonian fluid under stress. *Proc. R. Soc. Lond.* 225:49–63.
Bagnold, R. A. 1956. The flow of cohesionless grains in fluids. *Phil. Trans. R. Soc. Lond.* [Ser. A] 244:235–297.
Baraev, A. I. 1974. Wind erosion control. *Trans. 10th Int. Cong. Soil Sci.* 11:25–31.
Basiaux, P., Deltour, J., and Nisen, A. 1973. Effect on diffusion properties of greenhouse covers on light balance in the shelters. *Agric. Meteorol.* 11:357–372.
Bates, C. G. 1937. *The Windbreak as Farm Assets,* Farmer's Bull. 1405, USDA.
Bean, A., Alperi, R. W., and Federer, C. A. 1974. A method for categorizing shelterbelt porosity. *Agric. Meteorol.* 14:417–429.
Bergen, James D. 1976. Windspread distribution in and near an isolated, narrow forest clearing. *Agric. Meteorol.* 17:111–133.
Bisal, F., and Nielsen, K. F. 1962. Movement of soil particles in saltation. *Can. J. Soil Sci.* 42:81–86.
Black, A. L., and Siddoway, F. H. 1971. Tall wheatgrass barriers for soil erosion control and water conservation. *J. Soil Water Conserv.* 26:107–111.
Blundell, S. B. 1974. Evaporation to leeward of a shelterbelt. *Agric. Meteorol.* 13: 395–398.
Bocharov, A. P. 1986. *A Description of Devices Used in the Study of Wind Erosion of Soils,* A. A. Balkema, Rotterdam, Holland.
Bodolay, S. 1974. Wind erosion control in agricultural lands. *Trans. 10th. Int. Cong. Soil Sci.* 11:112–118.

Brown, K. W., and Rosenberg, N. J. 1971–72. Shelter-effects on microclimate, growth and water use by irrigated sugar beets in the great plains. *Agric. Meteorol.* 9:241–263.

Carter, D. J., and Houghton, H. J. 1982. Remote sensing of wind erosion in croplands. In papers selected for presentation at the 16th International Symposium on Remote Sensing of Environment, pp. 275–282, Vol. 1. Ann Arbor, Mich.

Catinot, R. 1974. (The forester's contribution to combating desertification in arid zones.) Contribution du forestier à la lutte contre la desertification en zone seches. *Bois et forets des tropiques* (France) 155:3–13.

CAZRI. 1985. *Research Highlights,* Central Arid Zone Research Institute, Jodhpur, India.

Chanduvi, F. (ed.). 1973. Evaluation and control of soil degradation in arid zones of Latin America. Santiago, Chile, FAO, Regional Office for Latin America, *Nature and Resources* 10(3), 33.

Chase, R. G., and Boudouresque, E. 1987. A study of methods for the revegetation of barren crusted Sahelian forest soil. *Proc. Soil, Water and Crop Management Workshop,* 11–17 January 1987, Niamey, Niger, US-AID.

Chepil, W. S. 1945a. Dynamics of wind erosion. I. Nature of movement of soil by wind. *Soil Sci.* 60:305–320.

Chepil, W. S. 1945b. Dynamics of wind erosion. II. Initiation of soil movement. *Soil Sci.* 60:397–411.

Chepil, W. S. 1945c. Dynamics of wind erosion. III. The transport capacity of the wind. *Soil Sci.* 60: 475–480.

Chepil, W. S. 1946. Dynamics of wind erosion IV. The translocating and abrasive action of the wind. *Soil Sci.* 61:167–177.

Chepil, W. S. 1951. Properties of soil which influence wind erosion, IV and V. *Soil Sci.* 72:387–401, 465–478.

Chepil, W. S., and Woodruff, N. P. 1959. *Estimations of Wind Erodibility of Farm Fields,* ARS, USDA, Prod. Res. Rep. No. 25.

Chepil, W. S., and Woodruff, W. P. 1963. The physics of wind erosion and its control. *Adv. Agron.* 15:211–302.

Chepil, W. S., Siddoway, F. H., and Armbrust, D. V. 1962. Climatic factor for estimating wind erodibility of farm fields. *J. Soil Water Conserv.* 17:162–165.

Chepil, W. S., Siddoway, F. H., and Armbrust, D. V. 1963a. Climatic index of wind erosion conditions in the Great Plains. *Soil Sci. Soc. Am. Proc.* 27:449–452.

Chepil, W. S., Woodruff, N. P., Siddoway, F. H., and Armbrust, D. V. 1963b. *Mulches for Wind and Water Erosion Control,* USDA-ARS 41-84.

Cole, G. W., Lyles, L., and Hagen, L. J. 1983. A simulation model for daily wind erosion soil loss. *Trans. ASAE* 26:1752–1756.

Cooke, R. J., and Warren, A. 1973. *Geomorphology in Deserts,* Batsford, London.

Craig, R. M., Smith, D. C., and Ohlsen, A. C. 1978. Changes occurring in coastal dune formation and plant succession along the Martin County coastline. *Proc. Soil Crop Sci. Soc. Florida* 37:14–17.

D'Almeida, G. A., and Schutz, L. 1983. Number, mass and volume distributions of mineral aerosol and soils of the Sahara. *J. Clim. Appl. Meteorol.* 22(2):233–243.

De Kesel, M., and De Vleeschauwer, D. 1981. Sand dune fixation in Tunisia by means of Polyurea Polyalkylene Oxide (Uresol). In: *Tropical Agricultural Hydrology* (R. Lal and E. W. Russell, eds.), pp. 273–281, Wiley, Chichester, U.K.

Dennett, M. D., Elston, J., and Prasad, P. C. 1978. Seasonal rainfall forecasting in Fiji and the southern Oscillation. *Agric. Meteorol.* 19:11–22.

Dennett, M. D., Keatinge, J. D. H., and Rodgers, J. A. 1984. A comparison of rainfall regimes at six sites in Northern Syria. *Agric. For. Meteorol.* 31(3/4):319–328.

Dhir, R. P. 1977. Soil degradation due to over exploitative human effort. *Ann. Arid Zone* 16(3):321–330. Central Arid Zone Research Institute, Jodhpur, India.

Dregne, H. E. 1976. *Soils of the Arid Regions,* Elsevier Scientific Publ., Amsterdam.

Dregne, H. E. 1978. Desertification: Man's abuse of the land. *J. Soil Water Conserv.* 33:11–44.

Dregne, H. E. 1982. *Impact of Land Degradation on Future World Food Production,* USDA-ERS 677, Washington.

El-Asswad, R. M., Groenevelt, P. H., and Nickling, W. C. 1986. Effects of polyvinyl alcohol on the threshold shear velocity and soil loss due to wind. *Soil Sci.* 141:178–184.
FAO. 1960. *Soil Erosion by Wind and Measures for Its Control on Agricultural Lands,* FAO Development Paper 71, Rome, Italy.
FAO. 1979. *A Provisional Methodology for Soil Degradation Assessment.* FAO/UNEP/ UNESCO report, Rome, Italy.
FAO, 1984. *Protect and Produce,* FAO, Rome, Italy.
Fleck, B. C. 1975. Dune erosion, sand drift control and community effort—Callaba Beach. *J. Soil Conserv. Serv. NSW* 31(1):19–24.
Floret, C. 1981. The effects of protection on steppic vegetation of the Mediterranean arid zone of southern Tunisia. Symposium "Dynamique de la vegetation dans les formations mediterraneennes lingneeses" Montpellier, 15–20 September 1980. *Vegetatio* 46:117–129.
Floret, C., and Le Floch, E. 1984. Agriculture and desertification in arid zones of northern Africa. UNEP/USSR Workshop on Impact on Agricultural Practices, Batumi, U.S.S.R., October 1984.
Framji, K. K., and Garg, B. C. 1976. *Flood Control in the World,* vol. 1, pp. 1–28, ICID, New Delhi, India.
Frankenberger, E. 1951. Quoted by WMO (1964).
Gillette, D. A., Blifford, I. H. Jr., and Fenster, C. R. 1972. Measurement of aerosol size distributions and vertical fluxes of aerosols on land subject to wind erosion. *J. Appl. Meteorol.* 11:977–987.
Gore, R. 1979. The desert: an age-old challenge grows. *Nat. Geogr.* 156–189.
Goudie, A., and Wilkinson, J. 1977. *The Warm Desert Environment,* Cambridge University Press, Cambridge, U.K.
Gupta, J. P., Aggarival, R. K., and Raikhy, N. P. 1981. Soil erosion by wind from bare sandy plains in western Rajasthan, India. *J. Arid Environ.* 4(1):15–20.
Gupta, J. P., Rao, G. G. S. N., Gupta, G. N., and Ramana Rao, B. V. 1983. Soil drying and wind erosion as affected by different types of shelter belts planted in the desert region of Western Rajsthan, India. *J. Arid Environ.* 6:53–58.
Hagen, L. J. 1989. *Wind Erosion Prediction System: Concepts to Meet Users' Needs.* Paper presented at the 44th annual meeting of the Soil and Water Conservation Society, July 30–August 2, 1989, Edmonton, Alberta.
Hagen, L. J., and Woodruff, N. P. 1973. Air pollution from dust storms in the Great Plains, *Atmos. Environ.* 7:323–332.
Hakini, A. H., Kachru, R. P., and Chakerabarti, S. M. 1976. Tillage in difficult soils: Dryland farming in Iran. *World Crops UK* 28(1):8–10.
Hamza, A., and El-Amani, M. 1977. *The Use of Satellite Imagery in Morpho-pedological Mapping, as Exemplified by the Map of Sbeitla, 1:200 000,* Pedologie et Teledetection, Rome.
Hogg, W. H. 1965. A shelter belt study—Relative shelter, effective winds and maximum efficiency. *Agric. Meteorol.* 2:307–315.
Hogg, W. H. 1965. Book view of windbreaks and shelterbelts by a working group of the commission for Agricultural Meteorology of the World Meteorological Organization. *Agric. Meteorol.* 2:411–415.
Hsu, S. A. 1973. Computing eolian sand transport from shear velocity measurements. *J. Geol.* 81:739–743.
Hyde, K. W. 1971. Conserving Israel's soil. 1971. *Turnoff* 3(2):23–30.
Hylckama, T. E. A. Van. 1970. Winds over saltcedar. *Agric. Meteorol.* 7:217–233.
ICRISAT Sahelian Center. 1985. *Annual Report,* Niamey, Niger.
IMD. 1985. *Frequency Distribution of Different Places in India,* India Meteorological Department, New Delhi, India.
ISC. 1985. *Improving Management of Resources,* pp. 28–31. ICRISAT Research Highlight 1985, ICRISAT, Patancheru.
Jackson, M. L., Gibbons, F. R., Syers, J. K., and Mokwa, D. L. 1972. Eolian influence on soils developed in a chronosequence of basalts of Victoria, Australia. *Geoderma* 8(2/3):147–163.

Jacobs, A. F. G. 1984. (Wageningen, The Netherlands.) Wind reduction near the surface behind a thin solid fence. *Agric. For. Meteorol.* 33(213):157–162.

Jenicks, R. 1979. Monitoring and critical review of the estimated source strength of mineral dust from the Sahara. In: *Saharan Dust: Mobilization, Transport, Deposition, Scope* (C. Morales, ed.), pp. 233–242, Wiley, New York.

Junor, R. S. 1978. Control of wind erosion on coal ash. *J. Soil Conserv. Serv. NSW* 34(1):8–13.

Kawatani, T., and Mewney, R. N. 1970. Turbulence and wind speed characteristics within a model canopy flow field. *Agric. Meteorol.* 7:143–158.

King, E. 1970. Ecological and meteorological investigations on shelter belts in the wet summer of 1968. *Agric. Meteorol.* 7:235–253.

Kishk, M. A. 1985. Desert encrouchment in Egypt's Nile Valley. In: *Soil Erosion and Conservation* (S. A. El-Swaify et al., eds.), pp. 15–23, Soil Conservation Society of America, Ankeny, Iowa.

Kowal, J. M., Kijewski, W., and Kassam, A. H. 1973. A simple device for analyzing the energy load and intensity of rainstorms. *Agric. Meteorol.* 12:271–280.

Krishnan, A., and Rao, G. G. S. N. 1979. Soil temperature regime in the arid zone of India. *Arch. Met. Geoph. Biokl., Ser. B.* 27:15–22.

Kutsenko, E. V. 1981. Use of polycomplexes for anchoring wandering sands and controlling deflation of coarse-textured soils. *Pochvovedenie* 36:58–61.

Lancashire, B. 1981. A micropower wind-run recorder. *Agric. Meteorol.* 24:291–296.

Le Houerou, H. N. 1970. North Africa: Past, present, and future. In: *Arid Lands in Transition* (N. E. Dregne, ed.), pp. 227–278, AAAS, Washington.

Lomas, J., and Schlesinger, E. 1971. The influence of a windbreak on evaporation. *Agric. Meteorol.* 8:107–115.

Long, S., Persaud, N., Gandah, J., and Ouattara, M. 1987. Influence of a neem (*Azadirachta indica*) windbreak plantation on millet yields and micro-climate in Niger, West Africa. *Proc. Soil, Water and Crop Management Workshop Sudano-Sahelian Zone*, 11–17 January 1987, Niamey, Niger.

Lorimer, M. S. 1984. Estimating the susceptibility of land components to wind erosion. *Proc. National Soils Conf.*, 13–18 May 1984, Brisbane, Australia, CSIRO.

Lowe, R. B., and Bergen, H. J. 1967. A mechanical wind speed and direction seven day recorder. *Agric. Meteorol.* 4:203–208.

Lyles, L. 1983. Erosive wind energy distributions and climatic factors for the West. *J. Soil Water Conserv.* 38:106–109.

Lyles, L., and Allison, B. E. 1980. Range grasses and their small grain equivalents for wind erosion control. *J. Range Mgmt.* 33:143–146.

Lyles, L., and Allison, B. E. 1981. Equivalent wind-erosion protection from selected crop residues. *Trans. ASAE* 24:405–408.

Lyles, L., and Krauss, R. K. 1971. Threshold velocities and initial particle motion as influenced by air turbulence. *Trans. ASAE* 14:563–566.

Lyles, L., Schmeidler, N. F., and Woodruff, N. P. 1973. *Stubble Requirements in Field Strips to Trap Wind Blown Soil*, Agricultural Exp. Stn. Res. Publ. 164, KSU, Manhattan, Kan.

Lynch, and Edwards, K. 1980. The analysis of limited climatic data with reference to wind erosion risk in semi-arid to arid regions. *Agric. Meteorol.* 21:37–47.

Maingnet, M., and Chemin, M. C. 1977. *Patterns of Aeolian Erosion in the Sahel of Niger on the Basis of Satellite Imagery and Aerial Photography,* Pedologie et Teledetection, Rome.

Malina, F. J. 1941. Recent developments in the dynamics of wind erosion. *Am. Geophys. Union Trans.* 22:262–264.

Mann, H. S. 1980. Soil erosion and soil movement: Technology for control. FAO/DANIDA training course on sand dune stabilization, shelterbelt and afforestation in the dry zone, 3–30 March 1980, CAZRI, Jodhpur.

Mann, H. S., and Rama Krishna, Y. S. 1980. Dust and dust storms in the Indian desert. International Workshop on Physics of Desertification, School of Theoretical Physics, Trieste, Italy, 10–28 November 1980.

Massman, W. J. 1983. The derivation and validation of a new model for the interception of rainfall by forests. *Agric. Meteorol.* 28(3):261–286.

Miller, 1973. Soybean water use in the shelter of a slat-fence windbreak. *Agric. Meteorol.* 11:405–418.

Miller, D. R., Rosenberg, N. J., and Bagley, W. T. 1974. Wind reduction by a highly permeable tree shelterbelt. *Agric. Meteorol.* 14:321–333.

Morales, B. 1979. Saharan dust. *Scope,* vol. 14, Wiley, Chichester, U.K.

Munro, D. S., and Oke, T. K. 1973. Estimating wind profile parameters for tall dense crops. *Agric. Meteorol.* 11:223–228.

Naegeli, W. 1953. The breaking effects of a forest on wind. Int. Union Forestry Res. and Organisation, 11th Congr. Rome, Section 11:12–17.

Pasak, V. 1974. Determination of the potential wind erosion of soil. *Trans. 10th Congr. Int. Soc. Soil Sci.,* 11:80–87.

Penning, Frits W. T., De Vries, P., and Iwam, S. 1986. A simple method for generating daily rainfall data. *Agric. For. Meteorol.* 36(4):363–376.

Peterson, T. C., and Parton, W. J. 1983. Diurnal variations of wind speeds at a shortgrass prairie site—a model. *Agric. Meteorol.* 28(4):365–374.

Plate, E. J. 1971. The aerodynamics of shelterbelts. *Agric. Meteorol.* 24:203–222.

Prospero, J. M., Glaccum, R. A., and Nees, R. T. 1981. Atmospheric transport of soil dust from Africa to South America. *Nature* 289:570–572.

Rahn, J. J., and Brown, M. 1971. Estimating corn canopy extreme temperatures from shelter values. *Agric. Meteorol.* 8:129–138.

Rama Krishna, Y. S. 1985. Microcrop shelterbelts for improving crop productivity under irrigated conditions. In: *Development and Management Training Course on Irrigated Agriculture in Arid Areas* (S. D. Singh, ed.), CAZRI, Jodhpur, India, pp. 661–674.

Rama Krishna, Y. S., and Sastri, A. S. R. A. 1980. Wind erosion. In: *Agroclimatic Features of Western Rajsthan,* CAZRI monograph, Jodhpur, India.

Randall, J. M. 1969. Wind profiles in an orchard plantation. *Agric. Meteorol.* 6:439–452.

Rapp, A. 1974. *A Review of Desertification in Africa: Water, Vegetation and Man,* Stockholm Report 1, Secretariat for International Ecology.

Riquier, J. 1982. A world assessment of soil degradation. *Nature Resources* 18(2):18–21 (UNESCO).

Rollin, E. M. 1983. The influence of wind speed and direction on the reduction of wind speed leeward of a medium porous hedge. *Agric. Meteorol.* 30(1):25–34.

Rosenberg, N. J. 1966a. Microclimate, air mixing and physiological regulations of transpiration as influenced by wind shelter in an irrigated beanfield. *Agric. Meteorol.* 3:197–224.

Rosenberg, N. J. 1966b. On the study of shelter effect with sheltered (screened) meteorological sensors. *Agric. Meteorol.* 3:167–177.

Rutter, N. 1968a. Geomorphic and tree shelter in relation to surface wind conditions, weather, time of day and season. *Agric. Meteorol.* 5:319–334.

Rutter, N. 1968b. Shelter effect of an old established shelter block of European larch on a slope of mean gradient 1 in 4. *Agric. Meteorol.* 5:335–349.

Rutter, N. 1968c. Tattering of flags at different sites in relation to wind and weather. *Agric. Meteorol.* 5:163–181.

Seginer, I., and Sagi, R. 1971–72. Drag on a windbreak in two-dimensional flow. *Agric. Meteorol.* 9:323–333.

Selby, M. J., Rains, R. B., and Palmer, R. W. P. 1974. Eolian deposits of the ice-free Victoria Valley, southern Victoria Land, Antarctica. *NZ J. Geol. Geophys.* 17:543–562.

Sen, A. K., and Singh, S. 1977. Significance of geomorphic factors on land use, planning and development in Bikaner. *Ann. Arid Zone* 16(1):13–24.

Sharp, R. P. 1964. Wind-driven sand in the Coachelle Valley, California. *Bull. Geol. Soc. Am.* 75:785–804.

Sidhu, P. S. 1977. Aeolian additions to the soils of North-west India. *Pedologia* 27(3):323–336.

Siemen, S. E. 1971–1972. A friction free potentiometer for determining wind directions by the use of reed contacts. *Agric. Meteorol.* 9:105–108.

Siguin, B., and Gignoux, N. 1974. An experimental study of wind profile modification by a network of shelterbelts. *Agric. Meteorol.* 13:15–23.

Singh, G., and Skidmore, E. L. 1974. Combating wind erosion and developing Great Indian Desert. *Soil Conserv. Dig.* 2(1):61–66.

Singh, H. P. 1977. A review: The problem of wind erosion and its control in western Rajsthan. *Soil Conserv. Dig.* 5:80–86.

Singh, H. P. 1982. Management of desertic soils. In: *Review of Soil Research in India*, ICAR, New Delhi, India, pp. 676–688.

Sinha, B. K. 1970. The effect of tillage operation and wheat stubble on wind erosion. *J. Res. Punjab Agric. Univ.* 7:178–182.

Skidmore, E. L. 1983. Wind erosion calculator: Revision of residue table. *J. Soil Water Conserv.* 38:110–112.

Skidmore, E. L. 1987. Wind erosion. In: *Soil Erosion Research Methodology* (R. Lal, ed.), Soil Management Support Services, International Society of Soil Science, Washington.

Skidmore, E. L., Fisher, P. S., and Woodruff, N. P. 1970. Wind erosion equation: Computer solution and application. *Soil Sci. Soc. Am. Proc.* 34:931–935.

Skidmore, E. L., and Hagen, L. J. 1970. Evaporation in sheltered areas as influenced by wind break porosity. *Agric. Meteorol.* 7:363–374.

Skidmore, E. L., and Woodruff, N. P. 1968. *Wind Erosion Forces in the United States and Their Use in Predicting Soil Loss*, Handbook 346, USDA-ARS, Washington.

Smalley, I. J. 1970. Cohesion of soil particles and the intrinsic resistance of simple systems to wind erosion. *J. Soil Sci.* 21:154–161.

Staple, W. J., and Lehane, J. J. 1955. The influence of field shelterbelts on wind velocity evaporation, soil moisture and crop yield. *Can. J. Agric. Sci.* 35:440–453.

Stevenson, C. M. 1969. The dust fall and severe storms of July 1, 1968. *Weather* 24:126–132.

Talbot, M. R., and Williams, M. A. J. 1978. Erosion of fixed dunes in the Sahel, central Niger. *Earth Surf. Processes* 3(2):107–114.

Thofelt, Lars, and Rufelt, Henry. 1984. A note on the influence of a windbreak on plant temperature. *Agric. For. Meteorol.* 32(1):1–11.

Thompson, D. F. 1981a. Wind erosivity indices for western New South Wales. *J. Soil Conserv. Serv. NSW* 37(3):157–165.

Thompson, D. F. 1981b. Wind flow in western New South Wales. *J. Soil Conserv. Serv. NSW* 37(2):79–90 (Soil Conservation Service, Gunnedal Research Centre, Gunnedah, New South Wales, Australia).

Thornthwaite, C. W. 1931. Climates of North America according to a new classification. *Geogr. Rev.* 21:633–655.

Ujah, J. E., and Adeoye, K. B. 1984. Effects of shelter belts in the Sudan savanna zone of Nigeria on microclimate and yield of millet. *Agric. For. Meteorol.* 33(2/3):99–107.

UNEP. 1977. *Desertification: An overview.* United Nations Conference on Desertification, 29 August–9 September 1977, Nairobi, Kenya.

USDA. 1972. *How to Control Wind Erosion*, Agricultural Information Bull. 354, USDA.

Wakhba, S. A. 1980. Use of polymers to control wind erosion. *Pochvovedenie* 35:52–54.

Walsh, B. N. 1968. *Some Notes on the Incidence and Control of Drift Sands along the Caledon, Bredasdrop and Riversdale Coastline of South Africa.* Bulletin, Department of Forestry, South Africa.

Williams, J. R., Jones, C. A., and Dyke, P. T. 1984. A modeling approach to determining the relationship between erosion and soil productivity. *Trans. ASAE* 27:129–144.

WMO. 1964. *Windbreaks and Shelterbelts*, WMO Tech. Note 59, Geneva, Switzerland.

Woodruff, N. P., and Armbrust, D. V. 1968. A monthly climatic factor for the wind erosion equation, *J. Soil Water Conserv.* 23:103–104.

Woodruff, N. P., and Siddoway, F. H. 1965. A wind erosion equation. *Soil Sci. Soc. Am. Proc.* 29:602–608.

Zachar, D. 1982. *Soil Erosion,* Elsevier Scientific Publishing, Amsterdam, Holland.

Zingg, A. W. 1953. Wind-tunnel studies of the movement of sedimentary material. *Proc. 5th Hydraul. Conf. Iowa Inst. Hydraul. Bull.* 34:111–135.

Zohar, Y., Schiller, G., and Karschon, R. 1971. Determination of the direction of the prevailing wind from the orientation of crescent-shaped Bedium shelters in Northeastern Sirai. *Agric. Meteorol.* 8:319–323.

Author Index

Abdou, F. M., 76
Abdujamin, S., 345
Abdurachman, A., 254, 328–329, 333–334
Adams, J. E., 63
Adeoye, K. B., 554, 556
Agbede, O. O., 464
Ahmad, N., 235
Ahn, P. M., 74, 77, 79
Ahuja, L. R., 426
Aina, P. O., 36, 41–43, 52, 89, 243, 325–326, 331–333
Alberts, E. E., 64, 514
Al-Durrah, M. M., 64, 135, 140
Alegre, J., 504, 508–509
Alexander, T. G., 464
Alles, W. S., 297
Alleyne, E. P., 442
Allison, F. E., 76
Amber, S., 251
Amphlett, G., 464
Ampt, G. A., 69
Anderson, D. M., 159
Anderson, H. W., 207
Anderson, M. G., 173–174
Anderson, M. J., 73, 82
Armburst, D. V., 540
Armiger, W. H., 328
Armon, M. N., 235
Arnoldus, H. M. J., 195, 234, 236, 242–243
Arulanoonden, K., 81
ASCE, 4, 240
Ashcroft, G. L., 69
Asselin, J., 210, 378, 506–507
Ateshian, J. K. H., 235
Atkinson, G., 447
Atlas, O., 36
Attaviroj, P., 429–431
Attle, J. R., 36

Babalola, O., 343
Babu, R., 243, 245

Baconguis, S. R., 465
Badarashi, B., 133
Baffoe-Bonnie, E., 380
Bagnold, D. V., 540
Bajpai, M. R., 343
Banasik, K., 260
Baraev, A. I., 531
Barber, R. G., 210, 243, 245, 248, 254, 266, 314–315, 364–365, 367, 369, 514
Barker, M. R., 384
Barnett, A. P., 335, 366
Bartelli, L. J., 113, 115
Bartholomew, M. V., 89
Baruah, P. C., 39
Barus, A., 438
Bates, C. G., 553
Bates, J. A. R., 89
Baumgartner, M. F., 195
Baver, L. D., 36, 71, 171, 226, 267
Bay, C. E., 117
Beasley, R. P., 11, 302
Becker, A., 36
Beer, C. E., 267, 269
Benge, M., 466
Bennett, H. H., 62, 312, 314
Bentley, W. A., 36
Bergsma, E., 64, 236, 496
Bernatti, R., 384
Bernollia, E. F., 425
Berry, L., 111
Bertoni, J., 119–120, 328, 378, 395
Bertrand, R., 311, 345, 514
Best, A. C., 41
Bhatacharya, B. B., 365
Bhatia, K. S., 62–63
Bhatt, P. N., 332
Bhola, S. N., 62
Bhumbla, D. R., 76
Birot, P., 425
Bisal, F., 140
Biscaia, R. C. M., 210, 245

565

Biswas, T. D., 62, 76
Black, A. L., 553
Blaisdell, F. W., 302
Blanchard, D. C., 36
Blandford, D. C., 216
Bligh, K. J., 313
Bocharov, A. P., 549
Bodman, G. B., 71, 159
Bodolay, S., 556
Bolline, A., 35, 145, 200–201, 236, 238, 251
Bond, J. J., 362
Bonell, M., 67, 426
Bonsu, M., 381, 384
Boon, W., 269
Borst, H. L., 117
Boudouresque, E., 550
Bourliere, F., 89
Bouyoucos, G. J., 61
Bowden, J. W., 79
Bradfield, R., 358
Bradford, J. M., 64, 135, 140
Brechner, E., 235
Brenneman, L. G., 269
Bridge, B. J., 72, 371
Brinkman, W. L. F., 425
Briones, A. A., 70
Brito, O. J., 243
Brito Chaves, I. de, 243
Brooks, F. L., 242
Brown, L. R., 15
Browning, D. R., 74
Bruce, R. C., 255
Bruce-Okine, E., 64, 134–135, 148, 216
Brundsden, D., 176
Bryan, R. B., 60, 63, 75, 84, 114, 136, 171
Bubenzer, G. D., 4, 140, 170, 225, 227
Budel, J., 110
Bureau of Sugar Experimental Station, 385
Burt, A. P., 75, 173
Butchbaker, A. F., 52

Cabeda, M. S. V., 244
Caine, N., 493
Campbell, B. L., 198
Carson, M. A., 104, 176
Carter, C. E., 36, 39–40, 42
Carter, D. J., 548
Casto Filho, C. de, 243
Cataneo, A., 243
CAZRI, 546
Celestino, A. F., 466
CENICAFE, 339–340
Cernuda, C. F., 64
Chadhokar, P. A., 453
Chakela, Q. K., 186
Chakrabarti, D. C., 62
Chan, C. C., 255, 305
Chandhray, T. N., 65

Chandra, S., 216, 342
Chang, H. M., 442, 507
Chang, T. P., 34
Changnon, S. A., Jr., 52
Chanphaka, U., 425
Chapman, G., 147
Chapman, P. E. V., 255
Charreau, C., 161, 236, 356, 411
Chase, R. G., 550
Chemin, M. C., 547
Chepil, W. S., 54, 65, 535–537, 540, 547–549, 558
Chheda, H. R., 342
Chiang, T., 172
Chiasson, T. C., 328
Chibber, P. K., 61–62
Childs, E. C., 69
Chinnamani, S., 464
Chisci, G., 133, 212, 272, 390, 404–405
Chong, S. P., 147
Chow, V. T., 141, 163, 185
Christiansson, C., 186
Ciesiolka, C. A. A., 203, 215
Clark, R. N., 402
Clegg, A. C., 425
Clinnick, P. F., 447
Cogo, N. P., 243
Cole, D. W., 423
Cole, G. W., 540
Cole, W. D., 206
Coleman, E. A., 71–72, 159
Collinet, J., 210
Collis-George, N., 71, 134, 159–160, 163
Colvin, T. S., 147, 366
Combeau, A., 138
Comerma, J. A., 303
Commonwealth Science Council, 451
Conaway, A. W., 63
Conway, G. R., 328
Cook, A. G., 90
Cook, H. L., 163
Cook, R. L., 255
Cooke, R. J., 534
Cooley, K. R., 240, 333
Cormack, J. M., 305
Coster, C., 465
Coutts, J. R. H., 200
Crouch, R. J., 80
Cruse, R. M., 135
CTFT, 236, 311, 466
Cuff, J. R. I., 193
Cummings, D. J., 409–410, 438
Cummings, R., 379
Cunningham, R. K., 89, 438
Curtis, W. R., 206

Dabin, B., 311
Dalrymple, B., 105–106

Author Index

Dandekar, M. M., 186, 189
Dangler, D. W., 242–243, 250
Das, D. C., 243
Datiri, B. T., 305
Daubenmire, R., 425
David, W. P., 269
Davidson, T., 89
Dawson, N., 384
De, S. K., 216, 343
Dearing, J. A., 186
De Boodt, M., 62, 64–65, 297, 299
Dedecek, R. A., 243–244
Defant, A., 35
Degani, A., 255
De Kesel, M., 559
De La Rosa, M. M., 466
De Leenheer, L., 62, 64–65
Delwaulle, J. C., 346
De Ming, S., 80
Denardin, J. E., 383
Denevan, W., 11
Dent, F. J., 15
de Ploey, J., 134, 136–137, 170, 199–200, 211
Derbyshire, F., 6, 106, 154–155, 157, 177
Derpsch, R., 380
Deshpande, T. L., 76, 80
DeTar, W. R., 322
De Vleeschauwer, D., 75, 296–297, 300–301, 559
Dhir, R. P., 545
D'Hoore, J. D., 90, 111
Dillon, J. L., 328
Disrud, L. A., 51, 142–143
Dissmeyer, G. E., 240
Donahue, R. L., 328
Donnollan, T. E., 210
Doornkamp, J. C., 107, 111
Doty, R. D., 426
Douglas, I., 188–190, 426
Doyle, A. D., 379–380
Dragoun, F. J., 267
Dregne, H. E., 14, 531
Dressler, J., 466
Drew, D. P., 174
Dubois, B., 332–333
Ducreux, A., 442
Dugain, F., 311–312
Dunne, T., 15, 188, 504, 506
Durand, P., 314
Dyrness, C. T., 426

Eckholm, E. P., 10
Edwards, K. A., 191–192, 255, 547
Edwards, W. M., 389
Ehlers, W., 412
Eigel, J. D., 36
Ekein, P. C., 138, 140

El-Amani, M., 548
El-Asswad, R. M., 548, 558
Elliott, W. J., 270
Ellison, W. D., 95, 131, 136, 140, 142, 200, 226, 267
El-Swaify, S. A., 79, 237, 242–243, 250, 333, 426
Eltz, F. L. F., 343
Elwell, H. A., 45, 50, 147, 237, 243, 262–264, 267, 324–325, 339
Embleton, C., 6, 141
Escolar, R. P., 76
Eswaran, H., 79
Evans, D. O., 367
Evans, R., 34
Eyles, R. J., 107, 109, 425
Ezaki, T., 135, 138

Fagi, A. M., 502, 517
Fahmy, M., 269
Falayi, O., 65
FAO, 15, 234–235, 248, 525, 543–544, 549
Farmer, E. E., 137, 142, 170
Fauck, R., 311–312, 330
Fausey, N. R., 406, 409
Fearenside, P. M., 210
Fekete, Z., 494
Felipe-Morales, C., 210, 301
Feller, C., 90
Fenster, C. R., 354, 390
Fetzer, K. D., 259
Finkner, S. C., 138, 140
Fissiha, T. W., 512
Fleming, G., 4, 8, 269
Fleming, W. M., 314, 493
Flenley, J. R., 425
Fonzen, D. F., 466
Foster, G. R., 112, 115–116, 130, 136, 138, 147–149, 154, 167–168, 231, 238, 240, 269–270, 272
Fournier, F., 35, 185, 188–189, 234, 311–313, 315, 324, 329, 440, 444
Foy, C. D., 328
Frankenberger, E., 553
Franzen, H., 205
Free, G. R., 117
Freebairn, D. M., 215, 269, 368, 379, 386, 388
Freeze, C. J., 195
Freire, O., 243, 245
Froehlich, W., 145
Fuchs, N., 36
Fullen, M. A., 33–34, 410

Gabriels, D., 65, 217, 297
Gaikwad, S. T., 89
Galabert, J., 236
Gallez, A., 80

Gananpin, D. J., 426
Garcia, A. S., 425, 430
Garg, K. C., 332
Geib, H. V., 208
Georges, J. E. W., 120, 334, 338
Gerasimenko, V. P., 204–206
Gerdel, R. W., 62
Gerlach, T., 207
Gerrard, A. J., 163, 173, 225
Ghadiri, H., 140, 216
Ghani, M. O., 76
Ghatol, S. G., 76
Ghildyal, B. P., 342
Gilbert, A. K., 154
Gillette, D. A., 541
Gilley, J. E., 138, 140
Gillspie, T., 35–36
Gilmour, D. A., 426, 493–494
Goel, K. N., 89, 332
Goel, S. K., 90
Golubev, G. N., 266–267
Goodlet, J. G., 104–105
Gorchichko, G. K., 200, 202
Gore, R., 531
Goudie, A., 545
Govindarajan, S. V., 62
Granger, O., 424
Greb, B. W., 401
Green, J. D., 119
Green, W. H., 69
Greenfield, J. C., 456
Greenland, D. J., 62, 76, 89, 91, 297, 438
Gregory, J. M., 322
Gregory, K. J., 104–105, 165–166, 185, 196, 204–205, 208
Grierson, I. T., 210
Grigorev, V. Y., 216
Guartsma, R., 236
Gumbs, F. A., 245, 337, 339, 400, 504
Gunn, R., 44–45
Gupta, J. P., 545, 553, 555, 557
Gupta, R. N., 62
Gupta, S. K., 305
Gupta, V. K., 191
Gwyer, G., 466

Haan, C. T., 112, 225
Hack, J. T., 104–105
Hadley, R. E., 89
Hagen, L. J., 51–52, 547
Hall, M. J., 35
Hallam, G., 346
Hamza, A., 548
Hanif, M., 448
Hansen, L. M., 36
Hansen, W. D., 328
Hardin, G., 12
Hardjowitjitro, H., 15, 186, 188

Haridasan, M., 62
Harpum, J. R., 111
Harris, R. F., 77
Harrison, P., 342
Harrold, L. L., 366, 388–389, 393–394
Hatch, T., 426
Haynes, J. L., 143
Heinonen, R. J., 287
Hempel, L., 191
Henin, N., 64–65, 234
Henry, J., 459
Hensch, B., 191, 195
Herrera, B. J. A., 342
Hibbert, A. R., 433
Hillel, D., 67, 69
Hills, R. C., 221
Hindson, J., 303
Ho, R., 425
Hock, P. C., 426, 428
Holtan, H. N., 71
Hoogmoed, W. B., 411–413
Hopkins, B., 425
Horton, R. E., 67, 70, 115, 163, 166, 226, 267
Hosegood, P. H., 454
Houghton, H. J., 548
Huberty, M. R., 69
Hudson, N. W., 35–36, 43, 112–113, 126, 143–145, 148, 203, 234, 263, 267, 302, 321, 323, 338–339, 446, 501
Hudspeth, E. B., 402
Hudugalle, N. R., 438, 443
Hughes, P. J., 173
Huke, S., 146–147, 325–327
Humphreys, G. S., 113
Hurni, H., 52
Hussein, M. H., 147
Hutchinson, J., 139
Hvorslev, M. J., 65
Hwang, Y. D., 61

ICAR, 313, 328
ICRAF, 465
ICRISAT, 397–398, 401–402, 551
IITA, 33, 112, 432, 438, 465, 471, 475–476
Imeson, A. C., 81, 174
Inoue, S., 138
ISSS, 61

Jackson, D. C., 112–113, 143–145, 321, 323, 339, 501
Jackson, I. J., 425
Jamal, T., 237–238
James, L. G., 158
Jarayam, N. S., 62
Jeje, L. K., 107
Jenicks, R., 546
Jenkinson, D. S., 89

Author Index

Jenny, H., 89, 91
Jha, M. N., 62
Jinze, M., 210
Johnson, C. B., 393
Johnson, H. P., 267
Johnston, A. E., 89
Johnston, J. R., 359
Jolly, J. P., 186
Jones, B. A., Jr., 140, 170
Jones, J. A., 175
Jones, M. J., 91, 161
Jones, P. A., 339
Jones, P. H., 430
Joseph, K. T., 323
Joshua, W. D., 243–244
Joss, V. J., 36
Junor, R., 546
Jurion, F., 459

Kadeba, O., 89, 91
Kalms, J. M., 411
Kaminsky, S. A., 195
Kampen, J., 359, 398–399, 401–402
Kamprath, E. J., 328
Kandiah, A., 66, 76, 81
Kang, B. T., 91, 342, 466
Kannegieter, A., 342
Kassam, A. H., 39, 42, 45, 47–48, 243
Keane, P. A., 342
Keersebilik, N., 33
Kellman, M. C., 462
Kemper, W. D., 65
Kerenyi, A., 216
Kerfoot, O., 425
Kerr, A. J., 343
Ketcheson, J., 388–389
Khalael, R., 69
Khan, S. U., 76
Khanbilvardi, R. M., 269
Khanji, J., 70
Khanna, M. L., 332
Khatibu, A. I., 365
Khybri, M. L., 395
Kimutai, J. N., 513
King, C. A. M., 107
Kinnell, P. I. A., 36, 45, 48, 50, 141
Kinzer, G. D., 44–45
Kirkby, A. V. T., 200
Kirkby, M. J., 104, 113, 115, 166, 176, 200, 233, 269
Kitanosono, T. A., 145
Klemes, V., 214
Klimes-Szmik, A., 63
Klinge, H., 89
Kneede, P. E., 173
Knisell, W. G., 272
Knowles, G. H., 193

Koenigs, F. F. R., 79
Koolhaas, M. H., 243
Kostiakov, A. N., 70
Koswara, J., 65
Kovda, V. A., 10
Kowal, J. M., 33, 39, 42, 45, 47–48, 89, 243, 330, 395–396
Kozachyn, J., 63
Kramer, L. A., 117, 122–123
Kramores, J. S., 77
Krauss, R. K., 143, 534, 536
Krishna, R., 544–545, 556
Krishnarajah, P., 465
Krusteva, V. S., 63
Kubota, T., 362, 388, 392
Kung, S., 173
Kurnia, U., 256–257, 305
Kuznetsov, M. S., 216
Kwaad, F. J. P. M., 200

Laban, P., 509
Lafever, H. N., 328
Laflen, J. M., 115, 145, 147, 269, 322, 366
Laguihon, W. Q., 525
Lal, R., 15, 34, 37–39, 48–52, 61, 64–66, 71–72, 74–77, 82–83, 87, 90–91, 112–116, 118–123, 125, 134–135, 139, 142, 144–148, 159–164, 199, 210, 216, 235, 243, 247, 252, 259–261, 265, 284, 291–295, 297, 305–306, 311, 325–327, 330–331, 333, 338, 340, 342, 365–366, 368–369, 371, 374, 376, 380–383, 387, 390, 397, 401–402, 404–405, 408–412, 423, 426, 436, 443, 460, 468–472, 476–477, 509
Lallmahomed, H., 425
Lam, Kin-Che, 202
Lamarche, V. C., 199
Lambert, A. M., 186
Lambour, J., 34
Landencia, P. N., 342
Lane, L. J., 270
Lang, R. D., 147
Larson, C. L., 69, 141
Larson, W. E., 135
Laskar, S., 62
Laursen, E. M., 169
Lawes, D. A., 76, 362–363
Laws, J. O., 36, 39, 41, 43–44, 227, 238
Lawson, T. D., 425, 434
Laycock, D. H., 507
Ledger, D. L., 109
Lee, J. H. N., 342
Lee, S. W., 34
Lehane, J. J., 553
Le Hourou, H. N., 546

Leigh, C. H., 429
Leite, J. A., 269
Lelong, F., 210
Lembaga, E., 427
Lembke, W. D., 142
Lenard, P., 35
Leopold, L. B., 203
Lettanzi, A. R., 149, 169
Lewis, L. A., 137, 200–201, 203–204, 206, 425
Li, R. M., 121, 124, 269
Li, Y. H., 15
Liao, M. C., 441, 447, 494–495, 505, 507–508, 510
Lieu, T. C., 426
Lima, F. A. M., 244
Lin, T. H., 62
Lindsay, J. I., 245, 337, 504
Lindstrom, M. J., 366, 410–411
Linville, A., 159
Lipman, E., 466
Lo, A., 242
Loch, R. J., 210, 467
Lockwood, J. G., 425
Lombardi, F. N., 119–120, 328, 335, 337, 378, 395, 447–448
Long, D. C., 120
Long, S., 538, 554
Longmore, M. E., 198–199
Lorimer, M. S., 531, 538
Louis, H., 108
Low, A. J., 412
Low, K. F., 234
Lowdermilk, W. C., 11
Lugo-Lopez, M. A., 62, 71
Luk, S. H., 62, 76, 84
Lundgren, B., 457, 465
Lundgren, L., 285–286
Lyles, L., 51, 65, 67, 142–143, 534, 536, 540

McCaffrey, L. A. H., 147
McCallan, M. E., 198
McCool, D. J., 36, 206
McCown, R. L., 375, 501
McCulloch, S. S. G., 425
McDonald, R. C., 161
McDowell, L. L., 389–390
McGregor, K. C., 147, 389–390
Machado, J. A., 381
Machado, S. A., 245
McHenry, J. R., 198
McIntyre, D. S., 84, 135
Mackie, C., 502, 517
MacLean, A. A., 328
McPhee, P. J., 249
Maduakor, H., 333
Maene, L. M., 147, 243, 249, 335

Mahmud, N., 237, 249, 255
Maignet, M., 547
Major, I., 63
Makepeace, P. K., 493
Malewar, G. U., 76
Malina, F. J., 534–535
Manipura, W. B., 365, 507
Mann, H. S., 544
Mannering, J. V., 354, 366, 390, 393
Manokaru, N., 425
Marques, J., 334, 336
Marston, D., 338, 377, 379–380, 388
Martin, G. L., 112, 136
Martin, W. S., 342
Martinez, M. R., 269
Mathews, A. A., 493
Mathews, H. L., 195
May, K. R., 36
Mazurak, A. P., 63, 170
Meeuwig, R., 77
Mehta, K. N., 62
Mein, R. G., 69
Meng, Q., 15
Menne, T. C., 313
Metwally, S. Y., 76
Meyer, D. L., 112, 117, 122–124, 130, 138, 140–141, 148–149, 155, 167–169, 210, 268–270, 366
Middleton, E. A., 61
Mielke, P. W., 187
Mihara, Y., 140
Miller, C. R., 267
Miller, R. W., 328
Millington, A. C., 195, 203, 214, 305, 463
Millington, T., 516
Millogo, E., 236
Mishra, B. K., 462–463
Misra, D. K., 365
Mitchell, H. W., 441
Mitchell, J. K., 4, 198, 225, 227
Moeyesons, J. R., 137, 170, 199
Mohammed, A., 339
Mokhtaruddin, A. M., 265, 335
Molchanov, A. A., 455
Moldenhauer, W. C., 65, 378, 391
Mondardo, A., 210, 245, 382
Monke, E. J., 169
Montecillo, L., 468
Mool, P. K., 493
Moore, T. R., 36, 342–343
Moormann, F. R., 77
Morales, B., 536
Moreau, R., 71
Morel-Seytoux, H. J., 69
Morgan, R. P. C., 137, 141, 145–146, 148, 200–203, 233
Morin, J., 400

Author Index

Morrison, J. E., Jr., 142–143
Mosher, P. N., 63, 170
Mosley, M. P., 137
Mota, F. O. B., 244
Mott, J., 411
Mou, J., 15
Mtakwa, P. W., 203–206, 248, 254, 260, 266
Muchena, F. N., 195
Muchow, R. C., 71
Mulllins, C., 384
Murphree, C. E., 47, 208, 238, 240
Musgrave, G. W., 113, 115, 226, 267
Mutchler, C. K., 36, 47, 119, 121–125, 141–142, 147, 208, 238, 240

Naegeli, W., 553
Nair, P. K. R., 457, 466
Naprakob, B., 461
Narain, B., 62
Nawaby, A. S., 36
Neal, J. H., 36, 138
Nearing, M. A., 64, 270
Negev, M., 269
Neumann, I., 466
Ngatunga, E. L. N., 248, 253, 263, 367
Nichols, M. L., 138
Nicou, R., 161, 356, 411
Niebling, W. H., 514
Nielsen, K. F., 536
Nieuwolt, S., 425
Nijhawan, S. D., 159, 332
Noer, Z. M., 465
Nordin, C. F., Jr., 186
Nortcliff, S., 34, 423
Northcote, K. H., 159
Nye, P. H., 91

Oades, J. M., 76, 210
Obeng, H., 381, 384
Oberholzer, E., 465
Obi, M. E., 342
Odum, H. T., 426
Ojeniyi, S. O., 464
Okigbo, B. N., 369, 383, 401
Olayemi, F. F., 210
Oliveira, J. B., 383
Ollagnier, M., 91
Olmstead, L. B., 159
Olofin, E. A., 62
Olson, G. W., 10, 12
Olson, O. C., 60, 82
O'Neal, A. M., 73
Onstad, C. A., 63–64, 171
Osamu, K., 524
Osborn, J., 77
Oschwald, W. R., 394
O'Sullivan, T. E., 255

Othieno, C. O., 441, 507
Overbeek, J. T. H. G., 297

Pacardo, E. P., 465, 468
Page, J. B., 159
Palmer, R. S., 141
Panabokke, C., 75, 133, 159
Pandey, A. N., 425, 493
Parfait, J. A., 425
Park, S. W., 44, 47, 132, 138, 141
Parker, G. G., 173
Parsons, D. A., 36, 39, 41, 43–44, 227, 238
Pasak, V., 542
Patel, N. A., 81
Paul, C. L., 386
Paulino, L. A., 19
Pauwels, J. M., 65
Payne, D., 140, 216
Peasley, B. A., 303
Peck, R. B., 73
Peele, T. C., 62
Pena, M., 365
Pena MacCaskill, I., 243, 255
Percy, M. J., 442
Pereira, H. C., 64, 339, 371, 401–402, 413, 429, 454
Perrens, S. J., 216
Pessotti, J. E., 245
Philips, J. R., 71
Phillips, R. E., 390
Pickup, G., 15
Piest, R. F., 269
Piest, R. G., 173
Pilgrim, D. H., 211
Pla, I. S., 297, 300
Plank, V. G., 36
Platford, G. G., 249, 272
Poesen, J., 141, 170
Politano, W., 195
Prentice, A. N., 401
Prihar, S. S., 362
Primavesi, A., 371, 382
Prospero, J. M., 547
Prove, B. G., 378
Pruppacher, H. R., 44
Punzalan, A. V., 465
Purnell, M. F., 489, 492, 498–499

Quansah, C., 137, 140, 380
Quantin, P., 138
Queensland Department of Primary Industry, 361, 367, 380
Quilty, J. A., 303
Quinn, N. W., 145, 322
Quirk, J. P., 75, 133, 159

Raghavan, G. S. V., 67
Raintree, J. B., 457

Ramaiah, R., 243
Ramakrishnan, P. S., 462–463, 553–555
Rania Rao, M. S. V., 333–334, 365
Rapp, A., 176, 186, 190, 531
Rathjens, C., 173
Rathore, R. K., 62
Rawitz, E., 402–403, 408
Raychaudhri, S. P., 89, 91
Reeve, I. J., 200, 216
Reid, L. M., 314
Renard, K. G., 240
Resck, D. V. S., 245
Resource Conservation Glossary, 352, 356
Richards, L. A., 67
Richey, C. B., 302
Richter, G., 193
Ripley, P. O., 368, 373
Riquier, J., 546
Ritchie, J. C., 196, 198
Robinson, A. R., 190
Robinson, D. O., 159
Roche, P., 331, 333
Rochette, C., 311
Rodda, J. C., 34
Rodriguez, G., 440
Rodriguez-Iturbe, I., 191
Roehl, J. W., 190
Rogers, J. S., 36, 40–41
Rogowski, A. S., 215
Rondilla, C. S., 465
Roose, E. J., 110, 112–113, 210, 236, 243, 247, 251–252, 259, 284–285, 297, 304, 311, 345–346, 365, 378, 435, 503, 506–507, 512, 514
Rose, C. W., 63, 140–141, 198, 269, 271
Rosewall, C. J., 377, 379
Rounce, N. V., 403
Rubber Research Institute of Malaysia, 80, 371, 445
Russell, J. R., 186
Russell, M. B., 64
Ruxton, B. P., 111
Ryan, K., 345

Sabhasri, S., 492
Sahi, B. P., 62–63
Salamin, P., 193–194
Sallaway, M. M., 388, 493
Sandhu, B. S., 76
Sarkar, S. K., 441
Sarmah, N., 63
Sastri, A. S. R. A., 545
Savat, J., 136–137, 269
Savigear, R. A. G., 103
Saxton, H. D., 138
Schleusener, P. E., 36
Schmidt, F., 196
Schottman, W. R., 145, 147

Schumm, S. A., 171, 203
Schwaertmann, U., 196
SCS, 276
SCSA, 3, 7–8, 130, 157, 425
Second National Soil Erosion Conference, Brazil, 382
Sen, A. K., 545
Senga, W. M., 512
Seubert, C. A., 91
Seubert, C. E., 428
Shah, R. K., 81
Shallow, P. G., 310
Shankar, V., 315
Shankranarayan, K. A., 315
Sharma, V. C., 62
Shaxson, F., 32, 186, 195, 365, 507
Sheng, T. C., 302–303, 305
Sherard, J. C., 62
Shi, D. M., 255
Shikula, N. K., 390–391, 394
Shin, J. S., 255
Siddoway, F. H., 541, 547, 553
Siderius, W., 193
Sidhu, P. S., 545
Sidiras, N., 382
Silva, J. R. C., 383
Silva, L. F. Da, 408
Sim, L. K., 426, 428
Simpson, L. A., 400
Sinclair, J., 161
Singer, M., 66, 81
Singh, G., 193, 254–255, 264, 544–545, 549
Singh, H. P., 11, 543, 551
Sinha, A. K., 342
Skaggs, R. W., 69
Skidmore, E. L., 55, 541
Slonecker, L. L., 67, 134
Slupik, J., 145
Smalley, I. J., 537
Smith, D. D., 45, 48, 89, 113, 115–117, 124, 142, 226–229, 231–232, 241, 267
Smith, G. E., 407
Smith, O. L., 64
Smith, R. M., 74
Smithen, A. A., 243
Soane, B. D., 87
Solorio, C. A., 312–313
Sood, M. C., 65
Soong, N. K., 298
Sprague, M. A., 376
Sreenivas, G. N., 243
Sreenivas, L., 200
Srikhajon, M., 236–237, 249, 254–255, 258, 264
Ssekabembe, C., 466
Stagg, M. J., 75

Staple, W. J., 553
Starkel, L., 494
Stehlik, O., 265, 267
Steichen, J. M., 407
Stennet, R., 305
Stephens, D., 513
Stevenson, C. M., 536
Stewart, B. A., 402
Stocking, M. A., 45, 81, 237, 243, 263, 324–325, 339
Strakhov, N. M., 185
Stromquist, L., 186–187
Stroosnjider, L., 411–413
Suarez de Castro, F., 139, 311, 313
Subramanian, V., 192
Sudjadi, M., 249, 472
Sukmana, S., 249, 466, 478
Sulaiman, W., 328, 335
Sutono, A., 377
Suwardjo, A., 251, 256–257, 305, 377
Swanson, A. P., 366
Swartzendruber, D., 69

Taiwan Agricultural Research Institute, 362
Takahashi, T., 426, 461
Talbot, M. R., 547
Tam, S. W., 311
Tavernier, R., 78
Taylor, S. A., 69
Temple, P. H., 203, 339, 513
Terwindt, J. H. J., 170
Terzaghi, K., 65, 73, 135, 176
Thaiutsa, B., 424
Thomas, D. B., 248, 364–365, 367, 369, 504, 512, 514–515
Thomas, M. F., 106, 108
Thomas, V. J., 492–493
Thompson, A. L., 158
Thompson, D. F., 547
Thompson, J. R., 267
Thornes, J. B., 141, 177
Thornthwaite, C. W., 540
Thorp, M. D., 107
Toky, O. P., 462–463
Torquebian, E., 423
Torri, D., 43–44, 48
Toth, A., 494
Touchton, J. T., 372, 377
Townshend, J. R. G., 207
Trimble, S. W., 450–451, 454
Triplett, G. B., Jr., 376
Troeh, F. R., 343
Truog, P. N., 378
Trustrum, N. A., 492–493
Tulaphitak, B., 400
Turchenek, L. W., 76
Turenne, J. F., 77

Turvey, N. D., 423
TVA, 455
Tyan, C. Y., 61

Ujah, J. E., 554,556
Umback, C. R., 142
Undang, K., 251
UNEP, 532
Unger, P. W., 361–362, 402
UNU, 432
USDA, 268, 401, 543, 549
USDA Agriculture Handbook, 196, 232
USDA Soil Survey Staff, 191
Utomo, W. H., 237, 249, 255

Valentin, C., 65, 67, 210
Van Asch, T. W. J., 147
Van Bavel, C. H. M., 62, 64
Van der Weert, R., 438
Van Doren, C. A., 113, 115, 208
Vanelslande, A., 75, 87–89, 216, 247–248, 253
Van Kampen, W., 346
Vann, J. H., 109
Van Vuuren, W. E., 240
Vardeni, B., 62
Vasques, J., 245
Venkobarao, K., 76
Vegara, N. T., 458
Verma, O. P., 328
Vieira, M. J., 210, 384
Virgo, K. J., 509
Virmani, S. M., 328
Voorhees, W. B., 410

Wakhba, S. A., 558
Waldvogel, A., 36
Walker, P. H., 141–142
Walling, D. E., 104, 165–166, 185–186, 191, 196, 204–205, 208
Walter, J. B., 305
Wandelaar, F. E., 259
Wang, J. Y., 44
Wang, S. T., 61–62, 85, 376
Warren, A., 534
Watnaprateep, P., 465
Watson, H. R., 525
Weber, F. R., 451
Webster, R., 109
Weil, R. R., 339
Weinstabel, P. E., 449–450
Weirich, F. H., 450–451, 454
Weismiller, R. A., 195
Wendelaar, F. E., 81
Wenzel, H. G., 141
Westin, F. C., 195
Whitaker, F. D., 340
Wiersma, J. L., 140, 167

Author Index

Wiersum, K. F., 251, 445, 460, 465
Wiesner, J., 35
Wigwe, G. A., 107–108
Wijewardene, R., 466
Wild, A., 161
Wilkinson, G. E., 33, 139, 243, 248, 323, 342, 371, 412–413
Wilkinson, J., 545
Willey, R. W., 331
Williams, C. N., 323
Williams, J. R., 240
Williams, L. S., 11, 50, 305
Williams, M. A. J., 63, 141, 425, 541, 547
Willis, W. O., 362
Wilson, G. F., 373
Winter, S. H., 315
Wischmeier, W. H., 4, 35, 45, 47, 60, 113, 115–117, 124, 138, 142, 226–234, 240–241, 243, 245, 248, 267–268, 364, 366
Wittmus, H. D., 353
WMO, 549, 551
Wockner, G. H., 368, 379, 386
Woodburn, R., 63, 117
Woodruff, N. P., 54–55, 65–66, 414–415, 537, 540–541, 547–549
Wu, H. L., 447

Wu, W. L., 494–495, 505, 507–508, 510
Wunsche, W., 383–384

Xiaoliang Experimental and Extension Station of Soil Conservation, 342

Yadav, R. C., 210
Yair, A., 75, 115, 174
Yalin, M. S., 171
Yamamoto, T., 63
Yang, Y. S., 255
Yariv, S., 132, 171
Yen, C. P., 451
Yeoh, C. S., 298, 305
Yoder, R. E., 62, 171
Yoshinori, T., 524
Young, R. A., 63–65, 104, 108–109, 120–125, 140, 142, 148, 167, 171, 201, 207
Ysselmuiden, I. L., 509

Zachar, D., 544, 546
Zanchi, C., 43–44, 48, 390, 404–405
Zech, W., 449
Zingg, R. W., 115, 226
Zoslavskig, M. N., 14

Subject Index

Africa, 17, 109, 147, 235, 237, 254, 390, 428, 449, 499, 531, 544
Aggregate size distribution, 63–65
Aggregate stability, 62, 76
Albizzia falcataria (tree), 428–429
Alfalfa, 327, 373
Alfisols, 16, 72, 76–77, 79–80, 83–84, 114, 123–124, 158, 204, 251, 289, 327, 363, 379–380, 389, 410–411, 500
Algeria, 546
Alpine forest soils, 84
Andean region, 489
Andosols, 249
Apple, 508, 514
Aridisols, 16, 244, 250, 374
Asia, 34, 126, 254, 262, 428, 499, 531, 544
Australia, 17, 34, 68, 72, 77, 81, 163, 192, 215, 314, 361, 371, 376, 378–380, 385–386, 400, 411, 426, 433, 445, 531, 544, 547–548

Baltic region, 547
Banana, 430, 475, 504–505, 508, 514, 518
Barbados, 311
Barley, 327
Belgium, 35, 145, 236
Black gram, 264
Brazil, 108–109, 118–120, 168, 176, 195, 210, 213, 243–246, 255, 313, 327, 334–337, 378, 380, 382–384, 395, 408, 447–448
Bulgaria, 63
Burkina Faso, 236, 311, 346, 522, 547
Burundi, 312, 314, 495

Cameroon, 311, 444, 475
Canada, 388–389, 558
Caribbean, 34, 376, 400, 489, 492
Cassava, 146, 252, 265, 288, 314, 325–326, 328–329, 331–332, 340, 355, 397, 428, 502, 509, 516, 519

Central America, 17, 126, 400, 499
Chad, 547
Chernozemic soils, 134
Chile, 365
Chili, 265, 335
China, 15, 61, 80, 210, 255
Citrus, 444, 508, 514
Clay minerals, 77–79, 148
  bulk density, 65, 87, 443
  exchangeable cations, 81
  hard setting/soil compaction, 88, 160, 408
  heat of wetting, 24, 72, 81, 133–134, 159, 163–164
  humus/organic matter, 67, 75, 80, 89, 91, 148, 478
  penetration resistance, 65, 87, 294, 442
  plow pan/clay pan, 407, 410
  puddling, 404–405
  soil temperature, 91
  surface seal/crust, 66, 158, 410–412
  total porosity, 65, 443
  transmission pores, 88
Climates/ecosystems:
  Antarctic, 535
  arid, 15, 29, 531, 546
  humid, 15, 29
  Mediterranean, 9
  Sahara, 546
  savanna, 16, 91, 264, 450, 554
  temperate, 450–451
  tropical rain forest, 16, 264, 424, 426, 432, 493
Cocoa, 252, 259
Coconut, 327, 426–427, 430, 432
Coffee, 252, 259, 314, 444, 447–448, 466, 502
Colombia, 7, 139, 246, 313, 372, 426, 440
Conservation tillage, 120, 296, 306, 322, 360
  concepts and definitions, 353

575

Conservation tillage (*Cont.*):
  types:
    broad-bed and furrow system, 397–401, 409
    chisel plow system, 390, 394
    conventional tillage, 112–113, 306, 340, 356, 381, 383, 389–391, 393, 415, 473–474
    intertillage, 362
    minimum tillage, 355, 390–393
    mound and hillocks, 357, 397
    no-tillage, 355, 376, 381–383, 389–391, 413–415, 473–474
    paraplow, 393, 408
    ridge tillage, 146, 326, 357–358, 396–397, 403, 406
    strip tillage, 394
    subsoiling, 362, 395
    tied ridges, 358–359, 395, 401, 403–404, 408
    vehicular traffic, 88, 340, 387, 409
Costa Rica, 441
Cotton, 12, 147, 259, 312, 314, 327, 330, 336, 403
Cowpea, 118, 252, 269, 327, 335, 354, 372, 469–470, 476, 504, 507, 509, 516, 519
Cropping systems:
  agroforestry systems, 375, 427, 432, 442, 456–458
  allelopathic effects, 475–476
  alley cropping, 375, 459–460, 464–466, 468, 473–474, 478, 510, 517
  buffer/grass strips, 344–345, 442, 463, 508, 510–514, 519
  burning, 337, 378, 445
  compound farm/home gardens, 427, 464
  cover crops/fallowing, 122, 337, 342, 368–373, 375, 377–378, 395, 437, 440, 442, 444, 447, 505, 510
  drainage/waterlogging, 404–405, 409
  grazing, 314, 446, 506
  land clearing, 87, 91, 199, 375, 409, 432–433, 436–440, 442, 461
  mixed/intercropping, 146, 331–332, 334, 337, 341, 375
  plantation/afforestation/trees/forestry, 423, 427, 431, 445, 447–448, 451–453, 455, 465, 475, 512, 525
  residue management, 347, 414–415
  rotations/cropping systems/farming systems, 259–260, 265, 325, 329–330, 334–335, 337, 459, 502
  shelterbelt/windbreak, 551, 553–556, 558
  shifting cultivation, 375, 428–429, 442, 445, 456, 459–462, 492
  soil conditioners, 558–559

Cropping systems (*Cont.*):
  strip cropping, 342–343, 345, 548
  taungya system, 428–429, 442, 445, 460, 464
Czechoslovakia, 193, 541

Dominican Republic, 15, 215, 344

East Africa, 17, 313, 331, 489
Egypt, 7, 558
El Salvador, 495
Engineering practices, 233, 449
  check dams, 303, 417, 522
    contour binding, 463
  chutes, 303
    road construction/urban setup, 446, 516, 518, 521
  drop structures, 303
  stone lines, 346
    stone terraces, 346, 516
  terraces, 126, 266, 301–302, 447, 449, 463, 467, 508, 513, 515–519
  waterways, 303, 447, 519
    hillside ditches, 447
Entisols, 16, 244
Erodibility, 60, 66, 148, 229, 243, 245, 248–254, 538–539
Erosion, 3–4
  agents, factors, and causes of, 5–6
  classes of, 191, 194
  climate, 26–27
  creep, 177, 424
  gully/ravine, 6, 82, 105, 171–175, 519
  hydrology, 26, 387–388
  land use, 26
  landforms, 26, 103, 169, 231, 538
    geomorphology, 103
    landscape model, 105
    lithology, 108
    position in landscapes, 105
    types, 104, 107, 109
  mass movement, 6, 175–176, 524
  modeling and prediction, 225, 267, 269
    common algebraic equations, 25
    CREAMS, 272
    erodibility nomogram, 230
    erosivity indices, 227, 234, 237–238
    Griffith University model, 269
    isoerodent maps, 227, 241–242, 244–249
    Musgrave's equation, 226
    MUSLE, 239–240, 519
    sediment transport model, 188–189
    SLENSA, 263
    upland erosion model, 269
    USLE, 226–227, 233, 239–241, 254–256, 324, 389, 519, 543
    WEPP, 262, 269
    WEPS, 543

Subject Index   577

Erosion, modeling and prediction (*Cont.*):
  WERM, 543
  wind erosion equation, 532, 539
  Zingg's equation, 226
 pipe or tunnel, 75, 82, 172
 processes, 23, 523
 rill, 6, 105, 167–168, 425
 sheet/splash/interrill, 6, 8, 105, 130–133, 137, 146, 157, 424
 slope, 26, 103, 149
  forms, 108
  geomorphologic processes, 105
  gradient, 105, 109, 112–114, 135–136
  length, 105, 114–118, 120–123, 541
  LS factor, 231–232
  shape, 105, 121–124, 366
  slope modification, 126
 soil, 26
  measuring (*see* Soil erosion measurement)
  stream bank, 177
  types, 6–7
  upland (interill/rill), 226
Erosion control, 283, 525, 548
 crop management, 320–321
  crop husbandry, 326
  crop stand, 338
  leaf area index, 327
  pest control, 338, 448
  time of planting, 338
  varieties, 327
 fertility management, 340, 509, 551, 558
  residue management, 551
  rooting systems, 451, 454
  weeding/hoeing, 507
 land use, 309, 493, 500–501, 506
  arable, 315
  forest, 310–311, 314, 325
  grazing, 314, 502
 land use capability, 316–317
 mulch, 284–286
  effects on evaporation, 286–287, 289–294, 388
  effects on soil conditions, 295–297, 299–301, 304–305
  principles, 254
Erosion-induced productivity decline, 521–522
Erosion-induced soil degradation, 526
Ethiopia, 452, 489
Eucalyptus, 456, 463, 554
Europe, 33–34, 262, 356, 365, 492, 544

France, 547

Germany, 193, 196, 547
Ghana, 91, 380–381, 384
Gliricidia, 468–470, 473, 475, 512, 516

Grasses/legumes, 514, 516–519
Green gram, 264
Guinea, 311, 440, 444
Guyana, 386

Haiti, 15
Hong Kong, 202
Hungary, 63, 193–194, 216, 493–494, 556
Hydrological properties, 67, 87
 evaporation, 361, 388, 540, 543
 soil water potential, 67, 90, 134, 159–160
 soil water storage, 361, 381, 388, 392, 400, 589
 water balance, 425
 water retention (p^F curves), 67, 87, 132, 162, 293, 478

Imperata/latana vegetation, 265, 428–429
Inceptisols, 16, 134, 250, 500
India, 7, 17, 31, 61–63, 65, 81, 110, 189, 192–193, 255, 264, 305, 313, 332–334, 362, 365, 386, 395, 397, 425, 441, 457, 461–462, 464, 495, 544–545, 555–557
Indonesia, 13, 19, 188, 213, 305, 327, 334, 377, 426–427, 429, 465, 477, 489, 495, 502, 516–517
Israel, 255, 400, 408
Italy, 43, 48, 390, 404–406, 547
Ivory Coast, 72, 90–91, 112, 210, 236–237, 284, 297, 304, 311, 365, 378, 435, 524

Jamaica, 305, 495

Kenya, 15, 72, 195, 246–247, 249, 266, 339, 344–345, 441, 454, 466, 495, 504, 506–507, 512–513, 515, 517
Korea, 255
Kraznozems, 134, 385

Latosols, 79, 336–337, 448
Leucaena, 429–430, 465–470, 473, 475, 511, 516, 519
Libya, 546
Litchi, 508, 514
Lithosols, 245
Luvisol, 247

Madagascar, 313, 324, 331, 333
Mahogany, 428–429
Maize/corn, 12, 118, 122, 145–146, 148, 252, 259, 265, 288, 313–314, 323, 326–327, 329, 331, 333, 335, 355, 372, 390, 396–397, 409, 428, 430–432, 468, 470, 475, 502, 504, 507, 509, 512, 514, 516, 519

Subject Index

Malawi, 109, 335, 369, 463
Malaysia, 62, 80, 108–109, 248–249, 297, 305, 327, 333, 335, 371–372, 426, 428–429, 440, 498
Mali, 413, 547
Mango, 504–505, 508, 514
Mechanical properties, 61–65, 82
Mediterranean, 235
Melon, 265
Mexico, 312
Microaggregation, 62, 77–78
Middle East, 19, 235, 543–544
Mollisols, 244, 500
Morocco, 246, 493, 497, 546
Mulberry, 504–505, 508
Mulch farming:
    ecofallow, 358
    live mulch, 337, 341, 358, 374, 376
    mulch factor, 260–261, 366–369
    planted fallow, 358
    plastic mulch, 362
    stubble mulch farming, 357–358, 380
    summer fallow, 358
Mung bean, 265, 328, 335

Nepal, 15, 259, 314, 489, 493–494, 496
New Guinea, 15, 113
New Zealand, 193, 491–493
Niger, 17, 109–111, 346, 446, 547, 549, 551, 554, 558
Nigeria, 18, 31, 33, 37, 42–43, 46, 48, 51, 72, 75–76, 90–91, 107–108, 112–114, 118, 120, 124, 137, 145, 161, 164, 171, 175, 203–204, 206, 210, 214, 235–236, 245, 247, 249, 252–253, 260, 297, 305, 323, 325, 327, 330, 339, 354, 362, 364, 368, 380, 395–396, 408, 434, 438, 464, 466, 495, 497, 519–520, 536, 554
Nitosols, 78, 80, 247
North America, 33–34, 356, 363, 388, 450, 492, 531, 544, 556

Oats, 122, 333, 373
Oil palm, 252, 259, 426, 430, 432
Okra, 556
Onions, 519
Oxisols, 16, 78–80, 244–245, 250–251, 384, 500, 516

Pacific, 34, 489
Pakistan, 7, 544
Papua, New Guinea, 113
Pastures, 313, 327, 329, 334, 343, 345, 373, 430, 432, 504–505

Peanuts/ground nuts, 259, 265, 327, 330, 333, 335–336, 428, 509, 514, 516, 519
Pearl millet, 259, 328, 558
Peru, 12, 210, 214, 301, 503–504, 509
Philippines, 327, 426, 429, 461, 465, 467, 489, 511
Pigeon pea, 264, 325, 332
Pineapple, 259, 264, 312, 333, 336, 362, 444, 504–505, 508
Podzols, 385, 391, 500
Potato, 333, 507, 509, 519
Puerto Rico, 62, 203, 425

Raindrop size, 31, 36
    hails, 51
    measurement, 35
    momentum, 25, 50–51
    size distribution, 35–44
Rainfall:
    amount, 235
    characteristics, 138
    erosivity, 6, 137, 146, 148, 227–228
        indices, 137
    intensity, 27, 30–32, 42–43, 144, 146, 387–388
        directional rains, 523–524
Red-yellow podzolic soils, 245
Restoration, 526–527
Rheologic properties, 73–75
    cohesion limit, 74
    consistency, 73
    liquid limit, 74
    plastic limit, 74
    shrinkage limit, 74
    sticky limit, 74
Rice (paddy), 264, 327, 329, 334, 336, 428, 430–431, 456, 516, 519
Rillability, 171
River watersheds:
    Amazon, 91
    Yellow River, 15–16
Rubber, 426, 428, 445
Rwanda, 491, 495–496, 503

Sahara, 195, 536
Sahel, 65, 68, 109, 411–412, 547
Salt-affected soils, 414
Scandinavian region, 547
Senegal, 90–91, 329–330, 411, 547
Sierra Leone, 463, 516–517
Sisal, 456
Soil:
    denudation, 5
    formation, 3
    sedimentation, 4–5
Soil chemical and mineralogical properties, 77, 89

Soil detachment, 4, 130, 137, 142, 166
  chemical erosion, 4
  delivery ratio, 190
  denudation, 5
  dissolved load, 186, 192
  physical erosion, 3
  raindrop impact, 6
  sediment yield, 4, 187, 506
  soil degradation, 8
  soil depletion, 9
  suspended/sediment load, 186, 192
  transport/entrainment, 4, 61, 130, 137, 166
  watershed (catchment/basin), 5
Soil erosion measurement, 200, 548–549
  aerial photography, remote sensing, and LANDSAT, 195, 548
  mapping, 191
  measuring structures:
    flumes, 185, 197–198, 460
    Geib multislot divisions, 208–210, 215
    Gerlach troughs, 205–207
    rainfall simulator, 216–217, 245
    runoff plots, 208–209, 211–214
    runoff samples, 198, 210–212
    sampling traps, 207
    splash measuring devices, 199–202
    V notch, 195–196
    weirs, 185, 195–196
  measuring techniques:
    buried nails, 203–204
    laboratory measurement, 216
    radioisotopes, 196–198
    rill measurement, 203–205
    soil level, 199
    tree roots, 200
  scales, 183, 185
Soil fauna:
  earthworms, 77, 85, 288, 290, 412–413
  termites, 86, 412
Soil losses, 239, 389
Soil orders, 253, 255–257, 380, 499–500, 517
Soil profile characteristics, 81–82
  laterite/phinthite, 109
  pedestals, 131
Soil properties, 60, 170
  particle size distribution and texture, 60–61, 82, 148, 170, 253, 258, 415
    clay, 67, 83
    gravels, 83
    sand, 83
    silt, 83
  structure, 60, 148, 547
Soil-ridge roughness, 540
Soil structure/aggregation, 62–63, 80, 288, 292
  model, 63

Soil water transmission, 68
  cumulative infiltration, 363, 374, 377, 382–383, 407, 409–411, 437–441, 443
  Darcy's law, 72
  infiltration curves, 68
  infiltration models, 67–69, 71
  infiltration process, 71
  infiltration rate, 65, 68, 88–90, 154, 158, 363, 375, 381, 406, 410, 412
  infiltration velocity, 68
  interflow, 435
  percolation, 61
  permeability and hydraulic conductivity, 72–73, 291
  permeability classes, 73–74
  sorptivity, 71, 363–364
  transmissivity, 71, 363–364
  wetting front, 71, 159, 169, 297–298
Soil wetting, 132
Sorghum, 259, 327, 330, 361, 378
South Africa, 313, 516
South America, 17, 19, 126, 254, 428, 499, 521
Soybean, 146, 148, 264, 325, 327, 332, 336, 354, 390, 516
Spain, 547
Sri Lanka, 67, 186, 243–244, 464, 507
Steeplands, 489–527
  Andean region, 12, 489, 500
  Ethiopian region, 15, 490
  Himalayan-Tibetan region, 15, 489, 494, 499
Strength properties, 84, 132, 171
  Coulomb-Hvorslev shear-strength equation, 67
  Coulomb's law, 66
  effective soil cohesion, 67
  pore water pressure, 66–67, 135
  shear strength, 24, 60, 66, 135
  shear stress, 25
  unconfined compressive strength, 60
Structural indices:
  dispersibility, 81
  dispersion ratio, 61
  erosion ratio, 61
  mechanical ratio, 61
  SiO_2/R_2O_3, 62
  slaking, 133
Structural methods:
  clod size, 65
  crusting, 65
  De Leerheer and De Boodt method, 64–65
  Henin's method, 64–65
  immersion method, 36
  soil tilth, 65, 73
  wash resistivity index, 80
  waterdrop technique, 64, 135

Subsaharan Africa, 19, 547
Sugar beet, 145, 327, 340, 355
Sugarcane, 312, 333, 336, 384, 386, 504–505
Sweet potato, 146, 326–327, 334, 336, 340, 355

Taiwan, 62, 255, 305, 362, 441, 451, 503–505, 508, 510, 514, 517–518
Tanzania, 107, 111, 186–187, 253, 263, 284, 286, 338, 365, 367, 403, 441, 465, 495, 513
Tea, 252, 504–505, 507, 518
Teak, 428–430, 465
Thailand, 69–70, 72, 236–237, 249, 255, 258, 337, 345, 362, 365, 388, 391–392, 400, 461, 489, 491–492
Tobacco, 12, 259
Tomato, 327
Trees, 500, 508, 510–512, 519, 548, 553–554, 557
Trinidad, 120, 246, 334, 337–339, 441, 503–504
Tunisia, 546, 548, 559

Uganda, 63, 107, 343, 513
United Kingdom/England, 33–34, 145, 206, 410
United States, 12, 40, 62, 68, 118, 171, 190, 215, 228, 231, 236, 241, 250, 312, 314, 335, 374, 388–390, 450–451, 454–455, 495, 541, 547
U.S.S.R., 14, 195, 202, 205–206, 216, 266, 390–391, 394
Utisols, 16, 77–80, 244, 250–251, 372

Vegetation/canopy cover, 538, 540
  crop management, 149, 231
    cover and management factor, 251–254, 259–261, 263–265, 334–335, 384
    crop rotations and erosion, 327
    ground cover, 364–365, 507
    growth stage, 325
    soil erosion, 322–324, 326
  height, 147
  rainfall erosivity, 143
Velocity, 52, 142, 545–555
  drag, 52–53, 533
  drift, 549
  lift, 52, 533
Venezuela, 237, 300, 303
Vertisols, 16, 134, 158, 163, 250–251, 327, 361, 365, 385–386, 398–399
Vetiveria zizanioides (khus grass), 347, 456, 515

Water management, 302, 446–448
  erosion, 366–367, 369, 371, 373, 386, 389, 391–392, 395, 397, 399, 403, 408, 414, 429, 436, 444–445, 448–449, 460–462, 467, 470, 473, 475, 502, 506, 508–510, 513–519
  hydrograph, 388, 392, 462
  nutrient loss, 366, 463–464, 557
  runoff, 365–367, 371, 373, 386–387, 436, 444–445, 448–449, 461–462, 467–469, 473, 475, 502, 506, 508–510, 513–519
  sediment lead/concentration, 367, 371, 387, 397, 461
  stream flow, 433–434, 450, 455
Watershed processes:
  hydrologic cycle, 155
  Manning's formula, 24, 169, 185
    fluid drag, 61, 167
    hydraulic radius, 169
    kinematic viscosity, 25
    Reynold's number, 25
    shear stress, 157
    surface roughness, 169
  overland flow/runoff, 106, 140, 161, 163, 173
    concentrated flow, 167
    hydrograph, 165
    types of flow, 106, 154, 156–157, 165
    laminar, 155, 165
    turbulent, 155, 165
West Africa, 235, 243, 252, 365, 378, 411, 444
Wheat, 327, 334, 379, 391
Wind-driven rain, 51, 142
  roughness height, 53
  wind profile, 53, 144
Wind energy, 545
Wind erosion, 414–415, 531–532, 538, 544, 550
  air turbulence, 534
  ballistic impact, 535
  Bernoulli effect, 533
  creep, 535
  desertification, 531–532
  dust storm, 536
  factors, 532, 537–538
  harmattan, 536
  measuring, 547–549
  saltation, 535–536
  suspension, 536–537

Yam, 146, 252, 265, 326–327, 340, 355, 397, 514

Zaire, 495
Zimbabwe (Rhodesia), 50, 112–113, 237, 243, 245, 259, 262–263, 321, 323–324, 338–339

ABOUT THE AUTHOR

Rattan Lal is Professor of Soil Physics at the Ohio State University in Columbus, Ohio. He is also President of the World Association of Soil and Water Conservation.